NSW Year 11

Finding Physics

a comic guide
with student activities

Wendy Pan
Illustrated by Cheylon Devine

First published 2024 by Finding Physics

Author:	Wendy Pan
Illustrator:	Cheylon Devine
Editor:	Harmoni Kwok
Answers checked by:	Harmoni Kwok
Typesetting:	Wendy Pan
Images:	All photographs were taken by or on behalf of the author. All diagrams were created by the author. Drawings were created by Cheylon Devine.
Cover design:	Cheylon Devine
Acknowledgements:	The author would like to thank Cheylon Devine for her joyous artwork and advice about typography, and Harmoni Kwok for her attention to detail. Special thanks to Neil and Lenora for their constant feedback, understanding and support.
ISBN:	978-1-7637082-0-4

Get in touch

To provide feedback on this book or support the author's next Finding Physics project, please visit:

www.findingphysics.com.au

Contents

Part 1: Motion and forces in 1D

Part 2: Dynamics

Part 3: Electricity and magnetism

Part 4: Waves and thermodynamics

Part 1: Motion and forces in 1D

F	D	H	E	A	F	H	Y	P	O	T	H	E	S	I	S	E	W	X	U	D
R	E	A	N	O	G	R	D	N	J	E	T	R	E	L	I	A	B	L	E	I
A	I	R	R	E	S	I	S	T	A	N	C	E	N	E	U	T	N	E	B	S
M	H	I	W	K	W	L	C	E	V	Q	Y	D	N	A	O	O	P	C	D	P
E	A	F	U	N	V	T	J	U	E	R	R	O	R	S	I	S	B	U	E	L
O	A	G	N	R	Y	J	O	C	R	G	E	U	G	T	I	C	M	E	S	A
F	L	R	N	P	H	E	N	N	A	C	O	V	C	A	R	O	H	I	N	C
R	T	A	G	I	L	O	V	E	G	U	C	A	X	A	S	N	N	G	A	E
E	P	V	H	Y	T	S	I	C	E	S	E	L	I	S	F	T	U	U	N	M
F	M	I	J	Y	G	U	V	C	A	R	W	I	E	R	U	A	O	N	P	E
E	H	T	E	I	B	D	D	J	H	L	R	D	I	R	E	C	T	I	O	N
R	E	Y	T	U	I	C	O	E	I	D	T	Y	N	A	R	T	G	F	G	T
E	B	C	N	M	E	N	U	N	C	E	R	T	A	I	N	T	Y	O	R	Y
N	F	O	K	I	N	E	M	A	T	I	C	S	E	C	O	P	A	R	A	N
C	O	N	C	L	U	S	I	O	N	Y	U	D	G	U	C	H	E	M	D	A
E	C	T	V	O												U	G	W	I	R
L	J	R	U	L												V	R	B	E	N
I	U	O	R	M												T	A	A	N	M
N	H	L	Q	U	A	N	T	I	T	A	T	I	V	E	N	L	P	P	T	V
E	Y	L	U	N	Y	O	C	O	C	I	M	N	O	E	C	K	H	L	X	E
A	G	E	A	F	U	R	R	P	I	M	L	B	D	P	X	S	O	K	C	L
R	T	D	L	A	O	M	O	L	L	W	K	N	I	L	U	J	U	J	V	0
R	F	D	I	S	T	A	N	C	E	R	E	W	U	O	Z	H	I	V	B	C
R	R	V	T	O	R	L	Y	M	N	P	J	V	E	K	A	G	U	E	A	I
E	S	C	A	L	A	R	L	R	E	T	H	N	Y	I	S	F	Y	C	N	T
G	F	M	T	R	N	V	P	D	T	Y	A	C	T	J	G	D	T	T	M	Y
R	F	K	I	P	I	F	E	D	H	T	G	X	R	F	D	H	R	O	A	E
E	D	I	V	U	T	A	D	S	N	U	F	Z	E	O	F	S	T	R	S	D
S	E	P	E	E	M	S	B	A	C	C	E	L	E	R	A	T	I	O	N	S
S	S	Y	B	L	O	A	T	L	W	O	D	A	W	C	G	A	E	H	S	C
I	W	F	C	H	D	S	A	C	E	P	S	I	N	E	R	T	I	A	D	F
O	A	E	X	I	N	D	E	P	E	N	D	E	N	T	H	W	M	G	F	G
N	Q	F	R	I	C	T	I	O	N	A	Z	Q	P	A	R	A	L	L	A	X

1.1 Describing Motion

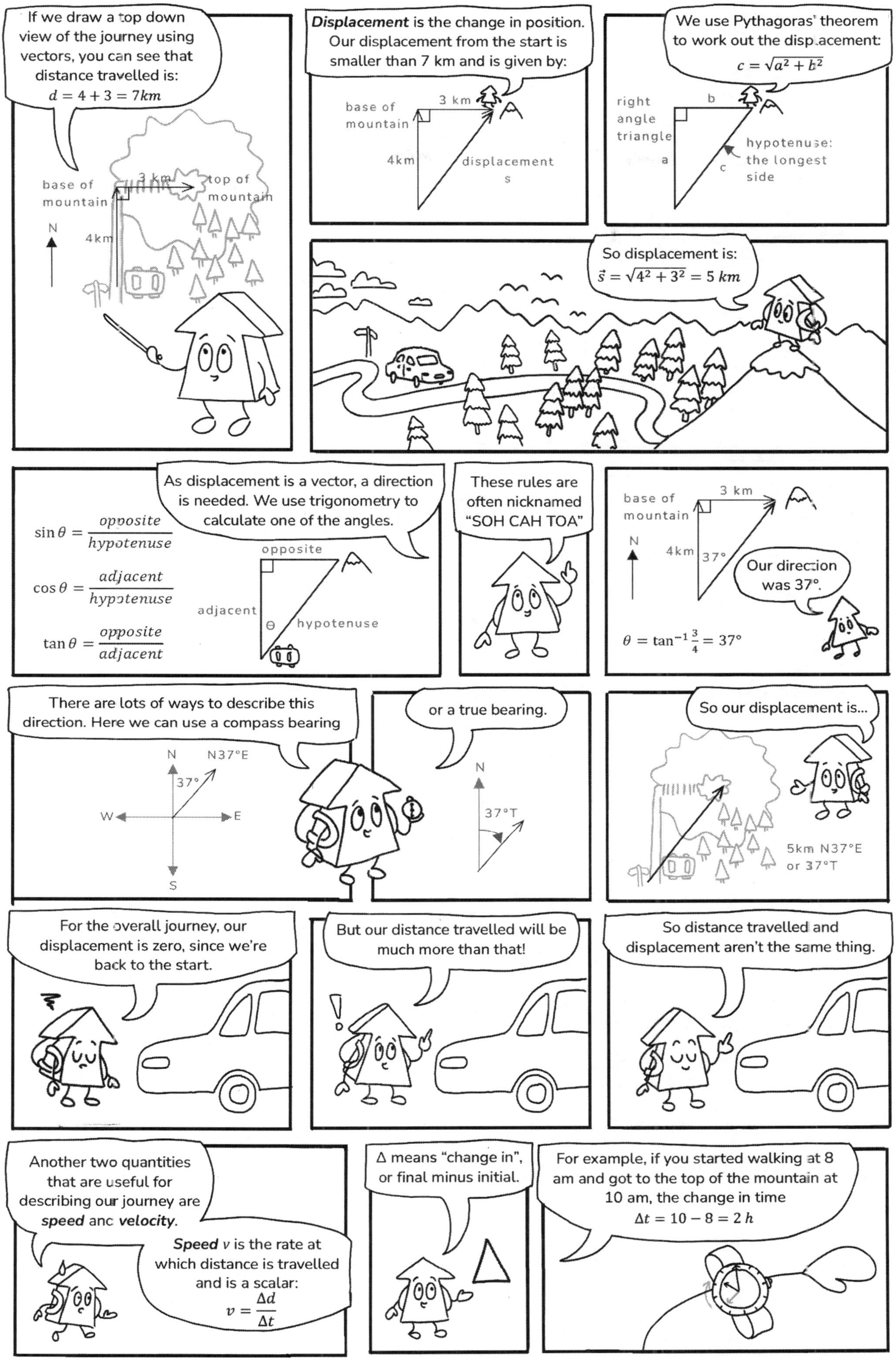

If we draw a top down view of the journey using vectors, you can see that distance travelled is: $d = 4 + 3 = 7km$
base of mountain
3 km
top of mountain
N
4km
Displacement is the change in position. Our displacement from the start is smaller than 7 km and is given by:
base of mountain
3 km
4km
displacement s
We use Pythagoras' theorem to work out the displacement: $c = \sqrt{a^2 + b^2}$
right angle triangle
b
a
c
hypotenuse: the longest side
So displacement is: $\vec{s} = \sqrt{4^2 + 3^2} = 5\ km$
As displacement is a vector, a direction is needed. We use trigonometry to calculate one of the angles.
$\sin\theta = \frac{opposite}{hypotenuse}$
$\cos\theta = \frac{adjacent}{hypotenuse}$
$\tan\theta = \frac{opposite}{adjacent}$
opposite
adjacent
θ
hypotenuse
These rules are often nicknamed "SOH CAH TOA"
base of mountain
3 km
N
4km
37°
Our direction was 37°.
$\theta = \tan^{-1}\frac{3}{4} = 37°$
There are lots of ways to describe this direction. Here we can use a compass bearing
N
N37°E
37°
W
E
S
or a true bearing.
N
37°T
So our displacement is...
5km N37°E
or 37°T
For the overall journey, our displacement is zero, since we're back to the start.
But our distance travelled will be much more than that!
So distance travelled and displacement aren't the same thing.
Another two quantities that are useful for describing our journey are *speed* and *velocity*.
Speed v is the rate at which distance is travelled and is a scalar: $v = \frac{\Delta d}{\Delta t}$
Δ means "change in", or final minus initial.
Δ
For example, if you started walking at 8 am and got to the top of the mountain at 10 am, the change in time $\Delta t = 10 - 8 = 2\ h$

3 km
4km
The speed at which you walked from the start to the top of the mountain would be: $v = \frac{7\ km}{2\ h} = 3.5\ kmh^{-1}$
Velocity $\vec{v}$ is the rate at which displacement changes and is a vector: $\vec{v} = \frac{\Delta\vec{s}}{\Delta t}$
v
Your velocity to the top of the mountain would be: $\vec{v} = \frac{\Delta\vec{s}}{\Delta t} = \frac{5\ km}{2\ h} = 2.5\ kmh^{-1}$
5km
The speed we just calculated, 3.5 km/h^{-1}, is ***average speed***.
We probably wouldn't walk at this speed the whole way though, since it would be tougher going up the mountain than walking on the flat ground.
Our speed up the mountain would be slower than the average speed overall.
3km in 1h
v_{av}=3kmh^{-1}
4km in 1h
v_{av}=4kmh^{-1}
This still doesn't give the ***instantaneous speed***, or speed at an instant in time, which would change from moment to moment as we encounter obstacles or just stop to catch our breath.
v=0
To calculate instantaneous speed, we need to measure how long it took to walk a very short distance up the mountain...
1cm
As we go down a zipline, our instantaneous speed would increase. The instantaneous speed at the bottom would be much bigger than the average speed.
This is because gravitational force would accelerate us. ***Acceleration*** is the rate of change in velocity: $\vec{a} = \frac{\Delta\vec{v}}{\Delta t}$
v
v
Acceleration is also a vector, with the same direction as the change in velocity.
v
a
So if our velocity increases, then we accelerated in the same direction as we moved.
$\vec{a}$
velocity increases
If our velocity decreases, then acceleration is in the opposite direction of our motion.
$\vec{a}$
velocity decreases
Scalars, vectors, distance, displacement, speed, velocity and acceleration are useful concepts to describe the motion of any bodies, be they planets or atoms. This branch of physics which studies motion is called ***kinematics***.

1.2 Describing motion, distance and displacement

1. Compare each pair of terms.

Terms	How are they similar? How are they different?
qualitative quantitative	
scalar vector	
distance displacement	
speed velocity	
instantaneous speed average speed	
velocity acceleration	

2. Circle the qualitative descriptions.

3 m s^{-1} fast N45°W one metre heavy

3. Circle the vectors.

distance mass acceleration time velocity

4. State if each sentence is describing distance, displacement, speed, velocity or acceleration.

a) Walk 5 m to the left. ____________

b) The ruler is 1 m long. ____________

c) If you drop a ball, gravity will make it go faster by 9.8 m s^{-2}. ____________

d) If you drop a ball, one second later it will be moving down at 9.8 m s^{-1}. ____________

e) You can travel at 60 km h^{-1} on the main road. ____________

f) The aeroplane is flying at 880 km h^{-1} at N45°W. ____________

5. The formulae for speed and velocity are shown below. Write what the symbols mean.

Speed: $v = \frac{\Delta d}{\Delta t}$

$v =$ ____________

$\Delta d =$ ____________

$\Delta t =$ ____________

Velocity: $\vec{v} = \frac{\Delta \vec{s}}{\Delta t}$

$\vec{v} =$ ____________

$\Delta \vec{s} =$ ____________

$\Delta t =$ ____________

6. You walk along a straight line 5 m to the left, then 8 m to the right, then 2 m to the left

Draw a diagram

a) What distance did you travel?

____

b) What is your final displacement from where you started?

____

7. The only way to get from the town of Congupna to Lemnos via the main road requires driving 6.8 km south, then turning left to go 2.7 km west.

Draw a diagram

a) What distance is travelled?

b) What is the displacement of Lemnos from Congupna?

Use Pythagoras' theorem and trigonometry.

8. The Borobudur temple in Indonesia has a base which is roughly square in shape, with each side around 123 m long. A tourist walks from one corner to the opposite corner around the base.

Draw a diagram

a) What distance did they walk?

b) What is the magnitude of their displacement from their starting position?

9. In a computer game called "Snake", you moved a hungry snake around a grid for a total distance of 15 units, but the snake's final displacement was 5 units to the right of where it started. Describe one way this could have happened, using a diagram.

Draw a diagram

10. A student walked 1 km from home to school, then 0.5 km from school to a fish and chip shop, then 2 km back home. They calculate their total displacement as 1 + 0.5 + 2 = 3.5 km. Explain whether their answer is correct.

11. The Earth spins once on its axis every day. The Earth's radius is 6371 km. A laboratory is located on the equator.

A circle's circumference is $2\pi r$.

a) What is the displacement of the laboratory after one day?

b) What distance has the laboratory travelled around the axis of the Earth in one day?

12. Just for fun: You travel 10 km north, then 10 km west, then 10 km south. You find yourself... back where you started. You have a working compass and know how to measure direction and distance carefully. Where on Earth could you be?

What are you assuming about Earth?

1.3 Speed and velocity

SI units

Scientists worldwide have agreed on an international system of units called ***SI units***. All units we use in equations are based on SI units.

However, prefixes are often used in front of a unit to suit the scale of the objects being measured. We often need to convert from these other units to SI units. Prefixes commonly used include:

Prefix	Symbol	Means	In scientific notation
milli	m	x 0.001	x 10^{-3}
centi	c	x 0.01	x 10^{-2}
kilo	k	x 1000	x 10^{3}

1. The SI unit for distance or displacement is the metre (m). Here are some other common units for length and where they might be used. Complete the table.

Distance	Value	What is this in metres?	Name the unit
Thickness of door	35 mm		millimetre
Height of A4 paper	29.7 cm		
Sydney to Canberra	286 km		

2. The SI unit for time is the second (s). Other common units are the minute, which is 60 seconds, and the hour, which is 60 minutes. Complete the table describing units for time.

Time	Value	What is this in seconds?	Name the unit
computer response time	10 ms		
cooking a soft boiled egg	6 min		
flight from Sydney to Perth	5 h		hour

Based on SI units, the unit for speed or velocity would be m s^{-1}. However, another commonly used unit for speed or velocity is km h^{-1}, for example for traffic on roads. Practise converting between these.

Example: The speed limit for cars in a school zone is 40 km h^{-1}. What is this in m s^{-1}?

Shortcut: Divide km h^{-1} by 3.6 to get m s^{-1}.

Answer:

$$\frac{40 \text{ km}}{\text{h}} = \frac{40 \times 1000 \text{ m}}{60 \text{ min} \times 60 \text{ s}} = 40 \div 3.6 = 11 \text{ m s}^{-1}$$

Try to visualise: this is about the width of a tennis court travelled in 1 s.

Example: On average, adults walk at a speed of 1.4 m s^{-1}. What is this in km h^{-1}?

Shortcut: Multiply m s^{-1} by 3.6 to get km h^{-1}.

Answer:

$$\frac{1.4 \text{ m}}{\text{s}} = \frac{1.4 \div 1000 \text{ km}}{1 \div (60 \text{ min} \times 60 \text{ s})} = 1.4 \times 3.6 = 5.0 \text{ km h}^{-1}$$

Example: Some elite swimmers can swim a 50 m freestyle race in 25 s. What is their speed in m s^{-1} and km h^{-1}?

Write the equation. Substitute data. Include units in the answer.

Answer:

$$v = \frac{\Delta d}{\Delta t} = \frac{50}{25} = 2 \text{ m s}^{-1} = 2 \times 3.6 \text{ km h}^{-1} = 7.2 \text{ km h}^{-1}$$

3. The speed of sound in air varies depending on temperature, but is around 340 m s^{-1}. What is this in km h^{-1}?

4. The speed of very fast jets is often compared to the speed of sound, using a Mach number. A jet travelling at Mach 1 is travelling at the speed of sound. A jet travelling at Mach 0.5 is travelling at half the speed of sound. Assume the speed of sound in air is 340 m s^{-1}. One of the fastest jets in the world can travel at Mach 9.6. What is this speed in m s^{-1} and km h^{-1}?

5. The fastest runner on Earth as at 2022 was Usain Bolt, who ran 100 m in 9.58 s.

   a) What was his speed in this race?

   b) Was his running speed faster than a car driving at 50 km h^{-1} on a suburban street?

Write the equation. Substitute data. Convert units to compare. Include a sentence to answer the question.

Rearranging equations

You often need to rearrange an equation to make the variable you are finding the subject. This means getting the variable on its own on the left hand side of the equation.

Example: Australia's longest road is the Eyre Highway, stretching 145.6 km across the flat, treeless Nullabor Plain. At the speed limit of 110 km $h^{-1,}$ how long would it take to drive the whole length of the Eyre Highway?

Write the equation. Substitute data. Make time the subject – whatever you do to one side of the equation, do to the other. Include the unit in the answer, consistent with the other units used.

Answer:

$$v = \frac{\Delta d}{\Delta t}$$

$$110 = \frac{145.6}{\Delta t}$$

$$110 \times \Delta t = 145.6$$

$$\Delta t = \frac{145.6}{110} = 1.32 \text{ h}$$

6. You walk from home to the train station in 10 min. Your average walking speed is 1.4 m s^{-1}. What is the distance you walked?

Since speed is in ms^{-1} you must convert time to s.

7. How far away do cars need to be so you can cross a street to the other side before they reach you? A typical suburban street is around 7.5 m wide. Assume cars obey the speed limit of 50 km h^{-1} and you walk at 1.4 m s^{-1}.

This is a two part problem. Work out the time. Then use this to work out distance.

Speed or velocity? Average or instantaneous?

8. Explain the difference between average and instantaneous speed, by filling in the blanks.

acceleration	average	average	distance	instantaneous
instantaneous	instantaneous	same	time	

The speedometer of a car shows you your speed at any one moment in ________. This is your ________ speed. The odometer shows the total ________ travelled. Dividing the odometer reading by the total time taken to travel it will give your car's ________ speed.

The ________ speed can vary within this time. For example, if you are driving at 50 km h^{-1} then brake for a red light, your ________ speed started off as 50 km h^{-1} and ended as 0 km h^{-1} but your ________ speed will be between these. However, if you drove at a ***uniform*** speed so there is no ________, then your average speed and instantaneous speed will be the ________.

9. Explain the difference between speed and velocity, by filling in the blanks.

average	circumference	direction	displacement	distance
instantaneous	tangent	time	time	velocity

Velocity differs from speed because it also has ________. If a child rides a merry-go-round that spins at a uniform speed, the ________ is not the same, because the child changes direction. You can work out the child's speed by measuring the ________ it takes for the child to spin once around and return to the starting position. You also need to measure the ________ the child sits from the centre of the merry-go-round, r. The speed throughout will be the ________, $2\pi r$, divided by the ________.

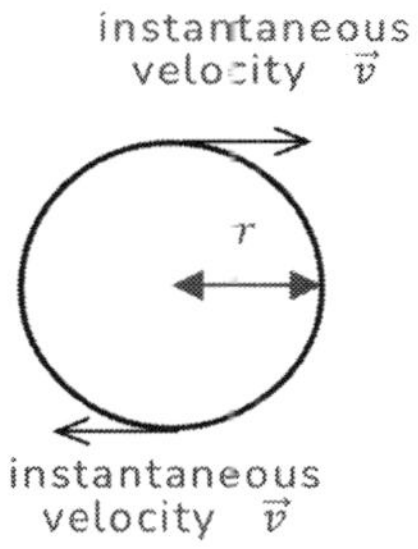

However, since the child returned to the starting point, there is no ________, so the ________ velocity will be zero. The ________ velocity at any point is the same magnitude as speed though, but the direction is at a ________ to the circular path travelled.

10. A student wants to drive from school to the theatre. They measure the distance between the school and theatre on a map to be 11.5 km. They estimate that if they drive at the speed limit on the road of 60 km h^{-1}, they will reach the theatre in 11.5 min. However, it actually takes 30 min to get there.

a) Explain what incorrect assumption the student made in calculating the time for the journey.

b) What was the average speed for the journey?

11. The Moon orbits the Earth once every 27.3 days, at an average radius of 384000 km. What is the average speed of the Moon around the Earth, in km h^{-1}?

12. A basketball court measures 28 m by 15 m. The coach makes a player jog from one corner to the opposite corner, around the sides of the court. The player jogs at a steady 2.4 m s^{-1} throughout.

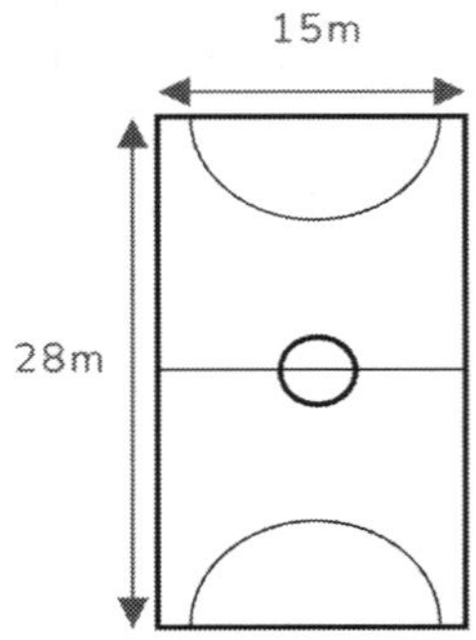

a) How long does it take for the player to reach the opposite corner?

b) Compare the player's average speed, instantaneous speed, average velocity and instantaneous velocity throughout.

Compare means explain similarities and differences. Include calculations and comment on comparisons eg. same, bigger, smaller.

1.4 Motion diagrams for constant velocity

Motion capture uses technology to record the position of moving objects at regular time intervals. Technologies used can range from basic ticker timers in school laboratories, to frame-by-frame video analysis. Motion capture is used in many areas, including sport, entertainment, quality control in machinery and the military.

1. Here is an example of a motion diagram, showing uniform velocity. Dots in the diagram represent an object travelling to the right, captured at a frequency, f, of 10 Hertz (Hz). This means 10 time intervals between dots per second.

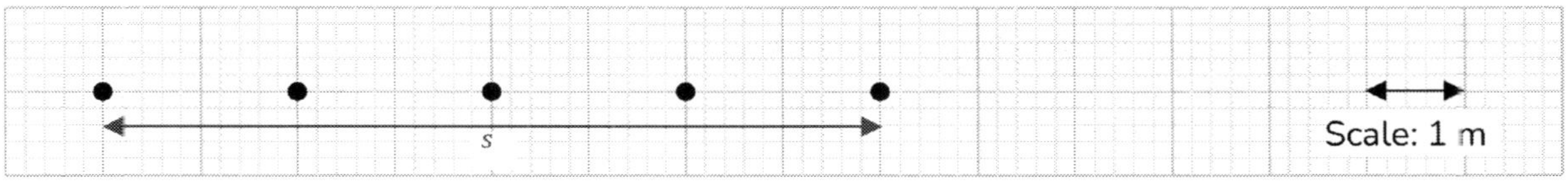

a) What was the total displacement, s?

> Count the intervals between major gridlines (dark lines).

Distance between first and last dots on the grid = ______________ units

Distance between first and last position in real life

= ______________ units × 1 m = ______________ m

b) What was the total time of the object's motion, t, recorded by the diagram?

> Count the dots after the first one to work out the number of periods shown.

The time interval between two consecutive dots is called the *period*, T:

$T = \frac{1}{f} =$ ______________ s

The number of periods shown in the diagram is ______________.

Total time for which the motion was recorded

$= T \times$ number of periods = ______________ s

c) Calculate the velocity of the object.

> Remember the unit and direction for vectors.

$v = \frac{s}{t} =$ ____

d) Explain why the diagram shows uniform velocity.

> Remember uniform means unchanging or constant.

____

____

2. A camera on a drone took a video of a car moving to the right. The dot represents where the car was in successive frames taken at 30 Hz.

> Highlight key information from the text and annotate the diagram.

> Calculate using larger measurements of distance and time for more accuracy.

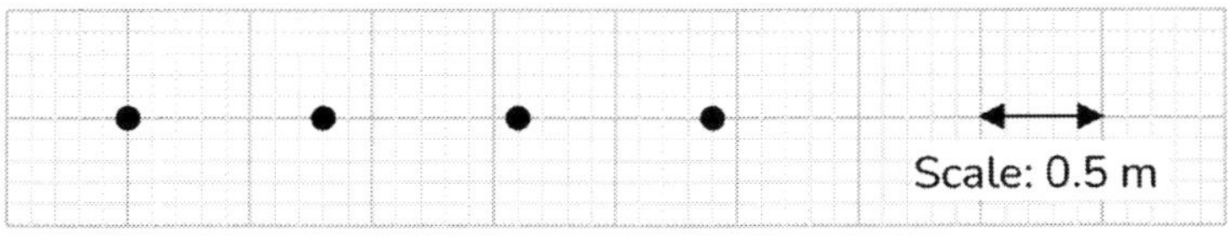

What was the velocity of the car?

> Remember the unit and direction for vectors.

Total displacement: $s =$ ____

Total time: $t =$ ____

Velocity: $v =$ ____

3. A runner on a beach runs at a constant speed of 2.3 m s^{-1} in a straight line. They take 120 steps per minute when they run.

a) What is the time interval between each footprint?

b) What is the distance in real life between each footprint (known as the stride length)?

c) Draw a motion diagram to represent the position of their footprints on the beach, showing 5 footprints. Use the scale shown.

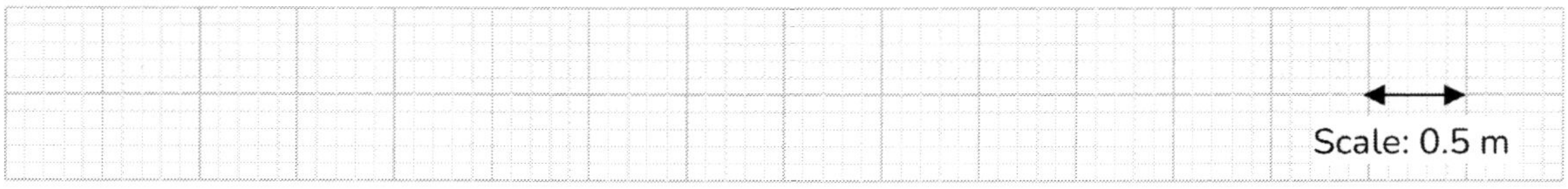

Convert the real distance between two consecutive footprints to distance on the scale:

Calculate each footprint position on the diagram by multiplying this by 2, 3, 4 and 5:

4. Stop motion is a style of animation where objects are moved small distances and photographed at every step. The photographs are played one after the other rapidly to make a video in which the objects look like they are moving on their own. For example, the "Shaun the Sheep" movies were made using stop motion.

An animator took a photograph of a tiny car as it was positioned on each dot on the grid below, from left to right, then played the photographs at 10 frames per second.

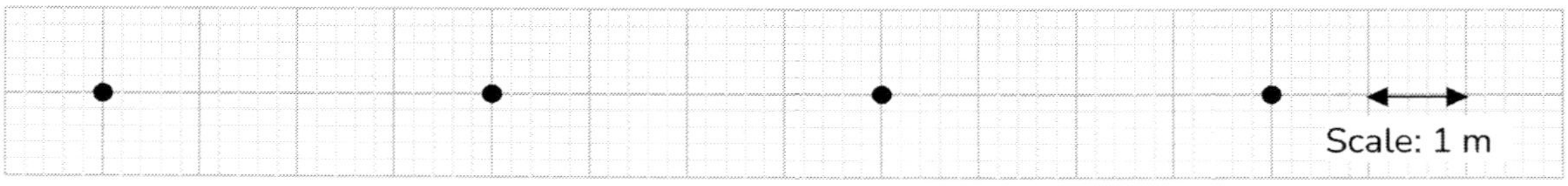

a) How fast would it look like the car was moving in the video?

b) They want to make a video of the car moving at the same velocity, but using 24 frames per second to play the photographs. On the same diagram as above, mark the positions they would need to place the car for the photographs to make a similar video. Show working below.

c) Outline an advantage and a disadvantage of making the animation in 10 frames per second versus 24 frames per second.

2.1 Uniform acceleration and frames of reference

When objects are travelling along a straight line, we can define one way as positive and the other way as negative. If we let up be positive, then down is negative.
+
−
This means initial velocity is positive, since we threw the ball up
$u=+5ms^{-1}$
and acceleration is negative, since gravity acts downwards.
$a=-9.8ms^{-2}$
Now we get $\vec{v} = (+5) + (-9.8) \times 0.5 = 0.1ms^{-1}$
The positive answer means the ball is still going up, but has almost slowed to a stop.
We get the same answer if we let down be positive too.
$u=-5ms^{-1}$
$a=+9.8ms^{-2}$
Now we get $\vec{v} = (-5) + (+9.8) \times 0.5 = -0.1ms^{-1}$
The negative answer this time means the ball is still going up.
Setting which direction is positive is part of defining a ***frame of reference***.
let north be positive
We can do it for any straight line motion, not just up and down.
let left be positive
so right is negative
From $\vec{v} = \vec{u} + \vec{a}t$ two other equations can be derived: $\vec{s} = \vec{u}t + \frac{1}{2}\vec{a}t^2$ $\vec{v}^2 = \vec{u}^2 + 2\vec{a}\vec{s}$
$\vec{s}$ means ***displacement***. But from where?
We need to choose a frame of reference that defines where the ***origin*** is as well as setting a direction as positive.
y
x
0
origin
In other words, the frame of reference is a coordinate system.
The ***origin*** is the point we will take as zero displacement. This can be anywhere, but we usually make the starting point the origin.
+
0
positive s
negative s
$\vec{v} = \vec{u} + \vec{a}t$ $\vec{s} = \vec{u}t + \frac{1}{2}\vec{a}t^2$ $\vec{v}^2 = \vec{u}^2 + 2\vec{a}\vec{s}$ are known as the ***SUVAT*** equations.
Guess why?
Now you have the tools to calculate everything about the motion of an object in a straight line, if it is undergoing uniform acceleration.

2.2 Uniform acceleration

Understanding acceleration

1. Recall the definition of acceleration, in words and in equation form.

2. Describe what is meant by uniform acceleration.

3. The following tables record data about the motion of objects in a straight line. Work out the change in velocity, Δv, each second. Then identify whether each table could be describing uniform acceleration and outline a scenario that the data could be describing. The first table is filled in as an example.

Example:

time (s)	velocity (m s^{-1})	Δv (m s^{-1})
0	0	-
1	5	+ 5
2	10	+ 5
3	15	+ 5
4	20	+ 5

Acceleration is uniform as the change in velocity per second is the same. The data could describe an object accelerating down a smooth ramp.

a)

time (s)	velocity (m s^{-1})	Δv (m s^{-1})
0	19.6	-
1	9.8	- 9.8
2	0	
3	- 9.8	
4	- 19.6	

Is the acceleration uniform?

Possible scenario:

b)

time (s)	velocity (m s^{-1})	Δv (m s^{-1})
0	3	-
1	1.5	
2	0	
3	-1.5	
4	-3.0	

Is the acceleration uniform?

Possible scenario:

c)

time (s)	velocity (m s^{-1})	Δv (m s^{-1})
0	0	-
1	0.4	
2	1.0	
3	0.4	
4	0	

Is the acceleration uniform?

Possible scenario:

SUVAT equations

4. The SUVAT equations can be used to calculate variables about motion involving uniform acceleration. Recall what each variable means in the SUVAT equations and the unit used to measure it.

Variable	Meaning	Unit
$\vec{s}$		
$\vec{u}$		
$\vec{v}$		
$\vec{a}$		
t		

5. Identify whether you can use the SUVAT equations and the reason why.

   a) A ball that is dropped. ....

   b) A ball that is bounced. ....

   c) A rollercoaster with hills and dips. ....

   d) A car braking steadily as it approaches an intersection. ....

   e) A skier gliding down a smooth, straight slope. ....

   f) An aeroplane flying in circles around an airport. ....

6. What variable is missing from each of the SUVAT equations?

Equation	$\vec{v} = \vec{u} + \vec{a}t$	$\vec{s} = \vec{u}t + \frac{1}{2}\vec{a}t^2$	$\vec{v}^2 = \vec{u}^2 + 2\vec{a}\vec{s}$
Missing variable			

To solve problems, use the equation with the variable you need to find as well as variables that you already know.

Example: When the acceleration of cars is compared, people often talk about the time taken to go from 0 – 60 mph (miles per hour), which is 26.8 m s^{-1}. A certain model of car does this is in 7.2 s. Calculate its acceleration.

Set the frame of reference: which way is positive? Identify the variables. Write the equation. Substitute data. Make acceleration the subject. Include the unit and direction in the answer.

Answer:

Let forward be positive	$\vec{v} = \vec{u} + \vec{a}t$
$s =$	$26.8 = 0 + \vec{a} \times 7.2$
$u = 0$	$\vec{a} = 3.7$ m s^{-2} forward
$v = 26.8$	
$a = ?$	
$t = 7.2$	

7. A student accidentally drops a coin from a second floor balcony. The coin hits another student 7.0 m below.

Acceleration due to gravity is $g = 9.8$ m s^{-2} towards Earth's centre.

   a) With what velocity did the coin hit the student below?

   Positive is: Equation: ....

   $s =$ Substitute: ....

   $u =$ Rearrange: ....

   $v =$ ....

   $a =$

   $t =$

   b) Compare this velocity with a car driving in a school zone at 40 km h^{-1}.

....

....

8. Michael Jordan the basketballer can jump 1.22 m vertically up.

What is the velocity at maximum height?

a) How fast would they need to leave the ground in order to do this?

Check whether direction is positive or negative when substituting.

Positive is: Equation: ______

$s =$ Substitute: ______

$u =$ Rearrange: ______

$v =$ ______

$a =$

$t =$

b) What height could they jump to if they were on the Moon? Acceleration due to gravity on the Moon is equal to 1.625 m s^{-2}.

Use part (a) to help. What variable stays the same?

Positive is: Equation: ______

$s =$ Substitute: ______

$u =$ Rearrange: ______

$v =$ ______

$a =$

$t =$

9. Rhythmic gymnastics often involves throwing apparatus, then doing acrobatic tricks while it is in the air. The official minimum ceiling height is 12.0 m.

Keep up with showing clear working!

a) What is the initial velocity the gymnast needs to throw vertically up with in order for the apparatus to reach this height?

______

______

______

b) How long will the apparatus spend in the air before returning to its starting point?

Displacement at starting point?

An equation in the form $ax^2 + bx = 0$ factorises as $x(a + bx) = 0$ giving $x = 0$ and $a + bx = 0$.

______

______

______

10. The world record for the highest trampoline jump is 6.73 m.

Sometimes two equations are needed.

a) How long did it take to reach the maximum height from the trampoline?

______

______

______

b) How long would it have taken to fall back down to the trampoline from the maximum height? Compare this time to your answer in part (a).

You can choose a new origin.

______

______

______

11. Show that for any object launched straight up, the time taken on the way up equals the time taken on the way down back to the starting point.

Use symbols to represent numbers: Let u_0 be launch velocity. Let acceleration be $-g$.

What variables are the same on the way up and down? Use this to find expressions for time on the way up and then down.

12. Olympic springboard diving involves jumping up from a flexible board 3.0 m above the water, then completing tricks in the air before diving into the water. A diver jumps up with a velocity of 5.44 m s^{-1}.

Draw a diagram

a) Calculate their displacement from the springboard and velocity at times 0.5 s, 0.7 s and 1.2 s after they leave the board.

time (s)	displacement (m)	velocity (m s^{-1})
0.5		
0.7		
1.2		

b) Sketch a diagram showing their path through the air and label roughly where they are in their journey at each time.

Visualisation is very helpful in problem solving.

c) What is the time taken for them to reach the water?

Put the equation in the form: $ax^2 + bx + c = 0$ then use the quadratic formula: $x = \frac{-b \pm \sqrt{b^2 - 4ac}}{2a}$

13. The reaction time of drivers in an emergency is on average 1.5 s. That is the time taken to observe a situation, think about it and press the brake. The deceleration of a typical car when the brake is fully pressed is about 7.0 m s^{-2}.

This problem has two parts – before and after the brake is pressed.

a) Estimate the average stopping distance for cars driving in an Australian suburban street with a speed limit of 50 km h^{-1}.

b) Before 2003, the speed limit used to be 60 km h^{-1}. What is the extra stopping distance required at this speed?

Pedestrian injuries were around 20% more frequent before 2003.

14. The table records data about the displacement from the starting point of an object in a straight line.

Assume initial velocity is zero.

a) Explain how the acceleration from t = 0 to the end of each time interval is calculated, assuming SUVAT equations can be used.

b) Calculated the missing values in the table.

c) Identify whether the object was in fact undergoing uniform acceleration.

time (s)	displacement (m)	acceleration (m s^{-2})
0	0	-
0.2	0.2	10
0.4	0.8	10
0.6	1.8	
0.8	3.2	

15. The table records data about the displacement from the starting point of an object in a straight line.

Assume initial velocity is zero.

a) Calculated the missing values in the table, assuming SUVAT can be used.

b) Identify whether the object was in fact undergoing uniform acceleration.

time (s)	displacement (m)	acceleration (m s^{-2})
0	0	-
1	100	200
2	300	
3	600	
4	1000	

16. An object is dropped from rest on Earth. It will accelerate towards the ground at 9.8 m s^{-2} (neglecting air resistance). This means every second, the speed increases by 9.8 m s^{-1}.

a) Complete the table by calculating the object's final speed and displacement at each time.

b) Calculate the object's average speed during the 4.0 s.

c) When does the average speed equal the instantaneous speed?

time (s)	speed (m s^{-1})	displacement (m)
0	0	0
1.0	9.8	
2.0		
3.0		
4.0		

17. Show that for uniform acceleration, the average speed equals the instantaneous speed at half time.

18. Show that for uniform acceleration, the average speed $\vec{v}_{av}$ is given by $\vec{v}_{av} = \frac{\vec{u}+\vec{v}}{2}$.

2.3 Visualising acceleration

1. The motion diagram shows an object dropped from rest. Images were taken at a frequency of 10 Hz.

   a) Describe how to read the motion diagram, by filling in the blanks.

change	dividing	increasing
multiplying	second	velocity

The motion diagram shows the position of the object at a rate of 10 images per ____________. We can find the time interval between positions by ____________ 1 by 10.

We can find the displacement during each time interval by ____________ the number of grid units travelled by 0.1 m. The displacement of the object during each equal time interval is ____________. This means the object's average ____________ must be increasing.

As acceleration is the ____________ in velocity over time, this means the object's acceleration must be downwards.

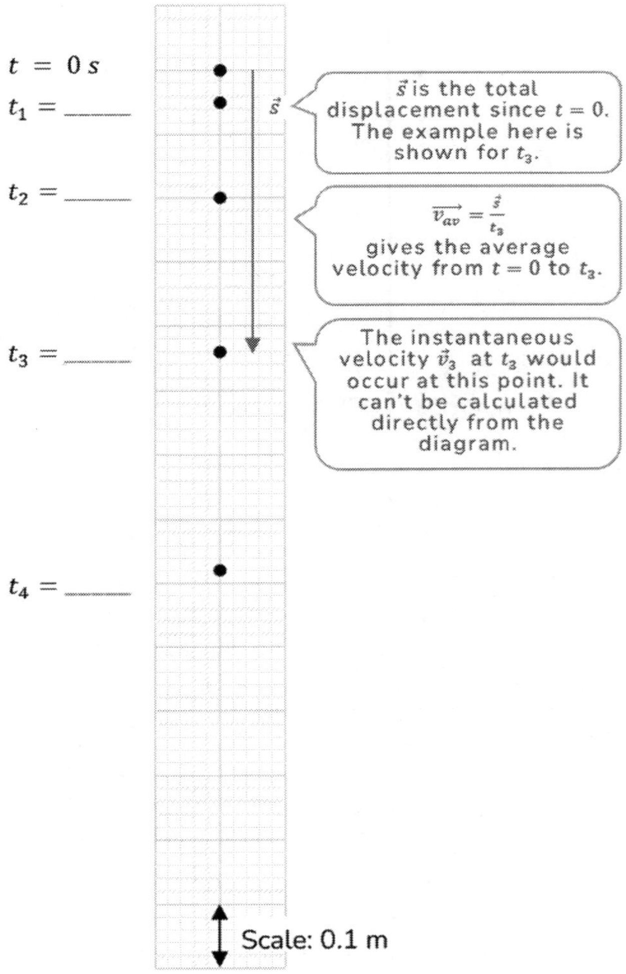

   b) Calculate the time interval between dots.

   ____

   c) Fill in the times t_1 , t_2, t_3 and t_4 at which each position on the diagram occurs.

   d) Fill in the table to record the displacements from the diagram for each time t_1, t_2, t_3 and t_4. Then calculate the average velocity. The first one is done for you. Leave the last column blank for now.

time (s)	displacement (m)	average velocity (m s^{-1})	instantaneous velocity (m s^{-1})
0.1	0.05	0.50	

   e) Identify the pattern in the average velocity.

   ____

   f) Velocity increasing by the same amount each equal time interval indicates that acceleration is constant. Was this the case here? ____________

   Yes means we can use SUVAT.

   g) Use the time and displacement for t_4 to calculate the acceleration.

   Using the biggest increases accuracy.

   ____

   ____

   h) Calculate the instantaneous velocity at times t_1 , t_2, t_3 and t_4 in the last column of the table.

i) Identify any pattern in the relationship between the average velocity and instantaneous velocity.

....

j) Determine where the next dot will be and mark it on the diagram.

....

2. True or false? Use the numbers from question 1 to help you.

a) If an object undergoes constant acceleration, its displacement increases by the same amount each equal time interval. ....

b) If an object undergoes constant acceleration, its instantaneous velocity increases by the same amount each equal time interval. ....

c) You can't always use SUVAT if an object is accelerating. ....

d) If acceleration is constant, the instantaneous velocity changes by at each time interval. ....

e) If an object is dropped from rest and undergoes constant acceleration, the average speed during a time period is given by $v_{av} = \frac{u+v}{2}$. ....

f) If an object undergoes constant acceleration, the average velocity during a time period will equal the instantaneous velocity halfway through the time period. ....

3. The dots on the motion diagram represent a racecar during a test drive, moving to the right with an initial velocity of 31.3 m s^{-1}. Two images were taken each second.

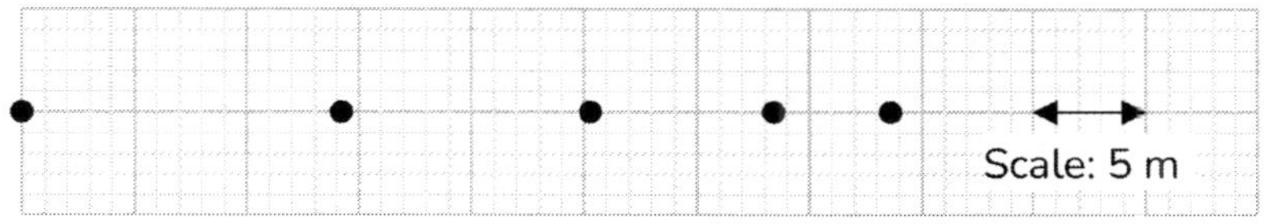

a) Explain what the driver was doing to the racecar.

Explain means describe giving a reason.

....

....

b) Calculate the acceleration, assuming it was uniform. Show that the diagram is consistent with uniform acceleration.

Choose one position and time. Calculate the acceleration as if it were constant. Does this value correctly predict the other positions?

....

....

....

....

4. A runner accelerates from rest at 6 m s^{-2}. If you capture their motion at 5 frames per second, draw what you would see on the motion capture diagram. Use a dot to represent the runner and mark three positions after the starting point on the grid. Show your working and state what scale you are using.

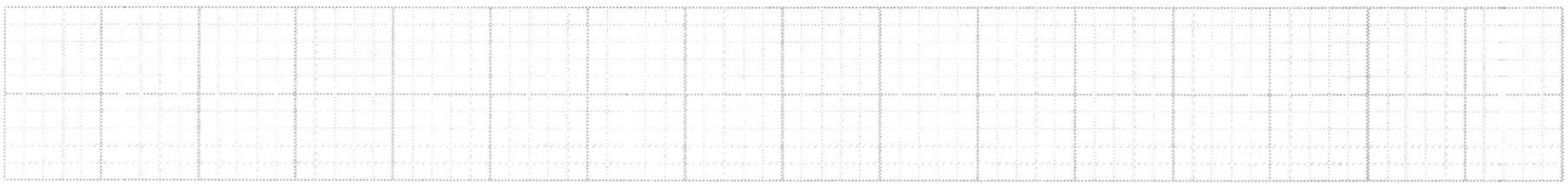

....

....

....

Choose a scale that is convenient to work with but takes up at least half the space.

3.1 What makes a good experiment?

A good method gives ***accurate*** results.
That means our measuring equipment and techniques allow us to observe the true value of a variable.
What measuring equipment shall we use for accurate results?
Measuring equipment with a smaller ***limit of reading*** is more accurate.
limit of reading ±0.5mm
mm
cm
limit of reading ±0.5cm
To prevent ***parallax error***, measurements need to be observed at eye level.
2
1
m
2
1
m
Human reaction time is another source of error.
The effect of human reaction time is reduced when timing something that takes longer
In 3... 2... 1... !
and more of a problem when timing something fast
Oops! Missed it!
To avoid human reaction time and increase accuracy, we could video the results.
00.0
Errors that occur consistently for every result are called ***systematic errors***, such as parallax error. Repeating the experiment won't help get closer to the correct value.
Other errors that aren't predictable are ***random errors***, such as the wind. Repeating the same measurement for many trials might give an average will be close to the true value.
A good method should obtain ***reliable*** results. That means repeat trials of the same measurement give similar results.

height (m)	time (s)		
	trial 1	trial 2	average
5.0	1.0	1.0	1.0

To see a ***trend*** or pattern in the results, a graph is useful.
time (s)
height (m)
Finally, our ***conclusion***! The results show that as the height above the ground increases, the time taken to reach the ground increases. Our hypothesis was correct.
But... is this a ***linear relationship***?
time (s)
height (m)
To increase validity, a wider selection of drop heights should be tested.
time (s)
height (m)
As the drop height increases, the time taken to reach the ground increases at a decreasing rate.
How do we verify the equation $s=\frac{1}{2}at^2$ though? Check out the part on Linear regressions...

3.2 Experimental skills

1. Write a definition for each term.

Term	Definition
variable	
independent variable	
dependent variable	
controlled variable	
fair test	
aim	
hypothesis	
conclusion	
validity	
accuracy	
reliability	

2. Here is an example of part of an experimental method:

Method:

1. Set up as shown in the diagram:

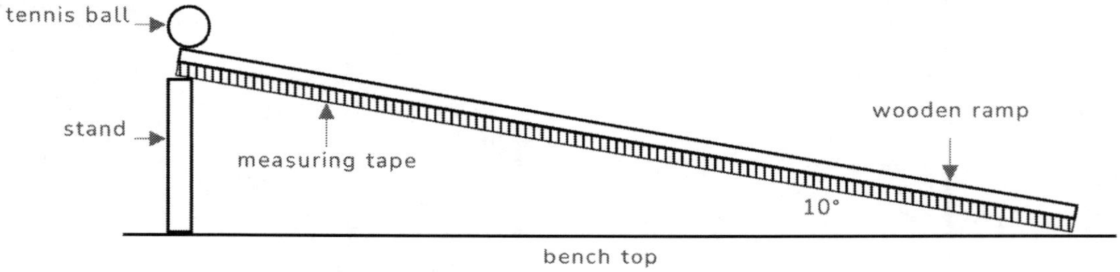

2. Release the ball from rest. Using a stopwatch, time how long it takes for the tennis ball to travel 0.5 m down the ramp and record the result.
3. ...
4. Repeat steps 1 to 3 for distances of 1.0 m and 1.5 m.

a) What is the independent variable? ____________________

b) What is the dependent variable? ____________________

c) Identify the controlled variables in the method.

____

d) Write an aim for this experiment.

Aim: To investigate the effect of ____________________ on the

____________________.

e) Write a hypothesis for this experiment.

Hypothesis: As the ____________________ increases, the

____________________ will increase / decreases / stay the same.

f) What steps have been taken in the method to get accurate results?

g) Suggest two improvements to increase the accuracy of results and explain why these are needed.

h) What step should be added at step 3 of the method to increase the reliability of results?

i) Apart from controlled variables, the method could be improved to obtain a more valid trend. Suggest an improvement to the method to achieve this.

Here is a table of results for the experiment in question 2. Use this to answer questions 3 to 6.

Distance (m)	Time taken (s)					
	Trial 1	Trial 2	Trial 3	Average	Absolute error	Percentage error
0.5	0.89	0.70	0.78	0.79	±0.1	±13%
1.0	1.18	0.98	1.08			
1.5	1.35	1.32	1.31			

3. The ***average*** is calculated by adding the results from each trial, then dividing by the number of trials. For the 0.5 m distance, the average time taken is:

$$\text{average} = \frac{0.89 + 0.70 + 0.78}{3} = 0.79$$

a) Calculate the average of the trials for the other distances in the table.

b) Write a conclusion for the experiment.

Calculated values cannot be more accurate than the actual results. Here, we average numbers with 2 significant figures so the answer is rounded to 2 significant figures too.

4. To determine how reliable these results are, we can calculate the ***uncertainty***, or how far from the average the trials were.

Errors only need to be given to 1 significant figure.

a) To calculate uncertainty in the reliability as an ***absolute error***, use:

$$\text{uncertainty} = \pm\frac{|\text{highest trial} - \text{lowest trial}|}{2}$$

For example, for the 0.5 m distance in the table:

$$\text{uncertainty} = \pm\frac{0.89 - 0.70}{2} = 0.1$$

Calculate the uncertainty as an absolute error for the other times.

b) An absolute error of ±0.1 s is big compared to a time of 0.5 s, but small compared to 50 s. It is therefore useful to calculate ***percentage error***.

$$\text{uncertainty} = \pm\frac{|\text{absolute error}|}{\text{observed value}} \times 100\%$$

For example, for the 0.5 m distance in the table:

$$\text{uncertainty} = \pm\frac{|0.1|}{0.79} \times 100\% = 13\%$$

Calculate the uncertainty as a percentage error for the other times.

Round percentage errors to the nearest whole.

5. Absolute and percentage errors can also be used to evaluate the accuracy of measuring instruments. The limit of reading of a measuring instrument gives an absolute error which is half of the smallest interval that it can measure.

For example, the ruler in the picture can read in millimetres, or 0.1 cm, so the limit of reading is 0.05 cm. The length of the pencil, together with the absolute error, is 7.7 ± 0.05 cm.

The percentage error for this is:

$$\text{uncertainty} = \pm\frac{|\text{absolute error}|}{\text{observed value}} \times 100\%$$

$$= \pm\frac{|0.05|}{7.7} \times 100\% = \pm 7\%$$

a) If this tape measure was used for the experiment in question 2, calculate the absolute and percentage error when measuring the distances.

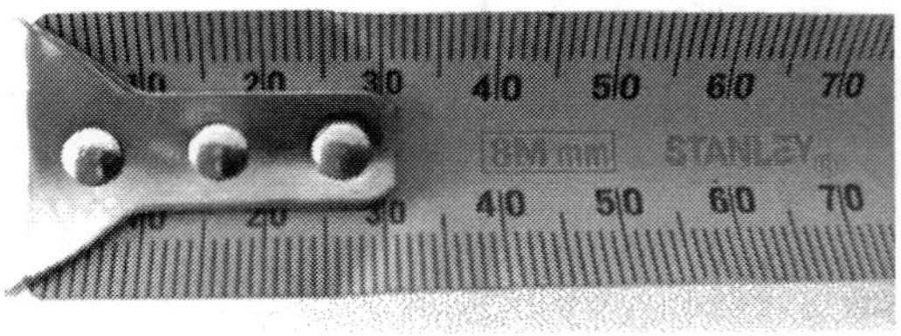

distance (m)	absolute error (m)	percentage error
0.5		
1.0		
1.5		

b) The average times for the experiment in question 2 are shown below. If this stopwatch was used to measure time, calculate the absolute and percentage error arising from the limit of reading.

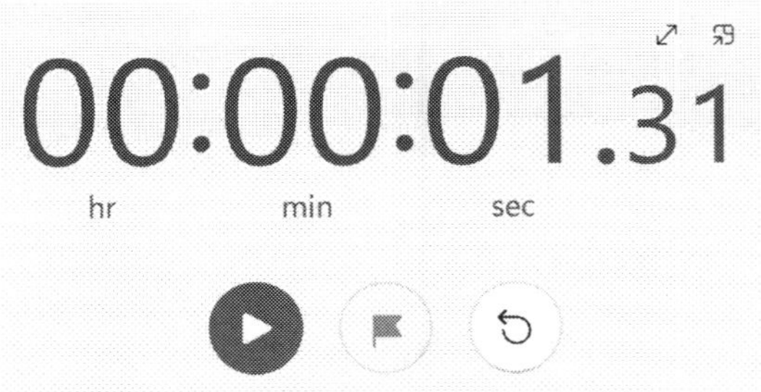

time (s)	absolute error (s)	percentage error
0.79		
1.08		
1.31		

c) Explain whether the limit of reading of the measuring tape and stopwatch are significant concerns in this experiment.

6. The average human reaction time to things we see is around 0.25 s. This time would be the absolute error due to reaction time for the times measured in the experiment.

a) Compare the percentage error due to reaction time when timing an event that takes 1.0 s and another that takes 5.0 s.

b) Look again at the data table for question 3. Suggest a reason for the difference in the reliability of the trials.

7. There are many errors that can lead to experimental results differing from the true value. These errors can be classified as ***random errors*** or ***systematic errors***.

An example of ***random error*** is the wind blowing. Random errors skew results unpredictably, but conducting many trials should give an average that is close to the true value.

An example of ***systematic error*** is zeroing error, such as using a scale which does not read zero even though there is no mass on it. Systematic errors skew results predictably the same way and conducting repeat trials will not bring the average closer to true value.

Classify the following errors as random or systematic.

a) Vibrations from people or traffic moving nearby.

____

b) Parallax error if a video camera is set up an angle to what is being observed.

____

c) The ball moving quickly so it is difficult to judge when it reaches a certain position.

____

d) A rough ramp, causing the ball to take slightly different paths each time.

____

Random errors can occur if motion experiments are conducted in windy conditions

Systematic errors occur if scales are not zeroed

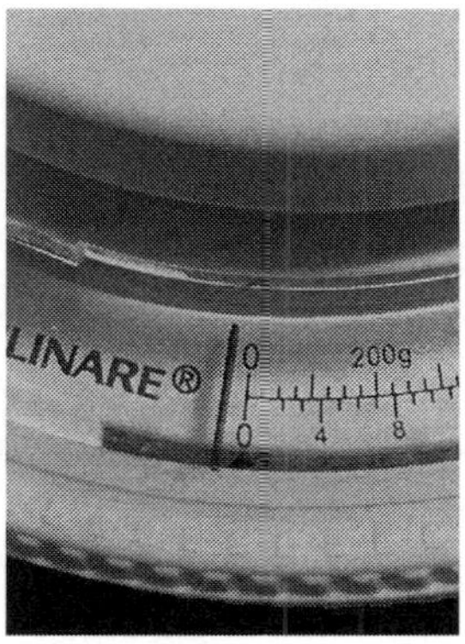

4.1 Motion graphs

Reading line graphs

The most common graph used in physics is a *line graph*. It has numbers on both axes and is used if data can be continuous – meaning that it is possible for data to have any numerical value. The advantages of a line graph are that it can show numerical trends in the data and can be used to predict results.

If the speedometer reading of a car is recorded every 0.5 s, the results can be graphed like this:

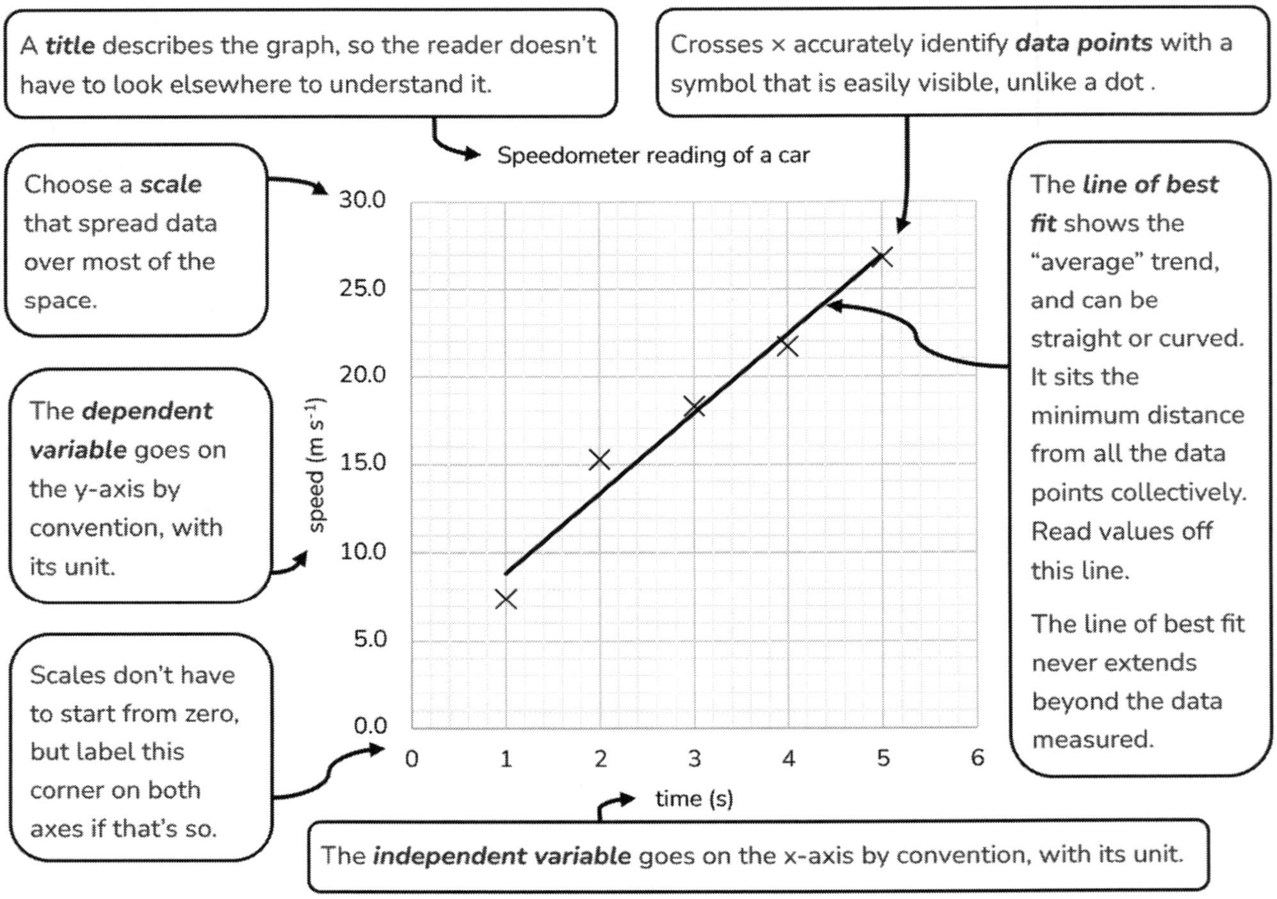

Use the information above to answer questions 1 to 8.

1. Recall when a line graph should be used and the advantages of using one.

   ____

2. Explain whether a line of best fit needs to touch any of the data points.

   ____

3. What was the speed of the car at 2.6 s? ____________________

4. What was the speed of the car at 4.6 s? ____________________

5. Estimate the speed of the car at 0 s. Line up a ruler with the line of best fit and draw a dotted line to extend it in order to predict this value. This is called *extrapolation*. ____________________

6. What was the driver doing to the car in real life? ____________________

7. Describe the trend in the graph.

   ____

8. How valid is the trend you stated in question 7? In other words, do all of the data points sit on the line of best fit? If not, suggest some real life reasons why this might be the case.

   ____

Gradient of a graph

The gradient of a graph often gives useful information. The gradient is given by the "rise over run". If we pick any two points (x_1, y_1) and (x_2, y_2) on the line of best fit:

$$\text{gradient} = \frac{\text{rise}}{\text{run}} = \frac{\Delta y}{\Delta x} = \frac{y_2 - y_1}{x_2 - x_1}$$

For a speed-time graph, as the y-axis shows speed and the x-axis shows time:

$$\text{gradient} = \frac{\Delta v}{\Delta t}$$

However, recall that:

$$a = \frac{\Delta v}{\Delta t}$$

So the gradient equals acceleration. Note its unit is the y-axis unit (m s^{-1}), divided by the x-axis unit (s), giving m s^{-2}.

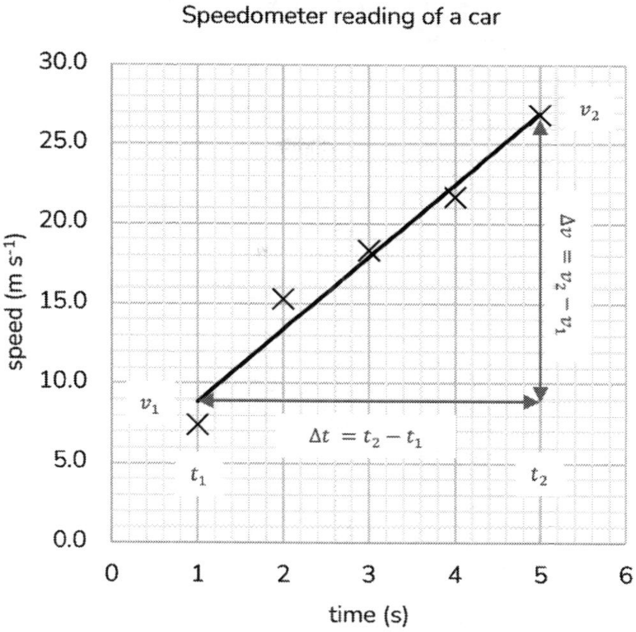

Use the graph above to answer questions 9 and 10.

9. Calculate the acceleration.

$a = gradient = \frac{\Delta v}{\Delta t} = \frac{v_2 - v_1}{t_2 - t_1} =$ ....

10. Explain whether the acceleration was constant.

....

11. The graph shows the speed of another car.

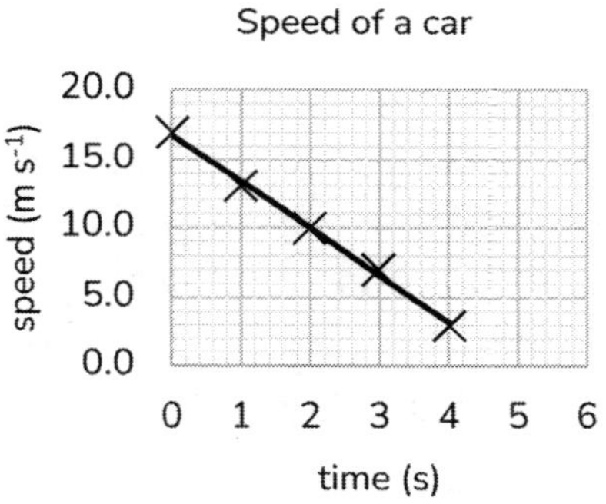

a) Calculate the acceleration of the car.

....

b) What is the driver doing to the car?

....

c) Predict the speed of the car at 5 s if it continues to be driven the same way.

....

12. Here's a different type of graph of an ice skater moving in a straight line.

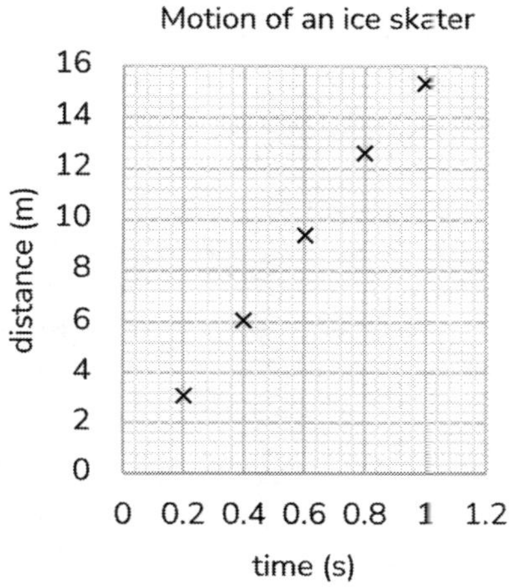

a) What is the dependent variable?

....

b) Draw a line of best fit. What does the gradient of the line of best fit represent? Calculate this.

....

....

....

> The gradient of any graph will be related to what's on the y-axis divided by the x-axis.

Gradient on curved graphs

13. Explain how to interpret curved graphs, by filling in the blanks.

4.9	accelerating	acceleration	changes	displacement	down	one
tangent	time	uniform	up	velocity	zero	

When the line of best fit is curved, the gradient ______________ throughout. However, even if a velocity-time graph is curved, the gradient still represents ______________. The acceleration will not be ______________ though. Likewise, for a curved displacement-time graph, the gradient still represents ______________, as both its gradient and velocity are given by change in ______________ over ______________. Velocity will not be uniform, meaning the object graphed is ______________.

The gradient at any one point can still be found by drawing a tangent to that point, which is a line that just touches the curve at ______________ point. The gradient of the line at that point can then be calculated using the "rise over run" of the ______________.

For example, in the displacement-time graph shown, at $t = 0$ the gradient of the tangent is zero, so the velocity of the ball is ______________.

At $t = 0.5\ s$ the gradient of the tangent is:

gradient = velocity

$= \frac{1.4-0.72}{0.54-0.4} =$ ______________ m s^{-1}.

______________ is taken as positive in the example, as the displacement increases as ball drops. Therefore a positive gradient, where the graph curves ______________ the page, means the velocity is downwards.

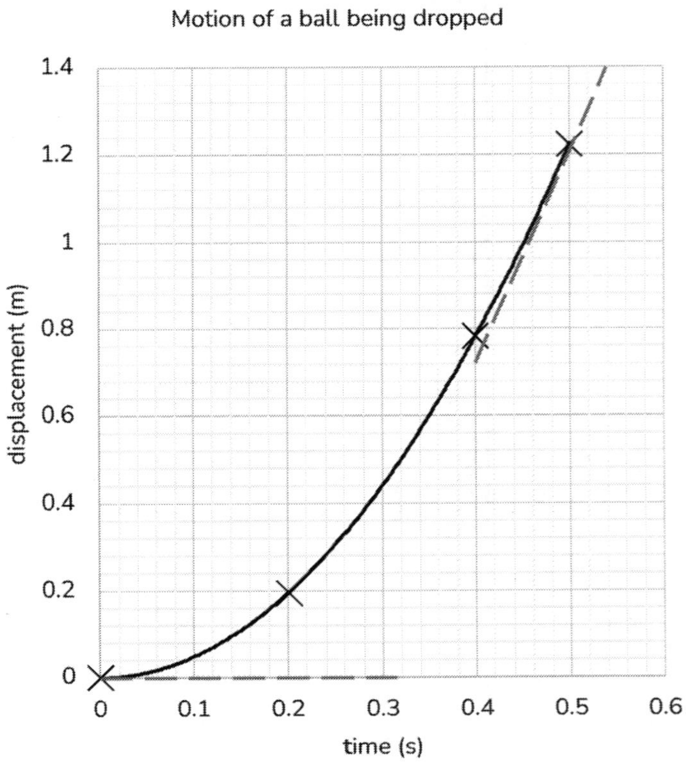

Time (s)	Velocity (m s^{-1})
0	0
0.2	
0.4	
0.5	4.9

14. Use the graph from question 13 for this question.

a) Complete the table by drawing the tangent at each of the times with missing data and using their gradient to calculate the velocity.

b) Graph your results.

c) Explain what the gradient of the graph you just drew represents. Calculate it.

____

____

____

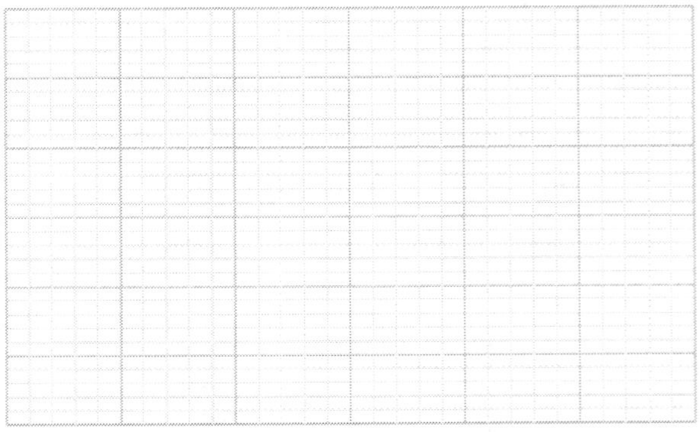

Area of graphs

15. Explain how to interpret the area of graphs, by filling in the blanks.

acceleration	area	distance	extra distance	initial speed	multiplied	rectangle
speed	total distance	trapezium	triangle	x-axis	y-axis	

In geometry, the ______________ of many shapes is related to height and width.

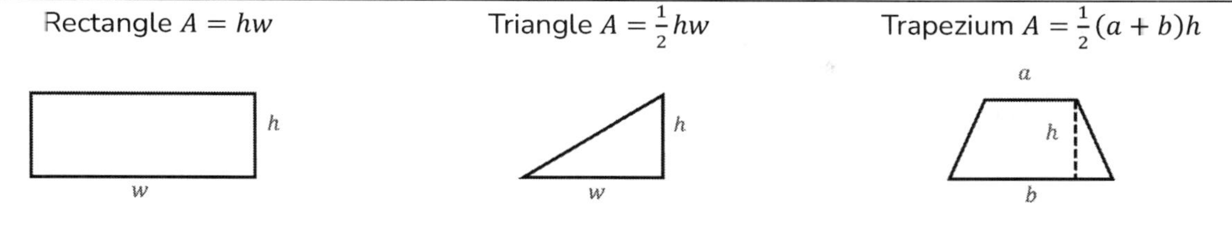

The area between a line graph and the x-axis is also calculated using the height, given by the change in the ______________, and width, given by the change in the ______________.

Let's revisit the example from question 1. Speed v is on the y-axis and time t is on the x-axis, so the area is related to speed v ______________ by time t. From the definition of ______________:

$$v = \frac{s}{t}$$

$$vt = s$$

So the area between the line graph and the x-axis on a speed-time graph gives ______________ travelled.

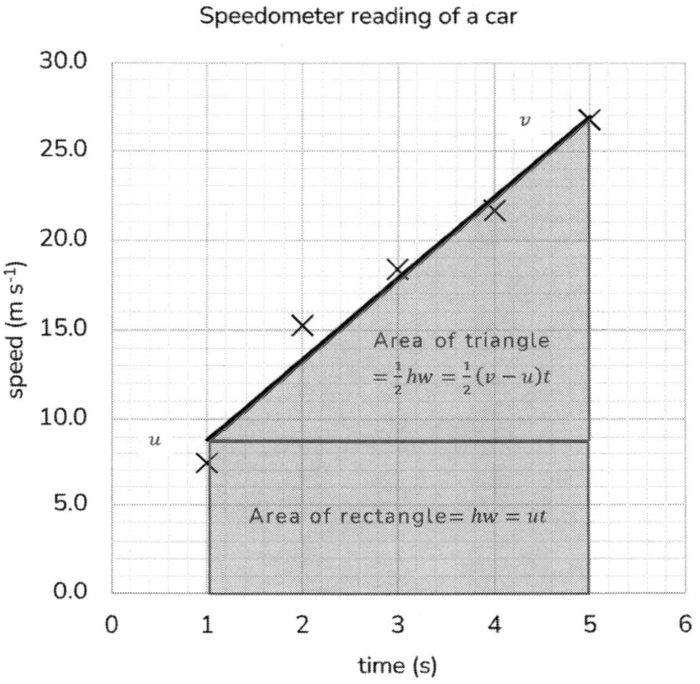

Looking closely at this particular graph, the area is in the shape of a ______________, made of a rectangle and triangle. The area of the ______________ is given by:

$$hw = 9 \text{ m s}^{-1} \times (5 - 1) \text{ s} = 36 \text{ m}$$

This is also the distance that the car would have travelled if it kept going between 1 and 5 seconds at its ______________ u of 9 m s^{-1}:

$$s = ut = 9 \text{ m s}^{-1} \times (5 - 1) \text{ s} = 36 \text{ m}$$

The area of the ______________ is given by:

$$\frac{1}{2}hw = \frac{1}{2}(v - u)t \ldots (1)$$

Rearranging the definition of ______________, $a = \frac{v-u}{t}$, gives $v - u = at$. Substituting into equation (1):

$$\text{area of triangle} = \frac{1}{2}hw = \frac{1}{2}(v - u)t = \frac{1}{2}at^2$$

Comparing this to $s = ut + \frac{1}{2}at^2$ shows the area of the triangle therefore represents the ______________ the car travelled because it accelerated, rather than continuing at its initial velocity u:

$$\frac{1}{2}hw = \frac{1}{2}(v - u)t = \frac{1}{2}(27 - 9) \text{ m s}^{-1} \times (5 - 1) \text{ s} = 36 \text{ m}$$

So the ______________ travelled by the car from 1 to 5 seconds was $36 + 36 = 72$ m.

16. The motion graph describes a car moving on a straight suburban road.

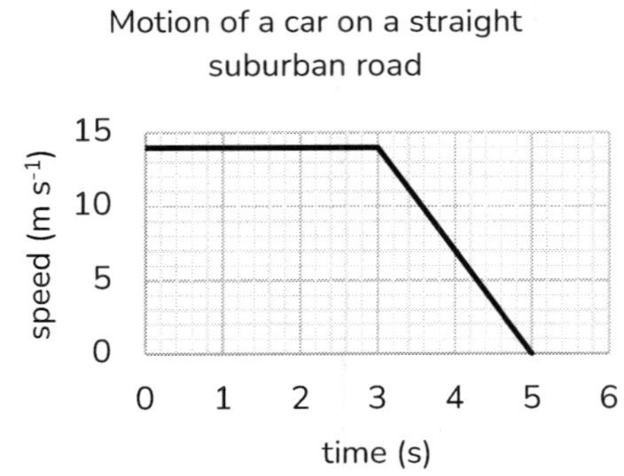

a) Describe the motion of the car in the first 3 s.

b) Calculate the distance travelled in the first 3 s.

c) Describe what happened to the speed and acceleration of the car between 3 s and 5 s.

Include supporting calculations if data is available, even if the question does not prompt you to.

d) Calculate the distance travelled between 3 s and 5 s.

Try this two ways – using the area under the graph and SUVAT.

17. The graph shows two cars, A and B, racing in the same direction along the same straight road. At 0 s, both cars start from the same position along the road.

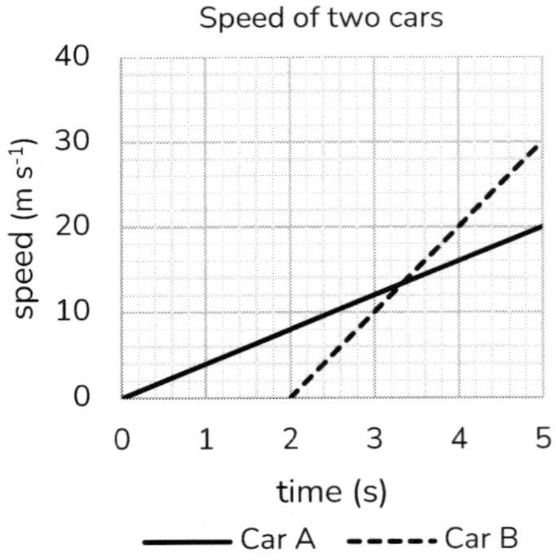

a) When do the cars have the same speed?

b) Compare the acceleration of the two cars.

c) At 5 s, how far apart are the two cars and which is in front?

18. Here is an acceleration-time graph of a rocket launch.

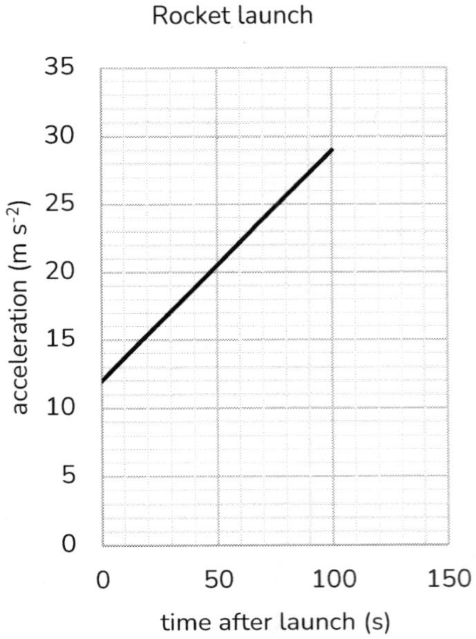

a) Explain why the area between the graph and the x-axis represents its change in velocity.

b) If the rocket was initially at rest, calculate the final speed of the rocket.

c) Explain whether the SUVAT equation $v = u + at$ can be used for this scenario.

Summary

19. Fill in the table.

Type of graph	What does the gradient represent?	What does the area between the graph and x-axis represent?	Sketch what this graph would look like for an object moving at constant speed
distance-time		nothing	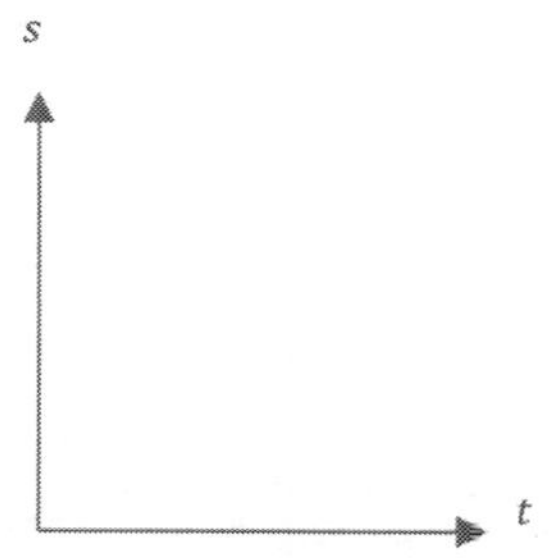
speed-time			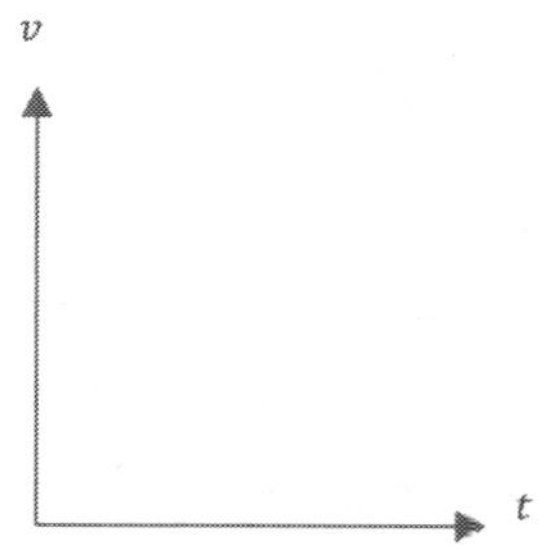
acceleration-time			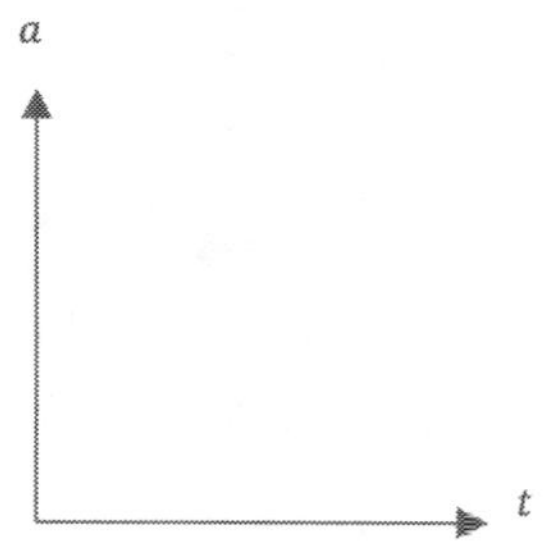

4.2 Motion graphs with vectors

Imagine a ball is thrown vertically upwards from a balcony, 4.00 m above the ground. The ball goes up for 2.00 m from its starting position, before it falls down to the ground. What will the acceleration, velocity and displacement graphs for this scenario look like?

As the ball moves both up and down, a frame of reference is needed. This also means that the motion graphs need a negative axis. Let's make the origin the starting point and let up be positive.

1. First use SUVAT to calculate all the data needed. Show that:

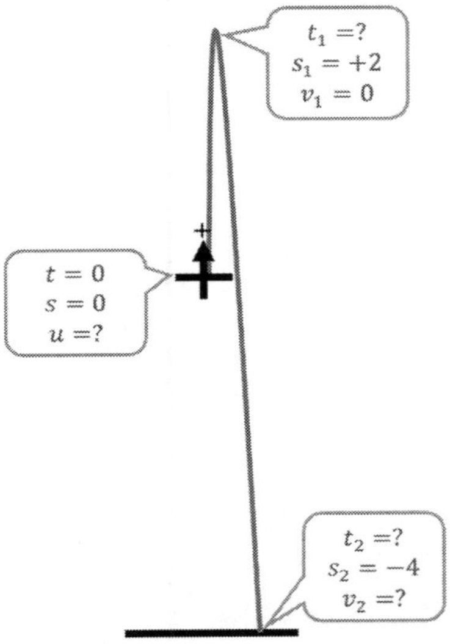

   a) the initial velocity u is 6.26 m s^{-1} up.

   b) the time taken to reach the peak t_1 is 0.639 s.

   c) time taken to reach the ground t_2 is 1.75 s.

   d) the final velocity as it reaches the ground v_2 is 10.8 m s^{-1} down.

Acceleration-time graph

The ball just undergoes acceleration due to gravity of 9.8 m s^{-2} for 1.75 s, when the ball reaches the ground, so the acceleration-time graph is just a horizontal line at -9.8 m s^{-2}.

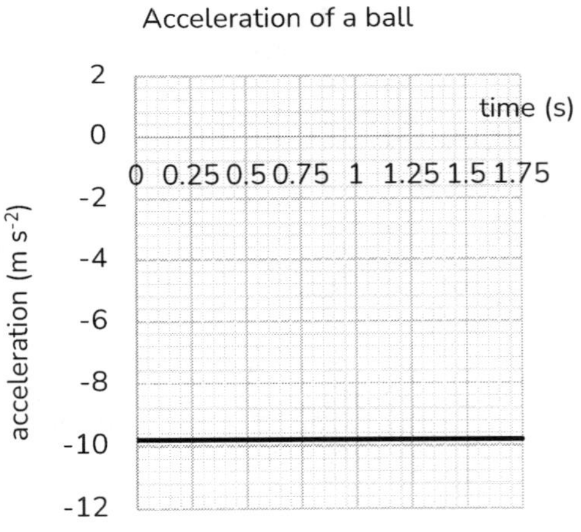

2. Calculate the area between the graph and x-axis.

3. Explain why the area between the acceleration-time graph and the x-axis gives the change in velocity Δv.

4. Explain why the change in velocity Δv is negative.

5. Recall the initial velocity u was 6.26 m s^{-1}. Use the area calculated in question 2 to determine the final velocity of the ball as it reaches the ground t_2. Compare this to the final velocity obtained in question 1(d).

An accelerometer is a device that records acceleration over time. There is one in every smartphone. Apps are available to display acceleration-time graphs for the phone. Apart from being a physics toy, it gathers information needed for navigation to work, to know when the screen is rotated, or to count steps while exercising. Accelerometers can also be found inside game controllers, car airbags to sense when there is a crash and laptops to protect the hard drive if it is dropped.

Velocity-time graph

6. Explain how to create the velocity-time graph for this scenario, by filling in the blanks.

acceleration	final	gradient	gravity	initial	negative
peak	positive	straight line	time	velocity	x-axis

The gradient of any velocity-time graph is found by dividing the rise, or change in ________________, by the run, or change in ________________, so it is equal to the ________________. Since acceleration is constant, the velocity-time graph will have a constant ________________ and look like a ________________. In this case, the gradient of the velocity-time graph will equal acceleration due to ________________ of -9.8 m s^{-2}. The gradient is ________________ because we set our frame of reference to take upwards as ________________.

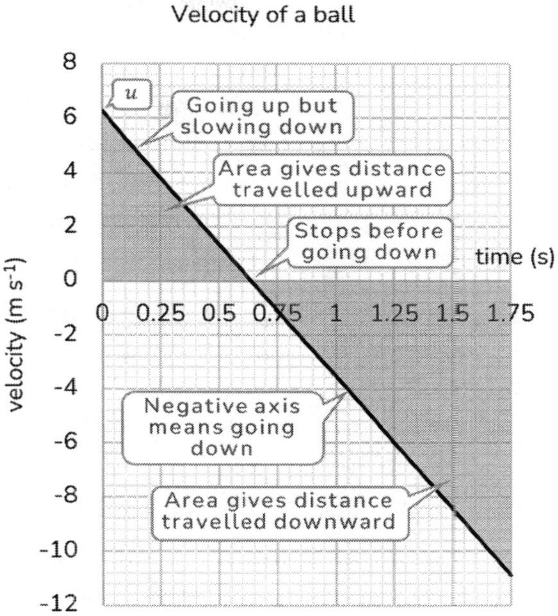

The line graph will start at the ________________ velocity of +6.26 m s^{-1} and end at the ________________ velocity of -10.8 m s^{-1}. The line graph will cross the ________________ when velocity equals zero, which occurs at the ________________ of the ball's path.

7. Calculate the gradient of the graph. Explain what it represents.

____

8. Explain what mistake has been made in the following statements:

a) "The ball returned to its starting point around 0.65 s."

____

b) "The velocity of the ball decreased after around 0.65 s."

____

9. The area between the graph and x-axis gives information about displacement and distance travelled.

a) Recall the maximum height of the ball above the starting point was 2 m. How is this related to the area of the triangle above the x-axis?

____

b) Calculate the area of the triangle below the x-axis and subtract it from the area of the triangle above the x-axis. What does the answer represent?

____

c) Add the areas of both triangles together. What does the answer represent?

____

Displacement-time graph

10. Explain how to create the displacement-time graph for this scenario, by filling in the blanks.

constant	decreases	flattens	gradient	increases	negative
negative	origin	peak	position	positive	positive

The displacement-time graph shows the ________ of the ball over time. Therefore we can start by plotting the positions at times already known, being the initial position, ________ of +2 m and final position of -4 m. The line graph will cross the x-axis where the ball returns to the ________.

The previous velocity-time graph helps us work out shape of the line connecting these points, because velocity is the ________ of the displacement-time graph. Since the velocity changed throughout the journey, the gradient of the displacement-time graph will not be ________.

As the ball goes up, the gradient of the displacement-time graph is ________, because velocity is ________. However, as it slows on the way up, the gradient ________, so the line ________.

As the ball falls down, the gradient of the displacement-time graph is ________, because velocity is ________. However, as it speeds up as it falls, the negative gradient ________.

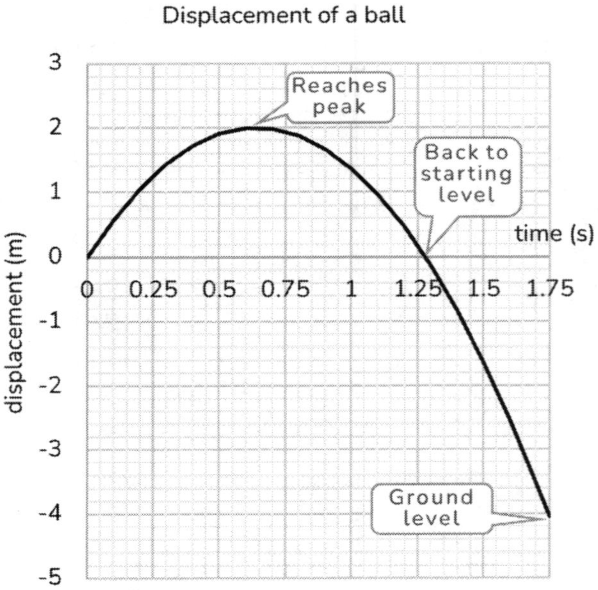

Calculating gradients of curved graphs is tedious. However, there's a mathematical technique called "differentiation" that makes this quick.
It is also difficult to calculate areas under curves. An advanced mathematical technique called "integration" can work out areas under graphs of any shape.
Integration and differentiation are tools in a branch of mathematics called calculus.

11. Explain why the gradient of a displacement-time graph represents the velocity.

____

____

12. Draw a tangent to the graph at the point representing the peak of the ball's motion. Explain what the gradient of this tangent represents.

____

____

13. Draw a tangent to the graph at a displacement of -4 m (in other words, ground level). Calculate the gradient of this tangent. Explain what this value represents.

____

____

More questions

14. A car is driven on a straight road that runs in a north-south direction. Sketch graphs that describe the following scenarios. Assume the car starts at the origin. Let positive be northward.

Scenario	Displacement-time	Velocity-time	Acceleration-time
Driving north at a constant velocity	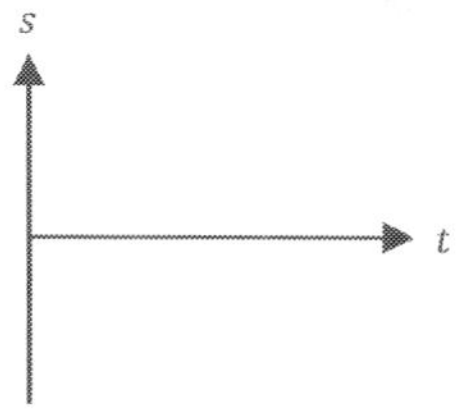	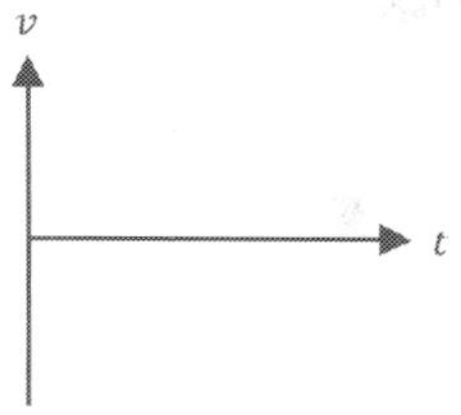	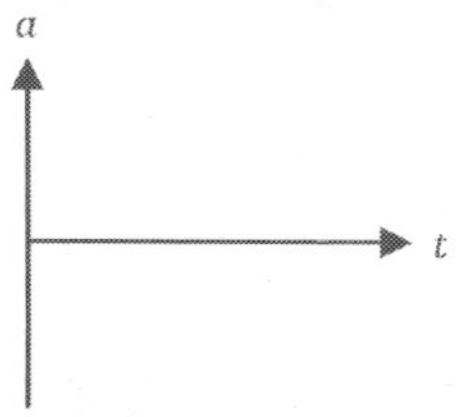
Driving south at a constant velocity	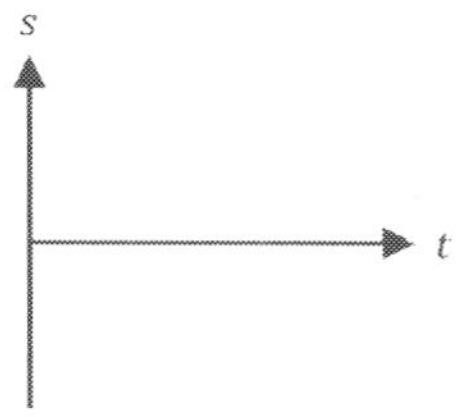	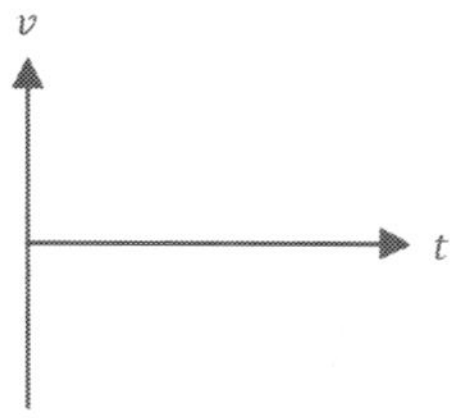	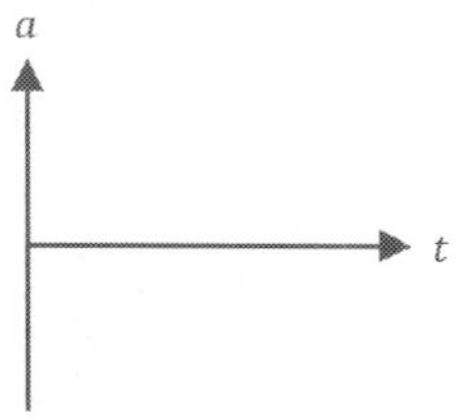
Constant acceleration from rest, moving northbound	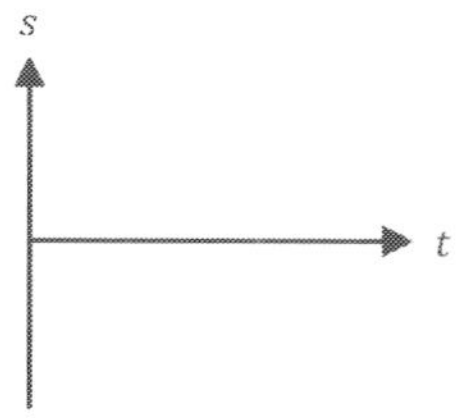	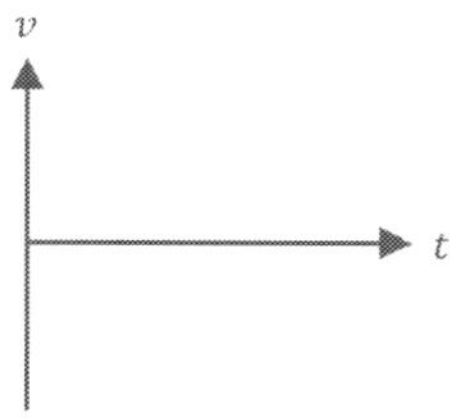	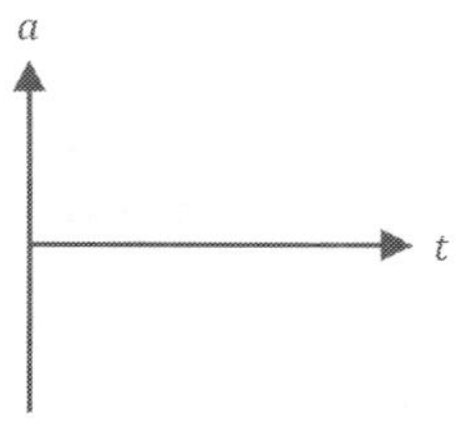
Constant acceleration from rest, moving southbound	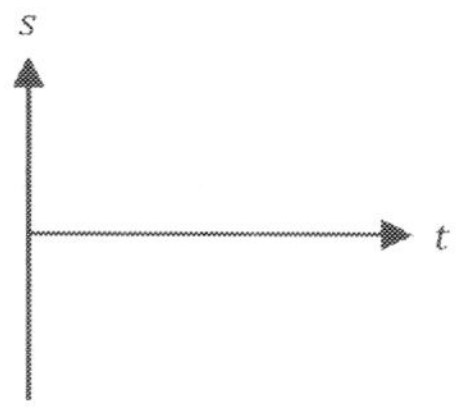	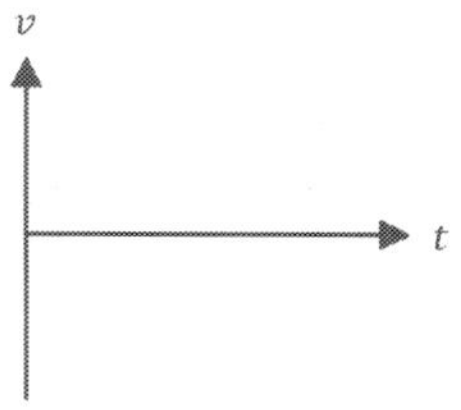	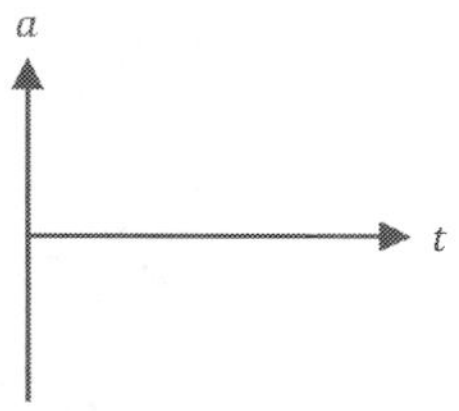
Driving north, braking with constant deceleration to stop	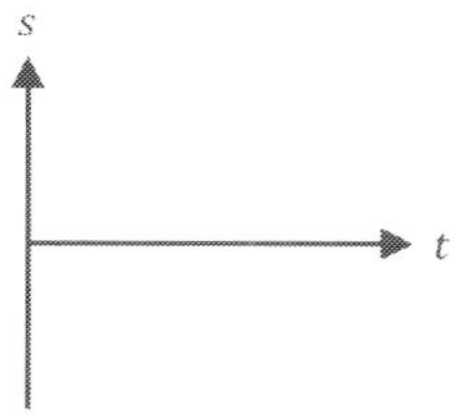	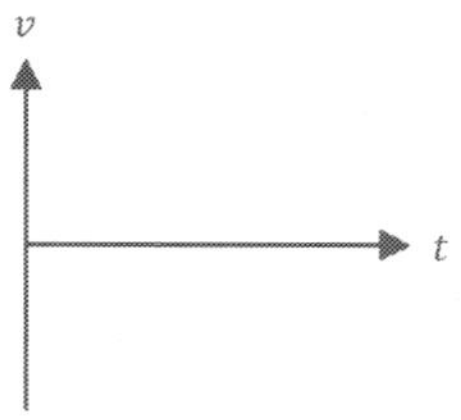	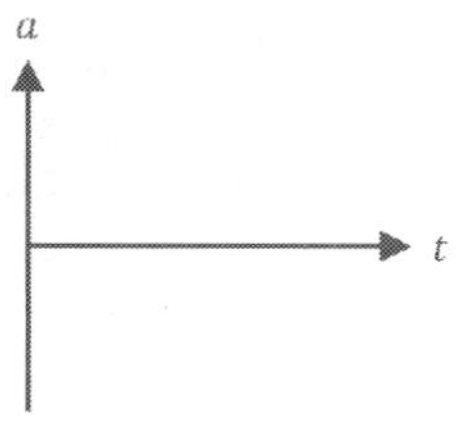
Driving south, braking with constant deceleration to stop	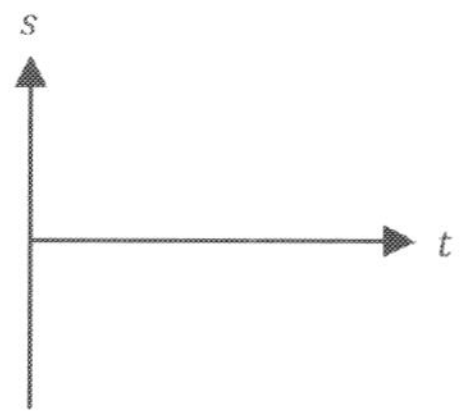	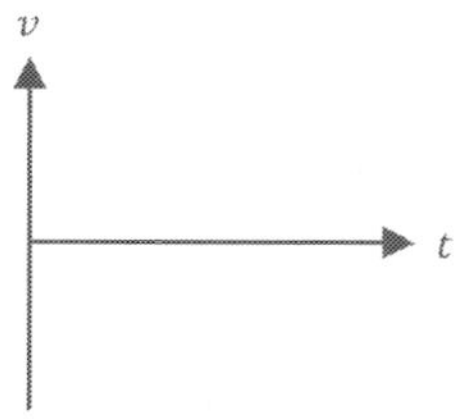	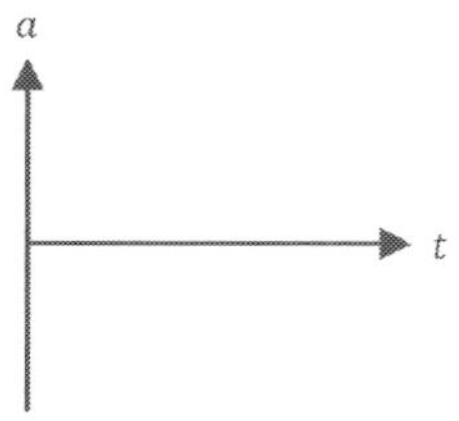
Driving north, braking with constant deceleration to U-turn, then accelerating at the same rate to go south	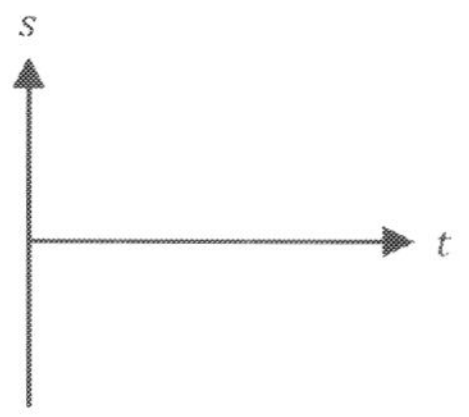	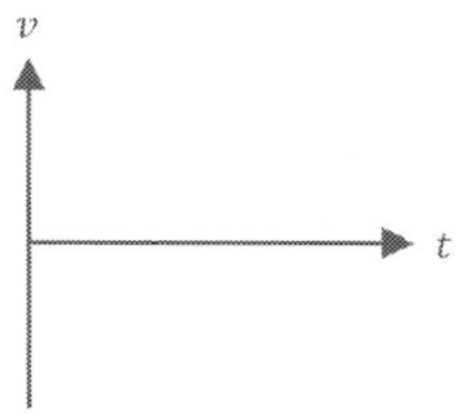	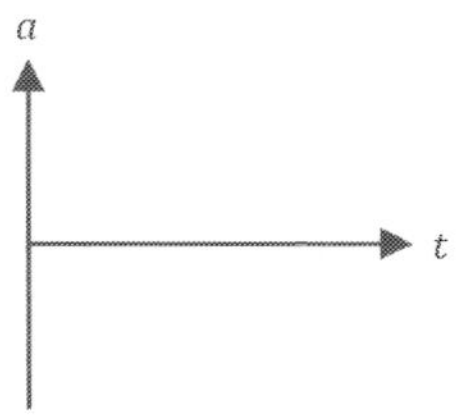

15. The graph represents the motion of an astronaut jumping on the Moon. The astronaut starts from the ground.

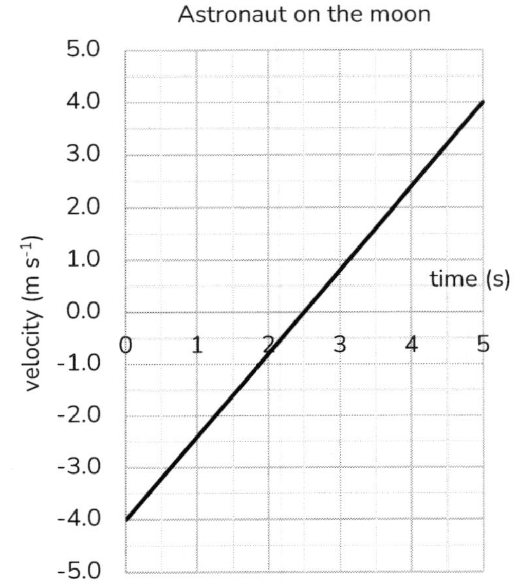

a) Explain which direction is taken as positive.

b) Describe the astronaut's motion at 2.5 s.

c) Determine the acceleration due to gravity on the Moon.

d) What was the distance travelled by the astronaut in 5.0 s?

e) Determine the position of the astronaut at 5.0 s.

16. The displacement-time graph represents the displacement of a pendulum, a mass on the end of a string.

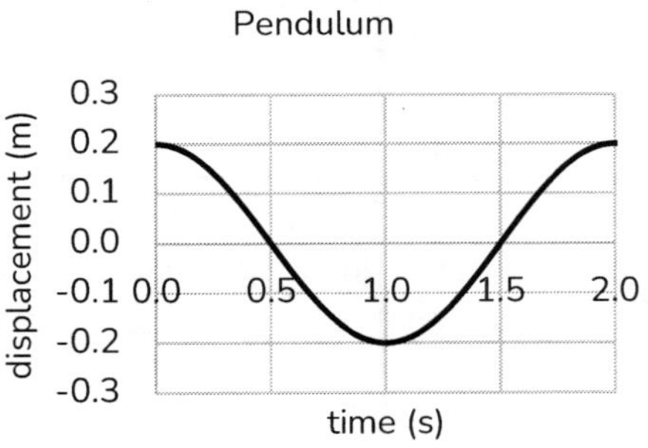

a) Describe the motion of the pendulum, as shown in the graph.

b) Sketch a velocity-time graph and acceleration-time graph for the pendulum.

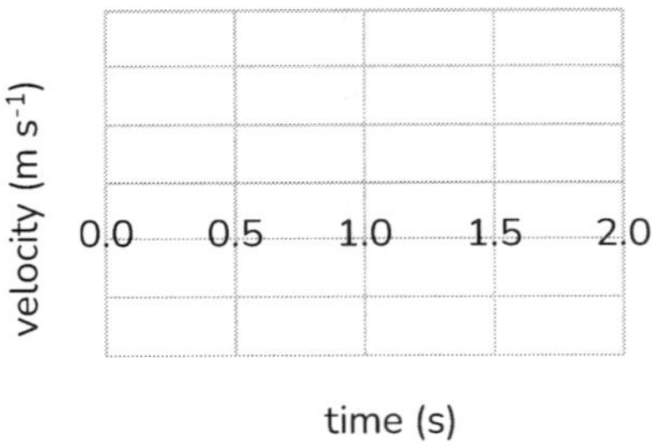

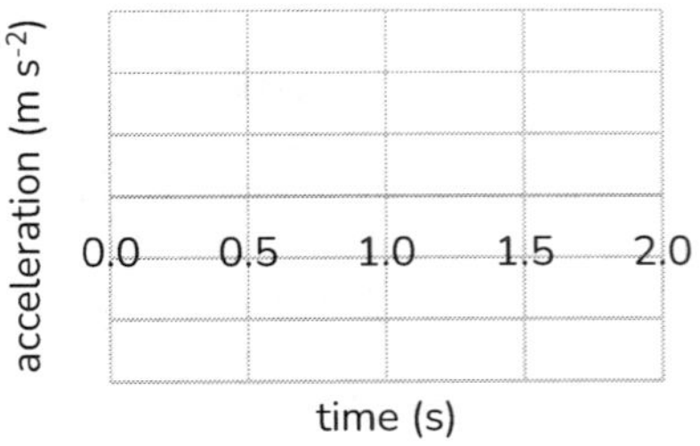

No scale is needed for a sketch.

5.1 Forces

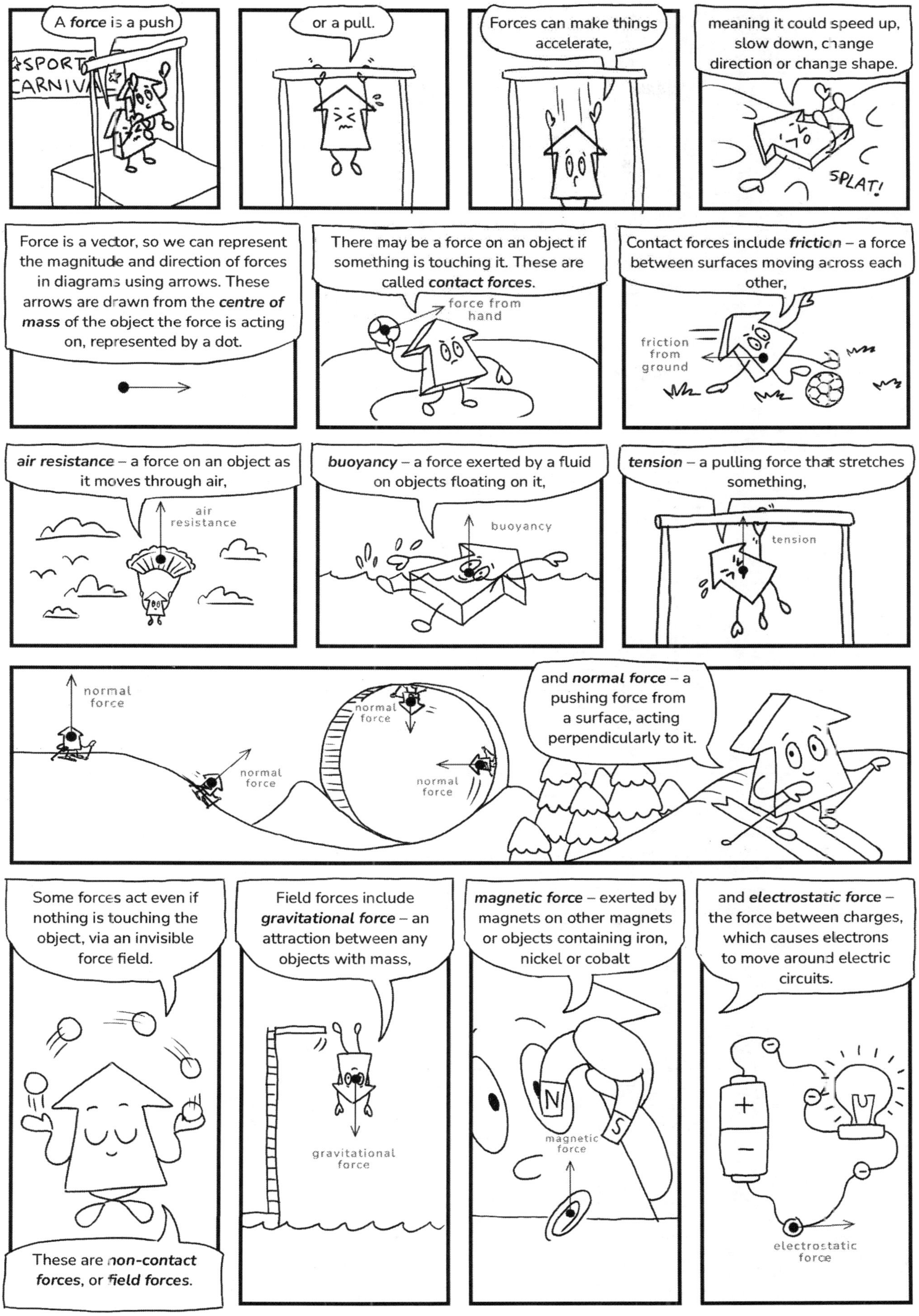
A force is a push
SPORTS CARNIVAL
or a pull.
Forces can make things accelerate,
meaning it could speed up, slow down, change direction or change shape.
SPLAT!
Force is a vector, so we can represent the magnitude and direction of forces in diagrams using arrows. These arrows are drawn from the centre of mass of the object the force is acting on, represented by a dot.
There may be a force on an object if something is touching it. These are called contact forces.
force from hand
Contact forces include friction – a force between surfaces moving across each other,
friction from ground
air resistance – a force on an object as it moves through air,
air resistance
buoyancy – a force exerted by a fluid on objects floating on it,
buoyancy
tension – a pulling force that stretches something,
tension
normal force
normal force
normal force
normal force
and normal force – a pushing force from a surface, acting perpendicularly to it.
Some forces act even if nothing is touching the object, via an invisible force field.
These are non-contact forces, or field forces.
Field forces include gravitational force – an attraction between any objects with mass,
gravitational force
magnetic force – exerted by magnets on other magnets or objects containing iron, nickel or cobalt
N
S
magnetic force
and electrostatic force – the force between charges, which causes electrons to move around electric circuits.
+
–
electrostatic force

Force is measured in newtons (N). It is named after Sir Isaac Newton.
1N
10N
There is often more than one force acting on an object.
force from person A
force from person B
If there is the same magnitude of force acting in opposite directions on an object, then the forces are balanced, meaning there is no net force.
No-one's winning...
force from person A
force from person B
If the net force is not zero, then the forces are unbalanced.
net force 100N
force from person A 100N
force from person B 200N
If we know the magnitude and direction of all the forces on an object, then we can predict its motion.
tension
gravitational force
Which way is the net force?
Newton's laws predict what happens due to forces on objects.
Should I get the credit for others' work?
The Law of Inertia states that a body maintains its state of rest or uniform velocity, unless acted upon by an external unbalanced force.
normal force
net force=0
gravitational force
Something moving at constant velocity also has balanced forces acting on it. It will keep moving at constant velocity unless the forces become unbalanced.
normal force
net force=0
gravitational force
The Law of Inertia means that anything moving in deep space will keep moving forever, unless an unbalanced force acts on it
OUT OF FUEL
and is why passengers on a car pitch forward if the car brakes suddenly.
SQUEEAK
The Law of Inertia is also called Newton's First Law. Inertia is the tendency to resist change in motion
I'm not moving
I'm not moving more
For linear motion, inertia and mass are the same thing.
But if you are spinning, your inertia also depends on your shape and radius.

Newton's Second Law is
$F = ma$
where F = net force (N)
m = mass (kg)
a = acceleration (ms^{-2})
So the force is directly proportional to the acceleration.
acceleration=a
If you double the force on an object, you double its acceleration.
acceleration=2a
The bigger the mass of the object, the bigger the force needed to achieve the same acceleration. So the force needed is directly proportional to the mass too.
acceleration=a
If an object has more mass, exerting the same force on it will produce less acceleration. In other words, mass and acceleration are inversely proportional.
acceleration=$\frac{1}{2}$a
If you double the mass, the same force will produce half the acceleration.
Newton's Third Law states that for every action, there is an equal and opposite reaction.
Ow!
force of bag on boxer
force of boxer on bag
This concept is about two objects exerting an action-reaction pair of forces on each other.
force from foot pushes boat back (action)
force from boat pushes person forward (reaction)
If the first object exerts a force (action) on the second, then the second object must always exert an equal and opposite force (reaction) on the first.
SPLASH
Action-reaction pairs are responsible for a lot of movement.
force of rocket on fuel (action)
force of fuel on rocket (reaction)
We need a force from something else in order to accelerate.
Oops...
SNAP!
Action-reaction and balanced forces are different concepts. Balanced forces are about forces on one object.
normal force
gravitational force
Action-reaction pairs are about two objects, acting on each other.
normal force (reaction)
force of apple on table (action)
gravitational force on apple from Earth (action)
gravitational force on Earth from apple (reaction)
Normal force isn't always the same magnitude as gravitational force either.
normal force
gravitational force
normal force
gravitational force
BOING
gravitational force
Knowing all the forces acting helps us predict what will happen to an object.
What forces will act on you today?

5.2 Introduction to forces

1. Describe forces, by filling in the blanks.

acceleration	balanced	contact	field	newtons	pull	unbalanced	velocity

A force is a push or a ____________, measured in ____________. Whenever there is a net force of zero on an object, the forces on the object are ____________, so the object's ____________ does not change. ____________ is caused by an ____________ force. Forces acting on an object must either be due to something in direct ____________ with it or act via a ____________.

2. The word bank lists types of forces.

air resistance	buoyancy	electrostatic force	friction
gravitational force	magnetic force	normal force	tension

Categorise these forces as contact forces or field forces in the table and define them.

Contact forces		Field forces	
Type	Definition	Type	Definition

3. A *freebody diagram* shows vectors representing all the forces acting on an object, where the arrows all start at the centre of mass. To identify all the forces acting on the object, think about everything touching it which could exert a contact force, as well as any fields acting on it.

A freebody diagram of the forces acting on a passenger in an elevator accelerating upwards is shown to the right.

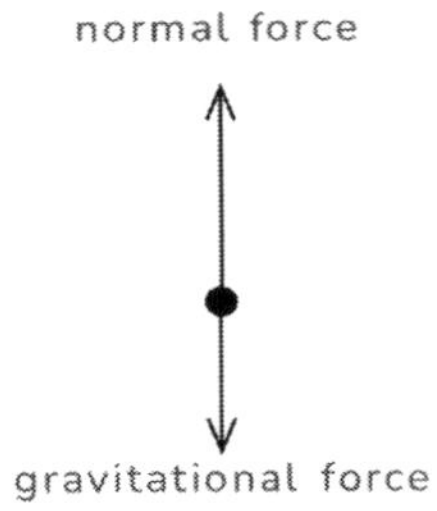

a) What exerts the normal force? What exerts the gravitational force? Which of these forces will stay the same as the passenger rides the elevator?

____

____

b) Draw a freebody diagram to describe forces the passenger in the elevator when it is:

moving up at constant velocity

moving up but slowing down as it reaches a higher floor

accelerating downwards as it returns to the ground floor

4. Draw a freebody diagram showing all the forces acting on the object in the following scenarios:

You sitting still on the ground

A ball bouncing up as it hits the ground

A skateboard sliding to a stop along a rough road

A diver sinking down into the water at constant velocity

A child accelerating down a smooth slope on a toboggan

A falling person being slowed by a bungee cord

5. Explain how forces allow us to walk, by filling in the blanks.

acceleration	backward	falling	forward	friction
inertia	mass	normal force	reaction force	

When you take a step forward, you must push your foot down and ________________ into the ground. ________________ between your foot and the ground opposes your foot moving backwards and pushes you ________________. If it were slippery, there would be less friction to stop your foot moving backward, so your foot would keep moving backward, due to ________________. This may result in you ________________.

Draw a freebody diagram of the forces acting on the foot when walking

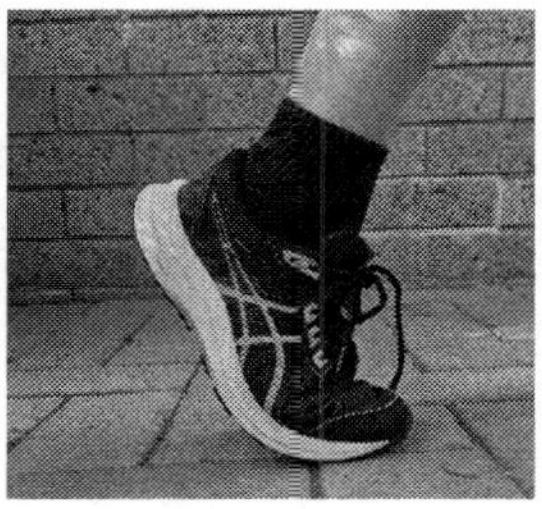

As you push your foot down into the ground, this causes an equal and opposite ________________ from the ground, pushing you up. This force from the ground on pushing on you is called the ________________, and it is the same magnitude as the force of you on the Earth. However, it causes a much bigger ________________ on you, since according to $F = ma$, the same force on a smaller ________________ causes a bigger acceleration.

You really make the Earth move! During COVID lockdowns, seismograms used for monitoring earthquakes detected a lot less movement.

6. Describe the forces on a rocket, by filling in the blanks.

accelerate	downward	gravitational	mass
reaction	upward	$F = ma$	

When a rocket is launched, gases produced from combustion of the fuel inside it are pushed ________________ and out. The gases exert an equal and opposite ________________ force ________________ on the rocket, called thrust. When the thrust exceeds the ________________ force on the rocket, the net force will be upward and it will start to ________________. The engine will exert a constant thrust on the rocket, but as fuel is used up, the total ________________ of the rocket decreases, leading to an increase in acceleration, according to ________________.

Draw a freebody diagram of the forces on the rocket just after lift off

7. A car cannot easily turn on a slippery, icy road. Instead, it will keep sliding forwards in a straight line, ending up on the outside of the turn. Explain why, and what the car needs in order to turn.

A moving car will keep ______________________ unless ______________________, according to ______________________.

To turn, there must be a force to ______________________.

This force is supplied by ______________________, pushing the car in the direction ______________________.

This force is the reaction force to ______________________.

8. Explain the following aspects of swimming.

a) Why must water be pushed backwards with each stroke in order to move forwards?

____

b) Why do you move forward faster if you push harder with each stroke?

____

c) Why do you glide forwards in the water momentarily, before needing to take another stroke?

____

9. A parent pushes a child sitting in a toy car, with a total mass of 40 kg, to move them forward with an acceleration of 1.5 m s^{-2}. What net force was needed?

____

10. Two ice skaters, Ash and Bee, are practising a move where they start standing and facing each other, then push each other apart. Ash's mass is 50 kg and Bee's mass is 70 kg. Ash pushes Bee away with a 200 N force.

a) Explain what force Bee pushes Ash away with.

____

b) Calculate the accelerations of both skaters.

____

____

11. True or false? If the statement is false, explain why.

a) If an object is moving, there must be a force acting on it.

____

b) Increasing the mass of an object will increase the force acting on it.

____

c) An action-reaction pair of forces occurs when forces are balanced.

____

d) Gravitational force pulls you down, but you pull the Earth up with the same magnitude of force.

____

e) Normal force is always equal to weight but upwards.

____

5.3 Mass and weight

1. Explain the difference between mass and weight, by filling in the blanks.

acceleration	distance	gravity	kilograms	mass	newtons
particles	same	surface	toward	weight	$F = ma$

Mass and weight are different concepts in science, although they might be used interchangeably in everyday language. ***Mass*** is the amount of matter in an object, measured in the SI unit ____________ (kg). In other words, your mass relates to the amount and type of ____________ (atoms and molecules) in you.

Weight is another name for gravitational force, which is the force of attraction between any two objects with mass, measured in ____________ (N). It is the force that produces acceleration due to gravity, which accelerates masses ____________ each other. The symbol g is used to represent acceleration due to ____________, and $g = 9.8$ m s^{-2} on average on Earth's surface. Acceleration due to gravity g varies on different planets and moons, because gravitational force increases as the ____________ of the objects increases. Gravitational force also decreases as the ____________ between objects increases.

The symbol W is often used to represent ____________ and F_G is often used to represent gravitational force. W and F_G are the ____________ thing, but the symbol W is more commonly used for gravitational force on Earth's ____________.

When the force accelerating an object is weight, the equation ____________ can be written as:

$$W = mg$$

where: W = weight (N)

m = mass (kg)

g = ____________ due to gravity (m s^{-2})

Mass relates to particles. This is a particle model of a solid.

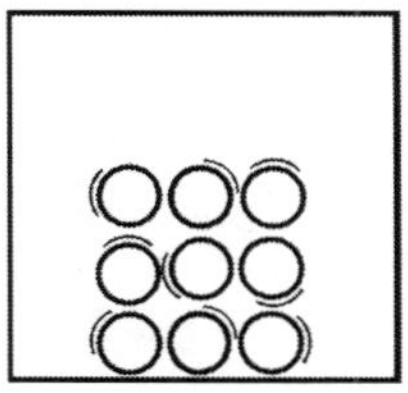

Gravitational force depends on mass and distance

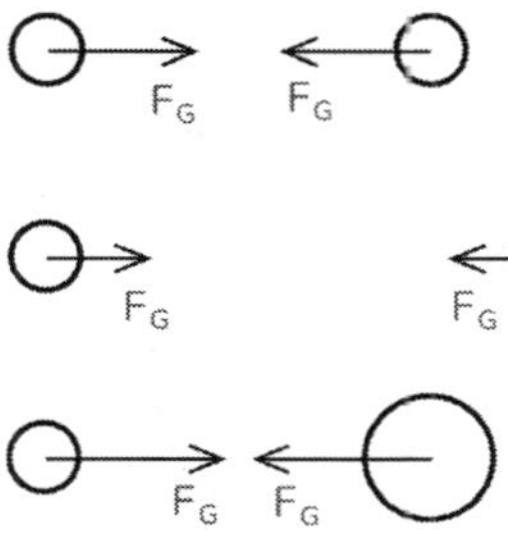

Notice the force vectors between masses are the same size, because they are action-reaction pairs.

2. Fill in the table to summarise mass and weight.

Concept	Mass	Weight
Definition		
Unit		

3. A person has a mass of 50 kg.

a) Calculate their weight on Earth.

b) Acceleration due to gravity on the moon is 1.625 m s^{-2}. Calculate their weight on the Moon.

4. "New weight loss program! Just go to the Moon!" Assess whether such a program would have satisfied customers. Note the Moon has less mass than Earth.

> Assess means explain the good, the bad and give your overall judgement.

____

____

____

____

5. At sea level on Earth, on average there is 1.03 kg of air above every cm^2 of area. Assume the average human head has a circumference of 57 cm.

> Circumference of a circle $= 2\pi r$
> Area $= \pi r^2$

a) Calculate the weight of air above a person's head.

____

____

____

b) Compare this to the weight of a washing machine with a mass of 90 kg.

____

____

6. Explain how scales work, by filling in the blanks.

acceleration	Earth	force	mass
newtons	normal	still	$W = mg$

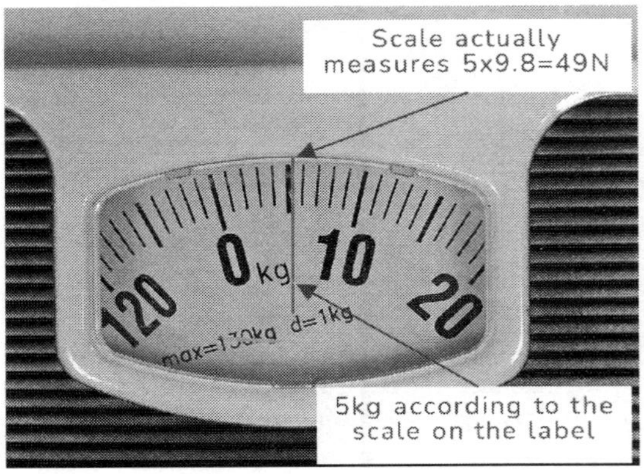

In everyday language, we often say we use scales to "weigh" objects, but the unit on the scales is grams or kilograms, not ____________. How scales actually work is they measure ____________, not mass. If you stand on a bathroom scale and jump up and down, the reading on the scale changes, even though your ____________ hasn't. So the bathroom scale is actually measuring the force you exert on the scale, which is equal and opposite to the ____________ force it exerts on you.

The labelling on the scale assumes that you are using it on ____________ to measure your mass, where ____________ due to gravity is 9.8 m s^{-2} and you are standing ____________ on it. It is labelled so that the force measured is converted to mass using ____________.

7. A world champion weightlifter with a mass of 50 kg holds a 100 kg mass still above them. What is the force they exert on the 100 kg mass?

> Which mass is relevant?

____

____

8. Some students were wondering what maximum force their index fingers could push down with. They measured this by pushing down on scales with their index finger as hard as they could and found on average, the scales read 4.8 kg. What is the average maximum force exerted by pushing down with the index finger?

9. An astronaut stands on bathroom scales on Earth and the scales read 75 kg. They take the scales to the Moon, where acceleration due to gravity is 1.625 m s^{-2}. What will the scales read when the astronaut stands on it on the Moon?

10. A rock is collected on Mars by a robot. When the rock is weighed on Mars using scales calibrated for use on Earth, the scales read 20 g. When the rock is brought back to Earth, the same scales read the rock's mass as 53 g. Explain the difference in reading.

11. Fill in the blanks to explain how a spring balance works.

balanced	gravity	kilograms	newtons
proportional	reaction	tension	weight

When a mass is hung at the end of a spring balance, the ________ of the mass stretches the spring down. The ________ in the spring pulls the mass up with an equal and opposite ________ force. The spring stops stretching further when the weight on the mass and tension are ________.

The length of the spring is ________ to the tension in it. To convert this tension to the mass of the object in ________, the tension, measured in ________, is ________ by acceleration due to ________ of 9.8 m s^{-2}.

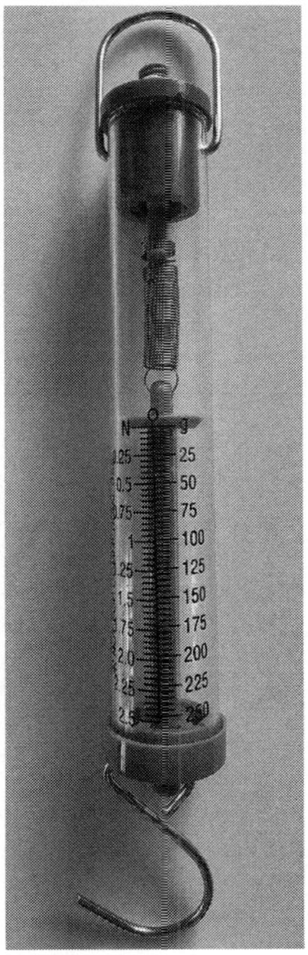

12. Astronauts floating in spaceships look and feel "weightless". This often leads to the misconception that there is no gravity in space. Explain what "weightlessness" actually is, by filling in the blanks.

absence	air resistance	dropped	Earth	free fall
inertia	launch	masses	normal force	satellites
small	straight	Sun	surface	twice
weight	zero			

There is gravity in space, keeping the Moon orbiting around the ______________. If this gravity disappeared suddenly, the Moon would continue moving in a ______________ line, tangential to its original orbit, due to inertia. Gravity also keeps the planets in our Solar System, including Earth, orbiting around the ______________, and human-made ______________ orbiting around Earth. The amount of gravity isn't ______________ either. For example, the International Space Station, which orbits Earth at an altitude of around 400 km, experiences an acceleration due to gravity of 8.7 m s^{-2}, compared to 9.8 m s^{-2} on Earth's ______________.

Hair Raiser ride at Luna Park

So the sensation of "weightlessness" doesn't come from the ______________ of gravity. At amusement parks on a "drop tower" ride, riders seated on chairs are lifted up high on a tower, then the chairs are ______________, plummeting down the sides of the tower. The riders experience "weightlessness" because their chair is no longer pressing up on them, so there is no ______________. In fact, the only force acting is gravity (if we ignore ______________)! The riders and the astronauts in the International Space Station are described as being in ______________, which leads to the sensation of "weightlessness".

In everyday life, we are used to experiencing the normal force from the ground or our seat, and this normal force is usually the same magnitude as our ______________. One way to describe this force is 1 g. If you are in a rocket where the normal force is pushing up on you ______________ as hard, you experience 2 g. In free fall, you would experience ______________ g. Using g allows us to compare the force we experience, even though we are different ______________.

Freebody diagram of you sitting normally

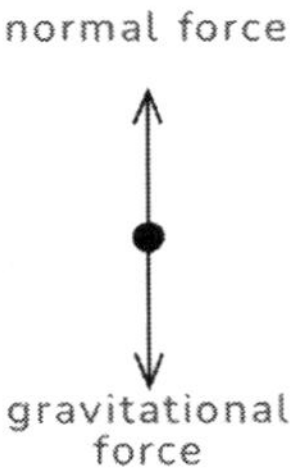

Astronauts typically experience 4 g during a rocket ______________, sometimes more. As our bodily fluids experience ______________ while our bodies accelerate, blood can pool in parts of the body and the heart has difficulty pumping, so shortness of breath, blurred vision and loss of consciousness are common effects, making higher g dangerous.

Freebody diagram of you in free fall

13. Draw a freebody diagram of the forces acting on you as you ride a drop tower at an amusement park:

when you start being lifted up

when you are dropped

when the ride brakes to a stop after you fall

5.4 Forces and constant acceleration

1. A small car with a mass of 1400 kg accelerates from rest to 27 m s^{-1} in 7.5 s. What force needs to be exerted by the engine to achieve this?

> Find acceleration using SUVAT. Then find the force using $F = ma$.

2. A racing car with a mass of 1500 kg has an engine which exerts a force of 11565 N. How long will it take to accelerate from rest to 27 m s^{-1}?

> Find acceleration using $F = ma$ first.

3. A sandboarder slides down a sand dune and arrives at a flat patch of sand at the base of the slope with a speed of 7.00 m s^{-1}. Friction from the sand brings them to a stop in 4.55 m. The mass of the sandboarder and board combined is 50 kg. What is the force of friction acting on them?

> Set a frame of reference. Include a direction in the answer for vectors.

4. The graph below shows the motion of a cyclist along a straight road. The combined mass of the bike and cyclist is 92 kg. For each stage of its motion, propose what the bike was doing and calculate the net force acting on the bike.

t = 0 – 3 s:

t = 3 – 6 s:

t = 6 – 10 s:

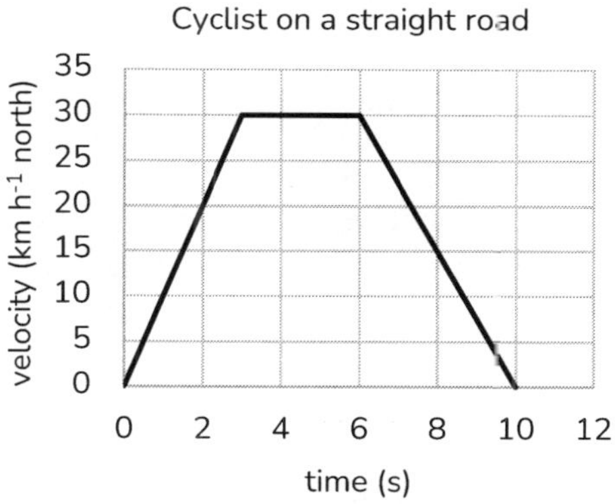

> Read the y-axis label for direction.

5. A stranded hiker of mass 75 kg is being air-lifted by a rope lowered from a helicopter. The rope accelerates the hiker upwards at 1.2 m s^{-2}.

> Draw a freebody diagram showing the forces on the hiker

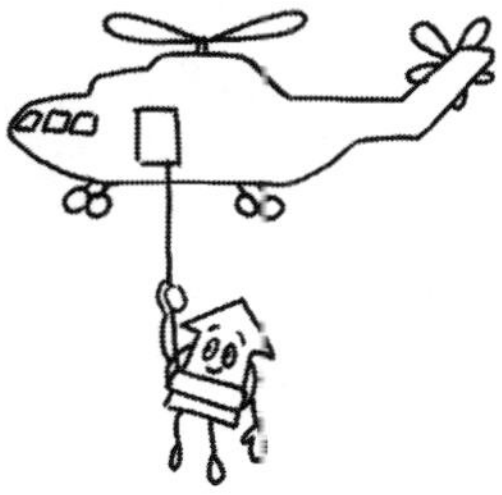

a) Calculate the net force on the hiker.

b) Calculate the weight of the hiker.

c) If up is taken as the positive direction, which equation below gives the net force F_{net}, in terms of the magnitudes of the tension of the rope T and weight of the hiker W?

$F_{net} = T - W$ $\qquad$ $F_{net} = W - T$ $\qquad$ $F_{net} = T + W$

d) Calculate the tension in the rope.

6. On an amusement park ride, seated riders are lifted up a tower, then dropped from rest to experience free fall for the first 30 m. The ride then slows down the falling riders to a speed of 4.0 m s^{-1} in 2.0 s after free fall.

Draw a freebody diagram

a) What is the normal force from the seat acting on a 60 kg rider during free fall?

b) Calculate the speed of the riders after free falling for 30 m.

c) Calculate the acceleration of the riders when the ride slows down.

d) What is the normal force from the seat acting on a 60 kg rider when the ride slows down? Compare this to the normal force on the rider during free fall, and explain whether the rider would feel heavier or lighter.

7. The dots on the diagram represent the position of a toy boat of mass 40 g, as it moves across water from left to right, starting from rest. The positions were recorded with a frequency of 10 Hz. The boat has a motor providing a constant thrust of 1.5 N to move the boat forwards. Assume there is a constant resistive force acting against the boat's motion, which is the sum total of friction and air resistance.

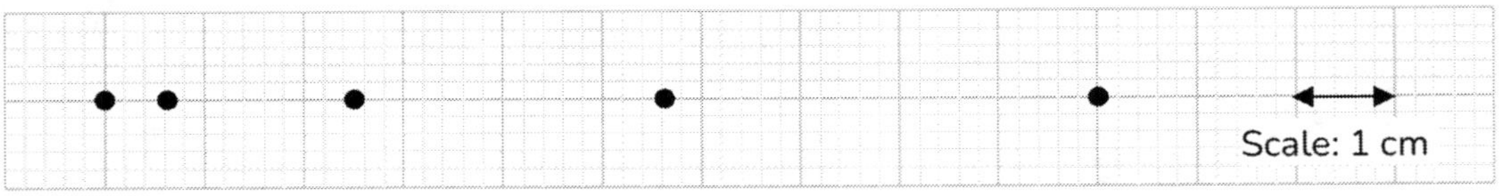

Draw a freebody diagram

a) What is the acceleration of the boat?

b) What is the net force acting on the boat?

c) Calculate the total resistive force acting on the boat.

8. Two objects, A and B, are thrown upwards with the same initial velocity. Object A has more mass than Object B. Compare their maximum heights, ignoring air resistance.

5.5 Harder force problems – Set 1

The Law of Inertia, $F = ma$ and action-reaction pairs can be used to solve a variety of physics problems.

1. A trapeze artist of mass 80 kg dangles by their legs from a bar. They hold onto a second trapeze artist of mass 60 kg with their arms, as shown. Both trapeze artists are stationary.

   a) Analyse the forces acting on the trapeze artists, by filling in the blanks.

action-reaction pair	arms	balance	bottom
equal	in contact	Law of Inertia	legs
move together	stationary	system	tension
top	trapeze bar	weight	zero

In physics, both trapeze artists together is called the ***system***, since they will ________________. As they stay at rest, there must be no net force on them, according to the ________________. Therefore the tension in the top person's legs, T_1, pulling them up, must ________________ the combined weight of the people, $W_1 + W_2$, pulling them down.

We can choose symbols to represent forces and subscripts to identify which force is which.

Freebody diagram of forces acting on the system

The net force on each person is also ________________, meaning the forces acting on each of them is balanced.

Between the people, the force the bottom person pulls the top person down with, T_2, with must equal the force the top person pulls the bottom person up with, since this is an ________________.

The forces on the top person are the ________________ from the trapeze bar, T_1, pulling them up, and their own weight, W_1, pulling them down. In addition, they are ________________ with the bottom person, who also pulls down on them with tension T_2. As the top person is ________________, the tension from the trapeze bar, T_1, must balance their weight, W_1, and tension from the bottom person, T_2. This means T_1 is greater than T_2, and the top person's ________________ must be stronger than their ________________.

Freebody diagram of forces acting on the top person

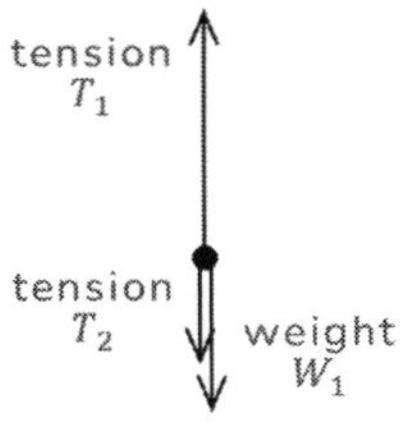

The forces on the bottom person are their own ________________, W_2, pulling them down, and tension from the ________________ person's arms, T_2, pulling them up. As there is no net force on the bottom person, W_2 and T_2 are ________________ in magnitude. The ________________ exerts no force on the bottom person, since it is not in contact with them.

Freebody diagram of forces acting on bottom person

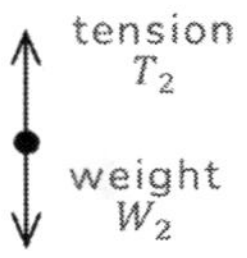

Another way to think about why T_1 is greater T_2 is that the tension from the trapeze bar, T_1, must support the combined weight of the ________________. On the other hand, T_2 only needs to support the weight of the ________________ person.

   b) Determine the magnitude of the tension in the top person's arms, T_2.

____

____

Set a frame of reference eg. let up be positive.

   c) Determine the magnitude of the tension in the trapeze bar, T_1.

____

____

2. A 60 kg acrobat stands on the shoulders of an 80 kg acrobat. What force must their legs exert so that they stand still?

a) Analyse the forces acting on the acrobats, by filling in the blanks.

action-reaction pair	balanced	normal force	system	weight

The forces acting on the ________ are the combined weight of the acrobats, pulling down, and normal force from the ground, pushing them up. These forces are ________ if they are stationary.

The force in the bottom acrobat's legs pushing down into the ground is equal in magnitude to the normal force pushing up from the ground, as this is an ________.

The forces acting on the top acrobat are their own ________, pulling down, and ________ from the bottom acrobat's shoulders, pushing up.

b) Calculate the force exerted by the top acrobat's legs.

Draw a freebody diagram

____

____

____

c) Calculate the force exerted by the bottom acrobat's legs.

Draw a freebody diagram

____

____

____

3. Road trains consisting of a truck pulling many trailers linked together is an important form of transport for goods across the remote Australian outback.

A 10 tonne truck is pulling two trailers, each with a mass of 30 tonnes. It accelerates uniformly from rest to 9.0 m s^{-1} in 18 s on a straight road.

a) Calculate the acceleration of the truck and trailers.

____

____

1 tonne (1 t) equals 1000 kg.

A diagram showing the forces on the truck and trailers is shown below.

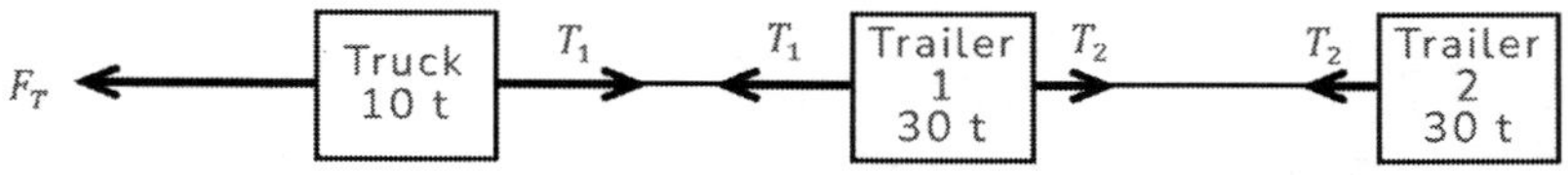

Set a frame of reference.

b) Calculate the force exerted by the engine of the truck, F_T, also known as thrust.

____

c) Consider the forces on just the 10 t truck. Thrust, F_T, pushes it forwards, but tension in the first coupling, T_1, pulls it back. The net force on the 10 t truck must produce the same acceleration as the whole road train. Use this information to calculate the tension in the first coupling, T_1.

We don't want the tension in the couplings to break them!

____

____

____

d) Trailer 1 also has T_1 in the first coupling pulling it forwards, with the reaction force of T_1 pulling the truck backwards. As well, the second coupling pulls Trailer 1 back with a tension of T_2. Calculate the tension in the second coupling, T_2.

Remember the mass in $F = ma$ is the one that is being accelerated by the net force.

e) The only force on Trailer 2 that is not balanced is the tension in the second coupling, T_2. Show that this tension gives Trailer 2 the same acceleration as the whole system.

Just considering the force on Trailer 2 would have been an easy way to calculate T_2.

4. The first passenger elevator was installed in New York in 1856, allowing modern high rise buildings to develop. Most elevators use a counterweight on the other side of a pulley to the elevator car. This means the motor that accelerates and decelerates the elevator does not have to work as hard, as gravity is also used.

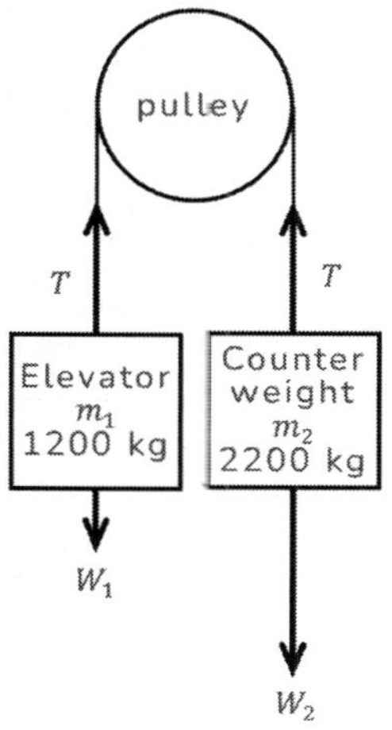

An empty elevator car of mass 1200 kg is attached via a cable and pulley to a counterweight of mass 2200 kg.

A pulley is a wheel with a groove that a cable can pass over, which just changes the direction of the force through the cable. This means we can think of the forces in this scenario as if they act in one dimension, like this:

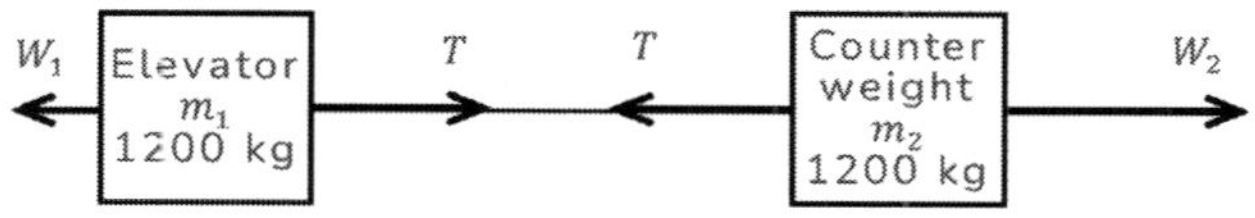

a) If the elevator car and counterweight are free to move, with no motor or brakes applying force on them, what is the acceleration of the elevator?

Set a frame of reference. Which way will the system move?

b) Using the forces on just the elevator, determine the tension in the cable.

c) Now consider the forces acting on the counterweight to determine the tension in the cable.

This should give the same answer as part (b)

5. Did you know you can measure the acceleration of an elevator using normal bathroom scale, using its measurements while you stand on it while the elevator is in motion?

A velocity-time graph for an elevator going from one floor to a higher floor is shown below.

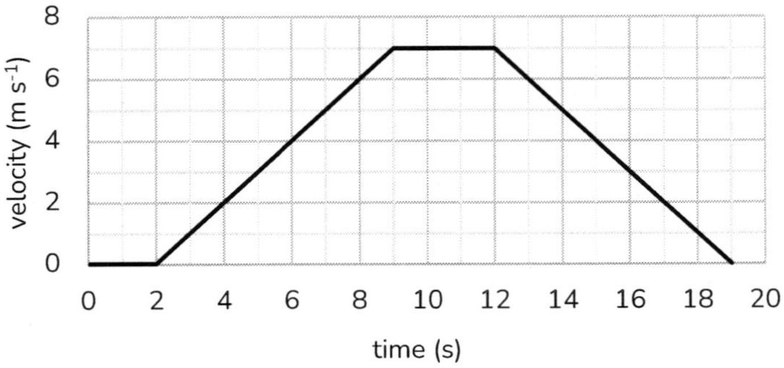

Use a small object to model the journey of the elevator to help visualise what is happening.

For each stage of the elevator's motion, describe its motion and determine what the scales would read if a 50 kg passenger was standing on it. Draw a freebody diagram of the forces on you for each stage.

Freebody diagram

From 0 to 2 s, the elevator is ______________________________.

The scales read ______________ because ______________________________

____

Freebody diagram

From 2 to 9 s, the acceleration of the elevator is ______________________________

____

Therefore the net force on the passenger is ______________________________

____

The normal force on the passenger is ______________________________

____

The reading on the scale is ______________________________

____

Freebody diagram

From 9 to 12 s:

____

____

____

Freebody diagram

From 12 to 19 s:

____

____

____

____

____

____

____

5.6 Harder force problems – Set 2

1. A rope is used to lift a 30 kg load so it accelerates upwards at 0.25 m s^{-2}. What is the tension in the cable?

Freebody diagram

Set a direction as positive.

2. A child of mass 40 kg is holding one end of a rope that is hung over a smooth tree branch. Their older sibling of mass 50 kg sitting on the tree branch thinks it would be fun to get down by holding on to the rope on the other side. Assume the tree branch is frictionless.

   a) What is the magnitude of their acceleration?

   b) What is the tension in the rope?

Freebody diagram

3. A car with mass 2220 kg has a maximum acceleration from rest to 97 m s^{-1} in 8.5 s.

   a) What is the acceleration of the car when it is not towing a trailer?

   b) This car is going to be towing a trailer of mass 1000 kg. What is the maximum acceleration of the car when towing the trailer?

What stays the same as part (a)?

   c) Draw a freebody diagram of the forces on each of the car and the trailer, when the car is towing the trailer at maximum acceleration. (Ignore any balanced forces.)

   d) Calculate the tension in the coupling between the car and trailer at maximum acceleration.

4. A supermarket uses shopping trolleys with a mass of 32 kg each. An employee pushes three shopping trolleys together, nested one in front of the other, so they accelerate at 1 m s^{-2}. Ignore friction.

a) Calculate the force the employee needed to exert on the trolleys.

b) For each trolley, draw a vector diagram of the forces. (Ignore any balanced forces.)

c) What is the force that the front and back trolleys each exert on the middle trolley?

Solve the parts you can first.

5. An actor of mass 50 kg sits on a giant prop of mass 70 kg, suspended in the air by a wire. The actor and the prop are lowered with an acceleration of 0.50 m s^{-2}.

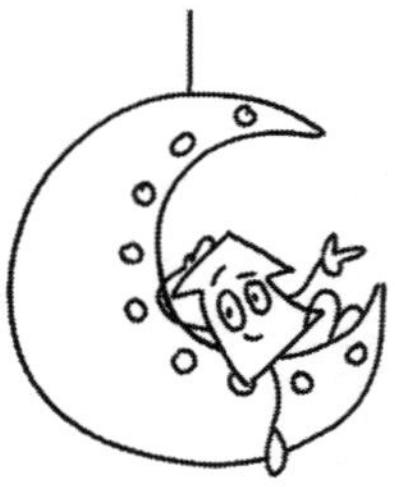

a) What is the resultant force on the actor?

b) What is the tension in the wire?

Freebody diagram

c) What force does the prop exert on the actor?

Where is it helpful to consider the system rather than individual objects?

6. A pulley looks like a wheel with a rope or cable passing around it, which changes the direction of the applied force on an object. They are used to lift heavy loads, from elevators and window blinds to weights at the gym.

Pulley on a train line

A 3.0 kg mass on a smooth bench is attached by a cable to a 2.0 kg mass, which is hanging freely over the bench.

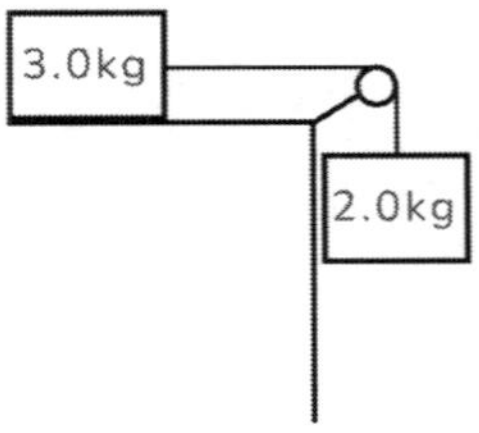

a) What is the acceleration of the system?

b) What is the tension in the cable?

7. To get to space, rockets need to carry a lot of fuel. The fuel typically makes up around 90% of the rocket's total mass. In contrast, what we actually want to get to space – the payload – is a small fraction of the mass of the fuel. The payload could be a satellite to put into orbit, equipment or even people.

To avoid carrying a big empty fuel tank once fuel runs out, every rocket ever launched into space so far has been a multi-stage rocket.

Each stage carries fuel and its own engine. The engine burns the fuel to produce exhaust gases that are pushed out, resulting in an upward force on the rocket called thrust.

The empty stage is jettisoned once the fuel has run out, and the next stage is ignited.

Multi-stage rocket

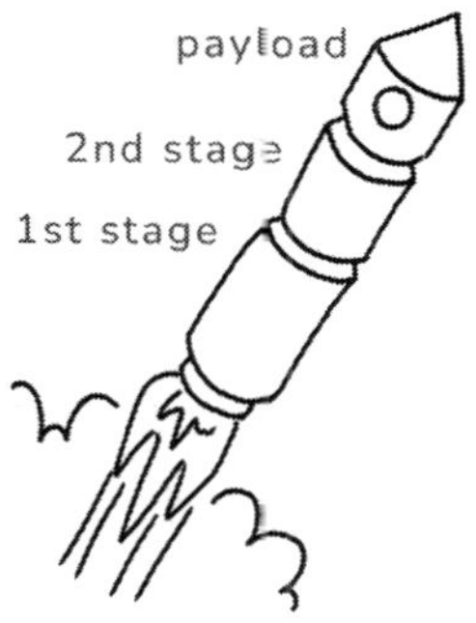

1st stage cut of fuel

a) Draw a vector diagram showing the forces acting on a rocket:

Just after ignition, but before the rocket leaves the ground	After ignition, when the rocket starts accelerating upward	When the rocket is in the air, just after the first stage has run out of fuel and before the second stage is ignited

b) The graph shows the motion of a rocket during its first stage. Explain the shape of the graph, assuming thrust from the engine, F_T, is constant.

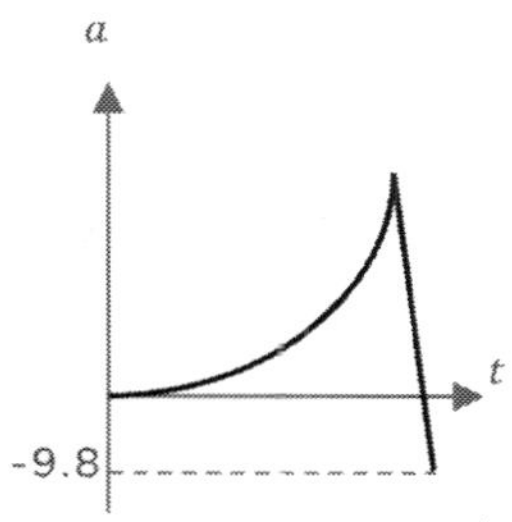

c) The second stage of the rocket fails to ignite, so the rocket eventually falls back to Earth. Sketch a velocity-time graph of the rocket's journey, starting from launch and ending just before the rocket crashes into Earth.

Explain the shape of your sketch.

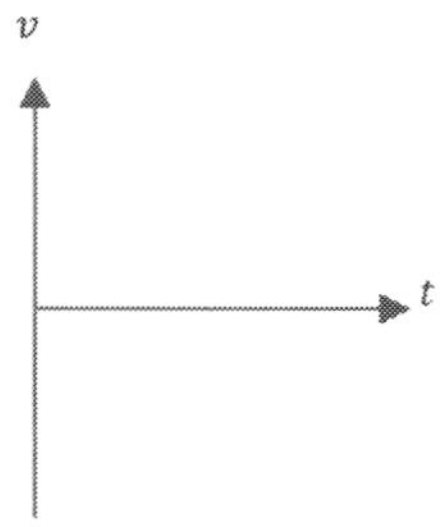

d) An astronaut of mass 80 kg is in the rocket. At the moment when the rocket's acceleration is 49 m s^{-2}, what is the normal force acting on the astronaut?

5.7 Linear regression

1. Explain how to use linear relationships to describe trends, by filling in the blanks.

acceleration	gradient	independent	inverse	mass	proportionally	qualitative
quantitative	science	straight	x-axis	y-axis	y-intercept	

The universe appears to follow mathematical rules, which ________ tries to discover.

Experiments can give us ________ conclusions like “as force on an object increases, its acceleration increases”. However, if we can also obtain an equation for the trend from the results of the experiment, we can use it to make ________ predictions.

Some real world trends are ***linear*** and their equation takes the form of:

$$y = cx + b$$

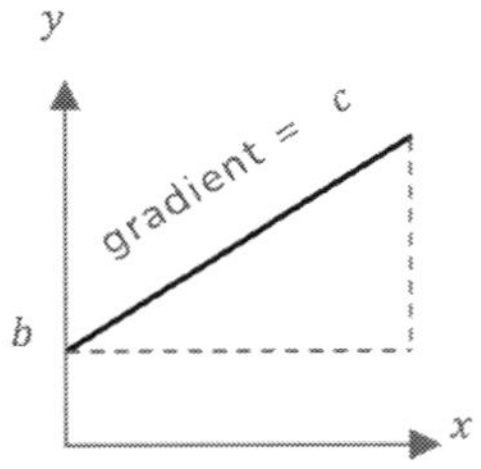

In words, this equation means that y is ***proportional*** to x.

Graphing this equation gives a ________ line, where c is the gradient and b is the y-intercept.

An example of a linear trend is obtained when changing the net force F on an object with mass m to see how it affects the ________ a produced:

$$F = ma$$

The graph for this is shown to the right, with the axes following the usual convention of putting the independent variable, net force F, on the ________ and the dependent variable, acceleration a, on the ________.

The graph shows that as the force on an object increases, its acceleration increases ________.

Rearranging the equation $F = ma$ gives:

$$a = \frac{1}{m} \times F$$

This is in the form of $y = cx + b$:

y	$=$	c	x	$+$	b
a	$=$	$\frac{1}{m}$	F	$+$	0

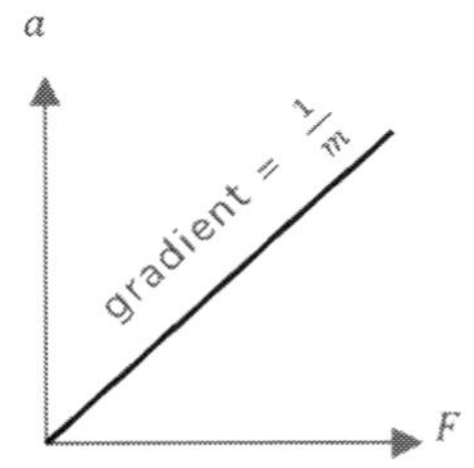

Now we can see that the ________ of the graph should be $\frac{1}{m}$, or the inverse of mass. The line of best fit also passes through the origin, since the ________ is zero.

To make it easier to see the relationship with mass, we can switch the axes and let go of the convention of putting the ________ variable on the x axis, as shown to the right. Graphing this would give the gradient as ________.

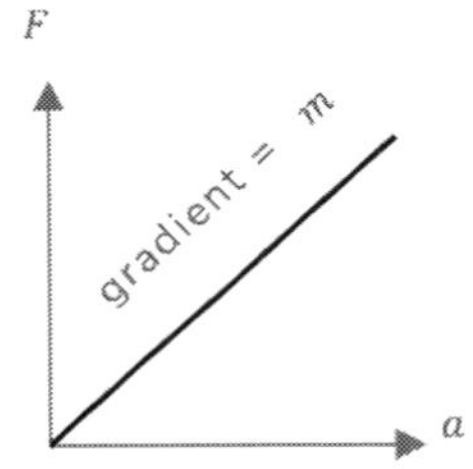

Note that when comparing graphs where the axes are switched, the gradient of one will be the ________ of the other.

2. Explain how to linearise other relationships, by filling in the blanks.

constants	gradient	gravity	increasing	linear	non-linear
rest	similar	squared	trend	uniform	variables

An example of a relationship that is ________ is a quadratic relationship, which takes the form:

$$y = ax^2 + bx + c$$

where a, b and c are constants. For example, the relationship between the displacement of an object undergoing ________ acceleration and time is given by:

$$s = ut + \frac{1}{2}at^2$$

Graph of s vs t

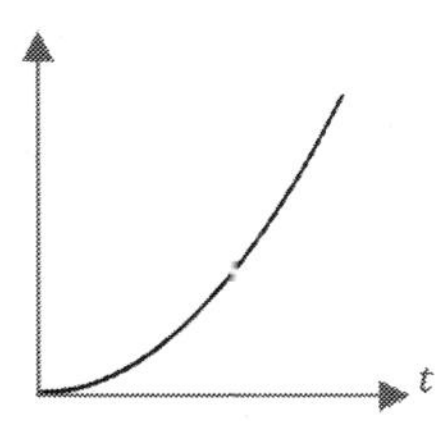

If the object starts at ________, this becomes $s = \frac{1}{2}at^2$, so graphing s on the y-axis and t on the x-axis gives the graph on the right. It shows that as time increases, displacement increases at an ________ rate.

If a trend looks like this, it is hard to tell whether the relationship is in the form of $y = x^2$, $y = x^3$ or some other power, as they have a ________ shape. It also hard to tell what ________ there are in the equation.

These look similar

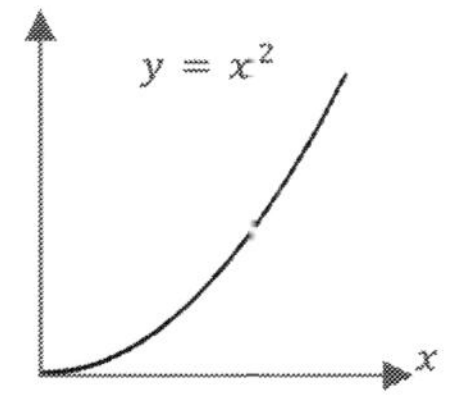

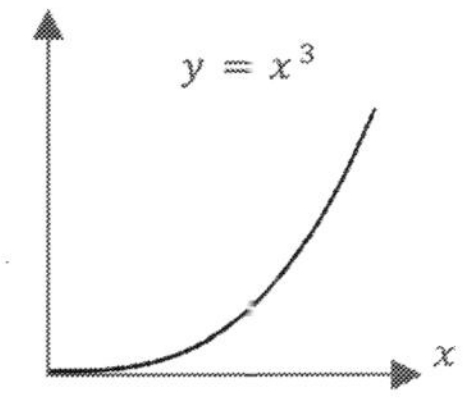

To determine an equation for the relationship, one technique used is a ***linear regression***. This involves choosing what to graph on each axis to obtain a ________ relationship. Something calculated from the independent variable goes on one axis, while something calculated from the dependent variable goes on the other, rather than the ________ on their own.

For example, if the displacement s of an object dropped from rest as time t increases is measured, to verify that $s = \frac{1}{2}at^2$ actually describes the ________, we linearise the equation by writing it in the form $y = cx + b$:

y	$=$	c	x	$+$	b
s	$=$	$\frac{1}{2}a$	t^2	$+$	0

If we graph s on the y-axis and t^2 on the x-axis, and the results are consistent with $s = \frac{1}{2}at^2$, then the graph will be a straight line with the ________ equal to $\frac{1}{2}a$. Since a dropped object accelerates due to ________, theoretically $a = 9.8$ m s^{-2} so the gradient should be 4.9 m s^{-2}.

Linear regression

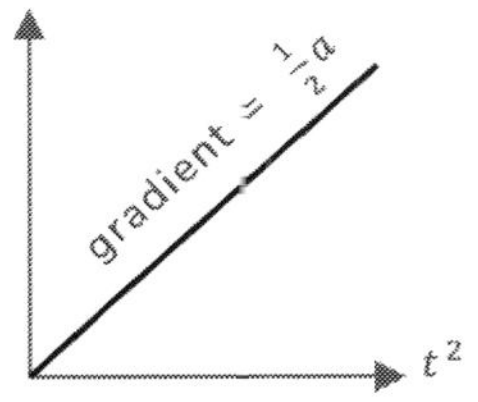

Note that since t^2 is graphed on the x-axis, the unit is also ________, to give seconds squared (s^2).

3. The density of a substance, D, is given by $D = \frac{m}{V}$, where m is the mass of the substance and V is its volume. You measure the mass m of several volumes V of the same liquid.

Rearrange the equation to make what is graphed on the y-axis the subject.

a) Express this equation as a linear relationship, by filling in the blanks.

y	$=$	c	x	$+$	b
	$=$		V	$+$	

Sketch the graph

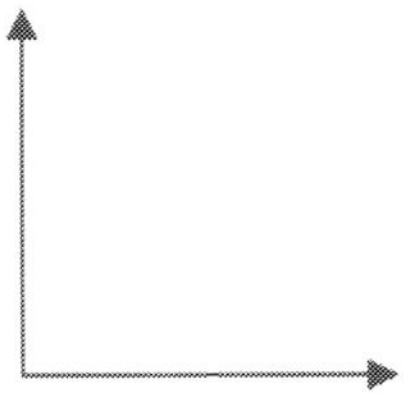

b) What should be graphed on the y-axis to determine whether your data matches the equation, if the volume V is graphed on the x-axis? Sketch the expected graph for the trend in the data.

c) What would the gradient of the graph represent? ________

4. You draw several circles with different radii r and measure the areas of the circles A. The area of a circle is $A = \pi r^2$.

   a) Express this equation as a linear relationship, by filling in the blanks.

y	$=$	c	x	$+$	b
A	$=$			$+$	

Sketch the graph

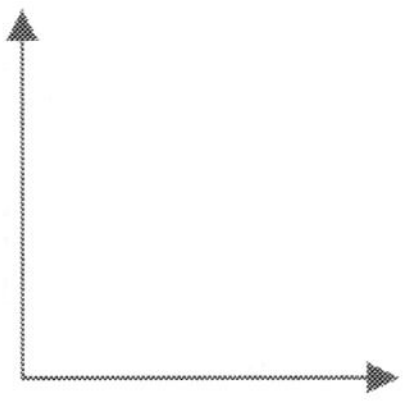

   b) What should be graphed on the x-axis to determine whether your data matches the equation, if the area A is graphed on the y-axis? Sketch the expected graph for the trend in the data. What would the unit for each axis be? ______

   c) What would the gradient of the graph represent? ______

5. The velocity v of an object undergoing uniform acceleration a is measured over time t, using a motion sensor.

   a) Write a hypothesis for the relationship between the velocity v and time t, in a linear form.

y	$=$	c	x	$+$	b
	$=$			$+$	

Sketch the graph

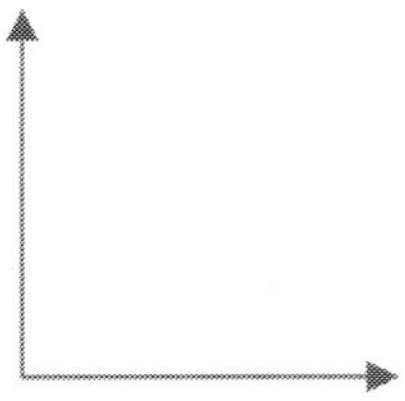

   b) Sketch the expected graph for the data, if the hypothesis is correct.

   c) What would the gradient of the graph and the y-intercept represent?

______

6. The final velocity v of an object undergoing uniform acceleration a is measured over several displacements s.

What you graph on both axes can be a value calculated from a variable.

   a) Write a hypothesis for the relationship between the final velocity v and displacement s, in a linear form.

______

Sketch the graph

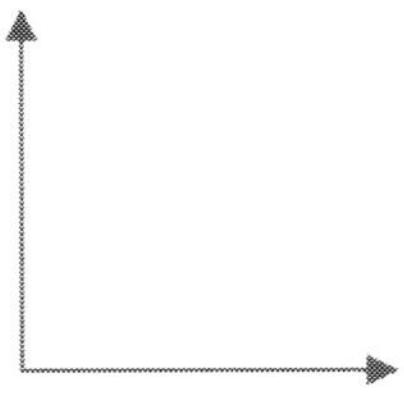

   b) Describe how you could graph the data to determine whether the data was consistent with your hypothesis. Sketch the expected graph.

______

______

7. A pendulum consists of a swinging mass. The period of a pendulum, T, is the time taken to swing back and forth once. A website states the period is related to the length of the pendulum, L, by:

A pendulum is a useful tool for keeping time, such as in grandfather clocks.

$$T = 2\pi\sqrt{\frac{L}{g}}$$

where g is the acceleration due to gravity.

You measure the period, T, when the length of the pendulum, L, is changed. Describe how you could determine whether your data was consistent with the equation.

Sketch the graph

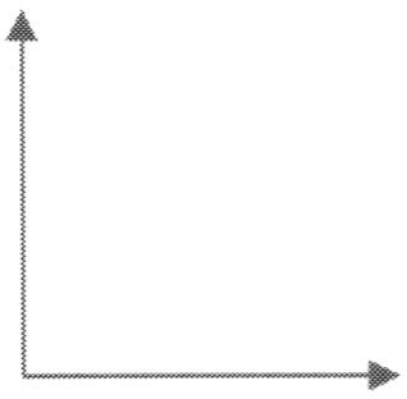

______

______

______

______

There is more than one correct choice of axes possible.

5.8 Experimental skills - Forces

Here is a method for an experiment to investigate forces.

Method

1. Set up as shown, starting with five 50 g slotted masses on a 500 g dynamics cart with a built-in accelerometer, and a hanging 50 g mass carrier with no extra mass. Hold the cart at rest.

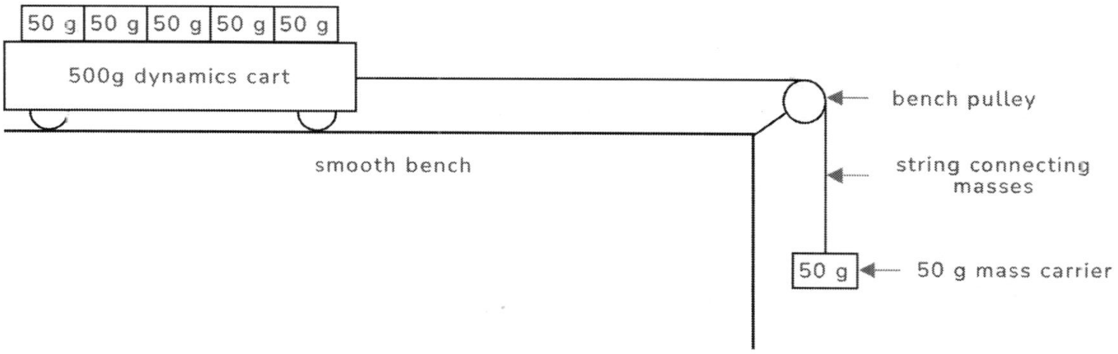

2. Release the cart. Measure its acceleration, using the built-in accelerometer.
3. Repeat step 2 two more times.
4. Repeat steps 1 to 3, after transferring one 50 g slotted mass from the trolley on to the mass carrier, in order to increase the mass hanging over the pulley.
5. Repeat step 4, continuing until all of the 50 g slotted masses on the cart have been transferred to the mass carrier.

1. What is the independent variable? ....

2. What is the dependent variable? ....

3. Identify all of the controlled variables in the method.

....

4. Write an aim for this experiment.

....

5. Write a hypothesis for this experiment. Support your hypothesis with a reason.

....

....

6. What steps have been taken in the method to achieve the following?

   a) Accuracy

   ....

   b) Reliability

   ....

   c) Validity

   ....

   ....

7. Discuss whether controlling the length of the string and the distance the trolley is pulled back will affect the dependent variable in this experiment.

....

....

8. Let the total mass of the cart and masses on it be m_1, the total mass hanging over the pulley be m_2 and the acceleration of the cart be a.

Results for the experiment are shown below, with the average acceleration of the trials calculated.

Total hanging mass m_2 (kg)	Acceleration a (ms^{-2})	Net force F_{net} ( )
0.05	0.61	
0.1	1.23	
0.15	1.84	
0.2	2.45	
0.25	3.06	
0.3	3.68	

a) Explain what provides the net force acting on the system and give an equation for this.

b) In the blank column, calculate the net force, F_{net}, exerted on the system for each value of m_2. Include the unit in the column heading.

c) Graph the acceleration on the x-axis and the net force on the y-axis.

d) Calculate the gradient of the graph. Explain what the gradient represents in this experiment.

6.1 Friction and the perfect shoe

However, the coefficient of friction also depends on electrostatic forces between particles on the surfaces.
Cling wrap looks smooth, but has a high coefficient of friction.
Ballet dancers sometimes dip their shoes in rosin, made of tree sap, to make them stickier on smooth wooden floors.
ROSIN
For every pair of surfaces, there are two values of the coefficient of friction.
When an object is stationary, it is harder to get it to move
than to keep it moving afterwards.
So the ***coefficient of static friction*** μ_s is used when the object is stationary.
Once the object is moving, the ***coefficient of kinetic friction*** μ_k is used, which is lower than μ_s.
Kinetic friction is also known as ***dynamic friction*** or ***sliding friction***.
The lower value of kinetic friction is due to the breaking of small bumps and weakened bonds on the surface as the object moves.
Table (microscopic)
Friction converts some kinetic energy into other forms, such as heat or even light.
So shoes that are too loose generate heat as they rub more against your skin.
But tight shoes exert more normal force on your skin, increasing friction. So the perfect shoe must fit right.
The right shoe also depends on your activity.
For example, different tennis shoe soles are needed for hard court, clay court or grass court.
Hard court
Clay court
Grass court
They can't cause too much friction, to allow tennis players to slide to a stop, rather than stop abruptly, as sliding reduces injuries.
So now you know about friction, what are you looking for in your "sole" mate?

6.2 Friction

1. Describe friction, by filling in the blanks.

decrease	electrostatic forces	increase	normal force	opposes	parallel
perpendicular	reaction force	roughness	surfaces	weight	

Friction is a force which resists movement between two ______________. Friction always ______________ the direction of motion and is ______________ to the surface.

The magnitude of friction on an object moving on a surface depends on the coefficient of friction and the ______________ between the object and surface.

Normal force is the force which is ______________ to the surface in contact with the object, and is the ______________ to the object pushing into the surface. If an object is sitting on a horizontal surface and is not pressed or lifted, the magnitude of the normal force is equal to its ______________.

The coefficient of friction μ depends on the ______________ of the surfaces in contact. If a surface is slippery, the coefficient of friction will ______________. If a surface is rougher, the coefficient of friction will ______________. The coefficient of friction also depends on the ______________ between the surfaces.

2. Identify what the symbols in the equation mean and their unit.

$$\vec{F_f} = \mu\vec{N}$$

Symbol	Meaning	Unit
$\vec{F_f}$		
μ		
$\vec{N}$		

3. Rubber is a common material used to make shoe soles. The table gives sample coefficients of friction of rubber on different surfaces.

Surface	μ_s	μ_k
rubber on dry concrete	1.0	0.80
rubber on wet concrete	0.30	0.25

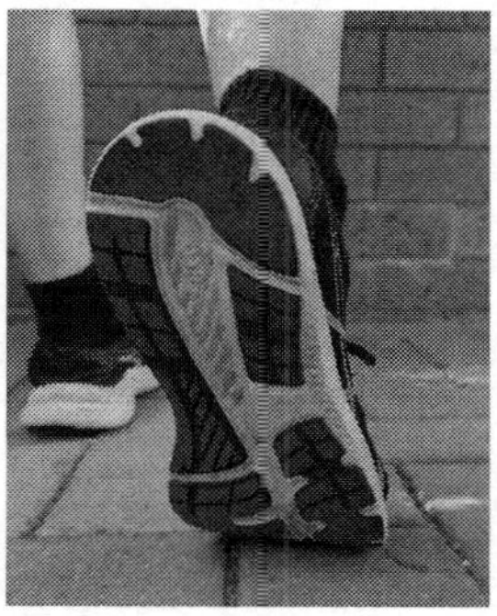

Contrast the coefficients of friction in the table and explain their impact on walking safely.

____

____

____

____

____

Contrast means explain differences. Specifically refer to information given in the question as part of your answer.

4. A small child and an adult slide down the same slippery dip, wearing similar clothing. Explain which will experience the larger frictional force.

____

____

____

____

5. The coefficient of friction of plastic packaging can be tested by wrapping a block in the plastic and pulling it along a surface covered with the same plastic. This is important, as friction during the manufacturing process can cause the plastic to drag along machinery, slowing down the process or damaging the goods being packaged.

A 500 g block wrapped in a certain type of plastic has a coefficient of static friction of 0.45 and coefficient of kinetic friction of 0.38.

a) What horizontal force needs to be exerted on the block to start it moving from rest?

b) What horizontal force needs to be exerted on the block to keep it moving at constant velocity?

6. The picture shows a 5.0 kg box being pushed to the right over a table with a force of 30 N. The coefficient of kinetic friction between the box and table is 0.3.

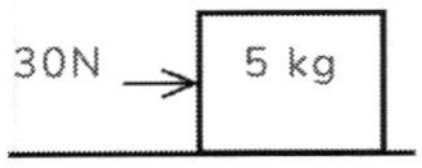

a) Draw a vector diagram showing all the forces acting on the box.

Vector diagram:

b) Calculate the frictional force on the box.

c) Calculate the acceleration of the box.

Write an equation for net force:

Remember to set a frame of reference.

7. A child is sitting on a sled on a horizontal surface. The combined mass of the child and sled is 40 kg. A friend needs to pull the sled with a horizontal force of 75 N to start them moving, and a 60 N force is required to keep them moving at a constant speed afterwards. What are the coefficients of static and kinetic friction?

What is the net force if velocity is constant?

8. A snowboarder with total mass of 80.0 kg glides across horizontal ground with an initial speed of 5.00 m s^{-1}. Assume the coefficient of kinetic friction between their board and the snow is a constant 0.030. What distance will the snowboarder travel before coming to a stop?

Find net force and acceleration first. Then calculate distance.

9. Parkour is a sport that involves running, jumping and climbing quickly, while negotiating obstacles. One parkour skill is called a wall run, which involves running short distances up walls to climb over them. The runner runs towards the wall, then plants a foot onto the wall, driving it downwards to change the direction of their velocity to upwards. However, there must also be some horizontal force into the wall from the foot, otherwise the runner will slip.

If the runner's mass is 60 kg, and the coefficient of friction between the wall and the runner's shoe is 0.70, with what horizontal force must the runner's foot press into the wall so they don't slip down?

Vector diagram:

10. A child slides down a 4.5 m tall pole in a playground, starting from rest and reaching the ground in 1.5 s. The child's mass is 45 kg. The coefficient of kinetic friction between the child and the pole is 0.46.

Assume the child slid down with constant acceleration.

With what force was the child clutching the pole?

6.3 Experimental skills - Friction

A student drags a box across a horizontal bench top, to test the aim using the equipment listed.

Aim: To investigate the effect of changing the total mass of a box and its contents on the force needed to move it at a constant velocity.

Equipment:

spring balance	100 g box with hook to attach to spring balance	4 x 100 g masses	bench top

1. What is the independent variable? ____________________

2. What is the dependent variable? ____________________

3. What variables need to be controlled? ____________________

4. Write a hypothesis for this experiment. Support your hypothesis with a reason.

____

____

5. Complete the method to investigate the aim.

Include steps to get accurate, reliable and valid results.

1. Set up as shown in the diagram below:

____

____

____

____

6. The student obtained the results below, after averaging the trials.

Total mass of the box and contents (g)	Force needed to move the box (N)
100	0.28
200	0.57
300	0.85
400	1.14
500	1.42

a) Graph the force needed to move the box on the y-axis and the total mass of the box and contents on the x-axis.

b) Explain what the gradient of the line of best fit represents. Hence calculate the coefficient of friction of the bench top and box.

____

____

6.4 Skydiving and air resistance

6.5 Air resistance

1. Describe air resistance, by filling in the blanks.

accelerate	acceleration	air resistance	balanced
decreases	greater	terminal velocity	velocity

When a skydiver jumps out of the plane, they ______________ downwards at 9.8 m s^{-2}. As their velocity increases, air resistance increases, so acceleration ______________. When acceleration is zero, and forces are ______________, the skydiver has reached terminal velocity.

Opening a parachute at this time will increase air resistance so that it is ______________ than gravitational force. This means net force is upwards, so ______________ is upwards. The ______________ is still downwards, but it decreases.

The decrease in velocity means that ______________ decreases, until it is the same magnitude as weight. At this point, the net force on the skydiver is zero and they reach a new, slower ______________.

2. Use a force vector diagram to show the forces acting on a skydiver at each stage of their jump.

t = 0 s Just stepped out of the plane	t = 1 s After falling for one second	t = 12 s Terminal velocity of 200 km h^{-1} reached	t = 68 s Parachute opens	t = 77 s Terminal velocity of 5 m s^{-1} with the parachute reached

3. Sketch a velocity-time graph for the jump, using the information in question 1 and 2.

4. At what stage of the jump is the net force:

   a) equal to weight? ______________

   b) upward? ______________

   c) zero? ______________

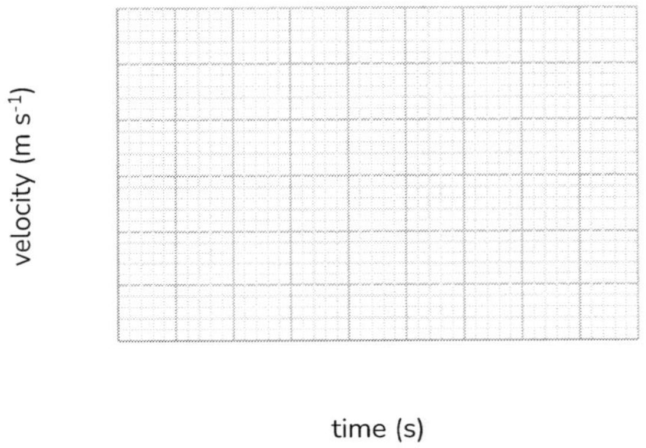

5. True or false? For each false statement, explain what is wrong with it.

   a) The instantaneous velocity of the skydiver can be calculated using $v = u + at$.

   ____

   b) The gradient of the velocity-time graph gives the acceleration of the skydiver.

   ____

   c) The area under the velocity-time graph gives the height of the skydiver above the ground.

   ____

   d) Air resistance can never be bigger than weight, otherwise the skydiver would move up.

   ____

   e) The higher the jump starts, the more force the skydiver will reach the ground with.

   ____

B	A	L	A	N	C	E	D	J	F	O	U	V	E	C	A	F	H	E	V	D
D	N	Y	J	L	V	D	G	E	G	E	H	A	N	I	X	R	O	G	M	U
E	M	I	L	I	E	J	D	U	C	H	A	T	E	M	L	I	E	R	T	E
O	T	E	N	S	I	O	N	N	W	M	E	N	A	P	U	C	G	A	V	Y
M	E	K	N	O	C	U	B	E	A	R	O	N	T	U	Y	T	O	V	I	B
R	S	R	T	E	E	L	A	S	T	I	C	C	O	L	L	I	S	I	O	N
U	T	E	H	G	O	E	K	E	T	W	E	N	A	S	I	O	U	T	X	O
L	E	L	O	H	F	E	E	I	Q	H	S	U	D	E	P	N	B	A	S	R
B	J	A	H	D	R	A	D	H	N	S	E	G	A	C	B	J	T	T	P	M
H	K	T	D	E	V	D	E	D	W	E	K	F	R	O	B	F	R	I	R	A
J	V	I	E	K	A	H	P	C	P	Y	T	L	C	D	E	D	A	O	F	L
P	N	V	D	V	W	L	T	N	E	E	U	I	M	O	T	C	C	N	H	S
T	L	E	C	E	H	K	N	E	T	F	O	R	C	E	W	O	T	A	L	M
L	O	M	P	S	L	D	L	P	O	P	L	E	T	H	K	C	I	L	N	Y
S	R	O	T	A	T	I	O	N	E	J	K	B	U	O	R	M	O	P	T	A
C	O	T	R	P												T	N	O	E	U
B	L	I	Y	N						**Part 2: Dynamics**						K	U	T	G	N
Y	L	O	W	F												L	H	E	A	T
R	I	N	E	L	A	S	T	I	C	C	O	L	L	I	S	I	O	N	D	E
O	N	N	T	H	E	R	I	N	O	U	L	P	W	E	I	G	H	T	X	A
U	G	F	C	E	B	N	P	U	P	I	B	D	R	T	A	N	D	I	E	R
D	R	W	Y	L	N	U	U	R	N	F	L	E	R	X	Y	R	O	A	T	H
M	E	A	Y	T	I	H	L	E	F	I	O	R	C	E	B	E	W	L	I	T
H	S	Y	O	U	K	N	L	F	R	A	T	Y	W	B	P	O	W	E	R	Y
W	I	Z	C	O	N	S	E	R	V	A	T	I	O	N	C	O	I	N	E	L
E	S	M	M	Y	N	E	Y	D	O	E	T	H	R	E	M	R	D	E	S	R
U	T	S	W	N	V	C	B	M	P	T	R	Y	K	I	O	N	G	R	U	S
G	A	H	J	L	C	U	R	I	O	L	S	I	T	Y	M	S	A	G	L	N
H	N	E	O	D	C	O	N	S	I	S	A	T	E	N	E	C	O	Y	T	Y
T	C	S	W	I	C	O	M	P	O	N	E	N	T	L	N	L	H	U	A	E
V	E	C	T	O	R	L	P	Y	O	U	S	U	E	C	T	C	E	E	N	D
R	H	U	N	K	E	D	X	Y	U	V	E	I	O	R	U	K	J	F	T	D
S	W	H	E	M	E	Q	U	I	L	I	B	R	I	U	M	D	Y	G	B	E

7.1 Vectors in 2D

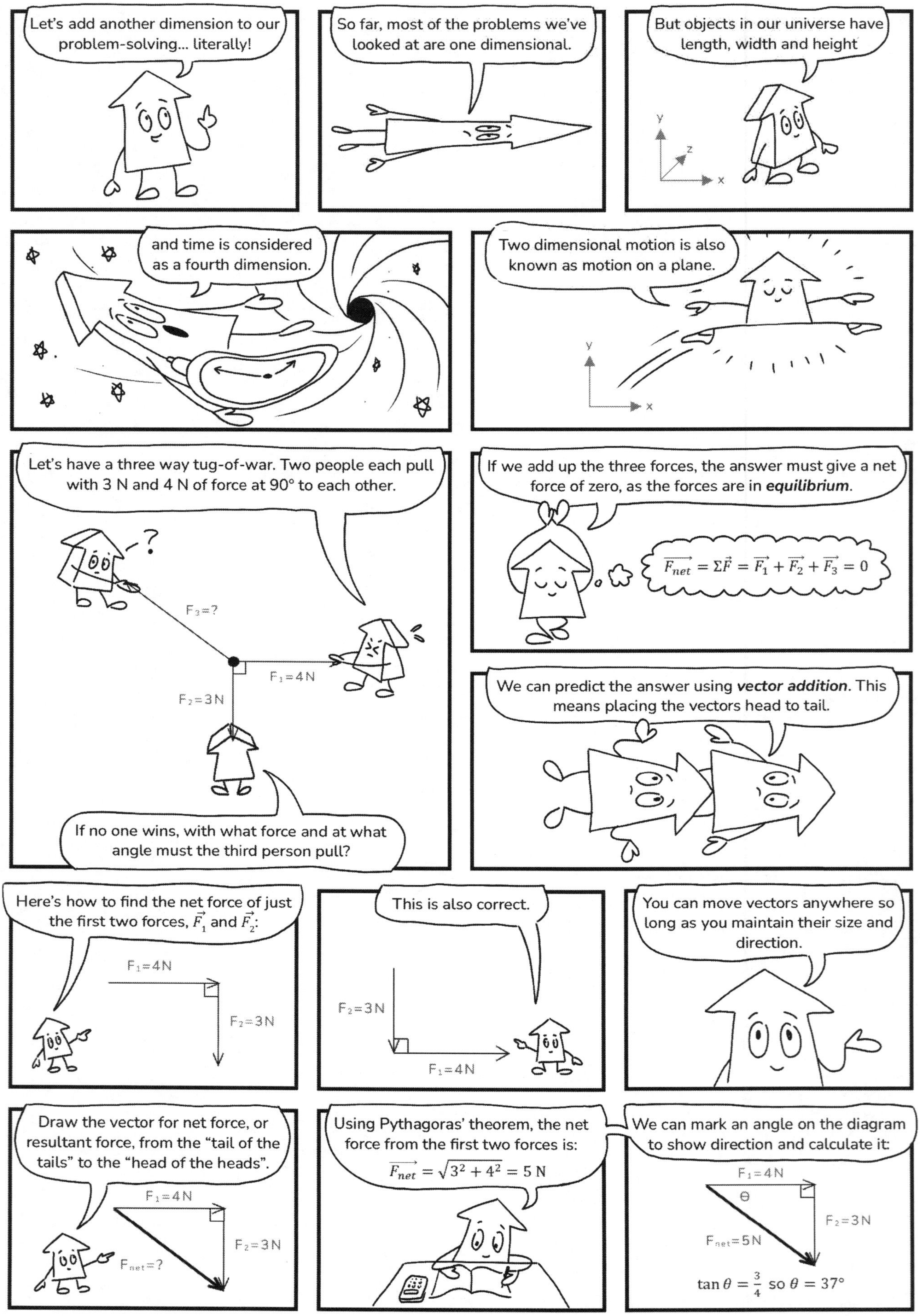

So the third person must exert a 5 N force the opposite way, so forces are balanced.
F₃ =5N
Adding all three forces, the arrows make a closed shape.
Fnet?
F₁=4N
37°
F₂=3N
F₃=5N
We can't draw a resultant vector, as net force is zero.
Let's see what this looks like on our tug-of-war diagram, as a ***freebody diagram***.
A straight line is 180°
F₃ =5N
143°
F₁=4N
37°
F₂=3N
Vector addition usually involves a little geometry, but it can get mathematically complicated if angles aren't 90°
Sine rule? Cosine rule?
F₂
Fnet
F₁
or if there are more than two vectors.
Fnet
It is possible to use a ***graphical method*** to draw the vectors to scale on graph paper, using a ruler and protractor, then measure the resultant.
Another method is to ***resolve*** each vector into ***components***. For displacement vectors, this is the same as breaking it up into x and y coordinates on a number plane.
resultant
y
x
In other words, vector addition of the x and y components gives the resultant.
We can resolve any vector into components, including force.
If you pull a box horizontally, you accelerate it horizontally
F
but if you pull the rope up vertically, you will only lift the box.
F
If you pull at an angle, the box will still accelerate, but there is a smaller component of horizontal force, since some of the force lifts the box instead.
F
Fy
Fx
We can use ***resolving into components*** to solve the same tug-of-war problem.
Because the net force is zero, all of the x components added together must give zero
and so must all the y components.
Resolving into components is really useful when adding a lot of forces!

7.2 Forces in 2D

1. Describe how to use vectors to determine net force, by filling in the blanks.

accelerate	arrows	balanced	closed polygon	direction	displacement	equilibrium
head to tail	magnitude	mass	resultant force	trigonometry	vector sum	velocity

A vector is a quantity that has both magnitude and ________________, such as force, acceleration, velocity or ________________. Vectors are represented by ________________. When several forces act on an object, the net force is the ________________ of all of these forces. Another term for net force is ________________. To add vectors, you draw them ________________. You can arrange the vector arrows any way to do this as long as you don't change their ________________ and direction. The net force is then drawn from the "tail of the tails" to the "head of the heads", creating a ________________. Rules of geometry, such as Pythagoras' theorem and ________________, can then be used to solve for the magnitude and direction of vectors.

If the forces are in ________________, then the net force is zero. This means forces are ________________ and the object is either not moving or moving at constant ________________. If the net force is not zero, then the object will ________________ in the same direction as the net force. The net force is equal to ________________ times acceleration.

Two forces at right angles

Example: A rock climber of mass 65 kg, strapped into a safety harness, rests halfway up a vertical wall with their legs pushing against the wall horizontally. The safety rope makes a 15° angle with the vertical.

a) Draw a freebody diagram showing the forces acting on the climber.

b) Draw a vector addition diagram showing the forces in equilibrium.

c) How much tension is in the rope?

d) How much force is exerted by the climber's legs on the wall?

Answer:

a)

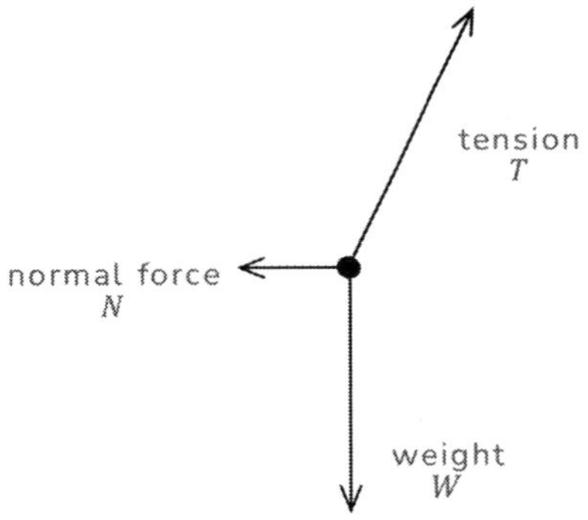

b)

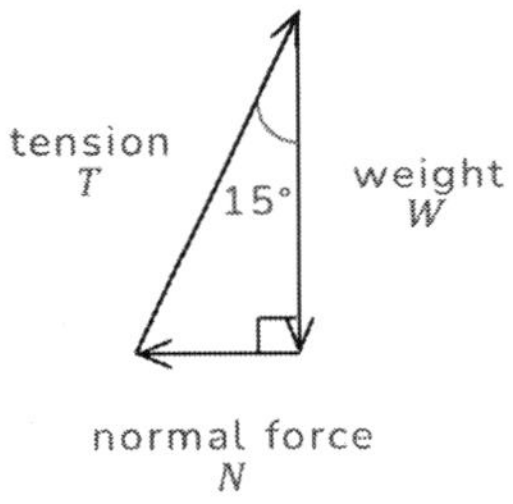

Always start your problem solving with drawing labelled diagrams

c) $\cos 15° = \frac{W}{T}$

$T = \frac{W}{\cos 15°} = \frac{65 \times 9.8}{\cos 15°} = 659.4 \ldots = 660 \text{ N}$

The tension in the rope is 660 N.

Use trigonometry and your vector diagram

d) $\tan 15° = \frac{N}{W}$

$N = W \tan 15° = 65 \times 9.8 \times \tan 15° = 170.6 \ldots = 170 \text{ N}$

The climber's legs exert 170 N on the wall.

2. A car of mass 1 000 kg is stuck in the mud. It is being pulled by two tow trucks. One pulls to the south with a force of 10 000 N, the other pulls to the east with a force of 5 000 N.

Vector diagram

a) What is the net force which the two tow trucks exert on the car?

Use Pythagoras' theorem to find net force

Direction can be given as a compass bearing

b) If the only forces were due to the tow trucks, what would be the acceleration of the car?

3. Tug boats are often used to help manoueuvre large ships into or out of harbour.

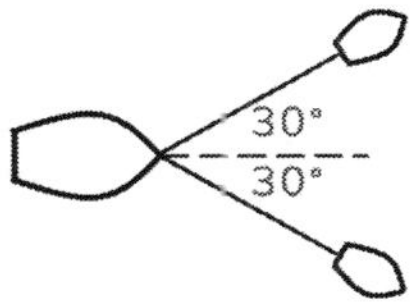

Vector diagram

A ship is being pulled by two tug boats with a net force 25 000 N. The tow ropes make an angle of 45° to either side of the direction the ship is moving.

What is the magnitude of force each tug boat exerts on the ship?

4. A parent holds a child horizontally back on a swing so the ropes the chair is attached to make an angle of 35° to the vertical.

Vector diagram

The child's mass is 28 kg.

a) With what force is the parent pulling the child?

b) What is the tension in each rope? (Since there are two ropes, one on each side of the swing, the combined tension will be twice the tension in each rope.)

Let F represent the force the parent exerts on the child. What other forces are there? What is the net force?

5. A child is hanging still from monkey bars in a playground. Each of their arms makes a 45° angle to the horizontal bars. The child's mass is 40 kg.

Vector diagram

What is the tension in each of the child's arms?

6. An acrobat performs a human flag move on a pole. Their arms each make a 45° angle to the pole, with the top arm pulling and the bottom arm pushing the acrobat up. The acrobat's mass is 60 kg.

   How much force is exerted by each arm?

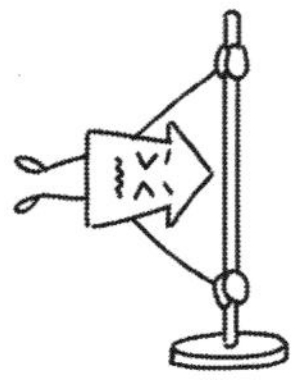

Vector diagram

Forces at any angles

If vector addition gives you a triangle without a right angle, then you need to use the sine rule or cosine rule.

	Sine rule	Cosine rule
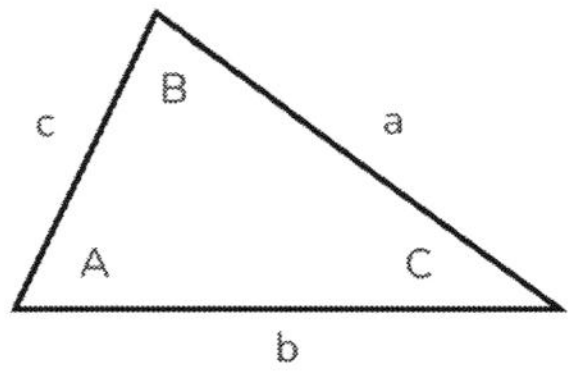	$$\frac{a}{\sin A} = \frac{b}{\sin B} = \frac{c}{\sin C}$$ Use this if the data you have and the unknown involve 2 angles and lengths of their opposite sides.	$$a^2 = b^2 + c^2 - 2bc\cos A$$ Use this if the data you have and the unknown give you 3 sides and 1 angle.

Example: A child travels horizontally along a zipline in a playground at constant velocity. The zipline makes a 150° angle at the point the child is hanging from. The child's mass is 50 kg. What is the tension in the cable?

Answer:

$$W^2 = T^2 + T^2 - 2 \times T \times T\cos 30°$$

$$(50 \times 9.8)^2 = 2T^2 - 2T^2\cos 30° = 2T^2(1 - \cos 30°)$$

$$2T^2 = \frac{(50 \times 9.8)^2}{(1 - \cos 30°)}$$

$$T = 946.6 \ldots N = 950\ N$$

The tension in the zipline is 950 N.

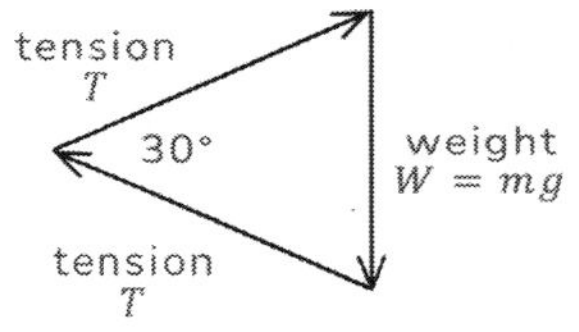

A straight line is 180°. Use this to work out angles as you rearrange the vectors into a closed polygon.

7. Two tug boats pull a ship with a net force of 25 000 N. The tow ropes make an angle of 30° to either side of the direction the ship is moving.

   What is the magnitude of tension in the rope connecting each tug boat exerts on the ship?

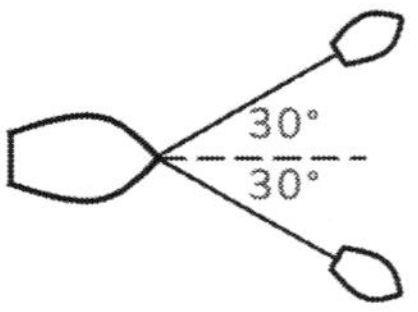

Vector diagram

A triangle's angles add up to 180°.

8. A 65 kg gymnast holds themselves in an iron cross position on the rings, with their arms each at 80° to their torso.

Vector diagrams

a) How much force is each arm pushing down with?

b) Compare the force in each arm in the iron cross to the support position, where the gymnast pushes up with their arms vertically by their sides.

c) Explain why it is impossible to do the iron cross with arms completely horizontal.

State roughly how many times bigger one is compared to the other.

9. A child of mass 35 kg is attached to a bungee jumper, consisting of two stretchy cords affixed to the ceiling.

Vector diagram

In a moment when the child is accelerating upwards at 2.0 m s^{-2} the bungee cords make an angle of 20° to the vertical.

How much tension is in each bungee cord?

Net force is not zero.

7.3 Graphical method of adding vectors

An alternative to mathematically calculating the magnitude and direction of vectors is the graphical method. This involves using a ruler and protractor to draw vectors to scale, preferably on graph paper, then measuring the answer. This method may give a quick answer when mathematical working is tedious, for example when adding a lot of vectors. However, it can be harder to draw and measure accurately.

Example: Two ants each exert a 3.0 N force on a breadcrumb. One pushes it east, while the other pushes it toward a 50° angle north of the first ant (or E50°N). What is the resultant force exerted by the ants?

Answer:

Step 1: Pick a scale that fits on the page and is easy to use. Let 1 cm = 1 N. So a 3.0 N force would be represented by 3 cm.

These forces might sound small, but some experiments show ants can carry around 5000 times their body weight. For this reason, engineers are interested in what makes their body structure so strong.

Step 5: Measure the angle of the resultant with a protractor.

Step 4: Draw the resultant, joining the "tail of the tails" to the "head of the heads". Measure the size in cm and convert it to N.

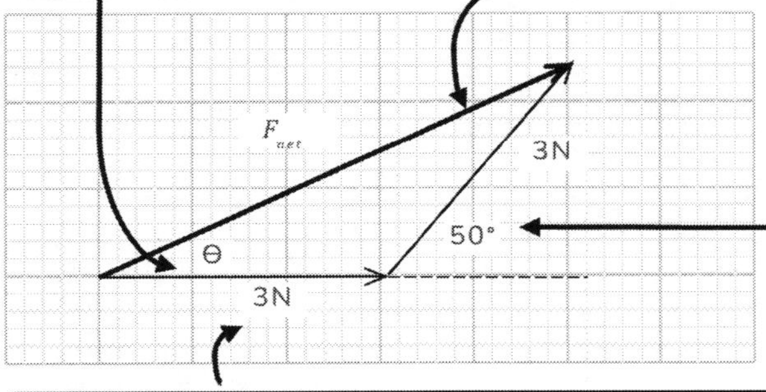

Step 3: Measure the angle for the next vector you are adding with a protractor. Draw the vector with a ruler.

If you need to, sketch a rough vector diagram first on the side, to help visualise what you need to draw to scale.

Step 2: 3 N = 3 cm. Draw this with a ruler on the grid line.

Step 6: The resultant force is 5.4 N at E26°N.

1. A gardener tries to hang a pot plant in the middle of a horizontal string. The maximum tension the string can take is 980 N. When the string is pulled so that it makes a 85° angle to the vertical on each side of the pot plant, the string breaks. What is the mass of the pot plant?

Step 1: Pick a scale. Let 1 cm = 100 N. How many cm would represent a tension of 980 N? ....

Step 3: The arrow from step 2 should end up somewhere here. Draw the second tension vector.

Step 4: The pot plant was in equilibrium, so adding a weight vector brings you back to the start. Measure the length.

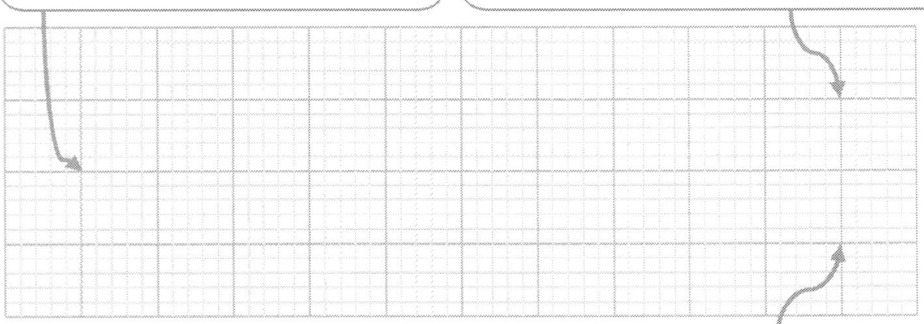

Step 2: Start here to draw the first tension vector, pulling the pot plant to the left. Measure the angle from the grid line. Label the vector.

Step 5: Convert the length for the weight into N, then calculate the mass.

....

2. A drone flies 40 m east, then 84 m in a direction S45°E, then 68 m in a direction N27°E, then 110 m in a direction S56°W. The whole journey took 25 s.

a) Use the graphical method to determine the final displacement of the drone from its starting position.

b) Determine the average speed of the drone.

c) Determine the average velocity of the drone.

These vector techniques work for any vectors, not just forces. Every vector in the diagram needs to be the same type though. Here they are all displacement vectors.

3. An archer shoots an arrow forward with a force of 90 N. At the instant the bowstring is released from rest, the angle between the strings at the back end of the arrow is 106°, with the arrow sitting horizontally halfway in between them. What is the magnitude of tension in the bowstring?

Have angles but no magnitude? Draw lines at those angles and see where they meet.

7.4 Subtracting vectors

As well as being added, vectors can also be subtracted. To subtract a vector, or make it negative, simply reverse its direction.

Then just add the vectors as you normally would, head to tail.

Here are vectors $\vec{a}$ and $\vec{b}$. $\vec{c} = \vec{a} + \vec{b}$ looks like: $\vec{c} = \vec{a} - \vec{b}$ looks like: $\vec{c} = \vec{b} - \vec{a}$ looks like:

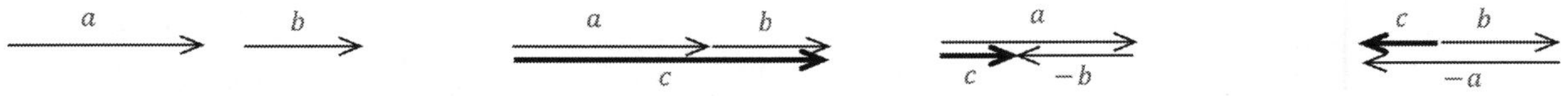

Here are vectors $\vec{a}$ and $\vec{b}$. $\vec{c} = \vec{a} + \vec{b}$ looks like: $\vec{c} = \vec{a} - \vec{b}$ looks like: $\vec{c} = \vec{b} - \vec{a}$ looks like:

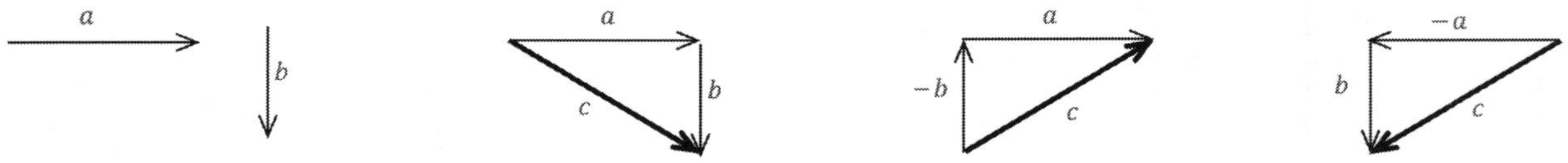

Vector subtraction is useful for determining the change in a vector quantity, for example change in velocity.

Example: The Australian Parliament House sits in the middle of a roundabout with a radius of 487 m. A driver takes 40 s to drive part way around the roundabout, so that they change direction from northbound to westbound. They drove at a constant speed of 70 km h^{-1}.

a) What was the change in velocity of the driver?

b) What was the average acceleration?

Answer:

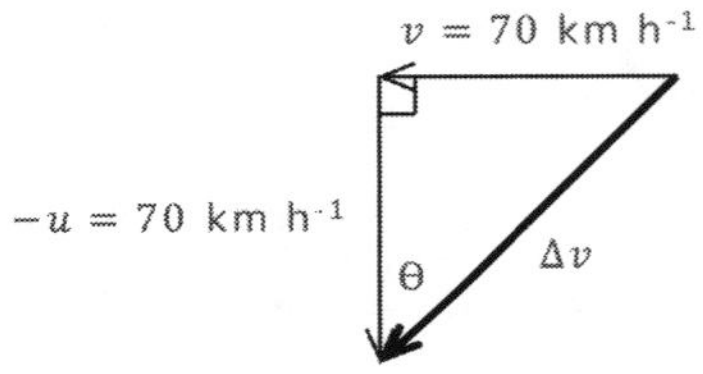

a) $\Delta\vec{v} = \vec{v} - \vec{u} = \sqrt{\left(\frac{70}{3.6}\right)^2 + \left(\frac{70}{3.6}\right)^2} = 27 \text{ m s}^{-1}$

$\tan\theta = \frac{70}{70} = 1$

$\theta = 45°$

Change in velocity is 27 m s^{-1} southwest.

b) $\vec{a} = \frac{\Delta\vec{v}}{t} = \frac{27}{40} = 0.68 \text{ m s}^{-2}$ southwest

1. Vectors $\vec{a}$ and $\vec{b}$ are shown below. Sketch the resultant vectors required.

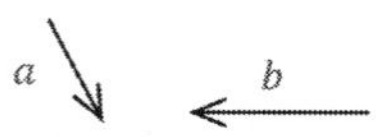

a) $\vec{a} + \vec{b}$ b) $\vec{a} - \vec{b}$ c) $\vec{b} - \vec{a}$

2. Vectors $\vec{a}$ and $\vec{b}$ are shown below. Sketch the resultant vectors required.

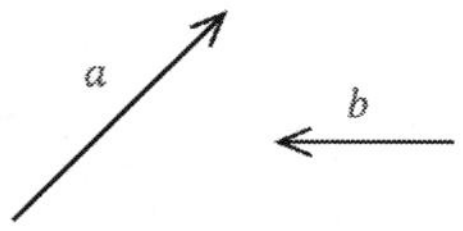

a) $\vec{a} - \vec{b}$ b) $\vec{b} - \vec{a}$

Compare $\vec{a} - \vec{b}$ with $\vec{b} - \vec{a}$ in all the examples on this page.

3. A yoyo is initially moving down at 1.0 m s^{-1}. After 0.50 s, it is moving up at 1.0 m s^{-1}.

Vector diagram

a) What is the change in velocity of the yoyo?

b) What is the average acceleration of the yoyo?

c) Identify the force causing this acceleration.

4. In hammer throw, an athlete spins a 7.26 kg metal ball on the end of a 121.5 cm wire in a circle, before releasing the wire to throw the ball as far as possible.

In one particular throw, the ball spins a quarter of the circle in 0.25 s at a constant speed of 7.63 m s^{-1}.

a) What is the magnitude of the average acceleration of the ball?

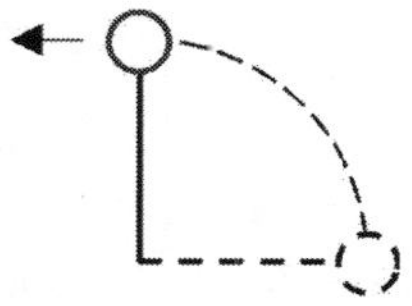

Vector d agram

b) What is the magnitude of the average force on the ball?

c) Identify the force causing the ball to spin.

7.5 Resolving vectors

1. Describe how to resolve vectors into components, by filling in the blanks.

component	coordinates	direction	distance	gravity
horizontally	magnitude	perpendicular	trigonometry	vertical

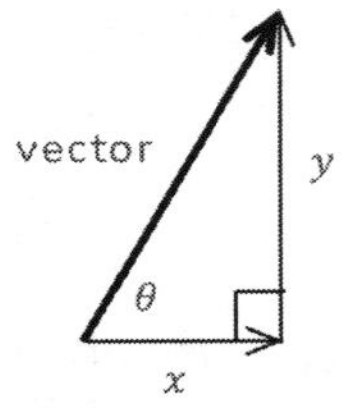

Resolving a vector into components means breaking it up into two parts that are ____________ to each other. It is like working out the x and y ____________ of a point on a number plane. If you use Pythagoras' theorem to calculate $\sqrt{x^2 + y^2}$, you get the ____________ from the origin to the point. Similarly, using Pythagoras' theorem on the components of a vector gives the ____________ of the vector.

To find the direction of a point on a number plane, you can use ____________ to find an angle. Similarly, if you know the components, you can use trigonometry to find the ____________ of a vector.

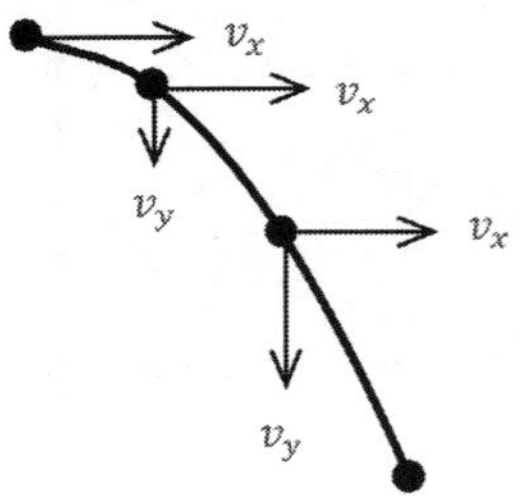

Resolving a vector is useful when there is something that only affects one ____________ of the vector. For example, if you throw a ball forwards, ____________ will pull the ball downwards. However, gravity will not stop the ball moving ____________, since this component of the ball's velocity is at right angles to gravitational force. Gravity will only change the ____________ velocity of the ball.

Example: A ship sails 100 km in a direction of N30°E. How far has the ship travelled:

a) towards the north?

b) towards the east?

Answer:

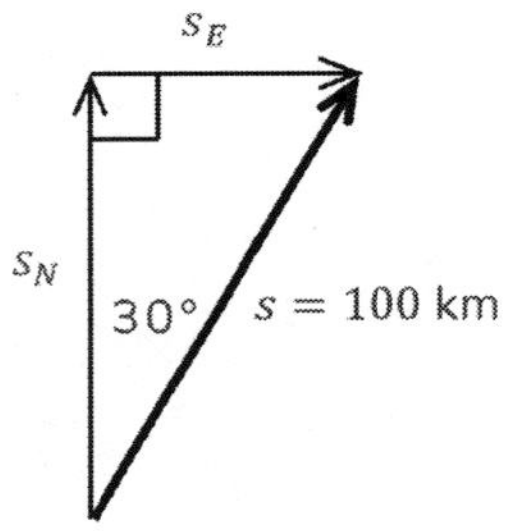

a) Let the north component of displacement be s_N.

$$\cos 30° = \frac{s_N}{100}$$

$$s_N = 100 \cos 30° = 87 \text{ km}$$

b) Let the east component be s_E.

$$\sin 30° = \frac{s_E}{100}$$

$$s_E = 100 \sin 30° = 50 \text{ km}$$

Resolve the displacement into north and east components.

You can identify components with a subscript.

2. An engineer is trying to determine the height of a vertical cliff, in order to install a zipline from it to a point on the ground 96 m away from the base of the cliff. They measure the angle of elevation of the top of the cliff from this point as 20°.

Vector diagram

a) What is the height of the cliff?

____

____

b) How long will the zipline need to be?

____

Resolve the displacement along the zipline into horizontal and vertical components.

Example: A traveller pulls a wheeled suitcase of mass 15 kg along the smooth horizontal ground at an airport. They pull on the handle with a force of 50 N at an angle of 60° to the ground.

a) What are the horizontal and vertical components of force they exert on the suitcase?

b) What is the normal force the ground exerts on the suitcase?

Answer:

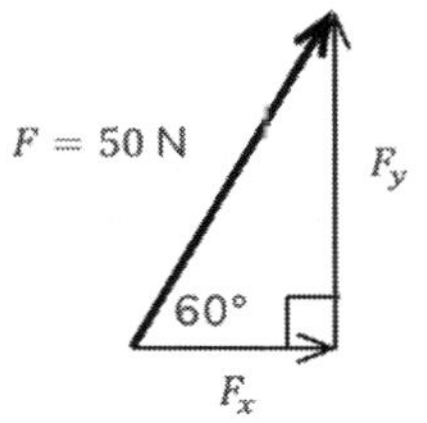

a) Let the horizontal component of force be F_x and vertical component be F_y.

$$\cos 60° = \frac{F_x}{50}$$

$$F_x = 50 \cos 60° = 25 \text{ N}$$

$$\sin 60° = \frac{F_y}{50}$$

$$F_y = 50 \sin 60° = 43 \text{ N}$$

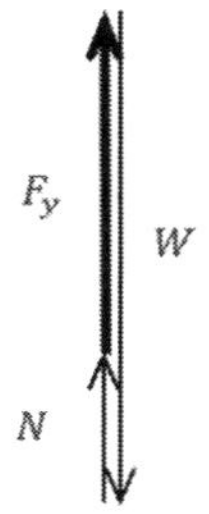

b) Suitcase is not accelerating vertically so vertical forces are balanced.

Only the vertical component of the force will affect the normal force

$$W = N + F_y$$

$N = 15 \times 9.8 - 43 = 104$ N upwards

3. You give a child a ride in a wheelbarrow on smooth horizontal ground. You push on the wheelbarrow with a force of 200 N, at an angle of 30.0° to the ground. The child and wheelbarrow have a combined mass of 60.0 kg.

a) What are the horizontal and vertical components of force you exert on the wheelbarrow?

Vector diagrams

b) What is the acceleration of the wheelbarrow and child?

Once you have the horizontal component of force, you can use it to calculate acceleration.

c) What is the normal force the ground exerts on the wheelbarrow?

Vector diagram

4. A ship travels at 40 km h^{-1} in a direction of N68°E.

In 10 minutes, how far north has the ship travelled?

Convert units! If the question doesn't state the unit needed, use the easiest one.

5. The Leaning Tower of Pisa is famous for its tilt (and the many attempts by engineers to stop it falling). A tourist tried to estimate the angle of the tilt by dropping a rock from rest over the edge of the roof. It took 3.346 s to reach the ground (ignoring air resistance) and landed 3.836 m from the base of the wall.

Vector diagram

a) At what angle to the vertical is the Leaning Tower of Pisa leaning?

SUVAT?

b) What is the distance from the place the rock was dropped to the base of the tower?

6. A hiker measures the length of a straight trail on a map, between their location and the top of a mountain, to be 1.0 km. The average walking speed of the hiker is 4.0 km h^{-1}, so they calculate it will take around 15 minutes to reach the mountain top. The elevation (vertical height above ground) of the mountain top is 500 m compared to their location. Evaluate the validity of the hiker's calculation, with reference to components.

Vector diagram

The hiker would be correct if

However, the hiker has not considered

This would give a total time of

Therefore the hiker's calculation is valid / invalid. (Circle the correct answer.)

Validity means correctness. Evaluate means assess the positives and negatives, with supporting calculations, and give a judgement. What has the hiker done correctly or incorrectly?

7. A runner of mass 65 kg pushes off the starting blocks in a race, with an average force of 882 N at a 45° angle to the ground. If the runner applies the force on the starting blocks for 0.35 s, with what horizontal speed do they leave the starting blocks?

Vector diagram

8. Compare the horizontal component of force acting on a runner starting a race, with and without starting blocks to push off.

Vector diagrams

With starting blocks, the normal force on the runner is ______________________

____

____

____

Without starting blocks, the normal force on the runner is ______________________

____

so horizontal force moving the runner forward is provided by ______________________

____

____

Key concepts: normal force, friction. Compare the vector diagram for each scenario.

9. A student vacuuming their floor discovers it is easier to pull the nozzle with the rod than to push the nozzle. Being a physicist, they do a fair test of this by using the same vacuum cleaner, on the same carpeted floor, with the rod at the same angle. Explain their observation, with the aid of vector diagrams.

Vector diagrams:

____

____

____

____

____

____

____

Try vacuuming! Compare the vector diagram for each scenario – what is different?

10. A 12 kg box is being pulled with a rope angled at 30° to the horizontal. With what force does the rope need to be pulled in order to just slide the block along a horizontal surface, with a coefficient of static friction of 0.75?

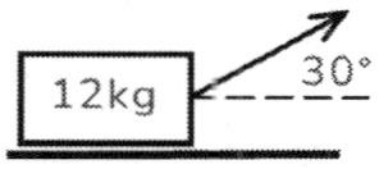

Vector diagrams

____

____

____

____

____

____

____

____

____

What forces are acting in the horizontal and vertical directions? You will need to solve two equations.

7.6 Adding more than two vectors

If more than two vectors need to be added, you can resolve each into components, then add the components together.

Example: Three people, A, B and C, are playing a three-way tug of war. Their ropes are knotted in the middle, and start at 120° to each other. A pulls on their rope with a force of 425 N, B exerts 450 N and C exerts 430 N. What is the resultant force on the knot in the middle?

Answer:

Freebody diagram

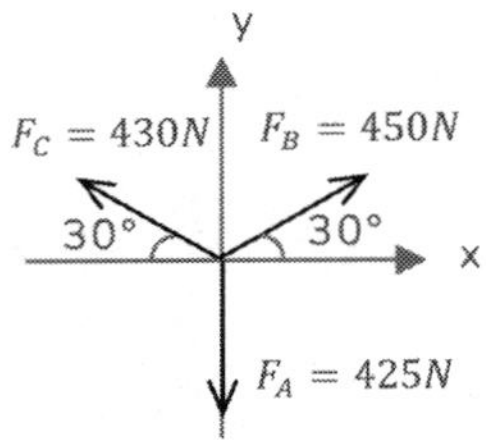

Force	x	y
F_A	0	−425
F_B	$450\cos 30°$	$450\sin 30°$
F_C	$-430\cos 30°$	$430\sin 30°$
F_{net}	17.32	15

Draw a freebody diagram on a number plane.

Resolve each vector into x and y components in table. Don't forget negative signs if a vector is pointed towards the negative axis.

Vector addition of components

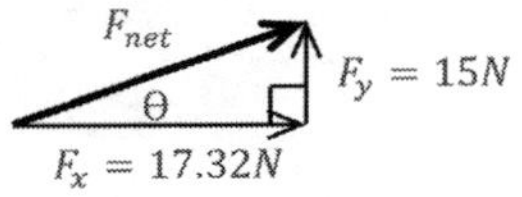

$F_{net} = \sqrt{15^2 + 17.32^2} = 22.9$ N

$\tan\theta = \dfrac{15}{17.32}$

$\theta = 40.9°$

The resultant force is 22.9 N, at 10.9° towards C from person B's rope.

Adding the total of the components using a vector diagram gives the resultant.

1. Complete each sentence.

   a) Breaking up a vector into two vectors at right angles that add up to give the original vector is called ______________________.

   b) If any number of vectors are added together, then the vector sum of their horizontal components will equal the ______________________ and the vector sum of their vertical components will equal the ______________________.

2. A person flying a 0.25 kg kite attaches a spring balance to the kite string. The tension in the string is 10.5 N and the string makes an angle of 45° to the horizontal when the kite is in equilibrium. What force is provided by the lift, F_L, (air resistance acting vertically upwards) and drag, F_D, (air resistance pulling the kite horizontally away from the string)?

Draw a freebody diagram then resolve each force into components:

Vector diagram

Force	x	y
F_{net}		

Calculate the lift: ______________________

____

If the kite is in equilibrium, what is the total force?

Calculate the drag: ______________________

____

7.7 Inclined planes

1. Discover how to analyse motion on inclined planes, by filling in the blanks using the word bank.

acceleration	along	horizontal	mass	net force
normal force	normal force	perpendicular to	ramp	reaction force
resolving	through	vertical	weight	weight

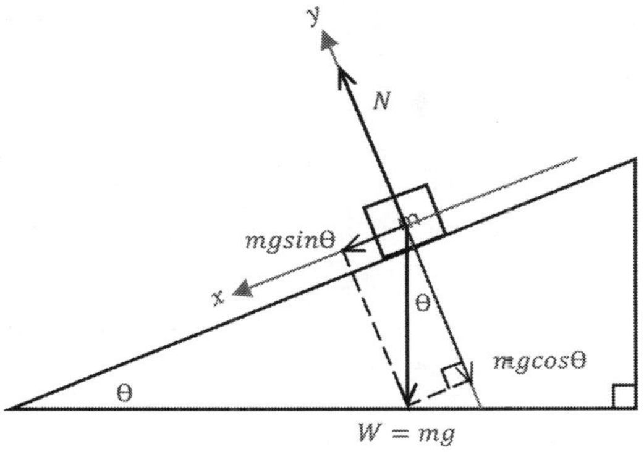

An inclined plane is a ________. An object on the ramp will tend to slide down, due to its ________. However, because the object cannot fall vertically ________ the ramp, its acceleration is restricted to being downwards ________ the ramp. So ________ weight into components, with one component along the ramp, is a useful way to find the acceleration of the object.

We can rotate our frame of reference so that the x-axis is along the ramp, in the direction of the ________ on the object, and the y-axis is perpendicular to the ramp, in the direction of ________ on the object.

If the angle the inclined plane makes to the horizontal is θ, then the component of ________ along the ramp is $mg \sin \theta$. If there is no other force with a component in the x-direction, such as friction, then the ________ of the object down the ramp can be found using:

$$F = ma = mg \sin \theta$$

$$a = g \sin \theta$$

This means that acceleration is independent of the ________ of the object.

The component of weight ________ the ramp is $mg \cos \theta$. If there is no other force with a component in the y-direction, perpendicular to the ramp, then the ________ on the object is:

$$N = mg \cos \theta$$

This is the ________ from the object pushing against the ramp. Normal force has a magnitude equal to weight if the ramp is ________ and zero if the ramp is ________.

2. Determine the sizes of the angles on the inclined plane shown, by filling in the table with the correct angle and supporting reason from the lists below.

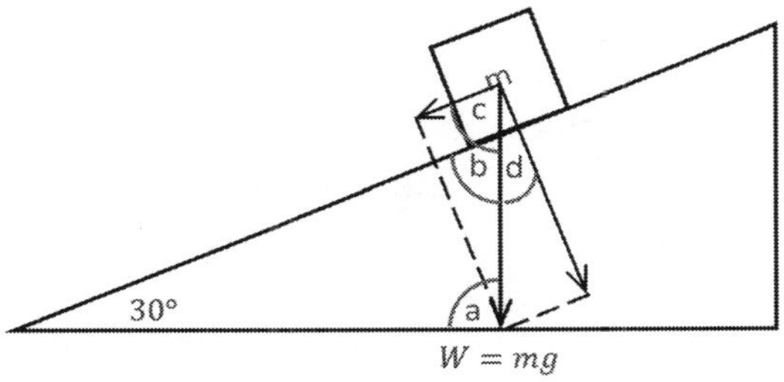

Magnitudes: 30°, 60°, 60°, 90°.

Reasons:

Components of weight are at 90° to each other.
Corresponding angles on parallel lines are the same.
Sum of angles in a triangle is 180°.
Weight is vertically down.

Angle	Magnitude	Reason
a		
b		
c		
d		

Example: A child of mass 25 kg slides down a slippery dip inclined at 30° to the horizontal. Assume there is no friction.

a) What would be the child's acceleration?

b) What is the normal force acting on the child?

Answer:

a) $F = ma = mg\sin\theta$

$a = g\sin\theta = 9.8 \times \sin 30° = 4.9$ m s^{-2} down ramp

b) $N = mg\cos\theta = 25 \times 9.8\cos 30° = 212$ N perpendicular to ramp

Vector diagram

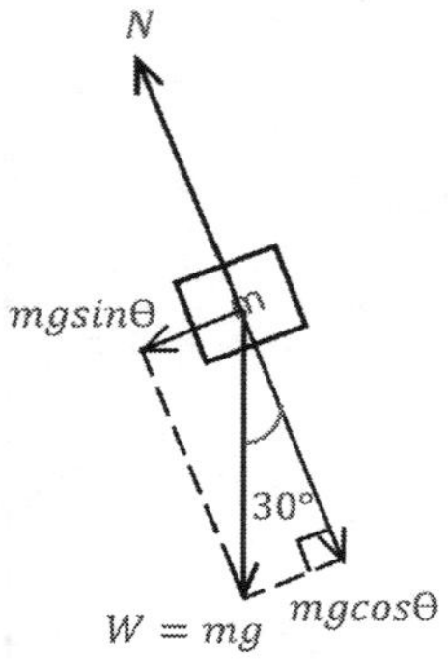

Example: An engineer is designing a ramp that boxes will slide down in a factory. The coefficient of kinetic friction between the ramp and the boxes is 0.45. If the angle of the ramp is 25° above the horizontal, what will be the acceleration of the boxes down the ramp?

Answer:

$F_{net} = ma = mg\sin\theta - \mu N$

$a = g\sin\theta - \mu g\cos\theta$

$a = 9.8\sin 25° - 0.45 \times 9.8\cos 25° = 0.14$ m s^{-2}

Vector diagram

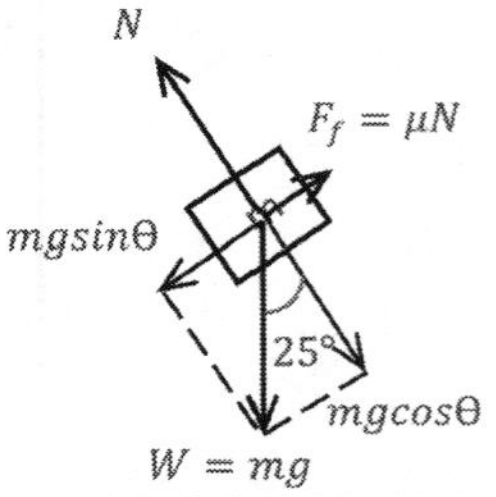

If it seems like information is missing, check if it cancels out.

3. An engineer is designing a water slide. For safety reasons, the acceleration of children on this water slide should not be greater than 6.8 m s^{-2}. However, the acceleration also needs to be at least 4.0 m s^{-2} to be fun. What is the maximum and minimum angle the water slide should make above the horizontal? Assume the slide is frictionless.

Vector diagram

4. A student uses a vector diagram to show the forces acting on an object on a frictionless inclined plane, tilted at an angle Θ. They draw the diagram shown. Discuss what the student did correctly and incorrectly.

Vector diagram

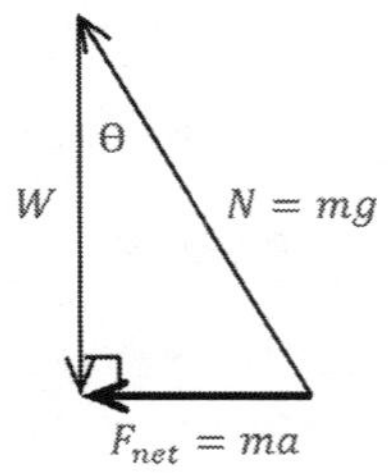

5. To determine the coefficient of static friction between a shoe with a leather sole and a wooden floor, the shoe is placed on a ramp made of the same material as the wooden floor. One end of the ramp is then raised, until the shoe just starts to slide down, when the ramp makes an angle of 14° to the horizontal. What is the coefficient of static friction?

Vector diagram

$\frac{\sin\theta}{\cos\theta} = \tan\theta$

6. A child of mass 25.0 kg slides down a slippery dip, inclined at 30.0° to the horizontal. The child's sibling thinks there is friction on the slippery dip. They measure the length of the slippery dip to be 10.0 m and the time taken for the child to slide down from rest to be 2.36 s.

Vector diagram

a) What was the child's acceleration?

b) What is the coefficient of friction between the slippery dip and child?

7. A skier skiing at 15.0 m s^{-1} is about to join the queue at a ski lift. There is a 30° upward slope approaching the ski lift, to slow skiers down. The coefficient of friction between the snow and skis is 0.065. The skier glides to a stop as they go up this slope in a straight line.

Vector diagram

a) What is the acceleration of the skier?

Which way is friction acting?

b) What distance does the skier travel up the slope before stopping?

Set a frame a reference.

7.8 Experimental skills – Inclined plane

Here is a method for an experiment to investigate an inclined plane.

Method

1. Set up as shown:

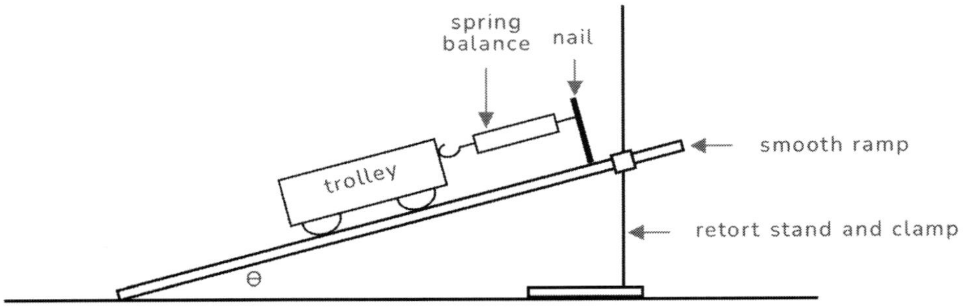

2. Using a protractor, set the ramp at an angle Θ=15°.
3. Record the tension in the spring balance.
4. Raise one end of the ramp and repeat steps 2 to 3, using angles of Θ=30°, 45°, 60° and 75°.

1. What is the independent variable? ....

2. What is the dependent variable? ....

3. Identify all of the controlled variables in the method.

....

4. Draw a vector diagram showing the forces acting on the trolley, then write a hypothesis for the experiment. Include an equation in the hypothesis.

Vector diagram

....

....

5. Here are the results for the experiment.

angle (°)	tension (N)	
15	0.2	
30	0.4	
45	0.6	
60	0.7	
75	0.8	

a) Using a linear regression, what would you graph to show the results are consistent with the hypothesis? What will the gradient be?

....

....

b) Use the blank column in the table to process data needed to graph one of the axes needed for the linear regression.

c) Graph your results using the axes you have chosen.

6. Use the graph to predict the angle at which the tension in the spring balance would be 0.5 N.

____

7. Explain two ways in which the mass of the trolley could be determined, using the graph. Find the mass of the trolley using both ways.

The line of best fit can be extrapolated ______________________________

____

The gradient of the line of best fit is ______________________________

____

8. The spring balance in the picture was used to take measurements in the experiment.

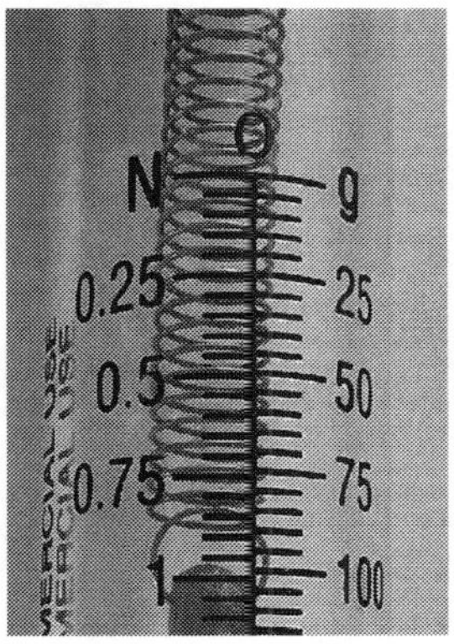

a) What is the absolute error of a 0.2 N reading using this spring balance?

____

b) What is the percentage error of a 0.2 N reading using this spring balance?

____

c) This scale can read up to 2.5 N. Discuss how a better choice of spring balance might improve the accuracy of this experiment.

The maximum reading on this spring balance is ________ but the maximum reading in the results is ________ so ______________________.

A better spring balance would have a ______________________________

____

9. Explain whether the results were reliable.

____

____

10. Discuss the trend shown in the results.

The trend shown by the results is that as tension ______________________________

____

The trend shows that the hypothesis ______________________________

11. Assess the validity of the trend.

Assess means discuss good and bad points and give an overall judgement.

The number of data points measured was ________. Compared to the line of best fit, most of these data points were ______________________________

Therefore, overall the trend was ______________________________

However, validity could be improved by ______________________________

____

because ______________________________

8.1 Relative motion – whose right?

If you are on a train moving at constant velocity and it is pitch black outside, you can't tell that you are moving
and if you see another train go by, you can't tell which train is moving.
This can mess with your brain sometimes.
Next station coming up!
The doors will open on your right
Huh? We're still moving?
!
Only relative to the other train...
In everycay life, we are used to thinking the ground doesn't move
but that depends on our frame of reference
and your frame of reference is only one of an infinite number of possible frames of reference.
Hi!

8.2 Relative motion in 1D

1. Describe relative motion, by filling in the blanks. Use the Relative motion comic to help you.

absolute	backwards	backwards	direction	forwards	ground
ground	ground	magnitude	observer	observer	opposite
police	police	police	police	relative	rest
same	truck	truck	truck	vector	

We are used to thinking about the ground as not moving and being the ________ frame of reference. So if someone is driving a truck at 60 km h^{-1}, we assume this is measured ________ to the ground. "Relative to x" means x is the frame of reference of the ________. So "relative to the ground" means as observed by someone at ________ on the ground. We can denote the velocity of the truck relative to the ground as $v_{t/g} = 60$ km h^{-1}, choosing "t" to represent "________" and "g" to represent "________".

Now an observer on the ground sees the police drive up behind the truck at $v_{p/g} = 70$ km h^{-1}, where "p" represents "________". The observer sees the police driving in the same direction as the ________ and catching up to it. However, the police would see the ________, and everything else at rest on the ________, moving backwards at $v_{g/p} = -70$ km h^{-1}, where $v_{g/p}$ is the velocity of the ground relative to the ________. In other words, in the frame of reference of the police, the ground is moving ________. Notice that the velocity of the police relative to the ground is the ________ magnitude but in an ________ direction to the velocity of the ground relative to the police.

Therefore to determine how fast the truck moves relative to the police $v_{t/p}$, we find the ________ sum of the velocity of the truck relative to the ________ plus the velocity of the ground relative to the ________:

$$\overrightarrow{v_{t/p}} = \overrightarrow{v_{t/g}} + \overrightarrow{v_{g/p}} = 60 - 70 = -10 \text{ km h}^{-1}$$

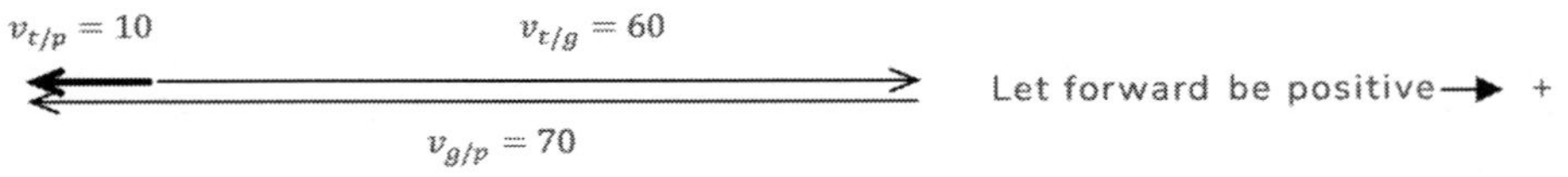

The resultant is negative, meaning that the police would see the truck moving ________ towards them at 10 km h^{-1}.

On the other hand, to determine how fast the police move relative to the truck $v_{p/t}$, we find the vector sum of the velocity of the ________ relative to the ground plus the velocity of the ground relative to the ________:

$$\overrightarrow{v_{p/t}} = \overrightarrow{v_{p/g}} + \overrightarrow{v_{g/t}} = 70 - 60 = 10 \text{ km h}^{-1}$$

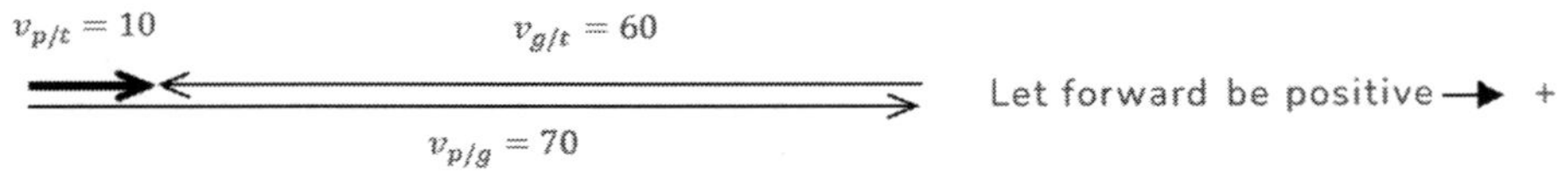

So the truck driver would see the police moving ________ towards them at 10 km h^{-1}. This is the same ________ as the velocity of the truck relative to the police, but in the opposite ________.

Example: A friend kicks a ball towards you at 20 m s^{-1}. At what speed do you observe the ball coming towards you, and how long does it take for you to intercept the ball, if you start off 25 m from your friend and you:

a) run towards the ball at 5.0 m s^{-1}?

b) run away from the ball at 5.0 m s^{-1}?

Answer:

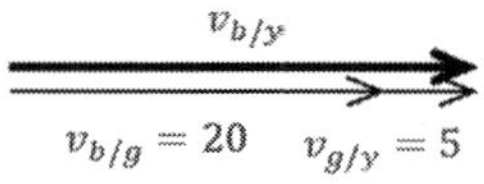

a) $\overrightarrow{v_{b/y}} = \overrightarrow{v_{b/g}} + \overrightarrow{v_{g/y}} = 20 + 5 = 25$ m s^{-1}

The ball comes towards you at 25 m s^{-1}. Time taken to intercept the ball is:

$$v = \frac{s}{t}$$

$$t = \frac{s}{v} = \frac{25}{25} = 1 \text{ s}$$

> Notice the ground is common to both velocities being added?

> Calculating time in the ground's frame of reference gives the same answer but is more work!

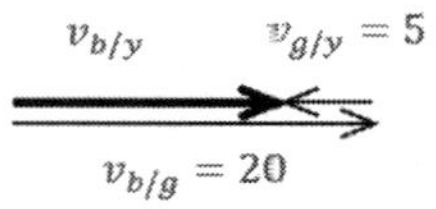

b) $\overrightarrow{v_{b/y}} = \overrightarrow{v_{b/g}} + \overrightarrow{v_{g/y}} = 20 - 5 = 15$ m s^{-1}

The ball comes towards you at 15 m s^{-1}. Time taken to intercept the ball is:

$$t = \frac{s}{v} = \frac{25}{15} = 1.7 \text{ s}$$

> How fast you perceive the ball depends on your own velocity. Get a ball and friend and try this out!

Vector diagrams

2. A white car travels on a highway at the speed limit of 110 km h^{-1}. It overtakes a black car in front which is travelling at 90 km h^{-1}. Both cars continue at the same speed.

   a) What is the velocity of the white car relative to the black car?

   ____

   b) What is the velocity of the black car relative to the white car?

   ____

   c) How many seconds are needed to safely overtake, if the white car needs to move forward by 100 m relative to the black car in order to do so?

   ____

   ____

> Notice (a) and (b) are the same magnitude but opposite directions?

Vector diagrams

3. A cricketer runs at 6.7 m s^{-1} then throws a ball. The ball moves at 28 m s^{-1} horizontally just after being thrown, as measured by a camera on the ground.

   a) Explain the advantage of doing a run up before throwing the ball.

   By doing a run up, the cricketer gives the ball an initial velocity of ____________, even if they drop the ball, due to the ball having ____________. The ball's final velocity of 28 m s^{-1} is the ____________ velocity of the ball relative to the cricketer plus the velocity of the cricketer relative to the ____________.

   b) How fast would the ball have been thrown by the cricketer, without a run up?

   ____

   ____

Vector diagrams

4. If you ride a bike, you might notice wind against you, even if it is not a windy day. That is because you are moving relative to the air particles around you, so you observe the air particles moving backwards relative to you. If you cycle along a straight path at 15 m s^{-1}, what is the velocity you would perceive the wind, if there was in fact:

a) a tail wind of 5 m s^{-1} (from behind you in the same direction you cycle)

b) a head wind of 5 m s^{-1} (against you opposite to the direction you cycle)

Vector diagrams

5. A river is flowing downstream at 1.0 m s^{-1}. Ash and Bee can each swim at a top speed of 2.0 m s^{-1} in still water. Cai can't swim and sits on the river bank. Ash swims upstream and Bee swims downstream at top speed. What is the velocity of:

a) Ash relative to the river?

b) Bee relative to the river?

c) Cai relative to the river?

d) Ash relative to Cai?

e) Bee relative to Cai?

f) Ash relative to Bee?

You can model this by walking on a treadmill, which represents the river.

g) Compare the time taken for Ash to swim in the river flowing at 1.0 m s^{-1} a distance of 30 m upstream from Cai, then back to the starting point, to swimming the same distance in still water.

6. While filming a movie, a drone flying above and eastwards along a street takes a video of an actor walking down the street. The diagram shows the motion of the actor, as captured by the drone:

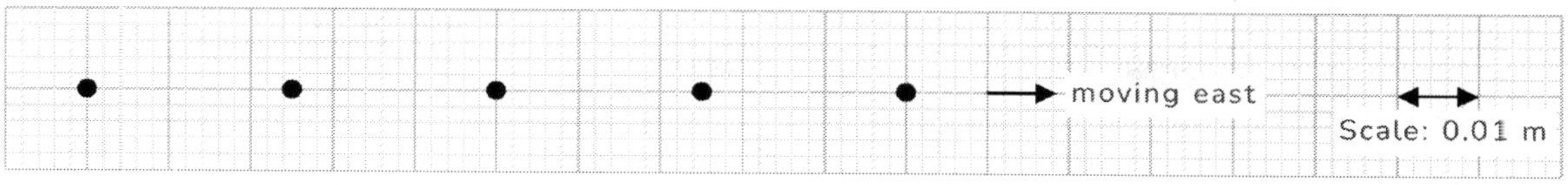

There is a fixed signpost on the street, which changes position in the video, as shown:

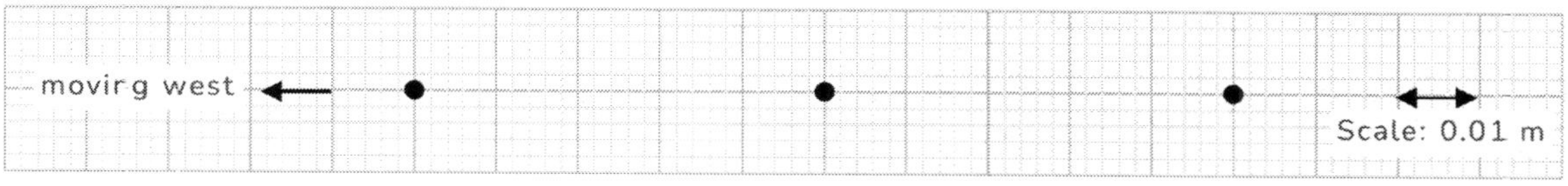

The drone takes videos at 24 frames per second.

a) What is the velocity of the drone relative to the ground?

____

Try this! Put your phone on a trolley and take a video of another moving object, against a plain background.

b) What is the velocity of the actor relative to the ground?

____

____

7. The graph shows the velocities of a train you are sitting in, travelling along a straight track, and a cyclist riding along a road next to the train track. Both velocities are measured relative to the ground.

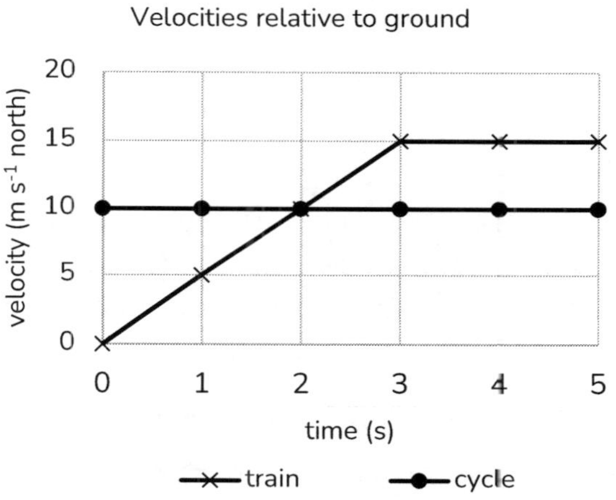

a) What is the initial velocity of the cyclist relative to you?

____

b) When does the cyclist look like they are stationary from your perspective?

____

c) What is the velocity of the cyclist relative to you after 3 s?

____

d) What is the acceleration of the cyclist relative to you, between 0 and 3 s?

____

____

If something is not accelerating relative to the ground, it will have the same acceleration as the ground relative to something else.

e) If you took a video of the cyclist from the train window, what would their look like in the video? Draw a velocity-time graph to describe the velocity of the cyclist relative to you, between 0 and 5 s.

Velocity of cyclist relative to you

8.3 Relative motion in 2D

Solving relative motion problems in two dimensions uses the same ideas as in one dimension, but also involves vector addition.

On land

Example: You are sitting on a dock and see the ship Alfa to the south of you, heading north towards you at 10 m s^{-1}. You see another ship, Bravo, heading towards the dock from the east, at 10 m s^{-1}.

a) What is the velocity of Bravo relative to Alfa?

b) What is the velocity of Alfa relative to Bravo?

Answer:

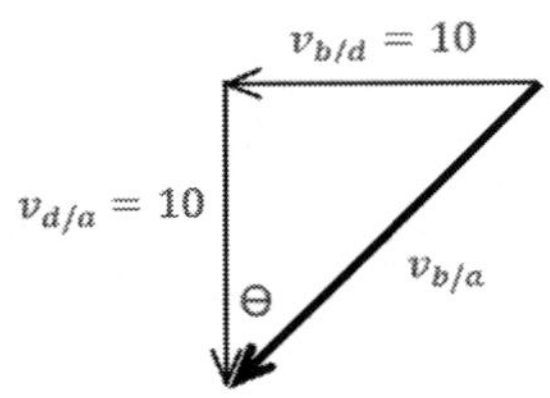

a) $\overrightarrow{v_{b/a}} = \overrightarrow{v_{b/d}} + \overrightarrow{v_{d/a}} = \sqrt{10^2 + 10^2} = 14 \text{ m s}^{-1}$

$\tan\theta = \frac{10}{10} = 1$

$\theta = 45°$

Bravo travels at 14 m s^{-1} from a north-easterly direction (heading south-west), relative to Alfa.

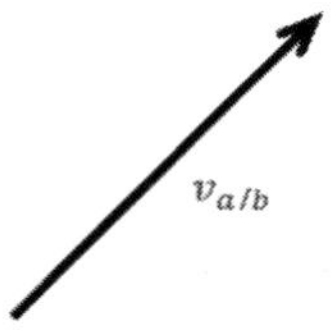

b) $\overrightarrow{v_{a/b}} = -\overrightarrow{v_{b/a}}$

Alfa travels at 14 m s^{-1} from a south-westerly direction (heading north-east), relative to Bravo.

Try modelling this with a friend. Walk to the same point, but from directions at 90° to each other. If you focus on your friend's head and ignore the background, they look like they head diagonally toward you.

1. Ash, Bee and Cai practise a figure skating routine. They need to start off at one point, then skate outwards away from each other at 120° to each other, at the same speed. They keep getting this wrong, complaining that it does not look like the other team members are skating at 120° to their direction.

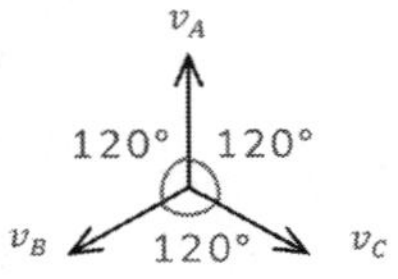

a) Draw vector diagrams representing the following. The first is done for you, as an example.

Bee's velocity relative to Ash | Ash's velocity relative to Bee | Ash's velocity relative to Cai

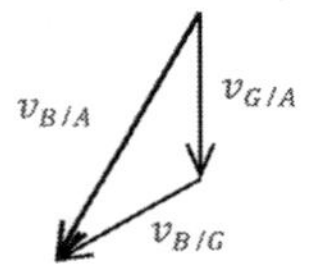

Cai's velocity relative to Ash | Cai's velocity relative to Bee | Bee's velocity relative to Cai

b) Explain qualitatively why Ash, Bee and Cai keep getting their skating wrong.

____

____

2. While sitting on a train moving at 15 m s^{-1}, you see rain pelting diagonally across the window at a 45° angle. When the train stops, the rain suddenly appears to fall gently and vertically down the window. If there was actually no change n the weather, why did the rain move faster when observed from the moving train, and what was the rain's velocity relative to the ground?

Vector diagram

____

____

____

____

____

____

3. A black car and a white car are heading towards the same intersection, each at 60 km h^{-1}. The black car is driving from north to south, while the white car is driving from north-west to south-east. What is the velocity of the black car, as observed by the white car?

Vector diagram

____

____

____

____

Recall the cosine and sine rules.

On the river

4. Ash sits on the west bank of a straight river 5.0 m wide, flowing southwards downstream at 1.0 m s^{-1}. Starting from Ash's position, Bee floats downstream in the river with the current at 1.0 m s^{-1}. Also starting from Ash's position, Cai swims eastward at 2.0 m s^{-1}, relative to the water.

a) Describe the motion of Ash, Bee and Cai, by filling in the blanks.

Ash	Bee	Cai	components	direction	east
magnitude	north	perpendicular	resultant	south	water

As Bee just floats on the water, their velocity is zero relative to the ______________. However, Bee's velocity is 1.0 m s^{-1} south relative to ______________. Ash's velocity relative to Bee is the same ______________ but the opposite ______________, 1.0 m s^{-1} ______________.

Since Cai's swimming velocity is directed eastward, their swimming will not affect their velocity north or south, as ______________ that are perpendicular have no effect on each other. Similarly, the river cannot slow down Cai's velocity to the ______________, since the current is ______________ to it. However, the current will make Cai drift ______________ at the same velocity as Bee. The displacement of Bee and ______________ downstream will therefore be the same as each other. However, Cai will have a velocity eastwards at 2.0 m s^{-1} relative to ______________.

Ash observes Cai moving diagonally downstream, with a total velocity that is the ______________ of their swimming velocity plus the velocity of the current.

b) Draw vector diagrams representing the following.

Bee's velocity relative to Ash	Ash's velocity relative to Bee	Ash's velocity relative to Cai
Cai's velocity relative to Ash	Cai's velocity relative to Bee	Bee's velocity relative to Cai

c) The river is 5.0 m wide. How long will Cai take to reach the east bank from the west bank?

You have both the velocity and displacement to the east.

d) In the time it takes to reach the opposite bank, how far will Cai have drifted downstream?

To find the downstream displacement, use the downstream velocity.

e) What is Cai's total displacement from Ash when they reach the opposite bank?

Vector diagram

It is convenient to give the angle to the bank.

f) Calculate Cai's velocity relative to Ash. Try this two ways – using the time and displacement you calculated in parts (c) and (e), as well as the resultant of velocities. Do you get the same answer?

If there is more than one way to get the right answer, it is a good idea to check both ways to test your understanding.

5. Ash is still sitting on the west bank of the river 5.0 m wide, flowing southwards at 1.0 m s^{-1}. Bee is still floating on the water. Cai goes for another swim, starting from where Ash sits, and still swims at a speed of 2.0 m s^{-1}, relative to the water.

This time, what if Cai wants to end up directly opposite Ash on the other bank, instead of downstream?

a) Describe the motion of Ash, Bee and Cai, by filling in the blanks.

Bee	component	diagonally	east	less
more	north	south	vector sum	water

Because Cai's displacement needs to be directly east of Ash, their resultant velocity also needs to be towards the ______________. This resultant velocity, relative to Ash, is the ______________ of their swimming velocity and the velocity of the water south. However, this time Cai is going to need a ______________ of their swimming velocity to be ______________ at 1.0 m s^{-1}, to counteract the river flowing ______________. So Cai will need to direct their 2.0 m s^{-1} swimming velocity ______________ upstream, relative to the water. This means the component of their swimming velocity to the east is ______________ than 2.0 m s^{-1}, so it will take ______________ time to cross the river, compared to allowing the river to carry them downstream while swimming across.

In addition, Cai appears to move diagonally upstream relative to ______________ at 2.0 m s^{-1}, who is in the same frame of reference as the ______________.

b) Draw vector diagrams representing the following.

Bee's velocity relative to Ash

Ash's velocity relative to Bee

Ash's velocity relative to Cai

Cai's velocity relative to Ash

Cai's velocity relative to Bee

Bee's velocity relative to Cai

c) Calculate Cai's velocity relative to Ash.

Use the vector diagram in (b).

____

____

____

____

d) How long will Cai take to reach the east bank from the west bank?

The velocity used must be in the same direction as the displacement.

____

____

In the air and ocean

6. A small plane flies at 30.9 m s^{-1} when there is no wind. There is a wind blowing from west to east of 12.5 m s^{-1}.

Vector diagram

   a) What is the velocity of the plane relative to the ground, if the pilot steers the plane northward relative to the air?

This is a similar problem to the river problem, except we have moving air instead of water.

Vector diagram

   b) The pilot needs to fly directly to a city 215 km north. In what direction should the pilot steer relative to the air, and what will be the resultant velocity of the plane relative to the ground?

   c) How long will the plane take to reach the city if the pilot flies the course described in part (b)?

7. Before swimming at the beach, you should check the ocean for rips. A rip is a channel of water moving away from the beach, formed by water returning to the ocean after pushing up on the beach. Rips often appear as deeper, darker patches of water, which may have fewer waves breaking or sandy water floating beyond the waves.

   The speed that water flows in a rip can be up to 2.4 m s^{-1}. Elite swimmers may have a swimming speed of around 2.0 m s^{-1}.

   a) If you are caught in a rip, explain why it is a bad idea to swim back towards the beach in the opposite direction of the rip current.

   b) Swimmers caught in a rip are advised to swim parallel to the shore, perpendicular to the rip current, to get out of the rip. Calculate the velocity of a swimmer who is caught in a rip current of 2.4 m s^{-1}, who follows this advice with a swimming speed of around 2.0 m s^{-1}.

Vector diagram

8. **Challenge**: A sailboat moves by using the force from a wind on the sail to propel the boat. The faster the wind on the sail, the more force there is to propel the sailboat.

a) Ash thinks that the sailboat will move the fastest if the boat is steered in the same direction as the wind is blowing – in other words, the wind comes from behind the sailboat. However, when Ash tries this, the sailboat only moves as fast as the wind. In addition, once the sailboat reaches this speed, the sail becomes slack, as if there is no wind pushing against it. Explain why.

Vector diagram

b) Bee knows a sailing technique called "tacking", in which a moving sailboat is steered at around 45° into the wind. This actually increases the speed of the wind hitting the sail, so the sailboat's speed can exceed the speed of the wind. Explain why.

Vector diagram

What is the velocity of the air relative to the boat?

9.1 Work, energy and power

Chemical potential energy is energy stored in chemical bonds, from the food we eat to fuel for vehicles.
The unit for energy is the ***joule***. A ***kilojoule*** is 1000 joules.
CHO
4888kJ
Elastic potential energy is energy stored when an object is stretched or compressed.
Kinetic energy always involves moving objects, and there are also many special types.
Heat, or ***thermal energy***, involves moving particles.
The hotter the air is, the more kinetic energy the air molecules have.
Sound involves vibrating particles.
TWANGGG
Electrical energy involves moving charges.
The curious thing about energy is that it does not seem to get created or destroyed. It just gets transferred from one object to another, or transformed from one type to another.
This is called the ***Law of Conservation of Energy***.
If we calculate the gravitational potential energy the ball has at the peak of its bounce, this is the same amount of kinetic energy it reaches the floor with
which is the same as the sum of the elastic potential energy, sound and heat produced on contact with the floor.
BONG!
The transformation of some energy to sound and heat means the ball doesn't quite bounce back to the same height,
but the total amount of energy stays the same.

As energy is just transferred from one object to another, or transformed from one type to another, we use the same unit for all types of energy, the ***joule.***
Work means the amount of energy transferred or transformed.
If we can see an object moving through a displacement due to a force acting in the same line, we can calculate work:
$W = \vec{F}.\vec{s}$
where:
W = work done (J)
$\vec{F}$ = component of force acting in the same direction as displacement (N)
$\vec{s}$ = displacement (m)
s
It doesn't matter what is doing the work, the equation is the same.
Gravity does the work
s
If you walk up a mountain, you need at least enough force to balance your weight to rise up the height of the mountain.
This is the same minimum force if you run up fast.
As you use the same force over the same displacement, you use the same amount of energy running or walking.
But your aching muscles tell you something is different. You needed more ***power*** from your muscles.
The rate we use or transform energy is called ***power***. In other words, ***power*** is the rate at which work is done:
$P = \frac{W}{t}$
where:
P = power (W)
W = work done (J)
t = time in which the work was done (s)
Power is measured in ***watts*** (W).
Watt is the unit of power?
It took humanity many years and minds to work out how to calculate energy and power, and that energy is conserved...
Emilie du Châtelet
... most of the time!

9.2 Work

Describing energy

1. Describe types of energy, by filling in the blanks.

attraction	charges	chemical	closed	conservation
efficient	elastic	electrical	electrostatic	food
force	gravitational	heat	higher	joules
kinetic	kinetic	position	potential	same
thermal	transferred	transformed	vibrating	work

Energy is measured in ________________ and is the ability to do ________________. Work can be calculated by multiplying the displacement an object travels through by the ________________ which acts on it in the same direction as the displacement.

The energy possessed by a moving object is called ________________ energy. Heat or ________________ energy involves moving particles with kinetic energy. Sound energy also involves kinetic energy, as it consists of ________________ particles. Electrical energy is also a special case of kinetic energy, as it involves moving ________________.

The kinetic energy of moving objects often comes from some type of ________________ energy, or stored energy. A ball that is lifted off the floor has ________________ potential energy. This gets transformed into ________________ energy when the ball is released. The ________________ the ball is lifted, the more gravitational potential energy it has, and the more kinetic energy it will reach the ground with once released. So gravitational potential energy is determined by the ________________ of an object in a gravitational field.

A battery has ________________ potential energy. The positive and negative terminals of the battery exert an ________________ force on the electrons in a circuit connected to the battery, giving those electrons kinetic energy. So electrical potential energy is transformed into electrical energy when there is a ________________ circuit allowing electrons to flow.

________________ potential energy is stored in the chemical bonds between atoms. This is the type of energy stored in ________________, as well as fuel combusted to power vehicles and generators. Actually, chemical bonds are made from electrostatic ________________ between negative electrons and positive nuclei, so they are a special case of electrical potential energy.

________________ potential energy is stored in objects that are stretched, such as a rubber band, or compressed, such as a spring.

If we calculate the total amount of energy we start with and end with, it appears that this total amount stays the ________________. This is the Law of ________________ of Energy, which means that energy is not created or destroyed. It just gets ________________ from one object to another or ________________ from one type to another.

Some types of energy may not be useful in a situation. For example, using your laptop transforms some electrical energy into ________________, but it is not the function of a laptop to be a heating device. The more wasted energy, the less energy ________________ a device is.

Equation for work

Work is calculated using the equation:

$$W = \vec{F}.\vec{s}$$

where: W = work done (J)

$\vec{F}$ = force acting in the same line as the displacement of object (N)

$\vec{s}$ = displacement of object (m)

Take care with symbols! Here W means work. But W is also used to mean weight. You can use an alternative symbol for weight or gravitational force, F_G.

$\vec{F}.\vec{s}$ means the "dot product of force multiplied by displacement". In other words, if the force moving the object is at an angle to the displacement, we only multiply the component of force that is in the same direction as the displacement. Any force that is perpendicular to the displacement cannot cause the displacement.

Example: You drag a suitcase on wheels for 500 m. The handle of the suitcase makes a 60° angle with the ground and you pull it with a force of 75 N. How much work did you do on the suitcase?

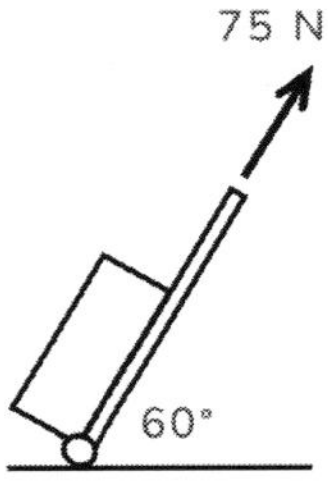

Note that substituting the component of force in the same direction as the displacement gives: $W = \vec{F}.\vec{s} = Fs\cos\theta$

Answer:

$W = \vec{F}.\vec{s}$

$= 75\cos 60° \times 500$

$= 18750$ J

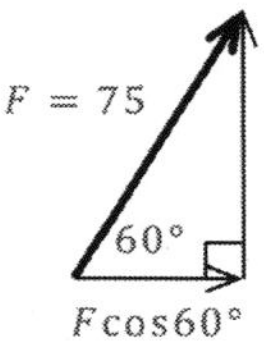

2. An ice hockey player pushes a puck for 25.0 m on the ice, using a hockey stick. The hockey stick is 45.0° to the ground and the player exerts a force of 6.50 N along the direction of the stick. What work is done on the puck?

____

____

Vector diagram

Example: A 1500 kg car travelling at 60.0 km h^{-1} brakes and slows to a stop in 12.0 s. What is the work done by the brakes? Express your answer in kilojoules.

A kilojoule (kJ) is 1000 J.

If there is a lot of information, list the known variables, then see what you can calculate from them.

Answer:

$v = 0$

$u = 60 \div 3.6$

$t = 12$

$m = 1500$

$v = u + at$

$0 = 60 \div 3.6 + a \times 12$

$a = -1.39$ m s^{-2}

$s = ut + \frac{1}{2}at^2$

$= 60 \div 3.6 \times 12 + \frac{1}{2} \times -1.39 \times 12^2$

$= 99.9$ m

$W = \vec{F}.\vec{s}$

$= mas$

$= 1500 \times -1.39 \times 99.9$

$= -208000$ J $= -208$ kJ

Work done by the brakes is 208 kJ.

Work is a scalar quantity – it has no direction. However, it is a ***signed scalar*** – it can be positive if the force acts in the same direction as the displacement, or negative if force acts in the opposite direction. Here the negative sign means kinetic energy is lost, as the force acts to oppose motion.

3. A cyclist on a bike accelerates from rest to 4.5 m s^{-1} on a horizontal straight road, over a 20 m distance. The mass of the cyclist is 75 kg. How much energy did they use to do this?

Energy transformed is the same thing as work done.

4. Heartbreak Hill is a stretch of road in Sydney around 2.0 km long, with an elevation of 80 m over this distance. A cyclist of mass 75 kg rides up Heartbreak Hill at a constant speed, exerting just enough force to oppose gravity. (Ignore friction.) Determine how much energy they used to ride up Heartbreak Hill, and compare this to the 300 kJ of energy contained in the apple they ate for breakfast.

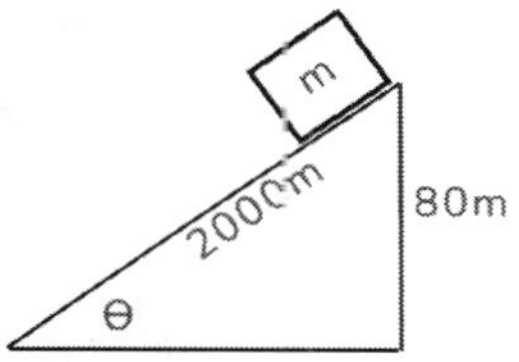

Even if you sit still, your body uses energy, because stuff inside you like blood is still moving.

5. You move a 52 kg sofa by 1.2 m, by pushing it at constant velocity across a carpeted floor with a coefficient of friction of 0.45. How much work did you do on the sofa?

Vector diagram

What is the net force if an object moves at constant velocity?

6. Resistance bands are often used in exercise and physiotherapy. The force required to increase the length of a particular resistance band was measured and graphed.

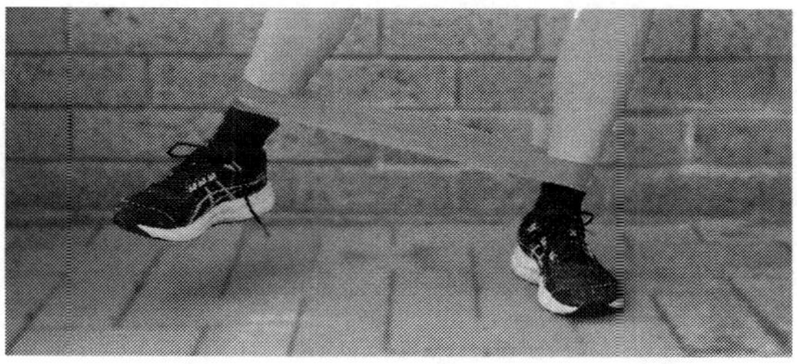

a) Describe the trend shown in the graph.

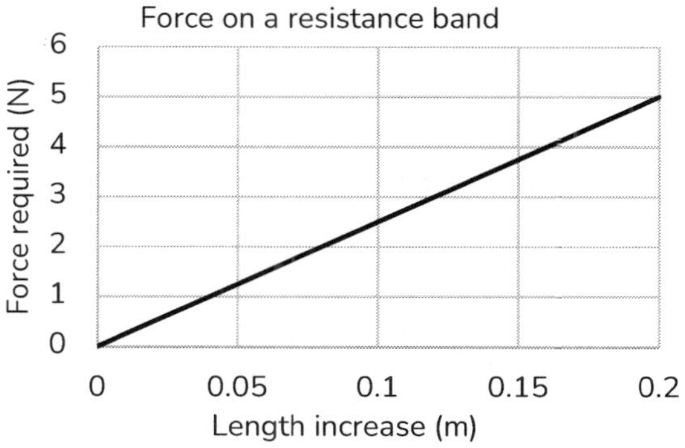

b) How much work do you need to do to lengthen this resistance band by 0.2 m? Explain why you cannot calculate this by merely multiplying 0.2 m by 5 N.

Remember area under a velocity-time graph gave displacement? Here, what do you get if you multiply the x and y axes?

9.3 Gravitational potential energy

1. Describe gravitational potential energy, by filling in the blanks.

decreased	drop	field	gravitational	gravity
height	increased	kinetic	mass	scalar
transformed	vector	we	weight	work done

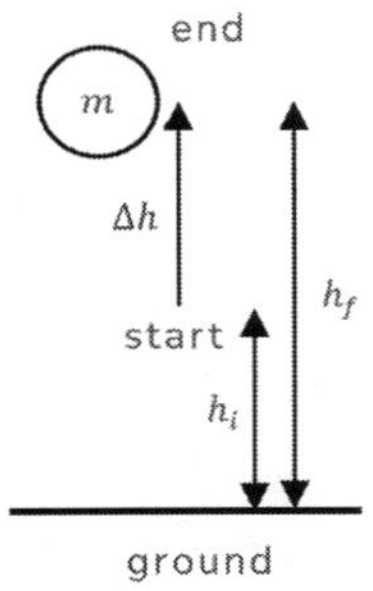

To lift an object of mass m upwards, we must apply a force equal to at least its ________, mg. If we change the object's height by Δh, we also change its ________ potential energy by ΔU. Since work means energy ________, change in gravitational potential energy is given by:

$$\Delta U = W = \vec{F}.\vec{s} = mg\Delta h = mg(h_f - h_i)$$

where: ΔU = change in gravitational potential energy (J)

W = ________ (J)

m = ________ of object (kg)

g = acceleration due to ________ (m s^{-2})

h_i, h_f = initial and final heights above ground (m)

Δh = change in ________ (m)

Remember change in anything is final minus initial.

Lifting 1 kg from 1 m to 2 m above the ground, the change in gravitational potential energy is:

$$\Delta U = W = mg\Delta h = mg(h_f - h_i) = 1 \times 9.8 \times (2 - 1) = +9.8 \text{ J}$$

Here, ________ do the work to lift the object, since our muscles exert the force to move it. The positive sign means we ________ the gravitational potential energy of the object. We have stored more energy in the object, so if we release it, it will have more energy to transform into ________ energy and hit the ground with more speed.

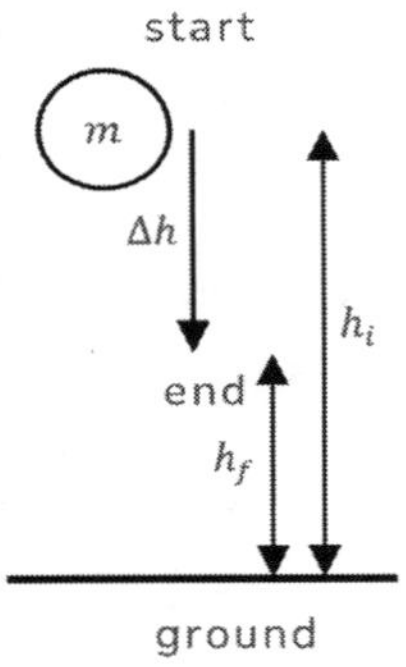

If we ________ the same object from 2 m to 1 m instead, the change in gravitational potential energy is:

$$\Delta U = W = mg\Delta h = mg(h_f - h_i) = 1 \times 9.8 \times (1 - 2) = -9.8 \text{ J}$$

The negative sign means the gravitational potential energy of the object ________. This time we aren't doing the work – the gravitational ________ is.

So energy and work can have a sign, even though they are ________ quantities. It's like a bank account can have positive and negative changes, even though money is not a ________.

2. In 2023, powerlifter Wendy Chan set a world record for deadlifting 150 kg. Her body mass was 47 kg. What work does she do if she lifts 150 kg by 1.2 m?

____

3. An Australian company called Green Gravity is storing renewable energy, by using it to lift large masses up abandoned mine shafts. When energy is needed, the masses are allowed to fall back down to turn a turbine and generate electricity. How much energy could be generated by a 5.0 tonne mass falling down a 1.0 km vertical mine shaft?

____

4. A wheelchair ramp has an incline of 15° and rises a vertical height of 1.2 m. What is the work done by a person of mass 65 kg sitting in a wheelchair of mass 18 kg, who moves themselves up the ramp at constant velocity?

Vector diagram

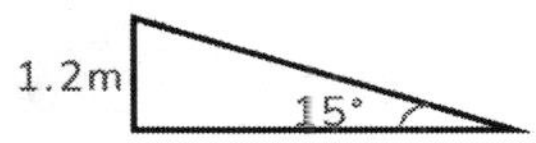

a) First try this by working out the force exerted by the person and the displacement up the ramp.

____

____

____

____

b) Now solve this the easy way, using the change in gravitational potential energy.

____

Analysing energy can provide an alternate way to solve problems.

5. Explain why a frame of reference is needed for potential energy, by filling in the blanks.

1 m	2 m	anywhere	basement	between	change	end
frame of reference	ground	height	length	lifting	positive	same

We often assume that something on the ground has zero gravitational potential energy because it has zero ________________. If so, ________________ an object off the ground by height h gives it gravitational potential energy U equal to:

$$\Delta U = mg\Delta h$$

$$U_f - U_i = mg(h_f - h_i)$$

$$U_f - 0 = mg(h_f - 0)$$

$$U = mgh$$

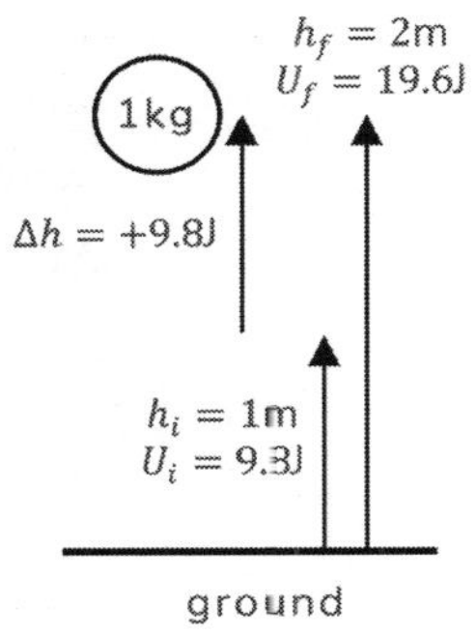

So lifting a 1 kg object ________________ above the ground gives it gravitational potential energy of +9.8 J. ________________ above the ground puts it at +19.6 J. The ________________ in gravitational potential energy ________________ these points is +9.8 J.

Actually, we can choose to set the point of zero potential energy ________________. If we set $h = 0$ to be the final position of the object, 2 m above the ground, and still let up be ________________, the change in gravitational potential energy required to lift the object from 1 m to 2 m above the ground now looks like this:

$$\Delta U = W = mg\Delta h = mg(h_f - h_i) = 1 \times 9.8 \times (0 - (-1)) = +9.8 \text{ J}$$

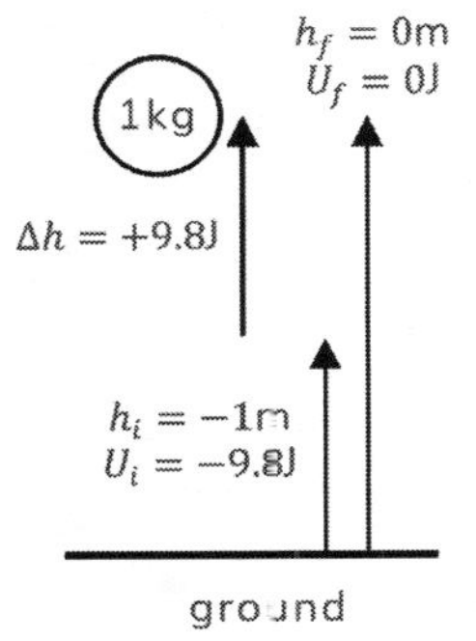

This is the ________________ change in potential energy, so it doesn't matter what frame of reference we use to calculate change. It's like measuring the length of a line - it's easier if you put your ruler with zero at one ________________ of the line, but you don't have to and you will still measure the same ________________.

Sometimes it is inconvenient to assume the ________________ has zero potential energy. If we drop something from the ground floor down the stairwell to the basement, it is easier to think of the ________________ as having zero gravitational potential energy. Spacecraft might not even return to Earth, so there is often a better ________________ to set as zero potential energy than Earth's surface.

6. A person of mass 92 kg takes an elevator from the fifth floor, 10 m above the ground floor, to the ninth floor, 22.5 m above the ground floor.

   a) How much energy is required to lift this person on this journey?

   ____

   ____

   b) What is the person's change in gravitational potential energy if they take the lift from the ninth floor back to the fifth floor, and what does the work?

   ____

   ____

7. The International Space Station has orbited at around 400 km above the Earth's surface for over 20 years, providing a space laboratory for scientists. Supply rockets with equipment and food are sometimes sent there.

   As the distance between a supply rocket and Earth increases, acceleration due to gravity decreases, as shown by the graph.

   A supply rocket of mass 9.5 tonnes is sent from the Earth's surface to the International Space Station.

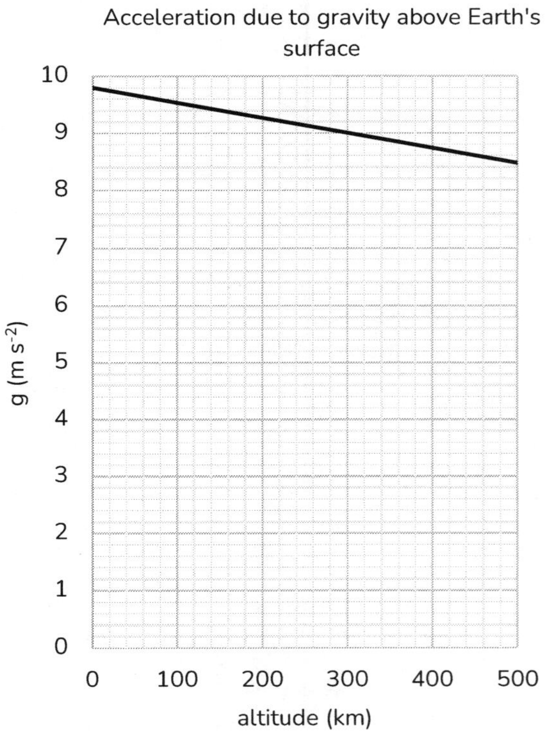

Does either the gradient or area of the graph mean something here?

   a) Determine the change in gravitational potential energy of the supply rocket.

   ____

   ____

   ____

   b) Compare the change in gravitational potential energy the supply rocket experiences during the first half of its journey, from Earth's surface to a 200 km altitude, to the second half, from a 200 km to 400 km altitude.

   ____

   ____

   ____

   ____

   ____

8. Pulleys can be used to change the direction of the force exerted by the cord, or make it easier to lift weights. For example, pulleys are used to open and close blinds, or in cranes to lift heavy construction materials.

A student conducts an experiment to see if the number of pulleys used really changes the force needed to lift a weight. Using one pulley, they gently pull the spring balance so a 250 g mass (load) just lifts at constant velocity, up a height of 0.500 m. They record the force required F (effort), as measured by the spring balance. Using a ruler, they also measure the distance the string was pulled (d).

This was repeated using two pulleys.

One pulley system | Load being lifted | Two pulley system

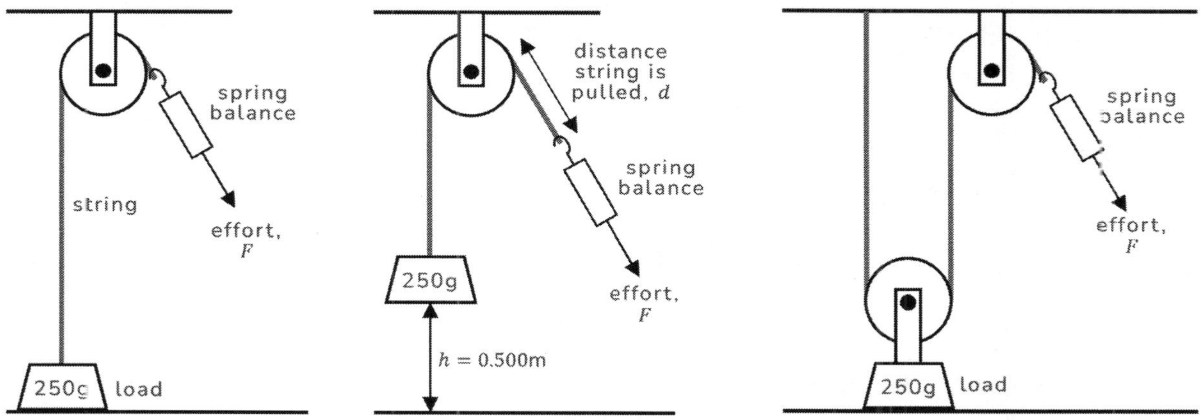

Here are the results:

Number of pulleys	Effort F (N)	Distance d (m)
1	2.45	0.500
2	1.23	1.00

a) What is the independent variable? ....

b) What is the dependent variable? ....

c) Identify all of the controlled variables. ....

d) The student infers that since adding a pulley to the system decreases the force needed to lift a weight by the same height, the Law of Conservation of Energy must not be true. Evaluate this inference.

....

....

....

....

....

....

e) What is the tension in each string, in each of the pulley systems?

....

....

....

....

9.4 Kinetic energy

1. Describe how to derive an equation for kinetic energy, by filling in the blanks.

constant	final	initial	kinetic energy	mass
moving	net force	substituting	velocity	work

Kinetic energy is energy possessed by a ________________ object. If a ________________ acts on an object, there will be a change in its kinetic energy. The net force also means ________________ is done on the object, equal to the change in kinetic energy:

$$\Delta K = W = \vec{F}.\vec{s} = m\vec{a}.\vec{s} \quad ...(1)$$

If the acceleration is ________________, we can find an expression for $\vec{a}.\vec{s}$ using:

$$\vec{v}^2 = \vec{u}^2 + 2\vec{a}.\vec{s}$$

$$\vec{a}.\vec{s} = \tfrac{1}{2}(\vec{v}^2 - \vec{u}^2) \quad ...(2)$$

________________ $\vec{a}.\vec{s}$ from equation (2) into (1) gives:

$$\Delta K = m \times \frac{1}{2}(\vec{v}^2 - \vec{u}^2)$$

$$K_f - K_i = \frac{1}{2}m\vec{v}^2 - \frac{1}{2}m\vec{u}^2$$

As both sides of this equation involve a ________________ quantity minus an ________________ quantity, we deduce that for any given mass and velocity, kinetic energy is:

$$K = \frac{1}{2}m\vec{v}^2$$

where: $K =$ ________________ (J)

$m =$ ________________ of object (kg)

$\vec{v} =$ ________________ of object (m s^{-1})

Example: What is the change in kinetic energy of a car of mass 1.0 tonne, if it accelerates from 40 km h^{-1} to 60 km h^{-1} after passing through a school zone?

Answer:

$$\Delta K = K_f - K_i = \frac{1}{2}mv^2 - \frac{1}{2}mu^2$$

$$= \tfrac{1}{2} \times 1000 \times \left(\tfrac{60}{3.6}\right)^2 - \tfrac{1}{2} \times 1000 \times \left(\tfrac{40}{3.6}\right)^2$$

$$= 77000 \text{ J} = 77 \text{ kJ}$$

2. A parent pulls a child on a wagon, with a combined mass of 50 kg, using a horizontal force of 295 N. The wagon is initially at rest, and travels 5.0 m along a smooth horizontal floor.

Remember work is energy transformed.

a) What is the change in kinetic energy of the wagon and child together?

____

____

b) What is the final speed of the wagon?

____

____

3. A slingshot is used to launch a rock of mass 25 g. The slingshot with the rock is pulled back by 10 cm. When released, the rock leaves the slingshot with a speed of 15 m s^{-1}.

   a) What is the final kinetic energy of the rock?

   b) What is the average force exerted by the slingshot on the rock?

Example: A truck driver wants to make more money, by either delivering a bigger load, or decreasing delivery times by driving faster. Which requires more work to accelerate from rest, and therefore more fuel – a truck of mass m accelerating to a speed v, or a truck of double the mass accelerating to half the speed?

> Algebraic expressions are often used for comparison. Give new symbols for the new scenario, then write an expression to compare each variable. Substitute the new expressions into what is being compared.

Answer: Work done by truck 1:

$W = \Delta K = \frac{1}{2}mv^2$

Let truck 2 have a mass $m' = 2m$ and speed $v' = \frac{1}{2}v$. Work done by truck 2:

$W' = \Delta K' = \frac{1}{2}m'v'^2 = \frac{1}{2} \times 2m \times (\frac{1}{2}v)^2 = m \times \frac{1}{4}v^2 = \frac{1}{2} \times \frac{1}{2}mv^2 = \frac{1}{2}W$

It takes half the work to accelerate a truck of double the mass to half the speed.

4. What happens to the kinetic energy of an object if you triple its speed?

5. If you thought you would die soon, what would you do? Emilie du Châtelet died giving birth in 1749. She thought this was likely, as she was 42, so she frantically spent her last months translating Newton's work from Latin to French. She added her own ideas, including deriving the Law of Conservation of Energy. Her work spread scientific ideas across Europe. Yet because she was female, she had to publish anonymously.

   Emilie du Châtelet suggested that kinetic energy was proportional to mv^2, and not mv as Newton had written. She concluded this from observing balls dropped onto a bed of clay and using the depth they sank to determine their final velocity.

   The table shows results from a similar experiment, where a ball of mass m is dropped from various heights h, and reaches the clay with velocities v.

h (m)	v (m s^{-1})	
1.0	4.4	
2.0	6.3	
3.0	7.7	
4.0	8.9	
5.0	9.9	

   Using a graph, explain whether the results support the hypothesis that the loss in gravitational potential energy equals the gain in kinetic energy, given by $mgh = \frac{1}{2}mv^2$.

   You may use the blank column to process results.

9.5 Energy transformations

1. Describe how energy can be transformed between gravitational potential energy and kinetic energy, by filling in the blanks.

conservation	final	final velocity	gravitational potential	initial
kinetic	rest	total	transformed	

When an object falls, it loses ________________ energy and gains ________________ energy. At any point in the object's journey, if we add up all the energy involved, the ________________ energy E must stay the same, according to the Law of ________________ of Energy. In other words, the total ________________ energy E_i, comprising the initial gravitational potential energy U_i plus initial kinetic energy K_i, must equal the total ________________ energy E_f, comprising the final gravitational potential energy U_f plus final kinetic energy K_f:

$$E_f = E_i$$

$$K_f + U_f = K_i + U_i$$

We can rearrange this formula to show that any gain in kinetic energy must equal the amount of gravitational potential energy ________________:

$$K_f - K_i = -(U_f - U_i)$$

$$\Delta K = -\Delta U$$

We can also use the formula to derive an expression for the ________________ of an object of mass m dropped from ________________, after travelling through a distance h to the ground:

$$K_f - K_i = -(U_f - U_i)$$

$$\frac{1}{2}mv^2 - 0 = -(0 - mgh)$$

$$\frac{1}{2}v^2 = gh$$

$$v = \sqrt{2gh}$$

2. A ball is released from rest at a height h above the lowest point on a smooth U-shaped ramp, as shown.

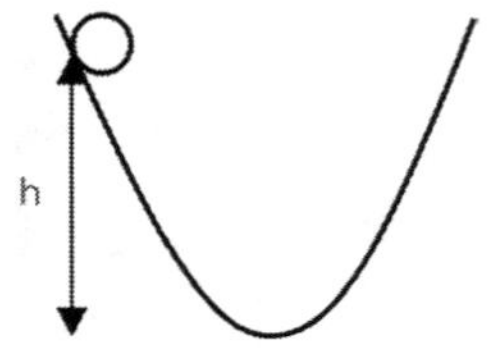

a) Explain where the ball will stop on the other side of the ramp.

____

____

____

Include relevant equations and key terminology.

b) The second half of the ramp is now flattened so that it is horizontal, and the same ball is released from the same position. Relate the motion of the ball to energy and inertia.

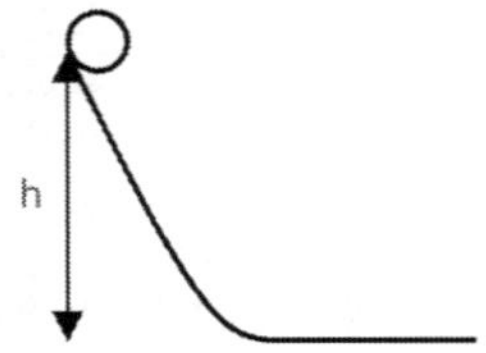

____

____

____

Relate means explain the link.

Example: A child of mass 45 kg rides a slippery slide, as shown in the diagram below. They are released from rest at point A. Assume there is no friction.

Notice the shape of the slide doesn't matter, only the change in heights.

Does it matter whether you use the data for point A or B to get the kinetic energy at point C?

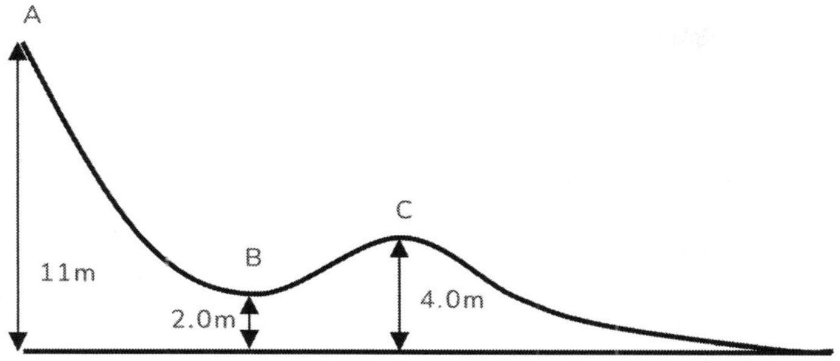

a) What is the child's kinetic energy and speed at point B?

b) What is the child's kinetic energy and speed at point C?

Answer:

a) $\Delta K = -\Delta U$

$K_B - 0 = -mg(h_B - h_A)$

$K_B = -45 \times 9.8 \times (2 - 11) = 3969$ J

$\frac{1}{2}mv_B^2 = -mg(h_B - h_A)$

$v_B = \sqrt{-2g(h_B - h_A)} = \sqrt{-2 \times 9.8 \times (2 - 11)} = 13$ m s^{-1}

b) $\Delta K = -\Delta U$

$K_C - 0 = -mg(h_C - h_A)$

$K_C = -45 \times 9.8 \times (4 - 11) = 3087$ J

$\frac{1}{2}mv_C^2 = -mg(h_C - h_A)$

$v_C = \sqrt{-2g(h_C - h_A)} = \sqrt{-2 \times 9.8 \times (4 - 11)} = 12$ m s^{-1}

3. A child of mass 45 kg on a toboggan reaches a section of a ski slope as shown, at a speed of 10 m s^{-1}. With what speed will child reach the peak of the crest? Assume there is no friction.

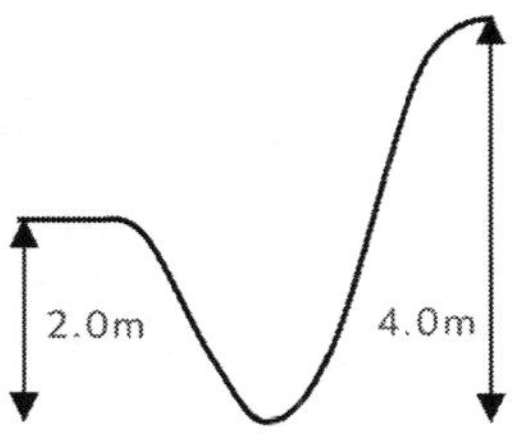

Underline key information.

4. A gymnast of mass 59 kg jumps on a trampoline and reaches a maximum height of 7.5 m above the trampoline. To bounce up, they compressed the trampoline to a depth of 50 cm.

   a) With what initial speed did they leave the trampoline?

   b) What is the average force the trampoline exerted on them?

5. Two blocks of mass 2.0 kg and 3.0 kg are connected by a string over a pulley. The 3.0 kg block sits on a smooth bench, while the 2.0 kg block hangs over the edge of the bench. The blocks are initially at rest. What is the speed of the blocks after the 2.0 kg block has dropped by 1.0 m?

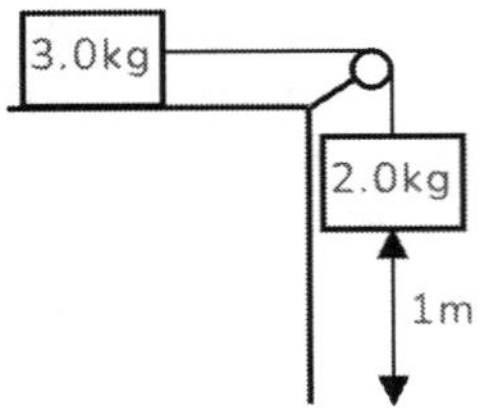

Calculate this two ways, as follows.

a) First calculate the acceleration of the blocks, by considering what force acts on them, then use this to calculate the final speed.

b) Then calculate the final speed by considering the energy transformed.

Which mass?

6. A student wants to slide down each of two straight, frictionless slippery slides from rest. Slide 1 is angled at 30° above the horizontal. Slide 2 is angled at 60° above the horizontal. Both slides start from 4.0 m above the horizontal ground. The student predicts that their speed and kinetic energy at the bottom of Slide 2 will be greater than Slide 1, because there will be more force acting on them down the slope. Discuss whether you agree with the student's hypothesis.

7. A trapeze swing is attached to the pivot point above by ropes 4.0 m long. A trapeze artist starts from rest with the ropes making a 60° angle to the vertical , holding onto the bar of the trapeze. When they reach the lowest point in the swing, where the rope is vertical, what will their speed be?

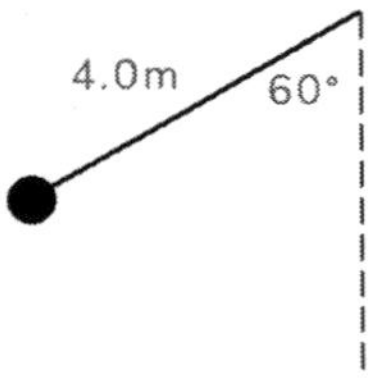

Some geometry will help get the data you need.

8. Ash is standing still on the ground floor and Bee is standing still in a glass elevator, with a 1.0 kg box on the floor. Ash watches the box rise up 10 m at a constant velocity of 2.0 m s^{-1}, as the elevator's motor makes the elevator rise. Bee observes that the box stays still.

a) What is the work done on the box from Ash's frame of reference?

b) What is the work done on the box from Bee's frame of reference?

c) Discuss whether these calculations are consistent with the Law of Conservation of Energy.

9. Describe the effect of friction and air resistance on energy, by filling in the blanks.

coefficient	conservation	air resistance	forever	friction	gravitational potential	kinetic
less than	light	transformations	moving	sound	shooting star	useful

If there s friction or air resistance acting on a ____________ object, some of its kinetic energy will be transformed to heat, possibly even sound or light.

For example, a meteorite falling through the Earth's atmosphere will transform ____________ energy into kinetic energy. However, air resistance from the atmosphere will transform some of the kinetic energy into heat and ____________. If we are lucky, we might get to see the meteorite as a ____________.

Friction often means the speeds measured in our motion experiments might be ____________ what we predict if we ignore friction. This is because some of the kinetic energy is lost as heat, possibly ____________ if you can hear the object moving. We can try to minimise the effect of friction by using smooth surfaces with a low ____________ of friction.

Friction and air resistance can sometimes waste energy by converting it into forms we do not find ____________. For example, a child on a swing is most interested in moving, therefore they want ____________ energy. The Law of ____________ of Energy predicts that the child would swing back and forth ____________, with energy being transformed from gravitational potential energy to kinetic energy, then back to gravitational potential energy. Instead, ____________ where the chains are anchored to the pivot, and ____________ as they move through the air means they eventually stop, because some of the kinetic energy is wasted as heat and sound. We can use an energy flow diagram to represent the energy ____________:

gravitational potential energy → kinetic energy → gravitational potential energy / heat / sound

Example: Two dance partners perform a dance move where one drags the other across the floor for 2.0 m at constant velocity. The dancer being dragged has a mass of 55 kg. The floor has a coefficient of kinetic friction of 0.38.

a) What is the work being done by the dancer doing the dragging?

b) Evaluate what happens to the energy used by the dancer doing the dragging.

Answer:

a) $W = \vec{F}.\vec{s} = \mu N.s = \mu mgs = 0.38 \times 55 \times 9.8 \times 2 = 410\ \text{J}$

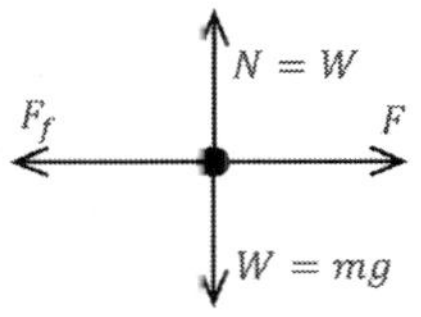

b) As the dancer being dragged is moving at constant velocity, their kinetic energy doesn't change. Therefore all the energy used by the dancer doing the dragging is being converted to 410 J of heat.

10. A toy parachute of mass 15.0 g drops silently from a height of 5.00 m above the ground. It reaches the ground with a constant velocity of 1.28 m s^{-1}. Evaluate the energy transformations that occurred.

11. In a chocolate factory, boxes of chocolates of mass 250 g slide down a ramp in the production line. The boxes pass the top of the ramp at the rate of 4 boxes per second.

A snapshot of the ramp with the moving boxes on them is shown below.

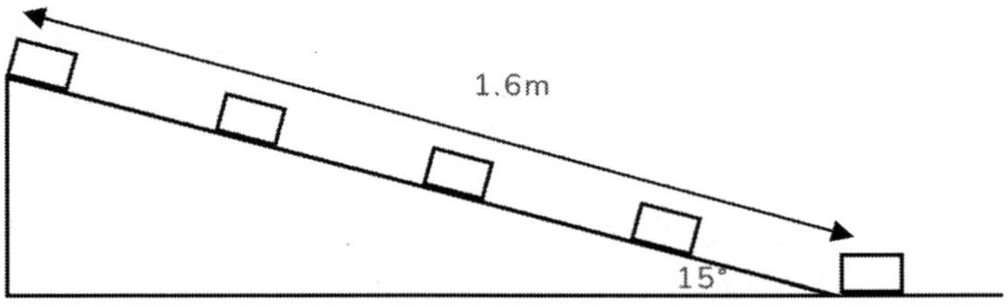

a) What is the kinetic energy of a box on the ramp?

b) Evaluate whether there is work done in this scenario and what energy transformations occur.

9.6 Power

1. Describe the difference between energy and power, by filling in the blanks.

conserved	electrical	energy	gently	gravitational potential	light
power	rate	steeper	time	work done	

It is easier to walk up a ________________ sloping ramp than climb up ________________ stairs to get to the same point. However, because you gain the same amount of ________________ energy, you do the same work. You do this work in less ________________ though, if you take the stairs rather than the ramp.

Power is the rate at which ________________ is transformed or used. In other words, it is the ________________ at which work is done. It is measured in *watts* (W):

$$P = \frac{W}{t}$$

where: $P =$ ____________ (W)

$W =$ ____________ (J)

$t =$ time in which the work was done (s)

Since energy is conserved, and power is energy transformed divided by time, that means for a given time period, power is also ________________. For example, if you ride a bike that generates electricity for a light bulb, the more power you exert with your legs, the more ________________ power you generate, and the brighter the light bulb glows, putting out ________________ energy at a faster rate.

Example: At the gym, while using the lat pulldown machine, you lift 25 kg up 1.8 m in 0.75 s.

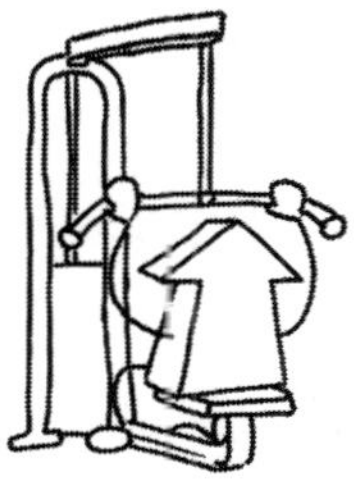

a) How much energy did you use?

b) What power did you exert?

Answer:

a) $W = mgh = 25 \times 9.8 \times 1.8 = 441$ J

b) $P = \frac{W}{t} = \frac{441}{0.75} = 588$ W

2. Compare work and power in the following table.

Feature	Work	Power
Unit		
Definition		
Formula		

3. Hospitals sometimes run a charity event called a bed race, in order to fundraise. In a bed race, each team must push a hospital bed with a person in it from the start to the finish line. In one such race, a team pushes a 100 kg bed with a 45.0 kg person it in, accelerating uniformly from rest at the starting line to 2.20 m s^{-1} in 2.00 s. How much power did the team exert?

____

____

4. A librarian who has the job of returning books from the ground floor of a library to shelves on the third floor decides to be smart and lazy. They install a motor which winds up a rope tied to a basket. The basket can be loaded with books on the ground floor and raised 12 m to the third floor. A motor with a power output of 63.5 W is used. How long would it take to raise a 20 kg load to the third floor?

5. A parent stops a runaway shopping trolley from hitting a small child. The shopping trolley's mass is 75.0 kg and it was moving at 3.00 m s^{-1}. The parent stops the trolley over a distance of 2.50 m. Assuming they decelerated the trolley uniformly, how much power did they exert on the trolley?

6. Show that if a force F moves an object in the same direction with an average velocity of v then the power required is $P = Fv$.

What is an equation for velocity?

Vector diagram

7. A conveyor belt 3.0 m long is angled at 30° to the horizontal. It moves a 2.0 kg box of flour up the ramp in 1.0 s at constant velocity.

a) What is the minimum coefficient of static friction that could exist between the conveyor belt and the box?

b) What power must the motor moving the conveyor belt exert?

Revise inclined planes

8. The dots in the motion diagram represent a child of mass 30 kg, on a toboggan sliding down a slope inclined at 11° to the horizontal. The images were recorded at 20 frames per second.

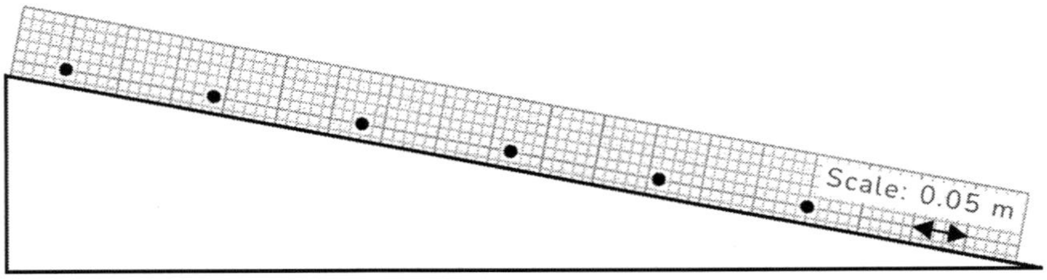

Discuss whether there is work done on the child and if so, at what rate.

9.7 SI units, dimensional analysis and prefixes

SI units

The International System of Units, or *SI units*, is the modern system of measurement established by international agreement.

There are seven base SI units, as shown in the table, from which all other units are derived. The way each of the base SI units is measured is internationally agreed. For example, until 2019, one kilogram was defined as the mass of a certain platinum-iridium cylinder, which was sealed in a jar and locked in a vault in France.

We now have more convenient definitions which scientists can easily measure from any country, based on quantities that appear to be constant in the universe, such as the speed of light.

Base SI units		
Quantity	Unit	Symbol
mass	kilogram	kg
length	metre	m
time	second	s
electric current	ampere	A
temperature	Kelvin	K
luminous intensity	candela	cd
amount of substance	mole	mol

1. Which of the following are base SI units?

g　　kg　　m　　km　　hour　　min　　s　　N　　J

Dimensional analysis

Dimensional analysis is working out the form an equation must take, based on what units are needed to make each side of the equation the same. We can also work out what each unit is equivalent to in base SI units.

Example: What is one newton equal to in SI units?

Answer:

A newton is the unit for force, and force can be calculated using the equation:

$F = ma$

Putting in the units for each quantity in the equation gives:

N=kg m s^{-2}

Dimensional analysis is a useful technique to check an equation is correct.

Example: Show that the units on both sides of the equation are the same for:

$s = ut + \frac{1}{2}at^2$

Answer:

Numbers, including fractions have no units, so we ignore them.

The left hand side is just in metres, m.

The right hand side has the unit: m s^{-1}×s+m s^{-2}×s^2=m

This is the same as the left hand side.

2. What is a joule equivalent to in base SI units?

What equation calculates the quantity measured in joules?

....

....

3. Show that $K = \frac{1}{2}mv^2$ gives a quantity that is measured in joules.

Use your answer to question 2.

....

....

4. What is a watt equivalent to in base SI units?

What is the unit of power?

SI prefixes

The SI prefixes are internationally agreed prefixes that can be put in front of a unit, to multiply or divide it by some power of ten. The ones used in this book appear in the table. More SI prefixes exist for very large or very small quantities.

SI prefixes		
Prefix	**Symbol**	**Multiplier**
giga	G	$\times 10^9$
mega	M	$\times 10^6$
kilo	k	$\times 10^3$
centi	c	$\times 10^{-2}$
milli	m	$\times 10^{-3}$
micro	µ	$\times 10^{-6}$
nano	n	$\times 10^{-9}$

5. Write the names of these units, their numerical value and the quantity they would measure.

   a) MJ

   b) GW

   c) nm

   d) µg

   e) ms

Example: A common bacteria found in intestines, Escherichia coli, is typically around 0.000002 m long. Write this using the most convenient prefix.

Count how many places you need to move the decimal point to the right.

Answer:

$0.000002 \text{ m} = 2 \times 10^{-6} \text{ m} = 2 \text{ µm}$

6. Express the following using a convenient prefix.

   a) The average adult needs about 8700000 J of energy per day.

   b) The Kangaroo Valley hydroelectric power station can generate 160000000 W of power.

   c) The wavelength of red light is around 0.00000065 m.

   d) The width of human hair is usually between 0.000017 m and 0.00018 m

7. The calorie is not an SI unit, but is still a commonly used unit to measure the amount of energy in food. One calorie is the same as 4.18 kJ. If a 200 g block of chocolate contains 1200 calories, how many joules of energy is this?

8. Horsepower (Hp) is a unit for power that is still used describe cars, but was originally used in the 1700s when horses worked to pull vehicles and machinery. One horsepower is the same as 746 W. The engine of one of the fastest cars in the world has a power output of around 1600 Hp. Express this in kilowatts.

9.8 Rolling

1. Describe rolling and relative motion, by filling in the blanks.

add	backwards	bottom	centre	conservation	forwards
friction	ground	reference	resultant	rim	rotational
tangential	top	translational	velocity		

If an object slides across a smooth surface, all the points on the object will have the same ________________. This motion in a straight line is called ***translational motion***.

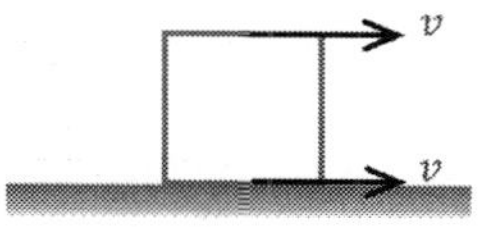

However, if a wheel moves at constant velocity v, but rolls without sliding, then only the ________________ of the wheel moves at constant velocity.

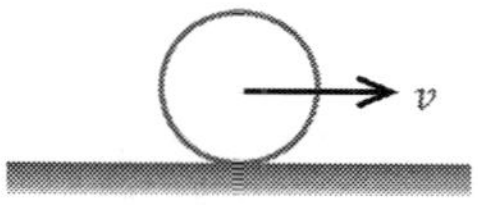

In the frame of ________________ of the wheel's centre, the rim undergoes ***rotational motion***. Each point on the rim has a velocity that is ________________ to the wheel. The bottom of the wheel moves ________________ relative to the wheel's centre at velocity $-v$, and experiences ________________ from the surface. Meanwhile, the top of the wheel moves ________________ relative to the wheel's centre at velocity v.

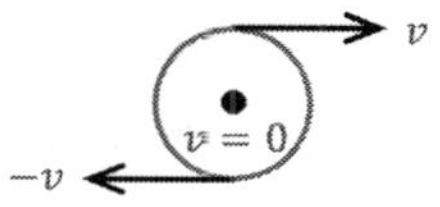

In the frame of reference of an observer on the ________________, the ________________ of the wheel is stationary for the moment it comes into contact with the ground. The ________________ of the wheel appears to move at velocity $2v$, which is the ________________ of its velocity relative to the wheel's centre and the velocity of the wheel's centre relative to the ground. Basically we ________________ v to the velocity of all the points on the wheel to change the frame of reference from the wheel's centre to the ground.

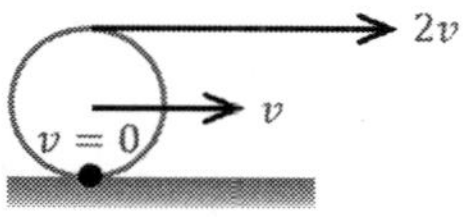

If we track the path of a point on the ________________, it looks semicircular from the ground's frame of reference.

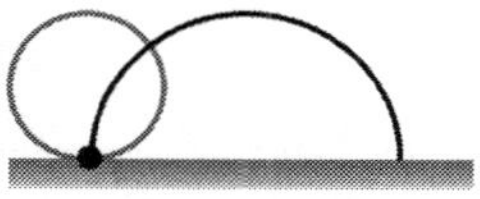

Try this by marking a dot on a jar or bottle.

Where does the extra energy for points on the rim come from? The Law of ________________ of Energy means some of the kinetic energy moving the wheel forward in a straight line, or ________________ kinetic energy, must be traded for ________________ kinetic energy instead.

2. Describe rolling resistance, by filling in the blanks.

accelerate	deformed	friction	gravity	opposing	rice grains
roll	rolling resistance	rotational	stationary	translational	variables

When a ball moves down a ramp, it may slide or ________________. So gravitational potential energy is transformed into a combination of ________________ kinetic energy and ________________ kinetic energy. If the ball rolls, its acceleration down the ramp will be less than predicted by the component of ________________ acting down the ramp, so it is as if there is a force ________________ its motion. This is called ***rolling resistance***.

________________ causes objects to roll. Cars on slippery ground experience less friction, so tyres can skid rather than roll, making it hard to ________________ by speeding up, braking or turning corners.

Rolling resistance also depends on ________________ such as the shape and size of the object, how easily ________________ the object and surface are, and the contents of the rolling object. For example, compared to an empty jar rolling, if the same jar is half full of rice, some energy must be transferred to move the ________________, with less energy available for the kinetic energy of the jar itself. If friction between the rice grains is not overcome, the rice may remain ________________, if the component of weight down the ramp is less than ________________.

3. Explain why a yo-yo descending down its string falls more slowly than the same yo-yo dropped and allowed to fall freely.

4. A 20.0 g ball is released from rest, from a vertical height of 45.0 cm above the bottom of a ramp. The ramp is inclined at 30°. The speed of the ball at the bottom of the ramp is 2.50 m s^{-1}.

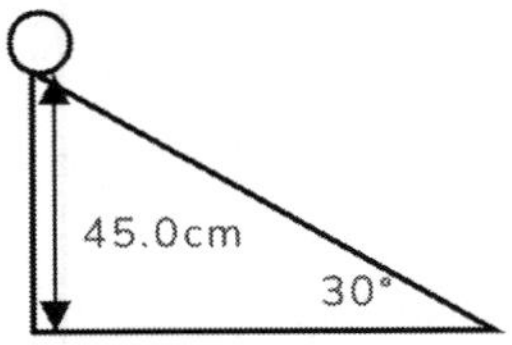

a) If the ball slid down the ramp rather than rolling, what speed should the ball have at the bottom of the ramp?

b) What percentage of the gravitational potential energy of the ball was transformed into rotational kinetic energy?

5. Pumping your tyres to the correct pressure can save fuel when driving a car. The graph shows the rolling resistance acting on the tyres of a 1200 kg car being driven at 100 km h^{-1} at various tyre pressures. The most common unit for tyre pressure in Australia is psi, or pounds per square inch, which describes how much force the air inside the tyre exerts per area of tyre. (Note psi is not an SI unit.)

Compare the power needed to keep this car moving at 100 km h^{-1}, if the tyre pressure is 20 psi and 50 psi, and explain which of these two tyre pressures you would choose to drive with. Assume rolling resistance is the only resistive force acting on the car.

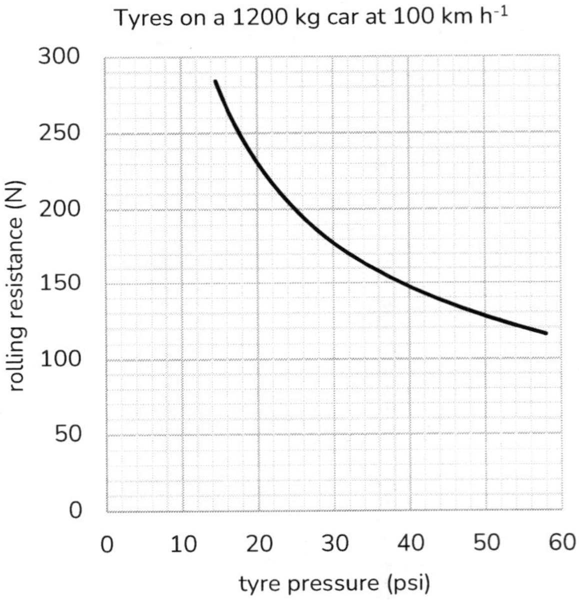

10.1 Momentum and impulse on skates

Oh! But my force acted for a different *time*!
Yes. Impulse is proportional to time. The longer the time, the bigger the impulse.
$\vec{I} = \Delta\vec{p} = \vec{F}t$
where:
$\vec{I}$ = impulse (kgms^{-1})
$\vec{F}$ = force (N)
t = time force acted for (s)
OK. I'm thinking back now to when you collided with me
you gave me more momentum, so we moved off faster...
p_i
p_f
and when I collided with you, I had less momentum to give so we moved slower.
p_i
p_f
Yeah, it's the ***Law of Conservation of Momentum:***
$\Sigma\vec{p_i} = \Sigma\vec{p_f}$
where:
$\Sigma\vec{p_i}$ = total initial momentum (kgms^{-1})
$\Sigma\vec{p_f}$ = total final momentum (kgms^{-1})
But isn't that the same thing as kinetic energy?
$K = \frac{1}{2}mv^2$
Well, not exactly.
Stand facing me, put your palms on mine
then let's both push away.
Huh. We didn't start with any kinetic energy, but we ended up with some.
Yes. So we can't explain why we moved by a transfer of kinetic energy.
So we *created* energy! We broke the Law of Conservation of Energy!
!
No... you got the energy you used to push me from the chemical potential energy in your lunch.
But we also created momentum! We didn't start with any, but then we moved.
Um... remember momentum is a vector?
Oh right! We went opposite ways, so the vector sum of our momentum is still zero!
p_A
p_B
$\Sigma\vec{p_f} = \vec{p_A} + \vec{p_B} = 0 = \Sigma\vec{p_i}$

But *why* is momentum conserved?
Or energy?
Good question... why does our universe follow these rules?
Well let me start by telling you how to *break* the Law of Conservation of Momentum...
We tried our experiments here on a flat, straight path
which looks the same as the flat straight path over there.
In other words, there is ***translational symmetry***.
translation
But if we tried our experiments on a slope
our momentum won't be conserved in collisions
because the gravitational potential energy at each point is different!
Precisely.
In 1918, Emmy Noether mathematically derived a beautiful theorem.
In simple terms, Noether's Theorem shows that whenever there is some form of continuous symmetry, there is a conservation law.
So translational symmetry means linear momentum is conserved
and rotational symmetry means angular momentum is conserved.
But what about the conservation of energy? What is symmetrical?
Ah... that's a tricky one.
Time is symmetrical.
If we try the same experiments now, in the same place, and again in a few minutes, we'll see the same results.
Duh... that's obvious.
Isn't it? How can time be different?
Well, I'll tell you about relativity some other time...
8 minutes ago

10.2 Momentum and impulse

Momentum and impulse

1. Describe momentum and impulse, by filling in the blanks.

change	final	impulse	initial	mass	minus
momentum	motion	multiplied	subtraction	vector	velocity

The momentum of an object is given by its mass ______________ by its velocity:

$$\vec{p} = m\vec{v}$$

where: $\vec{p} =$ ______________ (kg m s^{-1})

$m =$ ______________ (kg)

$\vec{v} =$ ______________ (m s^{-1})

It can be thought of as "mass in ______________" in classical physics. Since velocity is a ______________, momentum is too, with the same direction as the velocity.

Impulse is the ______________ in momentum. In other words, impulse is the final momentum ______________ the initial momentum:

$$\vec{I} = \Delta\vec{p} = \vec{p_f} - \vec{p_i} = m\vec{v} - m\vec{u}$$

where: $\vec{I} =$ ______________ (kg m s^{-1})

$\vec{v} =$ ______________ velocity (m s^{-1})

$\vec{u} =$ ______________ velocity (m s^{-1})

Impulse is a vector, with the same direction obtained by vector ______________ of the final and initial momentum.

Vector diagram

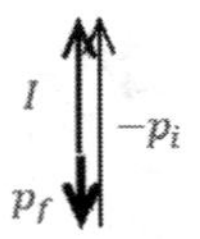

Example: A diver of mass 63 kg jumps off a diving board and reaches the water at 14 m s^{-1} downwards. Then they slow down underwater to 5.0 m s^{-1} downwards.

a) What is their initial momentum?

b) What is their final momentum?

c) What is impulse exerted by the water?

As momentum and impulse are vectors, remember to set a frame of reference.

Answer: Let down be positive.

a) $\vec{p} = m\vec{u} = 63 \times 14 = 882$ kg m s^{-1} down

b) $\vec{p} = m\vec{v} = 63 \times 5 = 315$ kg m s^{-1} down

c) $\vec{I} = \Delta\vec{p} = \vec{p_f} - \vec{p_i} = 315 - 882 = -567$ kg m s^{-1} (567 kg m s^{-1} up)

What you see

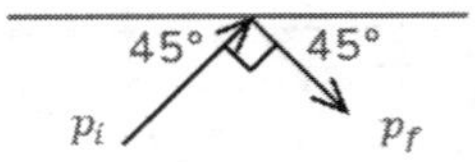

Example: A ball of mass 57.0 g hits a wall at 12 m s^{-1}, at an angle of 45° to the wall. It bounces off at 9.0 m s^{-1}, at an angle of 45° to the wall and 90° to its original direction. What change in momentum is exerted by the wall on the ball?

Vector diagram

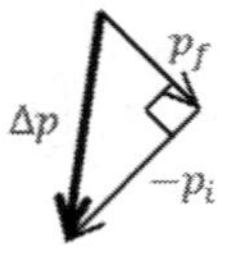

Answer:

$$\Delta\vec{p} = \vec{p_f} - \vec{p_i} = \sqrt{(0.057 \times 9)^2 + (0.057 \times 12)^2} = 0.855 \text{ kg m s}^{-1}$$

$$\tan\theta = \frac{0.057 \times 12}{0.057 \times 9}$$

$$\theta = 53°$$

Change in momentum is 0.855 kg m s^{-1} at 53° to the final momentum.

Reverse the vector's direction to subtract it.

2. A hockey puck of mass 170 g is gliding at 6.5 m s^{-1}. Calculate its momentum.

3. A car of mass 1200 kg accelerates from 40 km h^{-1} to 60 km h^{-1}. Calculate its change in momentum.

4. A child of mass m starts running on a flat track to reach a speed of v. An adult of twice the mass does the same but reaches half the speed. Compare their change in momentum and gain in kinetic energy.

5. A gymnast of mass 52 kg drops onto a trampoline with a downward velocity of 4.8 m s^{-1}. They bounce back up with a velocity of 4.2 m s^{-1}. What is the impulse exerted by the trampoline on the gymnast?

Vector diagram

6. A snooker ball of mass 165 g hits the side of a snooker table at 1.9 m s^{-1}, at an angle of 15° to the side. It bounces off at the same speed and same angle, as shown. What is the impulse exerted by the side of the snooker table on the ball?

What you see

Vector diagram

7. Sketch the momentum, force and kinetic energy graphs for an object:

a) with the following displacement-time graph.

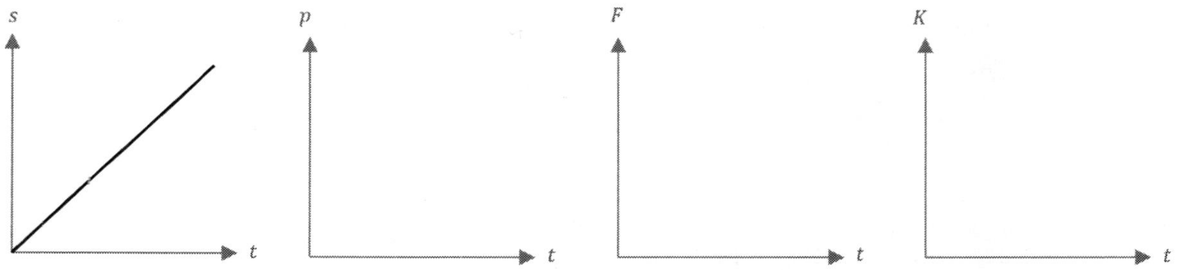

b) with the following velocity-time graph.

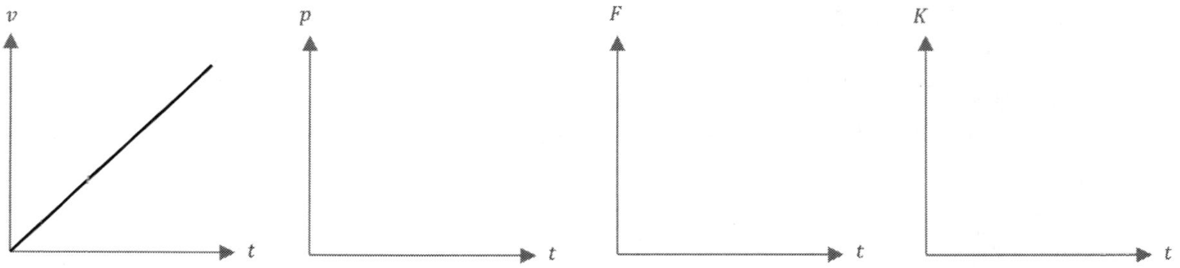

Force and impulse

8. Describe the relationship between force and impulse, by filling in the blanks.

acceleration	force	impulse	increase	longer	time	velocity

Recall that a net force acting on an object causes an ________________ that is proportional to the force, and that acceleration is a change in ________________ over time. Combining these gives:

$$\vec{F} = m\vec{a} = \frac{m(\vec{v} - \vec{u})}{t} = \frac{m\vec{v} - m\vec{u}}{t} = \frac{\Delta\vec{p}}{t} = \frac{\vec{I}}{t}$$

This formula tells us that impulse is also force multiplied by time:

$$\vec{I} = \vec{F}t$$

where: $\vec{I} =$ ________________ (N s)

$\vec{F} =$ ________________ applied (N)

$t =$ ________________ the force is applied to the object (s)

So impulse is increased if a larger force acts on an object, or if the force acts for a ________________ time. For example, pressing the accelerator of a car harder or for a longer time will ________________ the change in momentum of the car.

Example: A swimmer of mass 78.0 kg swims towards the wall at the end of the pool at 1.54 m s^{-1} perpendicular to the wall, turns then pushes off the wall with their feet to go back along the same lane at 1.83 m s^{-1}. Their feet were in contact with the wall for 0.32 s. What is the average force their feet exerted on the wall?

Vector diagram

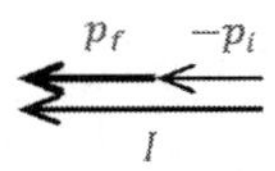

Answer: Let the direction of the swimmer's final velocity be positive.

$\vec{I} = \vec{F}t$

$m\vec{v} - m\vec{u} = \vec{F}t$

$78 \times 1.83 - (-78 \times 1.54) = \vec{F} \times 0.32$

$\vec{F} = 821$ N

Average force is 821 N away from the wall.

Only the average force can be calculated here. The instantaneous force will vary throughout the push.

9. The unit for impulse can be both kg m s^{-1} and N s. Show these are equivalent units, by converting each into SI base units.

____

____

10. Compare impulse, force and kinetic energy using the table.

Feature	Impulse	Force	Kinetic energy
Formula			
Unit			
Scalar or vector?			

11. Fill in the blanks to explain why a crash mat helps prevent injuries for acrobats when they fall on the ground.

decreases	final velocity	impulse	increases	initial velocity	stays the same	stop

When an acrobat falls, they come to a ______________. If the ______________ u they reach the ground at is kept the same, and they come to a stop with ______________ $v = 0$ after hitting the ground, the acrobat's change in momentum ______________, regardless of the surface they fall on. This change in momentum, or ______________, is given by the equation ______________ (write the equation).

However, using a crash mat ______________ the time taken t in which the impulse takes place. This ______________ the force acting on the acrobat from the ground, hence making it safer, according to the equation: ______________ (write the equation).

12. In soccer, the technique of following through means that a soccer player keeps their foot moving in contact with the ball in the direction of their target after their foot makes contact with the ball.

Complete the sentences to explain why a soccer player follows through after kicking a ball.

When a soccer player wants a ball to travel as fast as possible, the maximum strength of their muscles determines ______________.

However, by following through, they increase ______________.

Therefore, according to the equation ______________

______________.

13. A rower of mass 70.0 kg is rowing a rowboat of mass 16.0 kg. Before they drive the oars into the water, the velocity of the rowboat is 1.25 m s^{-1}. After one stroke of the oars through the water, the rowboat moves in the same direction, but at a faster speed. The oars push through the water for 1.18 s with an average total force of 18.2 N. What is the final speed of the rowboat?

14. The crumple zone on a car is a front end which is designed to crumple on impact with an obstacle. Ash and Bee compare crumple zones on two cars, each of mass 2000 kg.

Car A is an old car with a front end that will only crumple by 0.25 m if driven at 60 km h^{-1} into a sturdy wall. It comes to a stop after 0.030 s on collision with the wall.

Car B is a newer car that has a front end that will crumple by 1.0 m if driven at 60 km h^{-1} into the same wall. It comes to a stop after 0.14 s on collision.

Ash says: "Car A is safer, because it is stronger. It may also be cheaper to repair after a crash."

Bee says: "If you calculate the force acting on each car, Car B is safer."

Compare the impulse, work done and force acting on each car, and explain who is correct.

10.3 Impulse graphs

1. Explain how to analyse impulse graphs, by filling in the blanks.

decreases	gravitational force	momentum
multiplied	remains constant	time

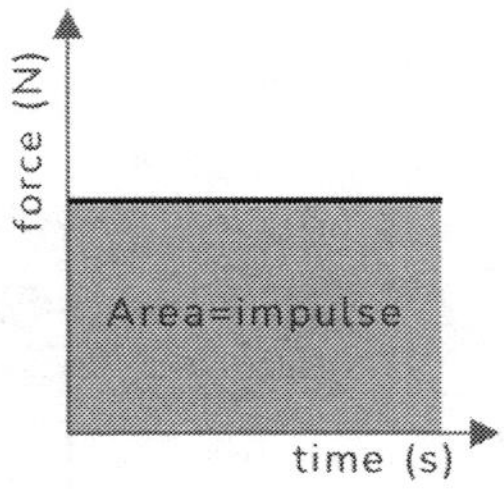

The equation $\vec{I} = \vec{F}t$ can be used to calculate impulse if the force acting ________________ over time, for example if the engine of a car applies constant thrust, or if the only force on a falling object is ________________.

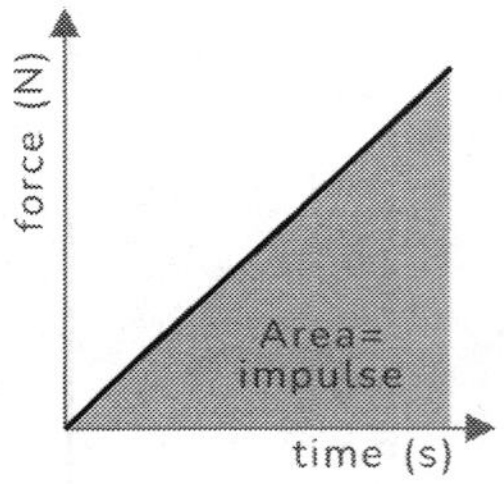

If the force acting on an object changes, the impulse can be calculated by finding the area between a force vs ________________ graph and the x-axis. The area will have units equivalent to the unit of the x-axis ________________ by the y-axis, in this case newton-seconds (N s). Whenever the force ________________, the area under the graph will be smaller too, representing a smaller change in ________________ during that time period.

Example: When a bow shoots an arrow, the more stretched the bowstring is, the more elastic potential energy is stored in it, and the greater the force the bowstring is able to exert on the arrow. The graph shows the force exerted by a bowstring on an arrow of mass 25 g, as the arrow is released.

a) What is the impulse applied to the arrow between 0 and 6 seconds?

b) What is the total impulse applied to the arrow?

c) What is the average force that the bowstring applies to the arrow?

d) If the arrow starts from rest, with what speed does it leave the bow?

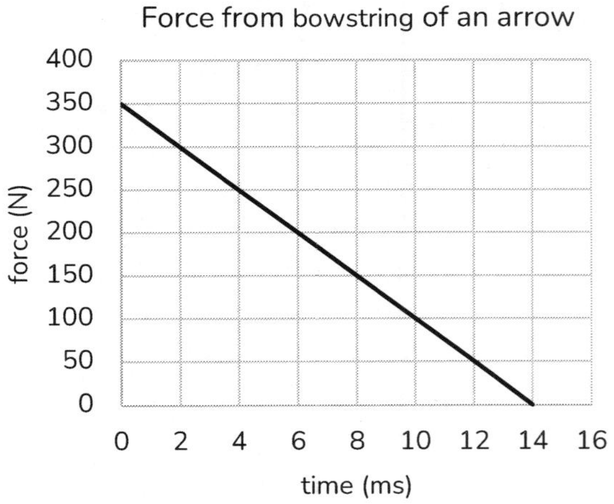

Pay attention to the units on the axes.

Answer:

a) $\vec{I}$ = area under graph from 0 to 6 s $= \frac{1}{2} \times (200 + 350) \times 6 \times 10^{-3} = 1.65$ N s

b) $\vec{I}$ = area under graph from 0 to 14 s $= \frac{1}{2} \times 350 \times 14 \times 10^{-3} = 2.45$ N s

c) The average force gives the same impulse in the same time.

$\vec{F} = \frac{\vec{I}}{t} = \frac{2.45}{0.014} = 175$ N

d) $\vec{I} = \Delta\vec{p} = \vec{p_f} - \vec{p_i} = m\vec{v} - m\vec{u}$

$2.45 = 0.025 \times \vec{v} - 0$

$\vec{v} = 98$ m s^{-1}

2. The graph shows the net force acting on a skydiver as they fall vertically out of an aeroplane from rest.

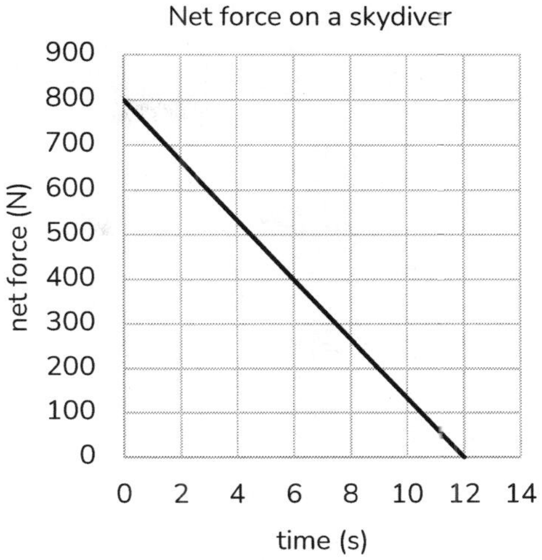

a) Draw a freebody diagram showing the forces acting on the skydiver at times 0 s, 6 s and 12 s.

0 s	6 s	12 s

b) Calculate the terminal velocity of the skydiver.

3. The graph shows the force acting on a 1250 kg car as it accelerates from rest.

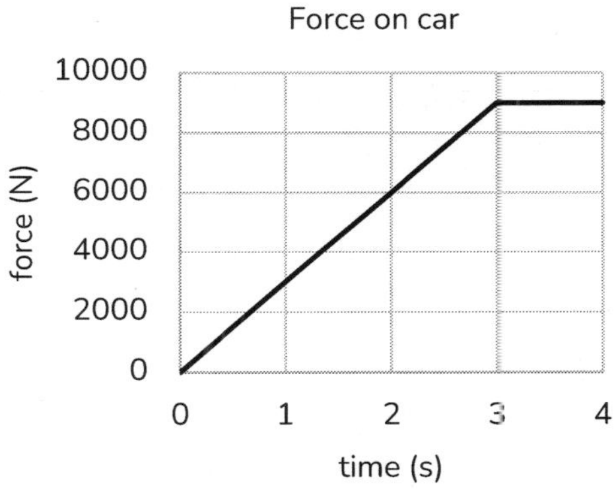

a) What is the change in momentum from 0 s to 4 s?

b) What is the car's final speed in km h^{-1}?

c) Assess the validity of the following statement:

"Between 0 to 1 s, you can calculate the impulse using: $\vec{I} = \vec{F}t = 3000 \times 1 = 3000$ N s.

Similarly, between 3 to 4 s, impulse is given by: $\vec{I} = \vec{F}t = 9000 \times 1 = 9000$ N s."

4. A 165 g ice hockey puck slides towards a player, who hits it back in the same direction at 32 m s^{-1}. The graph shows the force the hockey stick exerts on the puck.

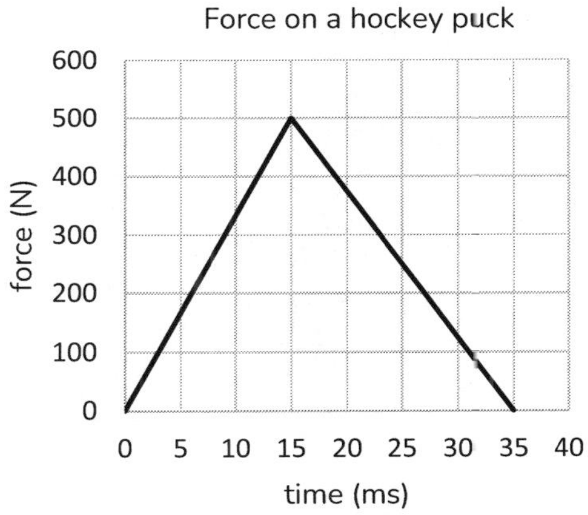

a) Calculate the initial velocity of the hockey puck.

b) What is the average force exerted on the puck by the hockey stick?

5. A force plate is often used to measure and improve sports technique. Like scales, it measures normal force exerted on it, but also records the time the force is detected. A force plate mounted on a wall is used to analyse two punches. The data is shown in the graph.

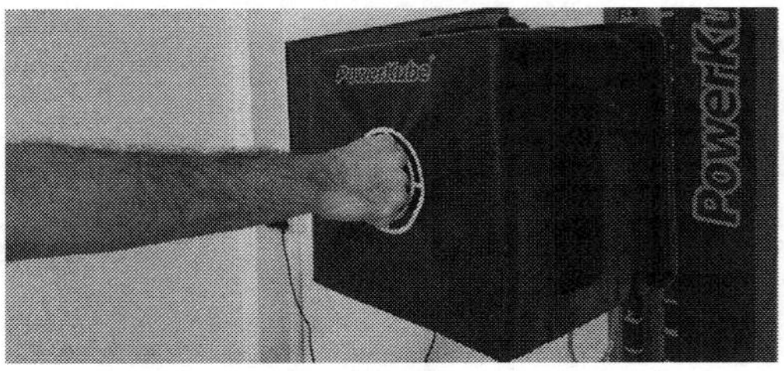

Compare the effect of the two punches on the same object being punched.

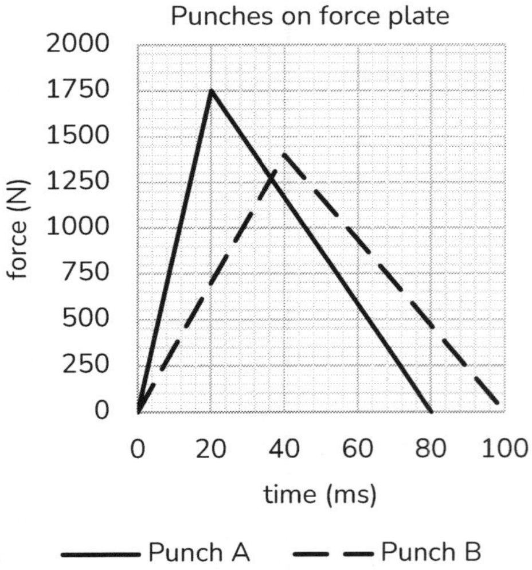

6. A squat jump is a useful leg exercise. You start from rest in a squatting position, rapidly push your feet into the ground to jump as high as you can, then land back in the same spot by gradually bending your knees into a squatting position.

An athlete performs a squat jump on a force plate. The graph shows the reading exerted by the athlete on the force plate, during a squat jump. In other words, the force plate records the magnitude of the normal force it exerts up on the athlete.

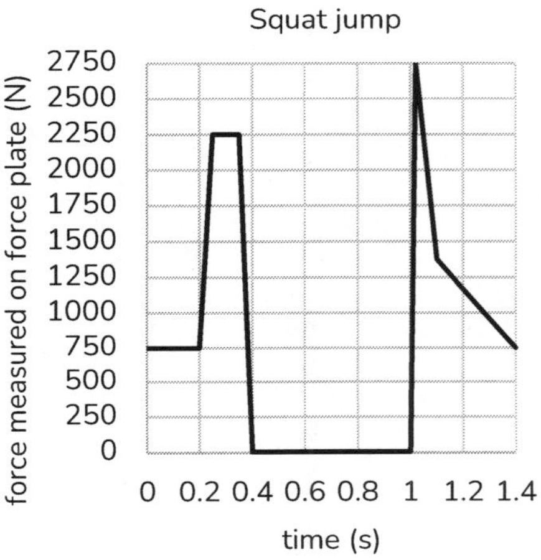

a) Identify what the athlete is doing in each of the time periods. Include a vector diagram showing all the forces acting on the athlete.

Time (s)	0 – 0.2	0.2 – 0.4	0.4 – 1.0	1.0 – 1.4
What is the athlete doing?				
Vector diagram		Draw this at 0.3 s		Draw this at 1.1 s

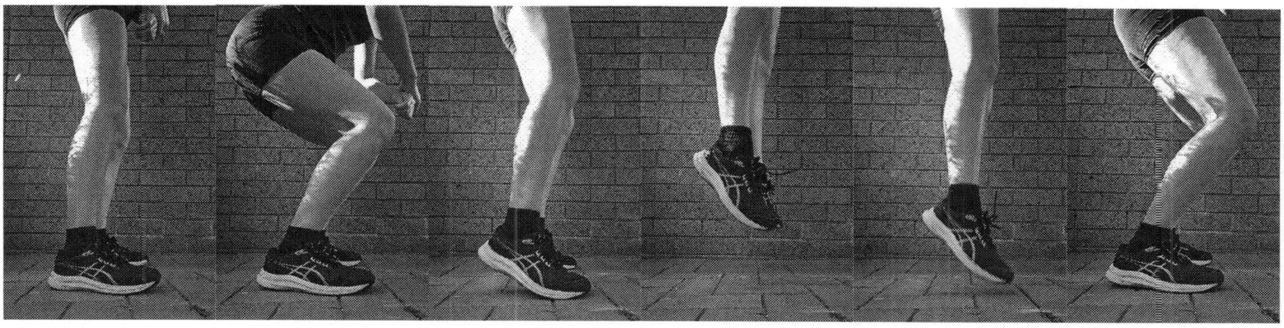

b) What is the mass of the athlete?

c) With what speed did the athlete leave the force plate when they jumped up?

What is the acceleration when they leave the force plate?

d) What impulse acted on the athlete in order to produce this speed?

e) What is the average net force acting on the athlete between 0.2 s and 0.4 s?

f) What is the work done by the athlete between 0.2 s and 0.4 s?

g) Estimate the distance the athlete squatted down before jumping. In other words, how far did their centre of mass move from the squatting position to just leaving the force plate?

h) **Challenge**: Analyse the change in momentum of the athlete between 0.2 s to 0.4 s and between 1.0 to 1.4 s, using the area under the graph. Discuss the relationship between your answers and the answer to part (d).

Approximate the areas

There is an impulse on the athlete due to the force plate, as well as due to gravity. So the net impulse is the area under this graph minus the impulse from gravity.

10.4 Conservation of momentum

1. Describe the Law of Conservation of Momentum, by filling in the blanks.

action-reaction	conservation	equal	field	final	impulse
initial	momentum	opposite	total	unbalanced	velocity

If two objects exert force on each other, either because they are in contact or because there is a ________________ exerting force between them, then the force the first object exerts on the second $\overrightarrow{F_{12}}$ will be ________________ in magnitude but in an ________________ direction to the force the second object exerts on the first $\overrightarrow{F_{21}}$. These forces are an ________________ pair:

$$\overrightarrow{F_{12}} = -\overrightarrow{F_{21}}$$

If these forces act for time t, then the ________________ is given by:

$$\overrightarrow{F_{12}}t = -\overrightarrow{F_{21}}t$$

Impulse means the same thing as change in ________________, so:

$$\Delta\overrightarrow{p_2} = -\Delta\overrightarrow{p_1}$$

Change in momentum is given by the ________________ momentum minus the ________________ momentum, and momentum is given by mass times ________________, so:

$$m_2\overrightarrow{v_2} - m_2\overrightarrow{u_2} = -(m_1\overrightarrow{v_1} - m_1\overrightarrow{u_1})$$

This equation can be rearranged to show that the total final momentum is equal to the ________________ initial momentum:

$$m_1\overrightarrow{v_1} + m_2\overrightarrow{v_2} = m_1\overrightarrow{u_1} + m_2\overrightarrow{u_2}$$

$$\Sigma\overrightarrow{p_f} = \Sigma\overrightarrow{p_i}$$

In other words, in a closed system, where there is no other ________________ force acting on the objects, total momentum is conserved. This concept is known as the Law of ________________ of Momentum.

Example: A truck is driving northbound at 30.0 km h^{-1}, having just pulled out of a carpark, when a car bumps into their rear, propelling the truck forwards to 58.0 km h^{-1}. Before colliding, the car was driving northbound behind the truck at 100 km h^{-1}. The truck's mass is 4.50 tonnes and the car's mass is 1.14 tonnes.

a) What is the final velocity of the car?

b) What impulse and force was exerted by the truck on the car, if the truck and car were in contact for 0.25 s?

Answer: Let north be positive.

As long as units on both sides of the equation are the same, you don't need to convert units.

a) $m_1\overrightarrow{v_1} + m_2\overrightarrow{v_2} = m_1\overrightarrow{u_1} + m_2\overrightarrow{u_2}$

$4.5 \times 58 + 1.14 \times \overrightarrow{v_2} = 4.5 \times 30 + 1.14 \times 100$

$\overrightarrow{v_2} = -10.5$ km h^{-1}

The car rebounds to travel 10.5 km h^{-1} southbound.

b) $\vec{I} = \Delta\vec{p} = m\vec{v} - m\vec{u} = 1140 \times -\frac{10.5}{3.6} - 1140 \times \frac{100}{3.6} = -35000$ N s

As force is in newtons, you need to convert to SI units for this part.

$\vec{I} = \vec{F}t$

$1140 \times -\frac{10.5}{3.6} - 1140 \times \frac{100}{3.6} = \vec{F} \times 0.25$

$\vec{F} = -140000$ N

The impulse was 35000 N s south and force was 140000 N south.

2. Box jellyfish have tentacles with a deadly sting. They also have a bell-like body, which can draw in and expel water to enable it to move by jet propulsion. A box jellyfish of body mass 0.25 kg propels itself from rest to a speed of 1.5 m s^{-1}. The jet of water expelled moves in the opposite direction to the box jellyfish at 0.21 m s^{-1}. What is the mass of water expelled?

3. A bowling ball of mass 4.5 kg strikes a bowling pin of mass 1.5 kg. The ball is initially travelling in a straight line at 6.2 m s^{-1} and the pin is standing still. After hitting the pin, the bowling ball continues travelling in the same direction, but at 3.1 m s^{-1}. What is the velocity of the pin?

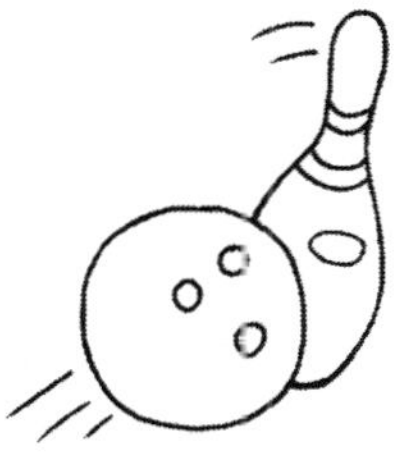

4. Ash and Bee cruise in a boat at 1.5 m s^{-1} northwards, under a bridge. Cai is on the bridge and decides to drop vertically down into the boat to join them. The combined mass of Ash, Bee and the boat is 170 kg. Cai's mass is 50 kg.

a) What happens to the velocity of the boat when Cai enters the boat?

b) Ash and Bee don't want Cai in the boat, so they drop Cai overboard, from rest relative to the boat. Explain what happens to the boat's velocity.

5. A toy gun of mass 100 g fires a foam bullet of mass 1.67 g. When the bullet is loaded, it compresses a spring inside the gun. Firing the gun releases the spring, shooting the bullet with a muzzle velocity of 21.0 m s^{-1}.

a) If the gun were fired horizontally at rest, and was free to move after firing, with what velocity would the gun recoil at?

b) How much elastic potential energy was in the spring, assuming all of this was converted to kinetic energy?

6. At a party, a dart is thrown horizontally into a piñata, initially at rest while hanging vertically. After being hit, the piñata with the dart embedded in it swings to rise up a 1.0 cm vertical height from its starting point. The dart's mass is 24 g and the piñata's mass is 450 g. With what speed did the dart hit the piñata?

Conservation laws are helpful.

Vector diagram

7. In a game of snooker, a white ball of mass 165 g moves at 3.2 m s^{-1} towards red ball of the same mass, which is at rest. The white ball hits the red ball so that the white ball is deflected at 45° to its original direction, while the red ball moves at 45° to the original direction of the white ball and at 90° to its final direction.

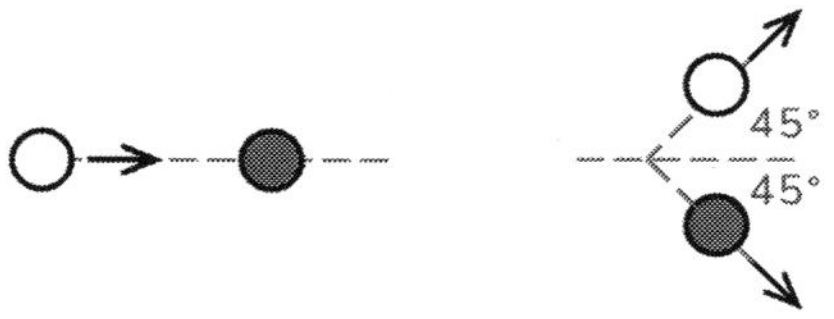

What are final speeds of the white and red balls?

Calculate the initial and final momentums, then use vector addition with the conservation of momentum.

8. A Taylor Swift concert in 2023 caused temporary seismic activity equivalent to a 2.3 magnitude earthquake. Discuss the effect on the Earth's motion, and estimate the change in velocity of the Earth, if all the people at the concert were to jump at the same time, then fall back to the ground. Assume there were 144,000 people of average mass 80 kg, who each jump 10 cm off the ground. The mass of the Earth is 6×10^{24} kg.

10.5 Elastic and inelastic collisions

1. Describe how conservation laws can be used to predict collisions, by filling in the blanks.

all four	bigger	conserved	deformed	elastic	energy
final	gravitational potential	half	inelastic	initial	kinetic
kinetic energy	loss	momentum	momentum	one	same
three	transferred	transformed	two	two	

A Newton's cradle is a toy made of identical balls hanging on strings in a line. If one ball at the end is pulled aside and released, only one ball at the other end moves out, and not the others. Why?

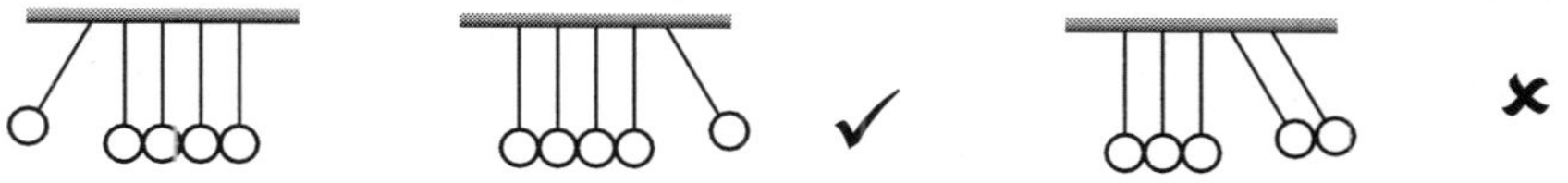

Let the mass of each ball be m. In a collision, momentum must be ________________, so:

$$\Sigma\overrightarrow{p_f} = \Sigma\overrightarrow{p_i}$$

Since $\vec{p} = m\vec{v}$, the Law of Conservation of ________________ is satisfied if one ball is pulled aside then collides with velocity $\vec{v}$ and causes any of the following scenarios:

- one ball at the other end exits with the ________________ velocity $\vec{v}$;
- ________________ balls at the other end exit with the velocity $\frac{1}{2}\vec{v}$;
- ________________ balls at the other end exit with the velocity $\frac{1}{3}\vec{v}$; or
- ________________ balls move with the velocity $\frac{1}{4}\vec{v}$.

However, if the Law of Conservation of ________________ is also true, then the total initial energy must equal the total final energy. The ball initially pulled aside starts with ________________ energy, which becomes kinetic energy when released. When it collides with the next ball, that energy must either be ________________ or transformed. If no energy is ________________ as heat, sound or deformation in the collision, then the total initial kinetic energy must equal the total final kinetic energy. This type of collision, involving only a transfer of ________________ energy, is called an ***elastic collision***.

The ________________ kinetic energy of the ball pulled aside, just before collision, is:

$$K_i = \frac{1}{2}m\vec{v}^2$$

If only one ball exits at the other end with velocity $\vec{v}$, then its ________________ kinetic energy is:

$$K_f = \frac{1}{2}m\vec{v}^2 = K_i$$

Since kinetic energy was conserved, this is an ________________ collision.

However, if ________________ balls exit at the other end with velocity $\frac{1}{2}\vec{v}$ then final kinetic energy is:

$$K_f = \frac{1}{2} \times (2m) \times (\frac{1}{2}\vec{v})^2 = \frac{1}{4}m\vec{v}^2 = \frac{1}{2}K_i$$

This implies that ________________ of the initial energy had to be lost as sound or heat, or the Law of Conservation of Energy would be violated. Since this much energy ________________ does not happen, two balls cannot exit. Repeating the same calculation while assuming that three or four balls exit would result in a ________________ loss in kinetic energy. Therefore, to satisfy both the Law of Conservation of Momentum and the Law of Conservation of Energy, one ball being pulled aside must lead to ________________ ball exiting the other side.

Collisions where kinetic energy is not conserved are called ***inelastic collisions***. In inelastic collisions, the initial kinetic energy is transformed into heat or sound, as the colliding objects are ________ instead of bouncing apart in their original form. A completely ________ collision happens when two objects collide and are stuck together. ________ is still conserved in these collisions, just not ________.

Example: Ash and Bee are ice skating. Ash skates forwards at 3.00 m s^{-1}, while Bee skates backwards in a straight line in front of Ash, at 1.50 m s^{-1}. Bee cannot see Ash approaching from behind, so they collide, with Ash hanging onto Bee to stop themselves falling. Both Ash and Bee have a mass of 40.0 kg.

a) What is the velocity of Ash and Bee immediately after collision?

b) Analyse whether this is an elastic collision and calculate the percentage of kinetic energy lost if it is inelastic.

Answer:

a) Let forward be positive.

$\Sigma\overrightarrow{p_f} = \Sigma\overrightarrow{p_i}$

$(m_A + m_B)\vec{v} = m_A\overrightarrow{u_A} + m_B\overrightarrow{u_B}$

$(40 + 40) \times \vec{v} = 40 \times 3 + 40 \times -1.5$

$\vec{v} = 0.75$ m s^{-1} forwards

b) $\Sigma K_i = \frac{1}{2}m_A u_A^2 + \frac{1}{2}m_B u_B^2 = \frac{1}{2} \times 40 \times 3^2 + \frac{1}{2} \times 40 \times 1.5^2 = 225$ J

$\Sigma K_f = \frac{1}{2}(m_A + m_B)v^2 = \frac{1}{2} \times (40 + 40) \times 0.75^2 = 22.5$ J

Percentage lost $= \frac{225-22.5}{225} \times 100\% = 90\%$ so this is inelastic.

Remember momentum is a vector but energy is not.

2. At an amusement park, Ash and Bee are riding dodgem cars. Bee is driving at 1.0 m s^{-1} when Ash bumps into Bee from behind, at 4.0 m s^{-1}. This slows Ash down to 1.0 m s^{-1}. The mass of each car and driver is 100 kg.

a) What is Bee's velocity after collision?

____

____

____

____

b) Analyse whether this is an elastic collision.

____

____

____

3. A bouncy ball of mass 75 g is dropped from rest at a height of 1.0 m. It reaches the ground and bounces back a height of 0.85 m.

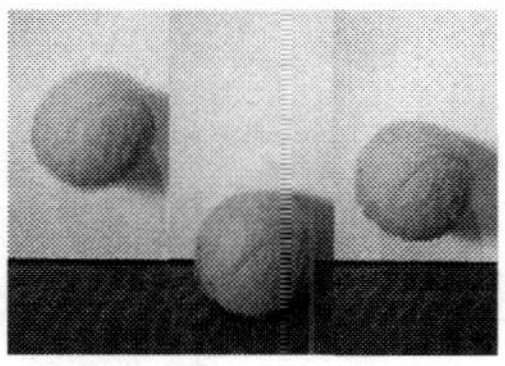

a) Calculate the change in momentum of the ball due to the collision with the ground.

Take care which velocities are used in the equations.

b) Analyse whether this is an elastic collision, by calculating the percentage of kinetic energy lost during the collision.

Analysing either kinetic energy or gravitational potential energy will give the same answer.

4. A driver of a car of mass 1.6 tonnes gets impatient with another driver in front, who is driving slow on a wet road. There is no overtaking lane, so the car pulls out illegally onto the wrong side of the road in order to overtake. The car is driving at 80 km h^{-1} when it hits a truck coming from the opposite direction in a head-on collision. The truck has a mass of 3.6 tonnes and was travelling at 60 km h^{-1} before the collision. Upon collision, the car and the truck are squashed into each other and move together.

a) What is the velocity of the wrecked car and truck?

b) What percentage of kinetic energy is lost during the collision? What might have happened to it?

5. Ash is skating eastwards at 3.0 m s^{-1} when they collide with Bee, skating northwards at 2.5 m s^{-1}. They fall and hold onto each other after colliding, slipping across the ice. Ash's mass is 35 kg and Bee's mass is 42 kg.

Vector diagram

a) What is their final velocity?

b) Analyse whether this is an elastic collision.

6. Ash and Bee are playing ice hockey, when they accidentally collide. Ash pushes Bee away during the collision. The graph shows the velocity of Ash and Bee before, during and after the collision.

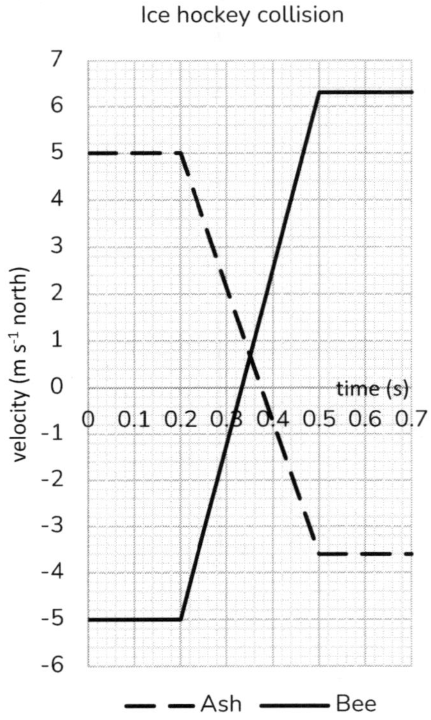

a) Ash's mass (with all their hockey gear on) was 85 kg. What was Bee's mass?

b) What force did Ash exert on Bee?

c) What force did Bee exert on Ash?

d) Determine whether this is an elastic collision.

7. **Challenge**: In a game of snooker, if a white ball hits a red ball with identical mass head-on, the white ball stops while the red ball moves with the same velocity as the white ball had initially, provided the white ball had no spin. Explain why, by deriving the relationship between the initial velocity of the white ball u_W and the final velocities of the white and red ball respectively, v_W and v_R.

Derive a relationship using each conservation law. Then solve them simultaneously.

8. A marble of mass 10 g is shot into a smaller marble at rest, on a piece of graph paper on a table. Images of the marbles were recorded at 40 frames per second, at the same scale.

The image below shows the positions of each marble before they collided:

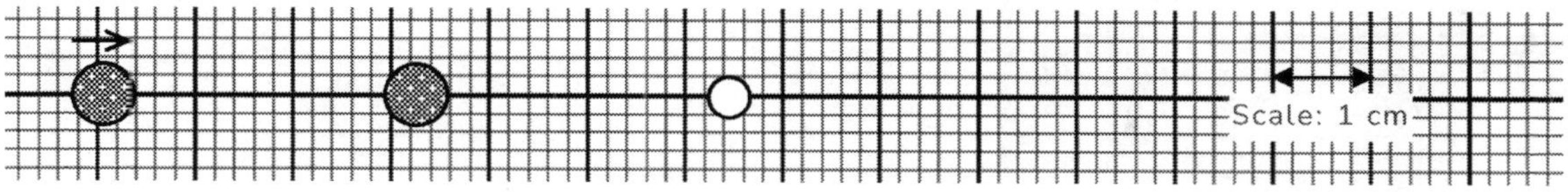

After they collided, their positions were recorded below:

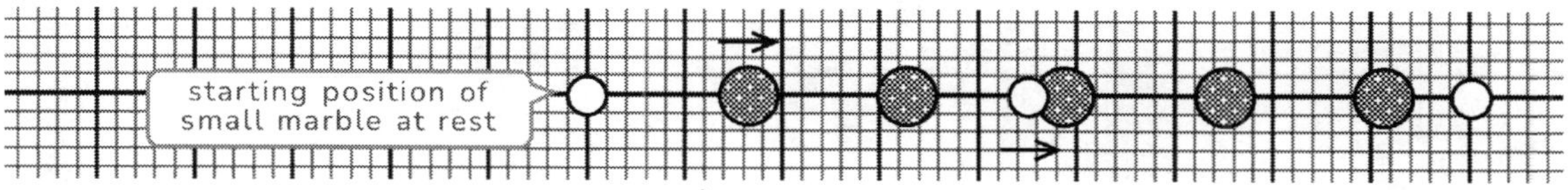

a) What is the time interval between each image?

b) In the first diagram, from the initial position of the large marble, how much time does it take before it collides with the small marble?

c) In the first diagram, from the initial position of the large marble, what distance does it travel before it reaches the small marble?

d) Calculate the initial velocity of the large marble.

e) What is the initial momentum of the large marble?

f) Calculate the final velocity of the small marble.

g) Calculate the final velocity of the large marble.

h) Determine the mass of the small marble.

i) Analyse whether this is an elastic collision.

9. A white snooker ball hit a black snooker ball, initially at rest. A video took images of this at 20 Hz, or 20 frames per second, as shown in the motion diagram below. The mass of each ball was 160 g.

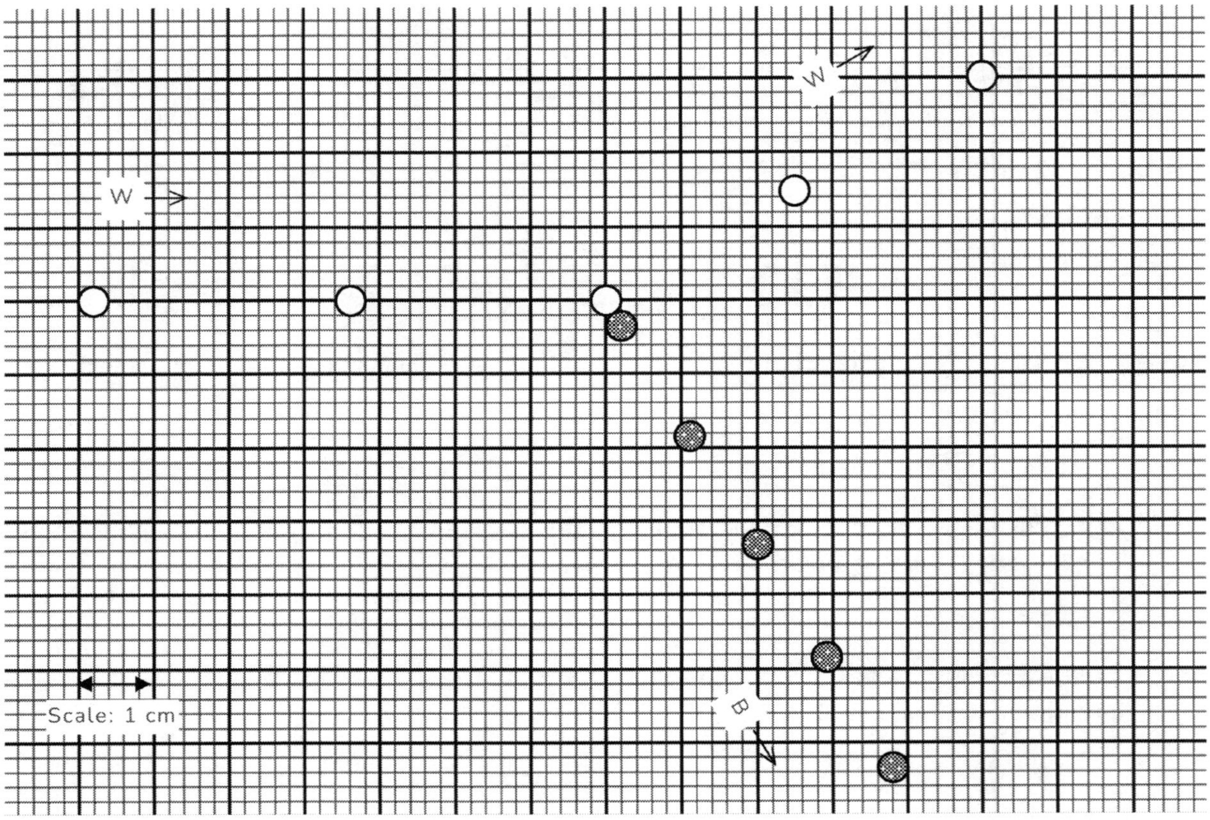

a) Determine the initial velocity and momentum of the white ball.

Use a protractor to measure the angles and a ruler to measure distance.

b) Determine the final velocity and momentum of the white ball.

Did you know if a ball hits an identical ball at rest, without spin, then the angle between the final velocities of the two balls is always 90°? Snooker players call this the 90 degree rule.

c) Determine the final velocity and momentum of the black ball.

Vector diagram

d) Analyse whether momentum was conserved in this collision.

Use vector addition to determine the total final momentum. Is it equal to the total initial momentum?

e) Analyse whether this was an elastic collision

Part 3: Electricity and magnetism

H	C	O	N	D	U	C	T	O	R	E	R	T	H	A	C	H	A	R	G	E
C	A	Y	R	S	O	U	T	H	A	R	D	I	R	O	N	T	T	O	N	A
O	K	H	P	E	R	M	A	N	E	N	T	M	A	G	N	E	T	E	V	R
U	T	C	O	I	L	D	A	J	F	J	V	E	S	Y	C	Q	R	J	Y	T
L	B	A	T	T	E	R	Y	I	V	U	G	N	I	Y	N	U	A	E	B	H
O	E	H	E	B	W	G	I	V	N	N	O	N	O	H	M	I	C	E	N	I
M	A	G	N	E	T	I	C	M	O	M	E	N	T	R	B	P	T	I	B	N
B	Y	N	T	C	G	R	R	O	B	R	D	F	J	V	N	O	M	S	C	G
E	V	N	I	I	N	E	L	E	C	T	R	O	S	T	A	T	I	C	P	L
M	U	P	A	M	M	E	T	E	R	W	Q	U	L	W	X	E	V	E	D	H
K	N	E	L	E	C	T	R	O	N	R	C	U	T	B	A	N	O	I	N	D
A	N	O	D	E	D	R	B	U	O	F	M	E	S	B	I	T	L	N	L	Y
O	H	M	I	C	C	L	O	S	E	D	C	I	R	C	U	I	T	S	H	Y
M	S	O	F	T	I	R	O	N	E	F	J	Y	V	R	M	A	M	U	D	M
A	E	D	F	V	Y	U	M	F	E	I	P	A	R	A	L	L	E	L	R	A
G	L	U	E														T	A	E	G
N	E	T	R														E	T	P	N
E	C	W	E														R	O	E	E
T	T	O	N	R	E	S	I	S	T	A	N	C	E	X	E	R	L	R	L	T
I	R	H	C	C	N	R	E	A	W	T	E	Y	O	N	U	R	E	A	O	I
C	O	P	E	D	O	A	N	N	A	I	E	L	B	E	E	P	L	R	T	C
F	M	R	E	S	I	S	T	I	V	I	T	Y	W	G	M	S	N	C	H	P
I	A	O	O	O	E	C	E	I	S	E	R	C	N	A	E	C	E	I	O	E
E	G	T	T	L	Y	N	V	P	N	O	E	V	H	T	X	O	R	T	P	R
L	N	O	F	E	R	R	O	M	A	G	N	E	T	I	S	M	C	J	E	M
D	E	N	P	N	L	T	L	R	C	U	C	M	R	V	P	P	A	F	N	E
U	T	E	O	O	E	H	T	P	T	J	O	U	L	E	J	A	T	R	C	A
G	O	H	S	I	L	O	A	W	E	H	S	E	R	I	E	S	H	V	I	B
V	J	U	I	D	N	R	G	E	H	A	Z	I	L	R	M	S	O	J	R	I
D	L	M	T	O	E	L	E	C	T	R	I	C	F	I	E	L	D	G	C	L
R	E	S	I	S	T	O	R	W	M	L	C	N	E	O	T	N	E	I	U	I
R	W	E	V	U	D	I	R	E	C	T	C	U	R	R	E	N	T	M	I	T
J	E	L	E	C	T	R	I	C	P	E	R	M	I	T	T	I	V	I	T	Y

11.1 Electrostatic – the force of life

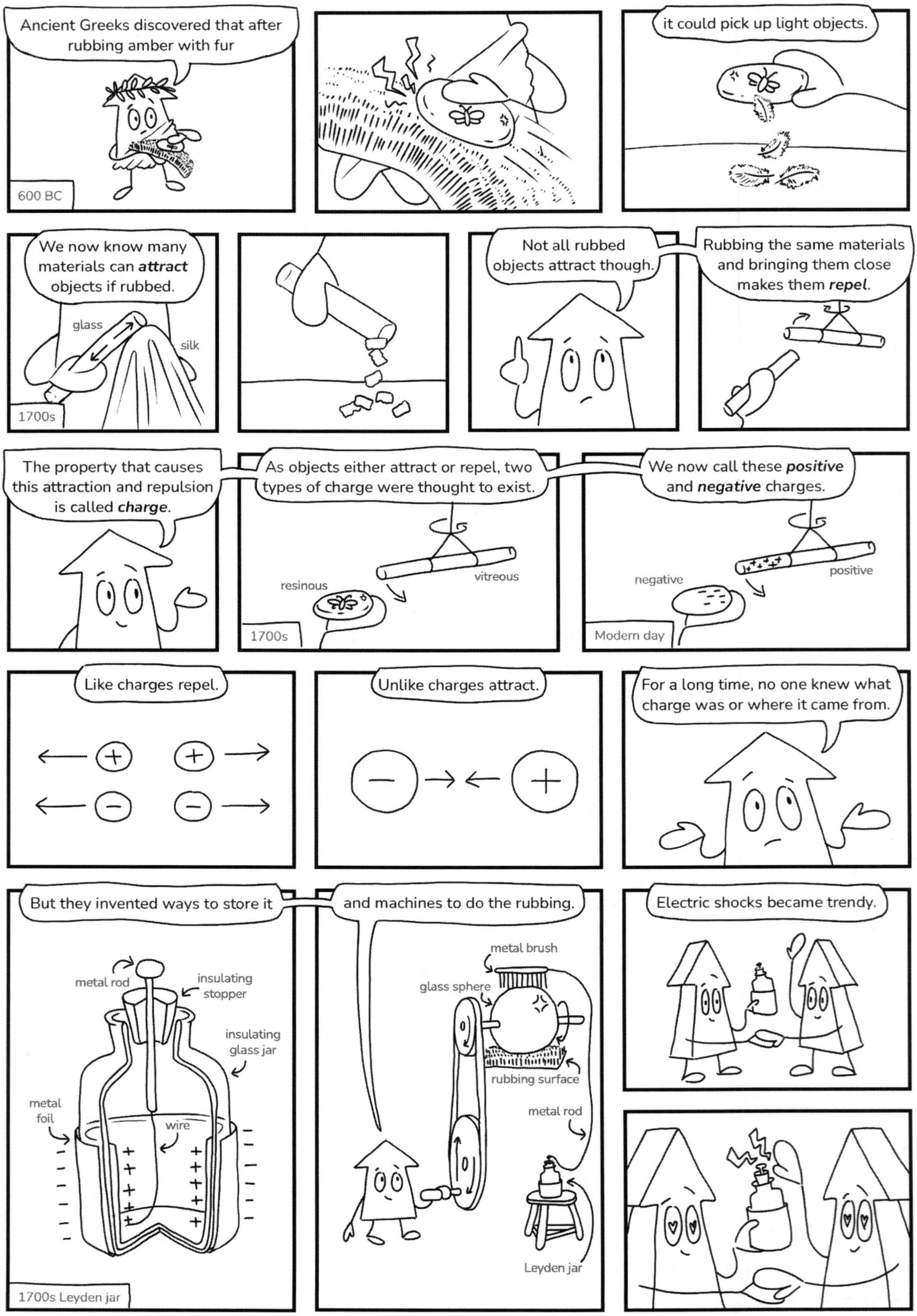

kite with metal wire
1752
Benjamin Franklin discovered that clouds became charged in storms
key
wet conducting hemp string
insulating silk string
and that lightning was a stream of charge
ZAP!
which could be safely conducted into the ground.
metal lightning rod
metal wire
After many experiments by many scientists, it was clear that charge is in everything,
but a lot of objects have equal numbers of positive and negative charges, making them *neutral*.
Also, charge is *conserved*.
It is not created or destroyed, it just moves around.
1897
Mysterious rays deflect in between charged plates
People only found out what charge was much later, when J. J. Thomson discovered a negative particle called the *electron*.
The electron is named after the Greek word for amber, elektron.
So when something is charged by rubbing it, electrons either get rubbed onto it
or off it.

The force between charges is called ***electrostatic force***.
The force between charges is not due to friction
– this is just one way of moving charges around!
Being close can be enough, as electrostatic force is a field force.
ZAP!
An ***electric field*** exists whenever there is a charge imbalance.
Electrostatic force and electric fields are everywhere.
They are responsible for chemical bonds.
Na+
Cl-
Our nerves send messages using electric fields.
nerve cell
Electric fields make our muscles work.
Electrostatic force is key to so many technologies.
N95
What would modern life be without it?
0%

11.2 Charging objects

1. Describe what charges do, by filling in the blanks.

added	attract	charges	conserved	electrons	electrons
electrons	field	flow	lightning	neutral	nucleus
opposite	protons	removed	repel	rubbing	two

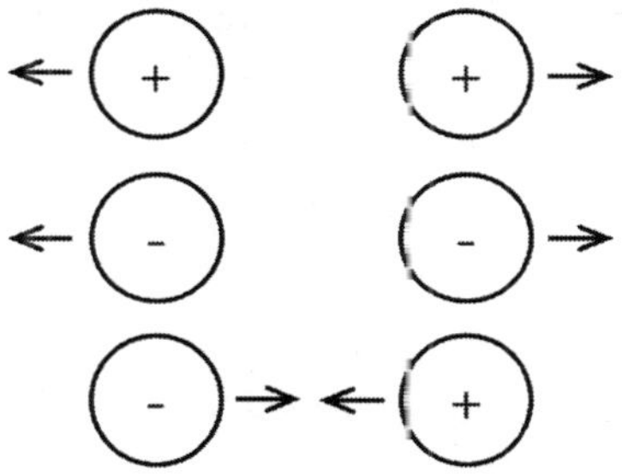

For a long time, people have known that ________ some materials together could leave materials with a property called charge, which results in either attraction or repulsion between them. The force between ________ is called *electrostatic force*. A charge can exert a force on other charges without touching them, due to the electric ________ around a charge.

To explain both attraction and repulsion, ________ types of charge are needed, called positive and negative. Like charges repel each other, so two positive objects will ________ and so will two negative objects. Unlike charges attract, so a positive and negative object will ________ each other.

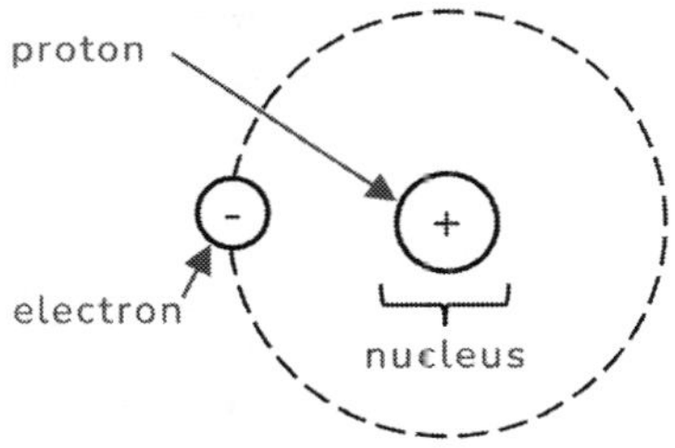

Where does the charge come from? Atoms in objects already have charges in them – positive ________ in the nucleus and negative ________ outside the nucleus. Protons and electrons have the same magnitude of charge, but ________ signs. A ________ atom has equal numbers of protons and electrons, so it has no charge overall.

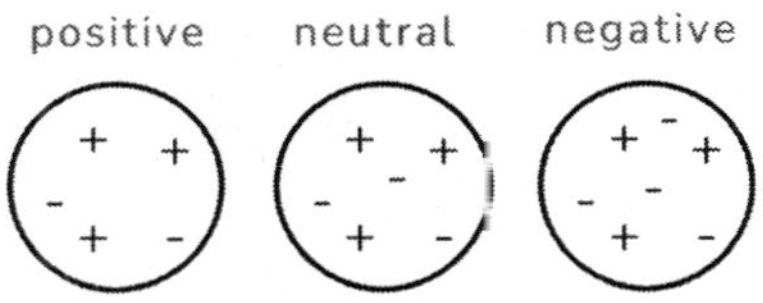

Charging objects only involves moving charges to different locations, not creating or destroying them, as charge is ________. Generally, only ________ are removed or added to atoms, as their mass is much lighter than protons and they are found on the outside of the ________. So if rubbing an object leaves the object with a positive charge, then the object has had electrons ________, leaving the object with more positive charge than negative charge. Similarly, a negatively charged object has had electrons ________, resulting in more negative charge than positive charge in the object.

If there are too many electrons on one object compared to another object nearby, the excess electrons can jump from one object to the other, resulting in sparks of ________. You may see a spark and feel an electric shock if you shuffle across carpet, then touch a metal doorknob – that is due to you building up charge as you shuffle, then ________ jumping between you and the doorknob!

The electricity running through the circuits of your house is actually a ________ of electrons.

2. Describe three methods of charging objects, by filling in the blanks.

attracted	conductors	Earth	electrons	excess	insulated	insulators	metal
negatively	no charge	positively	repelled	rubbed	safety	touching	transferred

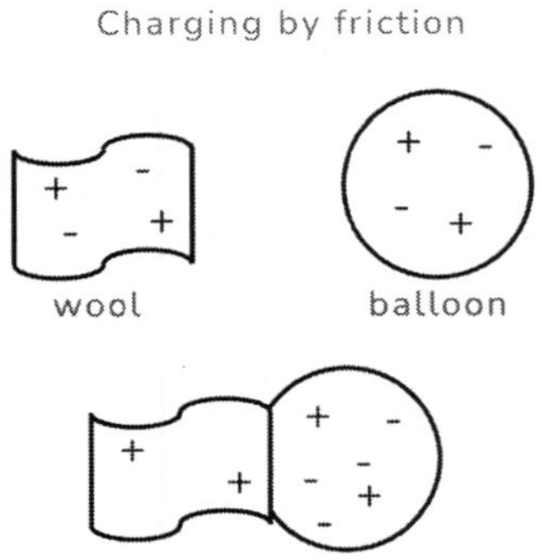

Charging by *friction* involves an object being ________ by a different material. This can either add or remove ________ from the object, depending on the nature of the material. Some objects have a greater tendency to gain electrons than others. For example, wool is more likely to lose electrons than rubber, so if you rub a rubber balloon with wool, the wool will become ________ charged, while the balloon becomes ________ charged. The more you rub, the more electrons are ________. However, since both wool and the rubber balloon are ________, charge is only transferred between the areas being rubbed, and does not flow throughout the whole object.

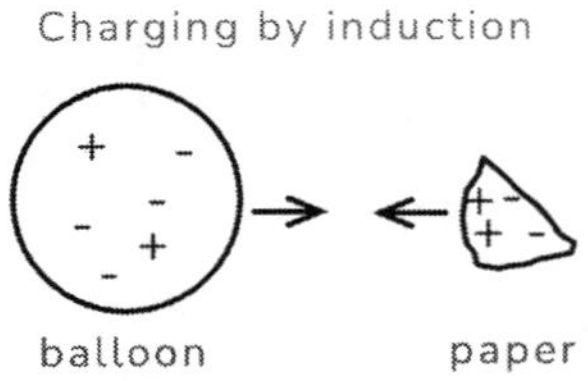

Charging by *induction* involves bringing a charged object close to a neutral one, for example a negatively charged balloon near some tiny pieces of paper with ________ overall. The electrons in the paper will be ________ by the balloon to the far side of the paper, leaving the near side of the paper positively charged. The near side of the paper is then ________ to the balloon.

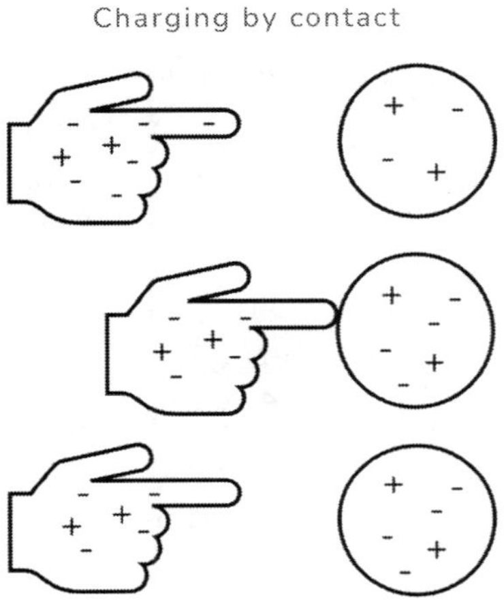

Charging by *contact* involves ________ a neutral object with a charged one. If one of the objects is a good electrical conductor, such as a ________, then charge can travel easily through the material and become evenly distributed throughout it. However, if the metal object is to retain its charge, then it needs to be ________ from its surroundings and supported by a material that does not allow charge to move through easily. Otherwise, if the metal is also in contact with other objects, especially ________, then it will not retain its charge, since it can easily flow to the other objects too.

A metal strip in contact with the ground can transfer excess charge through it to the ________. The process of removing ________ charge from an object by transferring it to another large object is called *earthing* or *grounding*. This process is used as a ________ feature in case of lightning strikes or electrical surges.

3. Four balloons, A, B, C and D, are charged with different materials. They hang on insulating threads. When brought close together, this is what is observed:

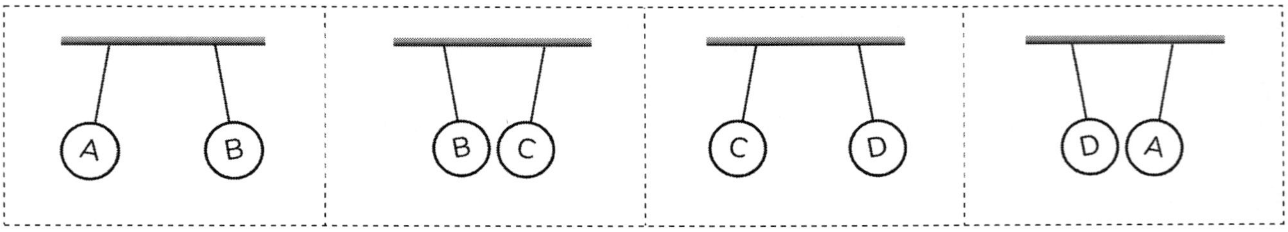

If A is positively charged, what charge do B, C and D have?

____

a) Explain what would happen if B and D were brought close together.

____

4. The diagram shows a model of the helium atom, with two protons, two neutrons and two electrons. Neutrons are neutral particles which are found in the nucleus of most atoms.

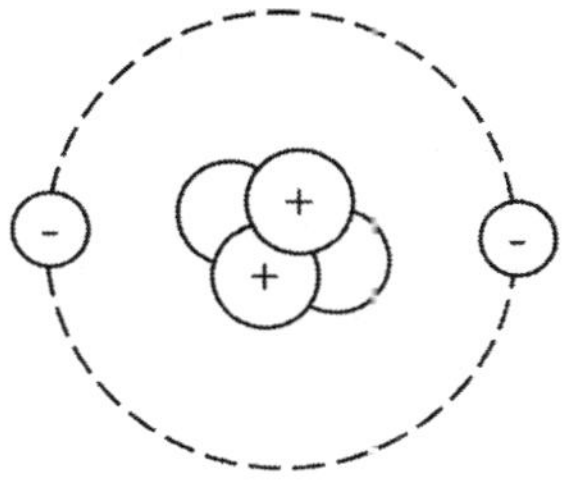

a) Explain what the overall charge on this atom is.

....

b) Explain what keeps electrons attracted to the nucleus.

....

....

c) If electrostatic force was the only force acting inside the nucleus, explain whether the nucleus would be stable and not break apart.

....

....

Look at the nucleus.

5. Above power lines, there is an extra conducting wire called an earth wire, which usually carries no current. The earth wire is attached directly to the metal support tower, while the power lines are only attached to the support tower via a stack of insulating discs. The earth wire's function is to protect the power lines from lightning. During a lightning strike, a stream of electrons is rapidly transferred from the sky to the ground. Explain how the earth wire and insulating discs protect power lines from a lightning strike.

....

....

....

....

....

....

What is the difference between conductors and insulators?

6. A PVC pipe is rubbed with a cloth. The PVC pipe is then brought close to an aluminium can, lying on its side on a table, so it is free to roll. The diagrams model what happens to the charges. Complete the caption for each diagram.

Charge is conserved, so the total number of charges stays the same.

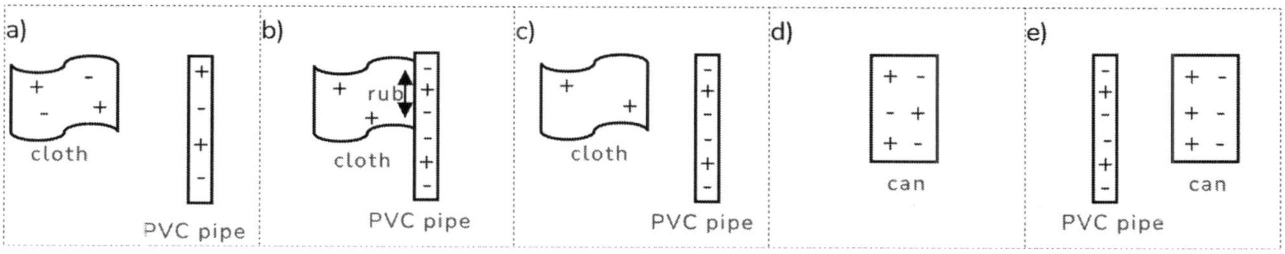

a) The cloth and PVC pipe initially.
Charge on cloth:

Charge on pipe:

Force between them:

b) They are brought together.
Type of charging:

Which way electrons moved:

c) They are separated.
Charge on cloth:

Charge on pipe:

Force between them:

d) Charge on can:

e) The PVC pipe is brought close to the can.
Type of charging:

Explain what happens to the can:

7. The *triboelectric series* ranks materials according to how likely they are to gain or lose electrons, compared to other materials. This ranking is often determined experimentally. Some materials in the series are shown in the table. Use it to answer the questions below.

hair	more likely to lose electrons
wool	↑
silk	↕
rubber	↓
plastic bag	more likely to gain electrons

a) Taking off a silk shirt by pulling it over your head can involve your hair sticking to the shirt, especially in dry weather. Then once your shirt is off, your hair might stand on end. Explain why.

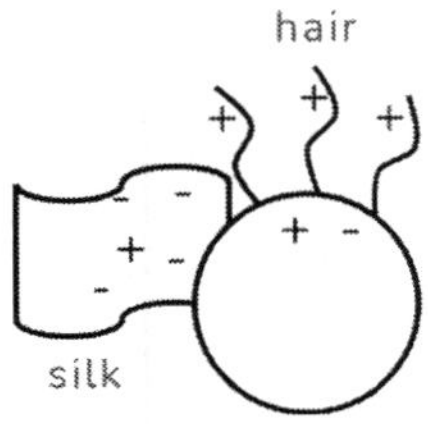

....

....

....

....

b) Complete the series of diagrams below, by adding charges on the objects.

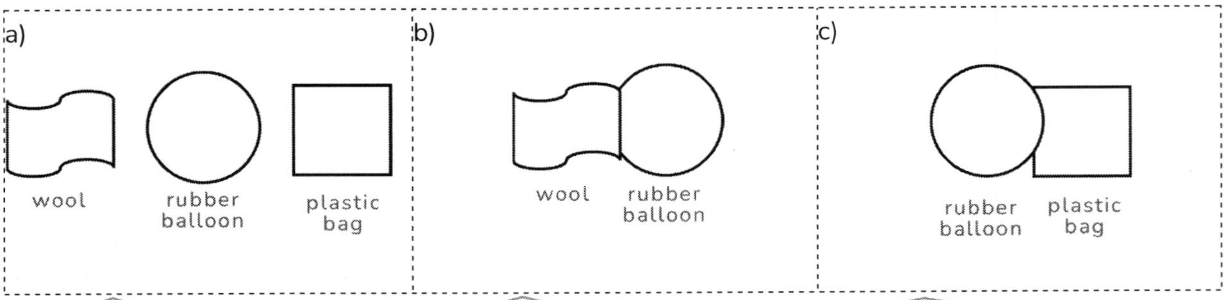

A piece of wool, a rubber balloon and a plastic bag have neutral charge overall initially.

The rubber balloon is rubbed with wool.

The balloon is then brought near a plastic bag. The plastic bag is attracted to the balloon.

8. An electroscope is a device that can be used to detect charge, made of two metal strips hanging in an insulated container. You can make a home-made version using aluminium foil rolled tightly into a ball, attached via a wire to two aluminium strips inside a jar. As aluminium is a metal, it is a conductor and electrons will move easily throughout it.

Complete the series of diagrams by adding charges to the diagrams. The first two diagrams are done for you.

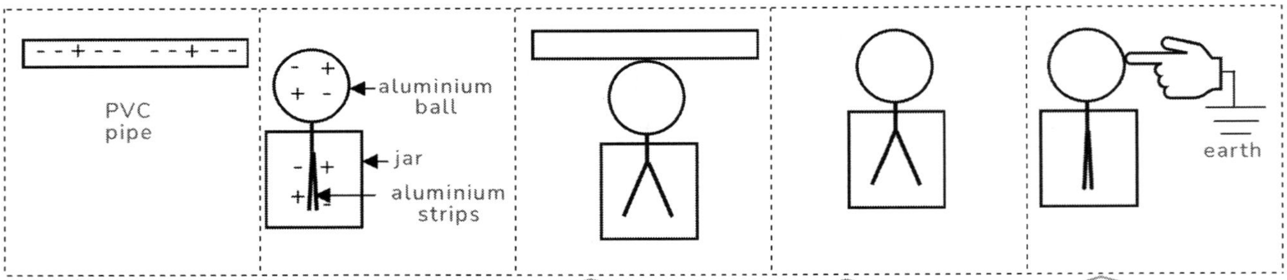

A PVC pipe is rubbed with a cloth, leaving it charged.

An electroscope is initially neutral.

The PVC pipe is brought in contact with the electroscope.

The PVC pipe is removed from the electroscope.

The electroscope is earthed, using a finger to touch it.

11.3 Measuring charge

1. Describe how charge and current is measured, by filling in the blanks.

charge	conserved	coulomb	electric field	proton	same
scalar	time	transferred	unit	whole	

Charge is a property of matter which causes objects to experience a force in an ______________. Charge is a ______________ quantity, but can have a positive or negative sign. The ______________ used to measure charge is the ***coulomb*** (C). This unit was named a long time before electrons were discovered, so it does not match a ______________ number of electrons. However, once electrons were discovered in 1897 the charge of one electron was found to be -1.602 x 10^{-19} C. This means the charge of one ______________ is +1.602 x 10^{-19} C.

Electricity experiments show that charge is ______________. That means that in a closed system, the total amount of charge stays the ______________. In other words, the total number of protons stays the same, and the total number of electrons too – they just get ______________ between objects.

Current is the flow of charge. In other words, it is the rate at which charge passes a point:

$$I = \frac{q}{t}$$

where: I = current (A)

q = amount of ______________ passing a point (C)

t = ______________ in which charge passes a point (s)

The unit for current is the ***ampere*** (A), which is one ______________ passing a point per second.

Example: How many electrons are there in 1 C?

Answer: Number of electrons $= \frac{1}{1.602 \times 10^{-19}} = 6.242 \times 10^{18}$ C

Example: When two conducting objects touch, electrons are able to distribute themselves evenly throughout the objects, so charging by contact can occur if either or both of the objects were initially charged. Two identical aluminium cans, standing on Styrofoam cups to insulate them from the table, are brought into contact. One aluminium can had a charge of +10.0 μC and the other was neutral, before touching.

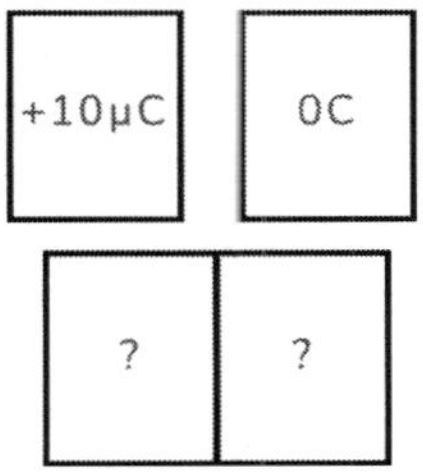

a) What is the charge on each can after contact?

b) How many electrons were transferred, and from which can?

Answer:

Since the objects have equal volume and surface area, halve the total charge.

a) Total charge $= +10.0$ μC

Charge on each can after contact $= +\frac{10.0}{2} = +5.00$ μC

b) Charge transferred from neutral to charged can $= +5.00 - 10.0 = -5.00$ μC

Number of electrons $= \frac{5.0 \times 10^{-6}}{1.602 \times 10^{-19}} = 3.12 \times 10^{13}$ electrons

2. Electrostatic spray painting is often used to paint cars. Droplets of spray paint are given a positive charge as they leave the paint gun nozzle. This means the paint drops repel each other, providing an even coating. A second advantage is that the paint drops are attracted to the neutral car body, minimising wastage.

   If one paint drop has a charge of $+1.5 \times 10^{-10}$ C, were electrons gained or lost, and how many?

3. Two identical metal balls, A and B, on the ends of insulated rods are touched together, then separated. Initially, ball A had a charge of +5.0 μC and B had a charge of -8.0 μC.

   a) Calculate the final charge on each sphere.

   b) How much charge was transferred, and from which ball?

   c) How many electrons were transferred?

4. The ampere (A) is an SI unit, but the coulomb (C) is not. What is a coulomb in SI units?

5. A toaster draws 4.0 A of current when turned on. How many electrons go past one point in the toaster's circuit per second?

11.4 Electrostatic force

1. Describe how the electrostatic force between two charges is determined, by filling in the blanks.

charged ball	directly proportional	charges	deflection	distance	air
electrostatic force	inversely proportional	material	vacuum	water	

In 1785, Charles de Coulomb worked out an experimental formula for the electrostatic force between two charges. He brought a charged ball close to another ________________ stuck on the end of a horizontal beam that could rotate. The electrostatic force could be measured using the ________________ of the beam. This force was ________________ to the product of the magnitudes of the charges and ________________ to the square of the distance between them:

$$F_E = \frac{1}{4\pi\varepsilon_0}\frac{q_1 q_2}{r^2}$$

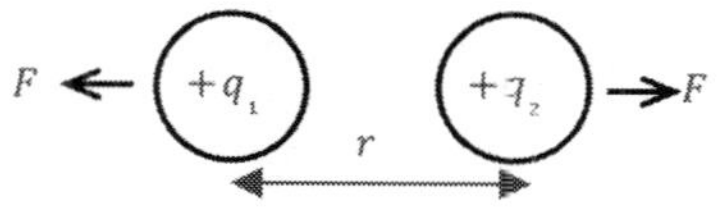

where: $F_E =$ ________________ between charge 1 and 2 (N)

q_1 and q_2 = charges 1 and 2 respectively (C)

$r =$ ________________ between charge 1 and 2 (m)

$\varepsilon_0 = 8.854 \times 10^{-12}\ \mathrm{A^2\ s^4\ kg^{-1}\ m^{-3}}$

= electric permittivity constant for a vacuum

ε is the Greek symbol epsilon. ε_0 is pronounced "epsilon nought".

Electric permittivity* ε** means how much a medium interferes with an electric field. A ***medium is the ________________ the electric field is going through. If the medium is a ________________, there are no charges in it, so it has the lowest possible electric permittivity, sometimes called the ***permittivity of free space* ε_0**. Any other medium will have atoms with ________________ in it, which will experience a force and therefore store some electrical potential energy.

To simplify calculations, we usually assume ________________ has the same permittivity as a vacuum. However, ________________ has a high electric permittivity compared to a vacuum, so on humid days, static electricity experiments don't work as well, since the force between charged objects is reduced.

Example: A hydrogen atom has one electron and one proton. The distance between the electron and proton is 5.3 x 10^{-11} m. What is the force between them?

Electrons in atoms have momentum which makes them orbit the nucleus, rather than collide with them.

Answer:

$$F_E = \frac{1}{4\pi\varepsilon_0}\frac{q_1 q_2}{r^2}$$

$$= \frac{1}{4\pi \times 8.854 \times 10^{-12}} \times \frac{+1.602 \times 10^{-19} \times -1.602 \times 10^{-19}}{(5.3 \times 10^{-11})^2}$$

$= 9.1 \times 10^{-18}$ N attractive

For any calculations, use: charge of an electron $= -1.602 \times 10^{-19}$ C

2. Look at the equation for electrostatic force.

Let the original force be F_E and the new force be F_E'. Write an equation for both then compare them.

a) What happens to the force if you double one of the charges?

____

____

b) What happens to the force if you double the distance?

____

____

c) Sketch a graph of the magnitude of electrostatic force as the distance between two charges is varied.

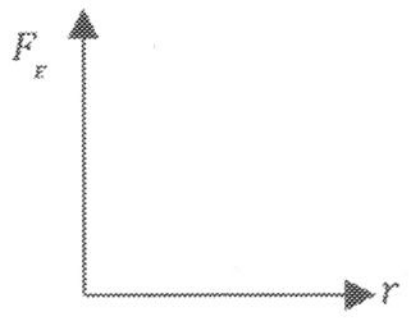

d) What choice of axes should be made in order to demonstrate that electrostatic force is inversely proportional to the square of the distance between the charges? Sketch this graph.

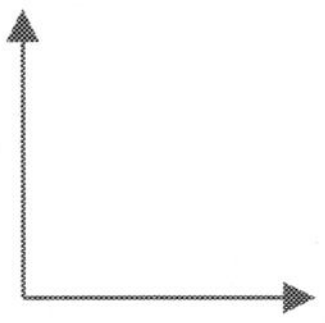

3. Two charges of +3.0 μC and +6.0 μC respectively are separated by 1.0 cm.

a) Determine the force between the charges.

---

b) Explain whether the +6.0 μC charge will exert a larger force on the +3.0 μC than the +3.0 μC exerts on it.

---

4. Three charges, each of equal magnitude, are placed as shown. Charge A has charge $-q$ and is a distance r to the north of Charge C, which has magnitude $-q$. Charge B has charge $+q$ and is a distance r to the east of Charge C. Which diagram best shows the resultant force on Charge C?

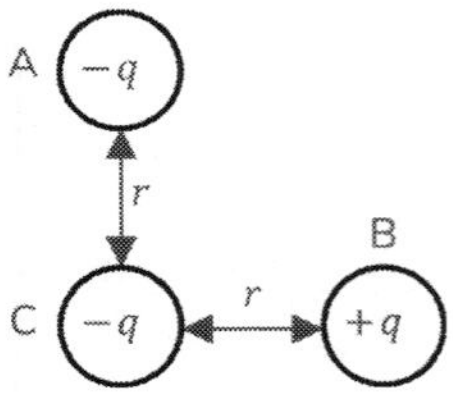

A.  B.  C.  D. 

5. Two identical balloons, A and B, each of mass 3.50 g, hang vertically on strings. They are both rubbed with a cloth so that they are equally charged. After rubbing, the two balloons are deflected so that the strings they are hanging from are 10.0° to the vertical and the centres of mass of the balloons are 30.0 cm apart.

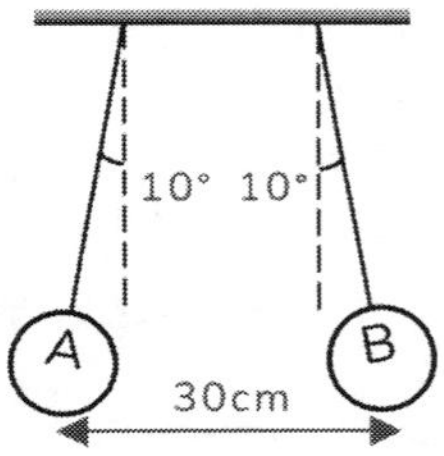

a) Draw a vector diagram of the forces acting on balloon A.

Vector diagram

b) What is the charge on each balloon?

---

6. SI units include the ampere (A), seconds (s), kilograms (kg) and metres (m).

1 A = 1 C s^{-1}.

Show that the unit for the electric permittivity constant is $A^2\ s^4\ kg^{-1}\ m^{-3}$.

Recall that units for quantities can be worked out from formulae that they appear in, and units on each side of an equation must be the same.

---

7. A water molecule is made of two hydrogen atoms chemically bonded to an oxygen atom.

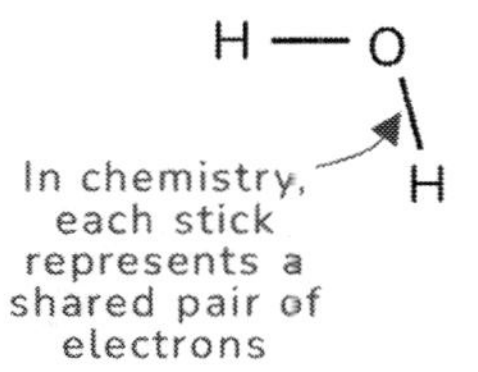

A hydrogen nucleus has one proton, while an oxygen nucleus has 8 protons.

The typical distance between a hydrogen and oxygen nucleus in a water molecule is 9.57×10^{-11} m.

The chemical bond between each hydrogen and oxygen atom consists of a pair of electrons they share. Assume that both these electrons are located halfway between the hydrogen and oxygen nucleus.

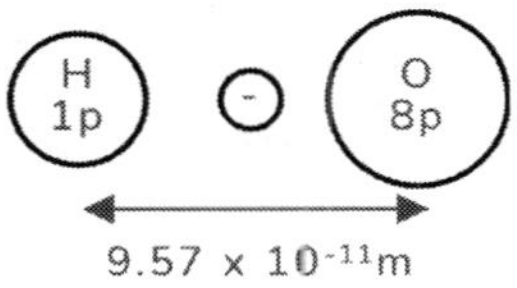

This calculation oversimplifies the bonding in water, in order to make it easier to understand. It ignores many factors, such as the other electrons in the atoms, and they aren't at rest. The electrons are also closer to the oxygen nucleus than the hydrogen nucleus.

a) Determine the force between the hydrogen nucleus and oxygen nucleus.

b) Determine the force between the hydrogen nucleus and one of the shared electrons, if this electron was 4.785×10^{-11} m away from the hydrogen nucleus.

c) Determine the force between the oxygen nucleus and one of the shared electrons, if this electron was 4.785×10^{-11} m away from the oxygen nucleus.

d) Compare the force between the hydrogen and oxygen nuclei and the forces exerted by the shared pair of electrons on them, to explain why the shared pair of electrons allows hydrogen and oxygen to chemically bond.

e) Explain why a water molecule is more positively charged near the hydrogen nuclei and more negatively charged near the oxygen nuclei.

11.5 Visualising electric fields

1. Describe how the direction of an electric field is determined, by filling in the blanks.

accelerate	balanced	charge	conductor
non-contact	positive	resultant	vector

A field is a region of space where a body experiences a ______________ force. An electric field is a field in which a ______________ experiences a force. Electric field is a ______________ quantity, since it has both magnitude and direction.

The direction of an electric field is defined by the direction in which a ______________ charge would experience a force in this field. In other words, each electric field line tells us the direction a positive charge would ______________, if it were placed on that field line and were free to move.

Electric field lines never cross – if there is more than one source of electrostatic force on a charge, then the direction of the electric field line shows the direction of the ______________ force on the charge.

Electric field lines are perpendicular to the surface of a charged ______________, since the charge is free to move throughout the object. This means that any components of the electric field that are not perpendicular to the surface will be ______________, making the resultant force at any point on the surface perpendicular to it.

Electric field lines around a positive charge

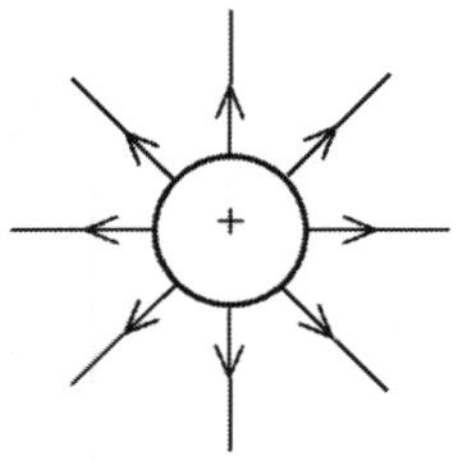

Electric field lines around a negative charge

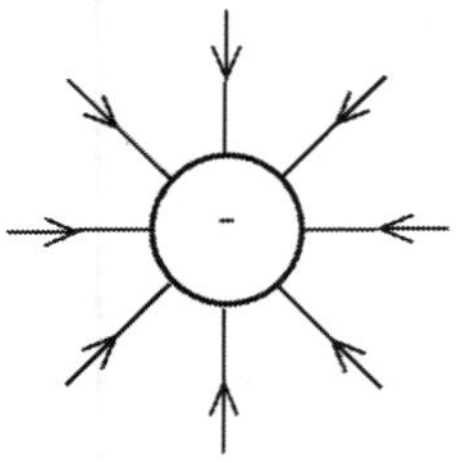

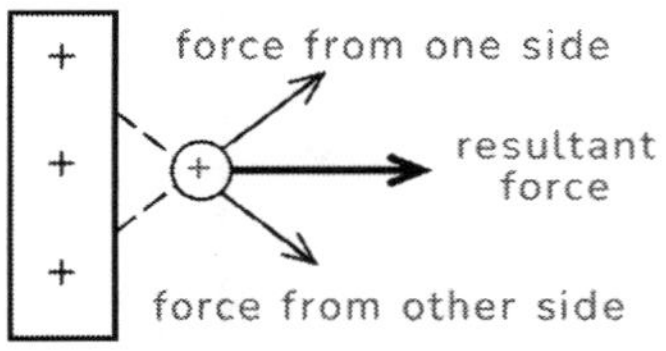

2. To map the electric field lines around a charged object, imagine putting a positive test charge at several positions around it and identify the direction of electrostatic force on the test charge at each position. The electric field lines must go in the same direction as these forces.

On each diagram below, draw a vector showing the direction of force from the charge in the centre, acting on each positive test charge, shown as a small +.

a)

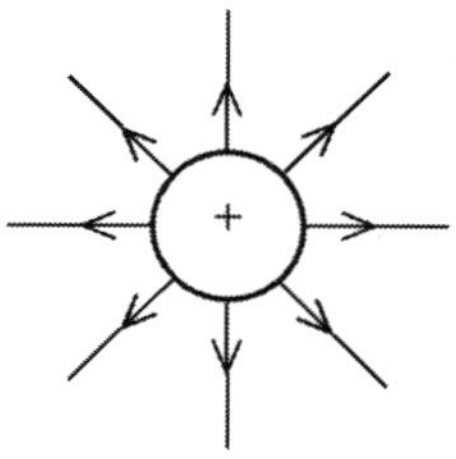

b)

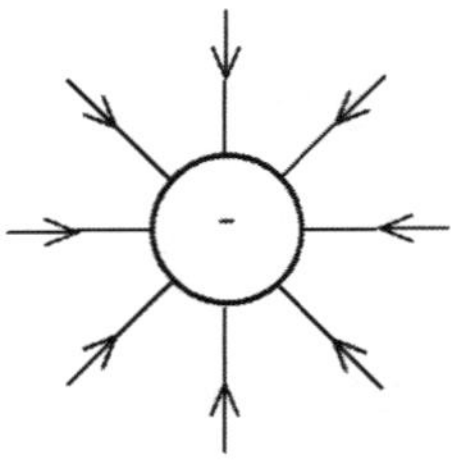

3. The diagrams show the electric field around each of two charges, A and B. Compare these charges.

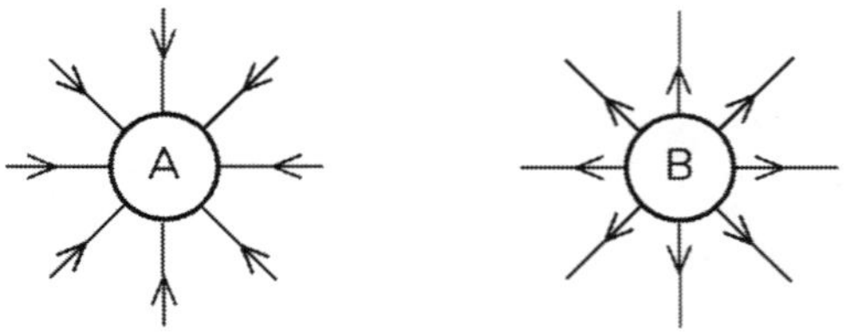

____

____

____

4. A and B are charges. State whether each is positive or negative.

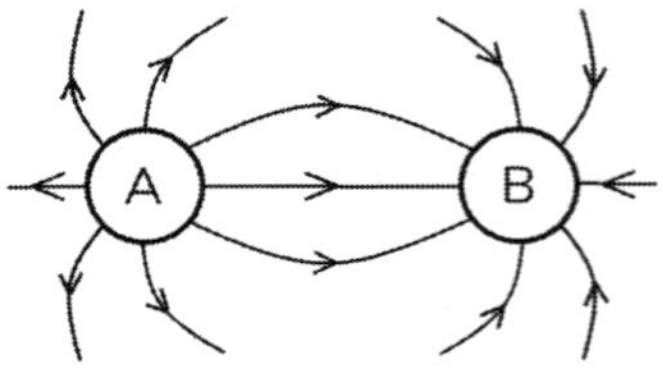

____

____

5. Describe electric field strength, by filling in the blanks.

apart	closer	inversely proportionally	magnitude
same	stronger	weaker	

Electric field around a point charge

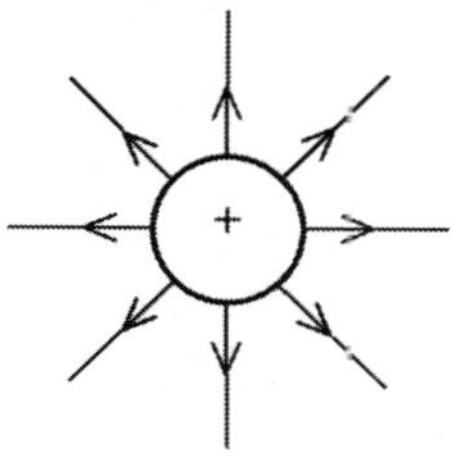

The ***electric field strength***, or ____________ of an electric field, tells us how much force per coulomb of charge an object would experience in the electric field. The distance between electric field lines represents electric field strength. Field lines closer together represents a ____________ electric field. So the field lines near a point charge (a single charge at a point) are ____________ together, and get further ____________ away from the point charge, representing that the electric field gets ____________. This is because the force F_E experienced by a charge q_1 due to another charge q_2 decreases ____________ with the square of the distance between the charges r^2, according to:

$$F_E = \frac{1}{4\pi\varepsilon_0}\frac{q_1 q_2}{r^2}$$

Uniform electric field

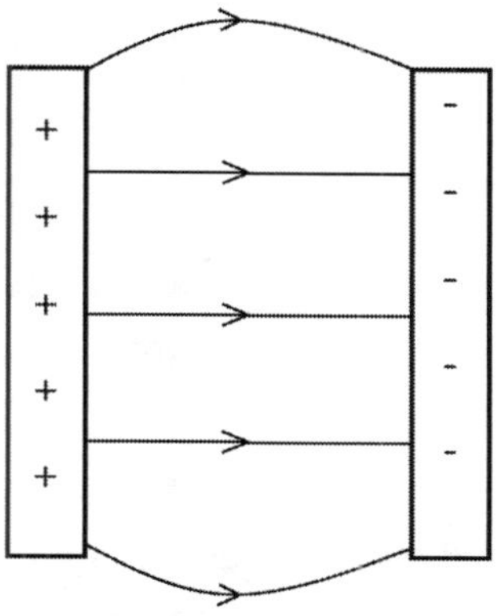

In a ***uniform electric field***, the field lines are the same distance apart, indicating that the electric field strength is the ____________ throughout.

6. A and B have the same magnitude of charge.

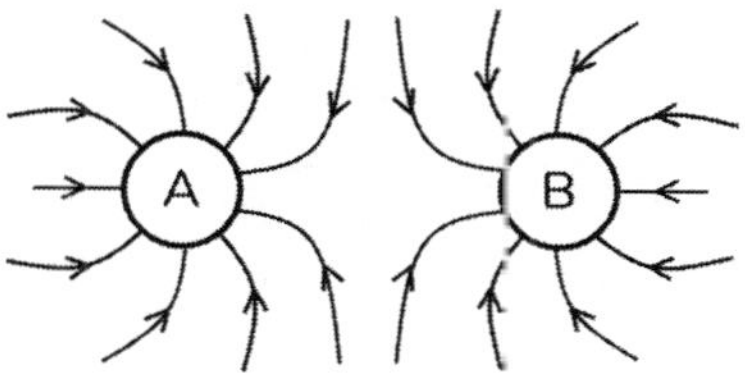

a) State whether each of charges A and B is positive or negative.

____

b) Explain what the electric field strength would be exactly halfway between A and B.

____

____

7. In each case below, draw the electric field lines between and around the charges. The charges have the same magnitude.

a)

b)

8. A negative charge is placed between two parallel charged plates.

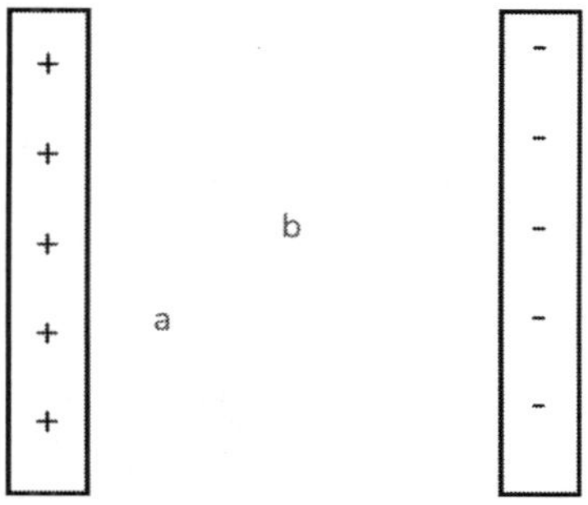

a) State the direction of force on the negative charge at position a.

____

b) Compare the strength of the electrostatic force on the charge at positions a and b.

____

____

____

9. If you put a conductor inside an external electric field, the inside of the conductor will have zero electric field. So a container made of metal will stop any electric field outside it from penetrating the container. This is why a mobile phone will not receive signal inside a closed metal box. This type of container is often known as a Faraday cage.

The diagram shows a neutral metal box, placed between two parallel charged plates. Explain why there is no electric field inside the box.

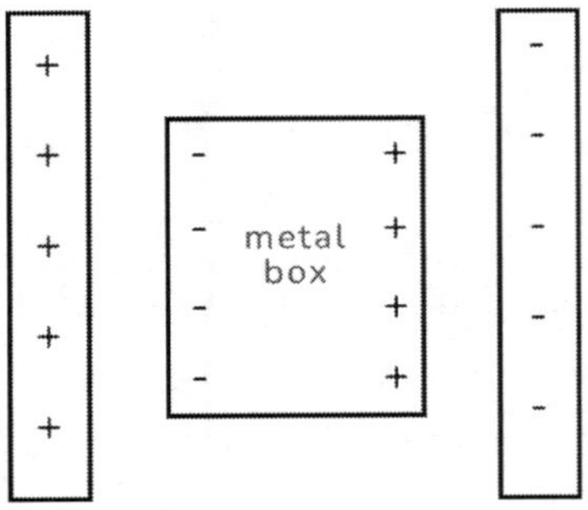

Key terms to use in your answer:
external electric field
conductor
induced electric field
resultant

10. An electric dipole is a pair of opposite charges that are separated by a very short distance. Dipoles are often seen in molecules. Dipoles will rotate to align with an external electric field.

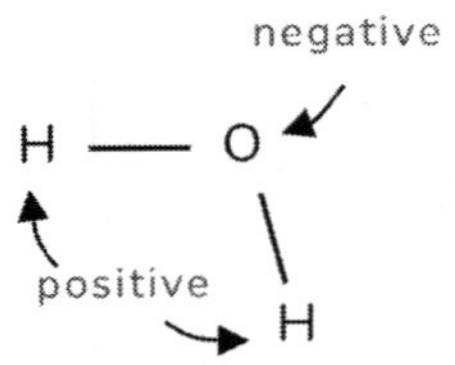

For example, the chemical bond between oxygen and hydrogen in water can be thought of as a dipole, with the oxygen end being negative and the hydrogen end being positive.

When a charged object, for example a balloon rubbed with wool so the balloon is positive, is brought close to a stream of water from a tap, the stream of water bends towards it. Explain why.

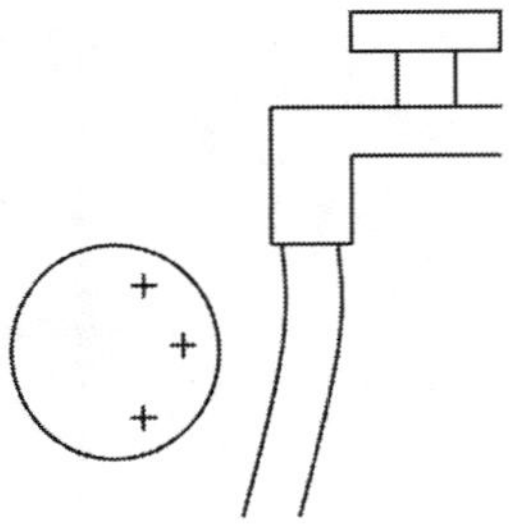

11.6 Electric field strength

1. Describe how to calculate electric field strength, by filling in the blanks.

acceleration	another	charge	electric field strength	electrostatic force
gravitational field strength	magnitude	mass	weight	

Electrostatic force between two charges increases as the ____________ of both charges increases. Likewise, gravitational force, or ____________, between two masses increases with mass. It is therefore useful to have some way to compare field strength due to an object in a way that does not depend on the properties, such as charge or mass, of ____________ object you put in that field.

For example, ***gravitational field strength*** is the gravitational force $\overrightarrow{F_G}$ per unit of ____________ m. Since gravitational force $\overrightarrow{F_G}$, or weight, is given by:

$$\overrightarrow{F_G} = m\vec{g}$$

we can rearrange this to see that numerically, ____________ due to gravity $\vec{g}$ is the same thing as ____________:

$$\vec{g} = \frac{\overrightarrow{F_G}}{m}$$

Similarly, ***electric field strength*** is the electrostatic force $\overrightarrow{F_E}$ acting on an object per coulomb of ____________ q:

$$\vec{E} = \frac{\overrightarrow{F_E}}{q}$$

where: $\vec{E} =$ ____________ at a point (N C^{-1})

$\overrightarrow{F_E}$ = electrostatic force exerted on a charge at that point (N)

q = charge (C)

This equation is often rearranged to make the ____________ the subject:

$$\overrightarrow{F_E} = q\vec{E}$$

Example: Before flat screen TVs, there were cathode ray TVs. These had a big tube at the back, which fired a beam of electrons at the screen, resulting in light being emitted wherever the screen was hit. Parallel electric plates on each side of the tube could be used to direct the electrons to the right position on the screen for each picture. If the plates on the left and right of the tube produced an electric field of 1.75 N C^{-1} to the right, as shown in the diagram, what is:

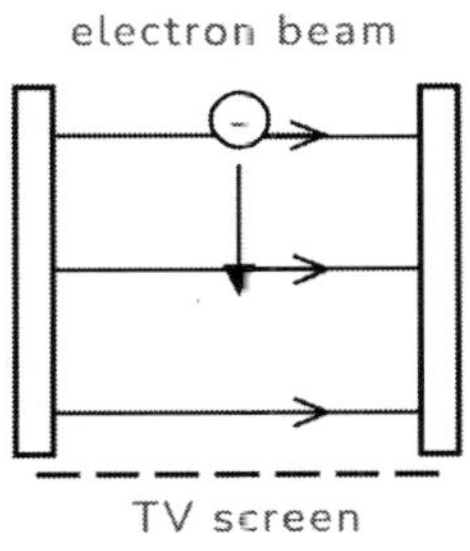

a) the force on an electron, travelling horizontally towards the screen; and
b) the acceleration produced?

Answer: Let positive be to the right.

Check the data sheet for useful values.

a) $\overrightarrow{F_E} = q\vec{E} = -1.602 \times 10^{-19} \times 1.75 = -2.80 \times 10^{-19}$ N to the left

Recall the electric field direction is the direction of force on a positive charge.

b) $m\vec{a} = q\vec{E}$

$9.109 \times 10^{-31} \times \vec{a} = -1.602 \times 10^{-19} \times 1.75$

$\vec{a} = -3.08 \times 10^{11}$ m s^{-2} (to the left)

For any calculations, use: $\varepsilon_0 = 8.854 \times 10^{-12}$ A^2 s^4 kg^{-1} m^{-3}

charge of an electron = -1.602×10^{-19} C

mass of an electron= 9.109×10^{-31} kg

2. Electric fields can be used to help identify what unknown particles might be. For example, after radioactive substances were discovered in the 1890s, people were very interested in doing experiments to find out what was being given off by them.

   Imagine in one such experiment, an electric field of 120 N C^{-1} to the right was applied to a particle. The particle was observed to accelerate to the right at 5.79 x 10^{9} m s^{-2}. The particle's mass was 6.645 x 10^{-27} kg. What was its charge?

Include whether the charge is positive or negative.

Vector diagram

3. Pith balls, made of a material like Styrofoam, are often used in electrostatic experiments, as they hold charge easily. When a pith ball of mass 1.2 x 10^{-4} kg and charge -1.5 μC is placed on a table, a charged plastic rod held above it is just able to pick it up.

   a) Draw a vector diagram showing the forces acting on the pith ball when it has just left the table.

   b) What is the minimum value and direction of the electric field due to the charged rod?

Which way is the force on a negative charge in an electric field?

4. When a pith ball of mass 4.8 x 10^{-5} kg and charge -3.0 μC is hung from a light string in a uniform electric field, the string deflects to the left, at an angle of 30° to the vertical.

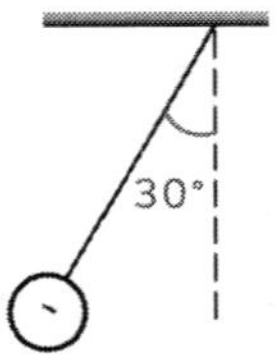

   a) Draw a vector diagram showing the balanced forces acting on the pith ball.

   b) What is the magnitude and direction of the electric field?

Vector diagram

For any calculations, use: $\varepsilon_0 = 8.854 \times 10^{-12}\ \mathrm{A^2\ s^4\ kg^{-1}\ m^{-3}}$

charge of an electron $= -1.602 \times 10^{-19}$ C

mass of an electron $= 9.109 \times 10^{-31}$ kg

mass of a proton $= 1.673 \times 10^{-27}$ kg

5. Explain why the electric field strength due to a point charge q_1 at a distance r from a second charge q_2 is given by:

$$\vec{E} = \frac{1}{4\pi\varepsilon_0}\frac{q_1}{r^2}$$

This is an example of an inverse square relationship, as the electric field is inversely proportional to the square of the distance.

6. Compare the effect of tripling the magnitude of a charge to tripling the distance away from the charge on the electric field caused by the charge.

Tripling the original charge q gives $3q$.

Let $\vec{E}$ be the original electric field, $\vec{E}_1$ be the electric field with triple the charge and $\vec{E}_2$ be the electric field with triple the distance.

$\vec{E}$ =

$\vec{E}_1$ =

$\vec{E}_2$ =

7. Hydrogen gas, or H_2, is made of two hydrogen atoms, each consisting of a proton in its nucleus. The hydrogen nuclei are 7.4 x 10^{-11} m apart. The two hydrogen atoms share a pair of electrons.

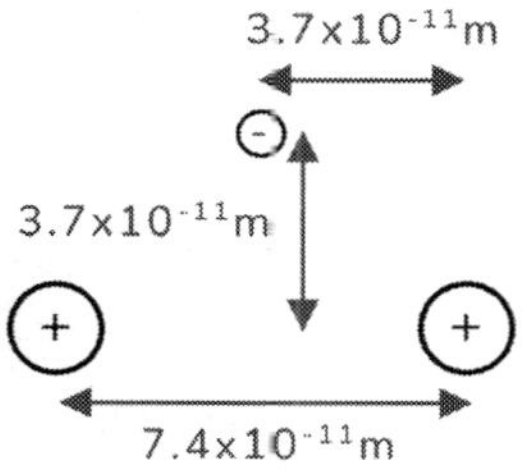

Let's model what would happen to one of these electrons due to the electric field from the protons. Assume the electron starts on the midline between the protons at rest, but 3.7 x 10^{-11} m from the midpoint, as shown in the diagram.

a) Draw a vector diagram showing the electric field due to each nucleus and the resultant of these electric fields, at the position of the electron.

Vector diagram

b) Determine the electric field at this position.

Use geometry to work out an angle.

c) Explain qualitatively the motion of this electron over time, due to the two hydrogen nuclei, if it is free to move.

11.7 Electrical potential energy

1. Describe electrical potential energy, by filling in the blanks.

change	chemical	circuit	displacement	electric field	force
gains	gravitational	gravitational field	gravitational potential	kinetic	losing
negative	opposite	same	stored	transformed	work done

Mass in a gravitational field

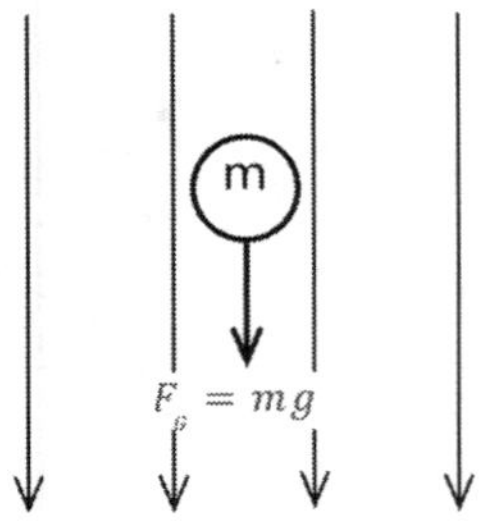

Recall that work is done whenever a force $\vec{F}$ causes a ____________ $\vec{s}$:

$$W = \vec{F}.\vec{s}$$

Energy is required to do work, so work is also the amount of energy ____________ from one type to another. For example, you do work to lift an object off the ground, as you need to apply a force at least as big as ____________ force to displace it upwards. In doing so, you transform kinetic energy into ____________ energy. The work you do can also be calculated from the ____________ in one form of energy, such as change in potential energy:

$$W = \Delta U$$

If you let the object go, it loses gravitational potential energy and gains ____________ energy. This time, the ____________ does the work, so the work you do is negative.

Charge in an electric field

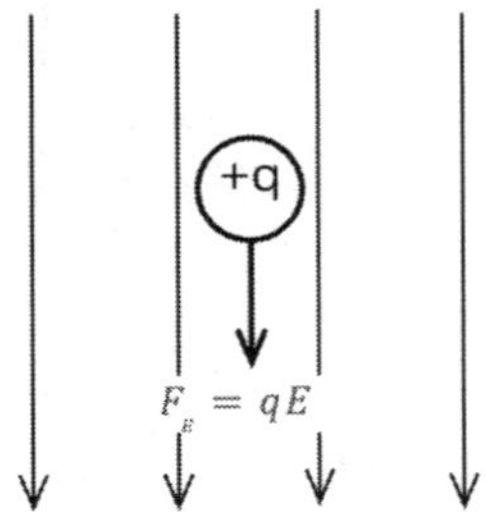

Similarly, an electric field exerts a force on a positive charge in the ____________ direction as the field. The charge gains kinetic energy by ____________ electrical potential energy, and the ____________ does the work. Since the electrostatic ____________ $\overrightarrow{F_E}$ on a charge q in an electric field $\vec{E}$ is given by:

$$\overrightarrow{F_E} = q\vec{E}$$

the ____________ on a charge in an electric field is given by:

$$W = \vec{F}.\vec{s} = qEd$$

where d is the displacement of the charge from its starting position.

On the other hand, to make a positive charge move in the ____________ direction to the electric field, something other than the electric field needs to do the work and the charge ____________ electrical potential energy. ***Electrical potential energy*** is the energy ____________ due to the position of charges relative to each other.

For a ____________ charge, work is done by the electric field to move it in the opposite direction to the field, whilst work is done by something else to move it in the same direction to the field.

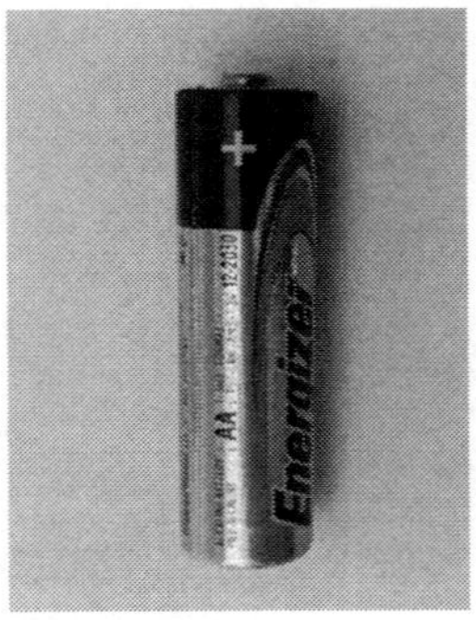

The terminals of a battery provide a source of electrical potential energy, due to a ____________ reaction inside the battery. The electrical potential energy is converted into kinetic energy if the battery is attached to an electric ____________ where charges are able to move, producing an electric current.

For any calculations, use: charge of an electron = -1.602×10^{-19} C

mass of an electron = 9.109×10^{-31} kg

mass of a proton = 1.673×10^{-27} kg

2. Proton beam therapy is a recent technology being developed to treat some forms of cancer. Assume that in one such beam, a proton is accelerated downwards by a uniform electric field of 1.0×10^{8} N C^{-1}. The proton starts from rest and travels 2.0 m before hitting the tumour. Ignore gravity.

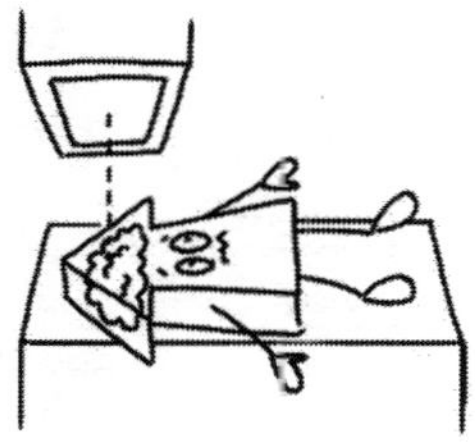

a) In which direction must the electric field be?

b) What is the electrostatic force on the proton?

c) What is the work done on the proton?

d) With what final velocity will the proton reach the tumour?

Note this is a simplified analysis. At high speeds, relativity needs to be considered, which we ignore here.

3. Electron microscopes use a beam of electrons to examine structures that are smaller than what a light microscrope can show, for example mitochondria in cells or defects in metal alloys.

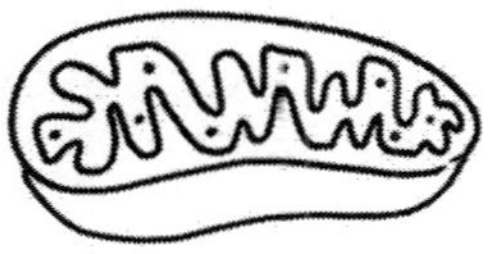

Assume that in an electron microscope, an electron is accelerated from rest by a uniform electric field over a distance of 15 cm. The electron needs to reach the sample being examined at a speed of 1.2×10^{8} m s^{-1}.

Determine the electric field required.

Again this is a simplified analysis, ignoring relativity at high speeds.

11.8 Electrical potential and potential difference

Electrical potential and equipotential lines

1. Describe electrical potential or voltage, by filling in the blanks.

accelerate	charge	does not	equipotential	ground	electrostatic force
highest	independent	inertia	kinetic energy	lowest	electrical potential
negative	negative	perpendicular	positive	positive	gravitational potential
same	sign	work			

Recall electric field strength is a measure of ________________ per unit charge, and it is a useful quantity to define because it is ________________ of the charge put into the field. Similarly, it is useful to define ***electrical potential***, or ***voltage*** at a point, as the electrical potential energy per unit charge, measured in volts (V):

$$V = \frac{U}{q}$$

where: V = electrical potential or voltage at a point (V)

U = ________________ energy (J)

q = ________________ (C)

Electrical potential is a scalar quantity, but it can have a positive or negative ________________, depending on whether voltage drops or increases.

Just as the ________________ is often the reference point for zero gravitational potential energy, the negative or earthed terminal that positive charges accelerate towards is taken as zero electrical potential.

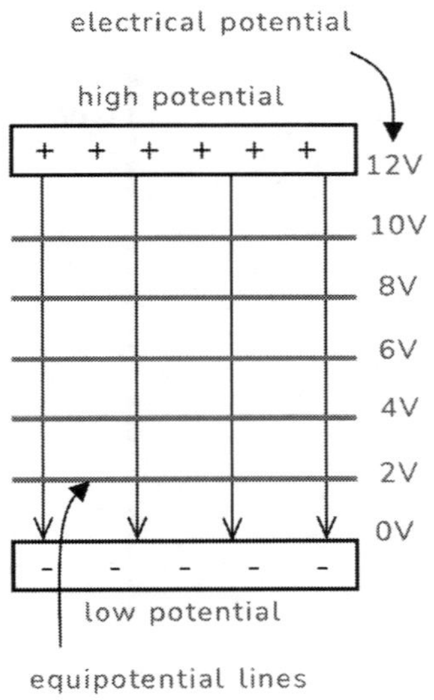

So between a positive and negative plate, the positive plate would be at the ________________ potential and the negative plate would be at the ________________ potential. In other words, a positive charge would have the most electrical potential energy at the positive plate and the least at the negative plate, because it would be lost as ________________.

Conversely, a negative charge would have the least electrical potential energy at the ________________ plate and the most at the ________________ plate. A negative charge would ________________ from low electrical potential to high electrical potential. The value and sign of electrical potential at each point ________________ change because the charge is negative. This is because for a negative charge, in the formula:

$$V = \frac{U}{q}$$

both U and q would be ________________, giving a ________________ V.

Equipotential lines are imaginary lines connecting points at the ________________ electrical potential. As an analogy, on a flight of steps, all points on each step are at the same ________________. No work needs to be done to move at constant velocity horizontally along a step, due to ________________. However, ________________ must be done to go up one step. Similarly, in an electric field, no work is done when a charge moves along an ________________ line. Therefore equipotential lines are always ________________ to electric field lines.

2. The diagram shows a uniform electric field, with three points A, B and C.

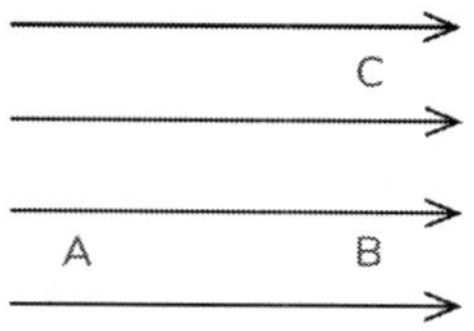

a) Explain whether the electrical potential energy of an electron increases in moving from B to A.

....

....

b) Explain whether the force on an electron at A is greater than at B.

....

....

Key words:
force
accelerate
kinetic energy
potential energy
electric field

Include relevant equations in explanations.

c) Explain whether A is at a higher potential than B? Does this change depending on the sign of the charge in the field?

....

....

d) Explain whether B is at a higher potential than C.

....

....

3. The equipotential lines and field lines around a positive charge are shown in the diagram below, as an example. Draw equipotential lines in the other two scenarios. The field lines are provided for you.

Example:

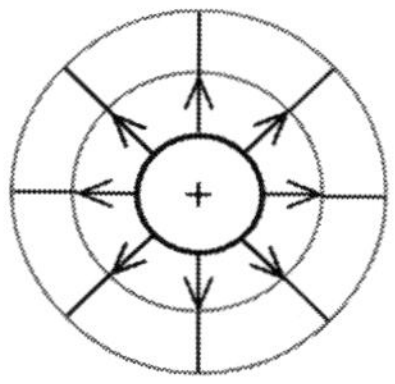

a)

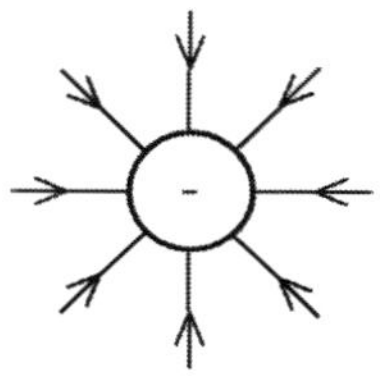

b)

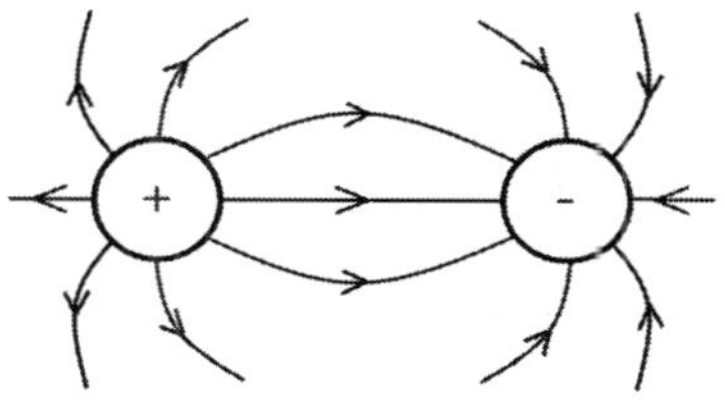

Potential difference

4. Describe potential difference, by filling in the blanks.

charge	magnitude	negative	potential difference	potential energy	work done

The change in electrical potential between two points is also known as ***electrical potential difference***, ***potential difference***, or ***voltage***.

Potential difference is related to work done on a charge by:

$$\Delta V = \frac{\Delta U}{q} = -\frac{W}{q}$$

where: $\Delta V =$ .... between two points (V)

$\Delta U =$ change in electrical .... (J)

$W =$ .... on the charge (J)

$q =$ .... (C)

The term "voltage" can mean:
- electrical potential (at one point);
- potential difference (between two points); or
- voltage drop.

The symbol "V" can be used to mean any of the above. All are measured with the unit "volt". Confusing? In reality, you can usually work it out from the context.

This formula is often rearranged to give:

$$W = -q\Delta V$$

If the electrical potential decreases between two points, then the change in electrical potential ΔV is .... The .... is also known as ***voltage drop*** V, so the formula is written as:

$$W = qV$$

Example: The diagram shows the electric field and equipotential lines between two charged parallel plates.

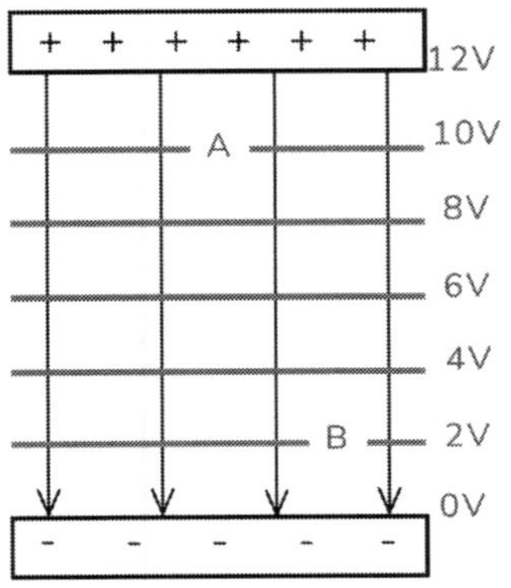

a) What is the potential difference between A and B?

b) What is the work done on a charge of +0.50 C in moving from A to B, and what does the work?

c) If the charge has a mass of 1.0 g, and starts at rest from A, what speed will it have when it reaches B if the electric field provides the only force on the charge?

Answer:

a) $\Delta V = V_f - V_i = 2 - 10 = -8$ V

b) $W = qV = 0.5 \times 8 = 4$ J done by the electric field

c) $W = \Delta K = \frac{1}{2}mv^2$

$\frac{1}{2} \times 0.001 \times v^2 = 4$

$v = 89$ m s^{-1}

5. For each pair of points, fill in the table to identify:

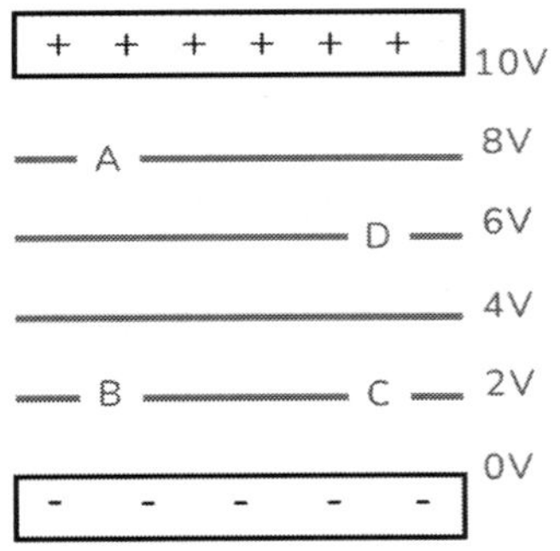

   a) the potential difference between them;

   b) the work done moving a charge of +0.25 C between them; and

   c) what does the work – the field, an external force or neither?

Points	A to B	B to C	A to C	A to D	B to D
Potential difference					
Work done on +0.25 C					
What does the work?					

6. Ion beams have many uses in scientific research as well as industry. For example, a beam of ions can be used to probe the surface of materials, or implant elements into materials. A hypothetical ion beam accelerates an ion of mass 2.2 x 10^{-25} kg and charge 1.6 x 10^{-19} C, by applying a voltage of 20 kV. What is the final speed of the ion, if it started from rest?

11.9 Potential difference and electric field strength

1. Relate potential difference and electric field strength, by filling in the blanks.

charge	electric field strength	electrostatic force	metre	parallel
potential difference	ruler	uniform	voltmeter	work done

Recall that ***electric field strength*** E is defined as the ________________ F_E acting on an object per coulomb of ________________ q:

$$E = \frac{F_E}{q}$$

Recall that ________________ W moving a charge through a distance d in an electric field E is:

$$W = \vec{F}.\vec{s} = qEd \qquad ...(1)$$

Also, work done on a charge can be calculated using the ________________ V between the two points that the charge moves between:

$$W = qV \qquad ...(2)$$

Combining equations 1 and 2 gives:

$$qV = qEd$$

$$V = Ed$$

This gives a second way to define ***electric field strength***, as the potential difference per ________________:

$$E = \frac{V}{d}$$

where: $E =$ ________________ (V m^{-1})

$V =$ potential difference between two points (V)

$d =$ distance between the same two points, ________________ to the electric field (m)

This equation can only be used for a ________________ electric field, since it is based on an assumption that the force was the same throughout the displacement over which work was done. This equation provides a practical way to determine electric field strength experimentally, since the voltage between two plates can be measured with a ________________ and the distance between them can be measured with a ________________.

Example: A potential difference of 12 V is applied between two parallel plates, which are 12 cm apart.

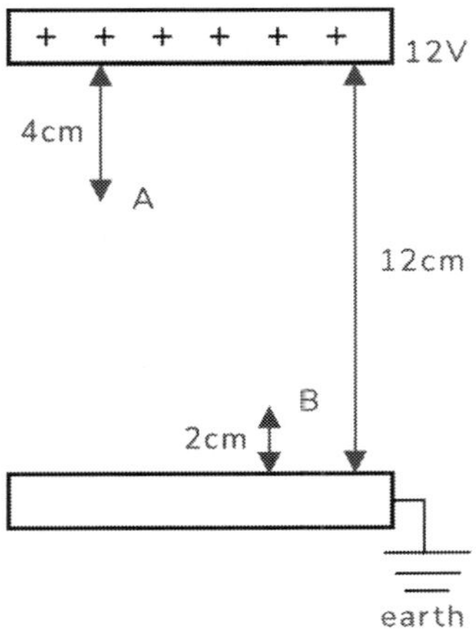

a) Determine the electric field between the two plates.
b) Point A is 4.0 cm from the positive plate. What is the electrical potential at point A?
c) Point B is 2.0 cm from the earthed plate. What is the electrical potential at point B?
d) Determine the voltage drop between points A and B.

Answer:

a) $E = \frac{V}{d} = \frac{12}{0.12} = 100$ V m^{-1} from the positive plate to earthed plate

b) Voltage drop between positive plate and Point A is:

$V = Ed = 100 \times 0.04 = 4$ V

so electrical potential is: $12 - 4 = 8$ V

c) Voltage drop between Point B and earthed plate is:

$V = Ed = 100 \times 0.02 = 2\ \text{V}$

This is also the electrical potential at Point B, since the electrical potential at the earthed plate is 0 V.

d) $8 - 2 = 6\ \text{V}$

2. In the example above, the distance between the plates is reduced to 6.0 cm. Analyse the effect of this on:

   a) the electric field between the plates;

   b) the electrostatic force acting on a charge between the plates; and

   c) the gain in kinetic energy of a positive charge as it reaches the earthed plate, after leaving the positive plate from rest.

3. Explain why there are two possible units for electric field strength.

4. Compare the quantities in the table below.

Quantity	Definition	Formula	Unit	Vector or scalar?
Electrostatic force				
Electrical potential energy				
Electric field				
Potential difference				

Which of the above quantities are independent of charge?

5. Which option correctly shows the relationship between the electrical potential at a point as a function of the distance from a charged plate to that point, in a uniform electric field?

A.

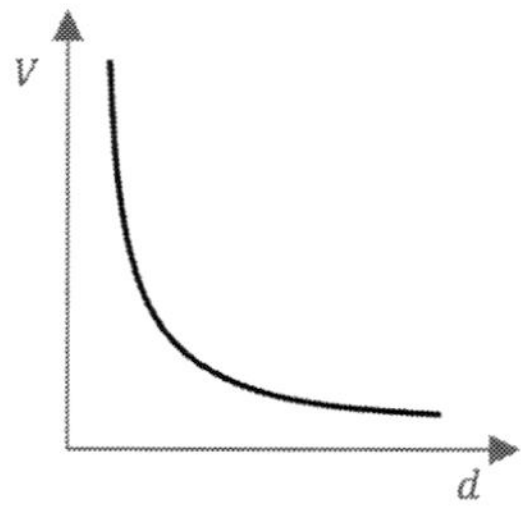

B.

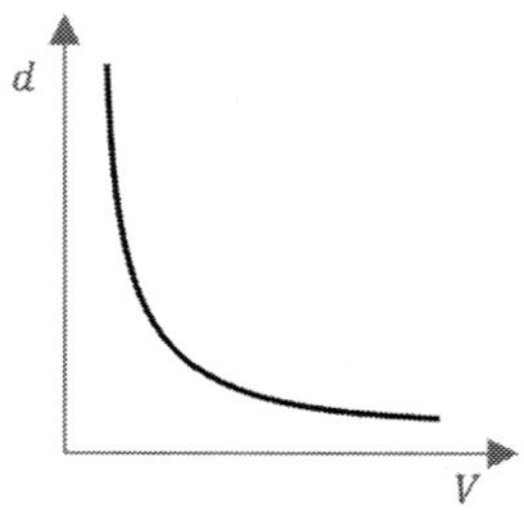

C.

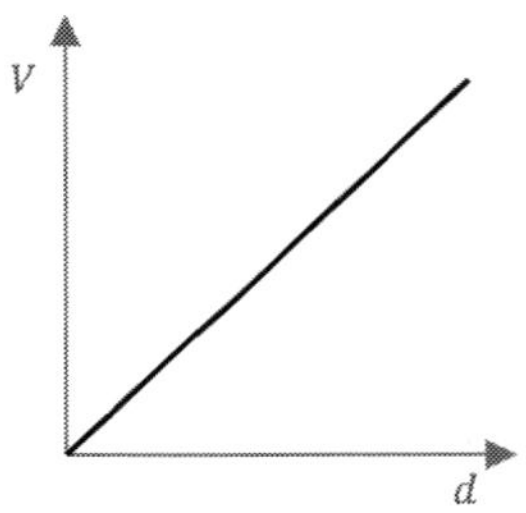

D.

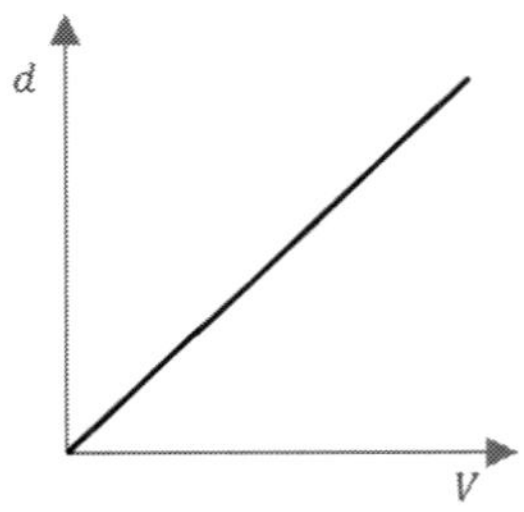

6. A device called a capacitor is made of parallel plates, which become charged if the plates are attached to a battery. After the battery is switched off, the plates still have charge for a short time and therefore store electrical potential energy. The gap between the plates can be air, but is often filled with a material called a dielectric. Capacitors have many real life applications.

a) Defibrillators apply a voltage across the chest to stimulate a failing heartbeat, and use capacitors to store the charge needed to produce this voltage. If the voltage between the charged plates is 500 V and they are held parallel to each other, 25 cm apart, what is the electric field between the plates?

Dipoles between uncharged plates

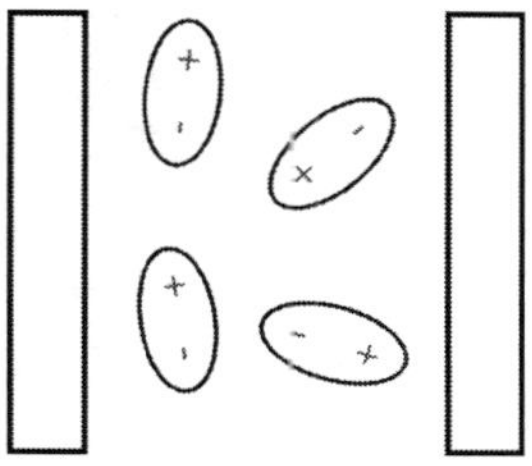

b) The first diagram shows a capacitor made of two parallel plates with a material between them that contains molecules which behave as dipoles, when no voltage is applied.

Complete the second diagram to show what happens to the dipoles when a voltage is applied.

Explain what happens to the dipoles and the electric field between the plates.

Sketch what happens when a voltage is applied

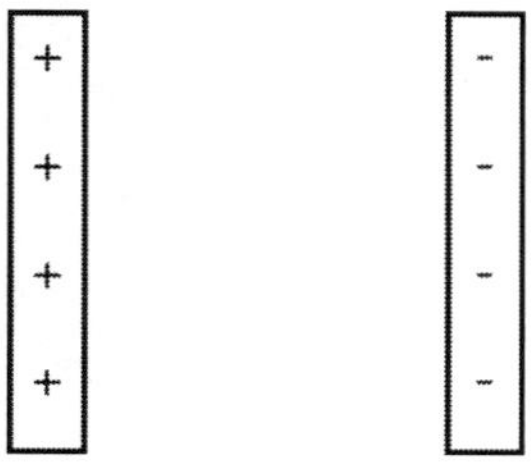

c) Discuss whether you can use the formula $E = \frac{F_E}{q}$ where $F_E = \frac{1}{4\pi\varepsilon_0}\frac{q_1 q_2}{r^2}$ to calculate the electric field strength between the plates of a capacitor.

7. A voltage of V is used to accelerate a beam of electrons across a distance d, starting from rest.

The symbol for velocity is v and voltage is V. Write clearly so you don't mix them up.

a) Derive an expression for the final speed of the electrons.

b) Analyse what happens to the final speed of the electrons if the voltage is doubled to $2V$.

8. The ionosphere is a part of Earth's atmosphere which starts from around 80 km above the ground. The ionosphere consists of charged atoms and molecules, which have become charged due to radiation from the Sun. In fair weather, the electric field between the positively charged ionosphere and negative ground can fluctuate between 100 V m^{-1} and 300 V m^{-1}. Assume that at a particular time, this electric field is 100 V m^{-1} over a small area of ground, and the ionosphere and ground act like charged parallel plates producing a uniform electric field.

a) Sketch a diagram showing the electric field and equipotential lines between the ground and this height.

Diagram

ground

b) Determine the potential difference between the ground and a point 1.5 m above the ground.

c) Recent studies suggest that some spiders can use the Earth's electric field to fly long distances, even across oceans. The silk that spiders eject from their bodies has a negative charge, which experiences a force in Earth's electric field. This allows spiders to fly using a technique called "ballooning".

Consider a spider that spins a thread with a charge of -20 nC. The combined mass of the spider and thread is 0.20 mg. Assume the Earth's electric field is 100 V m^{-1}. Taking into account electrostatic and gravitational force, what would be the resultant acceleration of the spider?

12.1 Electric circuits

The unit for voltage is the ***volt***. It is measured using a ***voltmeter***.
V
Conductors, like metals, have low ***resistance***, so current flows easily through them.
Insulators, like rubber and plastic, have high ***resistance***, leading to lower current.
Air is also an insulator with a high resistance.
The unit for resistance is the ***ohm***.
Ω
Resistance converts the kinetic energy of the charges to some other form, such as heat or light.
This can be useful
or wasted.
A component designed to add resistance to a circuit is called a ***resistor***.
This can be useful to prevent a circuit overheating, due to a high current.
Adding resistance along the same path, or in ***series***, increases the total resistance of the path,
reducing current.
YOU PLACED
99th
TIME
10 : 17

If the path is broken at any point, all current stops.
So adding light bulbs in series makes them dimmer
and if one breaks, there is no more current in the circuit.
Adding an alternate path by placing resistance in ***parallel*** decreases the total resistance,
increasing total current.
The alternate path doesn't change how fast current moves through any one path though.
Therefore turning a light bulb on
and off
won't affect another circuit in parallel.
However, the more light bulbs on in parallel
the more total current is drawn and the more electrical energy is used.
BILL $$$
Using a ski run to model electric circuits has limitations though.
Charges travelling along one path in a circuit will experience the same electrostatic force, moving them at the same rate.
We have neutral charge overall
and so we can choose our pace!

12.2 Circuit basics

Components in electric circuits

1. The table lists some common components in simple circuits. Practise drawing the symbol for each component.

Component	Function	Symbol	Copy the symbol
cell	Source of electrical potential energy, created by a pair of different metals chemically reacting in a conducting solution. One metal becomes the positive terminal, the other becomes the negative terminal.	 The long line represents the positive terminal. The short line represents the negative terminal.	
battery 	Source of electrical potential energy, created by two or more cells connected together.	 This can be drawn with more cells.	
power pack 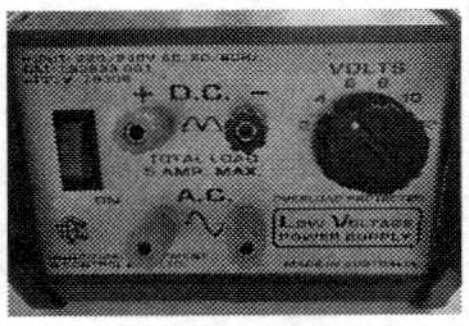	Source of electrical potential energy, where the voltage can be adjusted.	 The arrow means "variable".	
wire 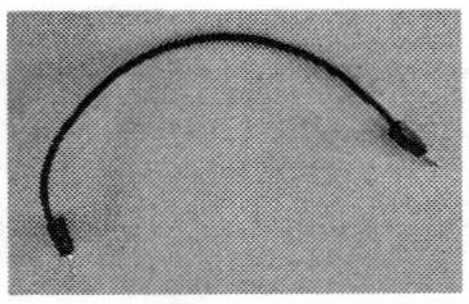	Conducting material, usually copper, to connect components together and to form a closed circuit. Often protected with insulating material.		
light bulb 	Transforms electrical energy into light.	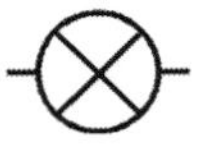 or 	
resistor 	Transforms electrical energy into heat. This can be useful if a circuit requires less current.		
variable resistor 	Transforms electrical energy into heat, but the resistance can be varied.	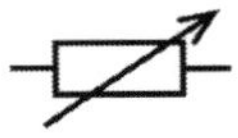	
open switch 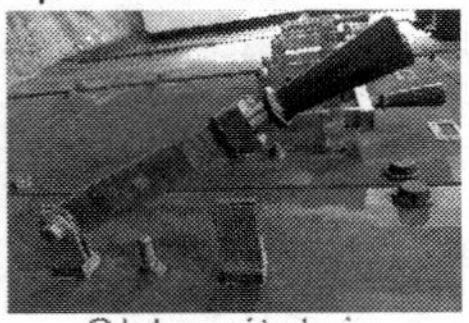 Old switch in World War 2 military bunker	Turns a circuit off by creating an open circuit, with a gap between conducting material.		
closed switch 	Turns a circuit on by allowing an unbroken, conducting path across the ends of the switch.		

Component	Function	Symbol	Copy the symbol
voltmeter 	Instrument used to measure voltage, the electrical potential energy per unit charge.	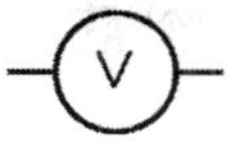	
ammeter 	Instrument used to measure current, the amount of charge flowing past a point per second.	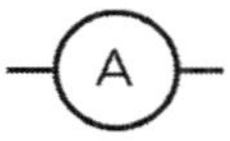	

2. Describe features of a simple circuit, by filling in the blanks.

ammeter	ampere	broken	charge imbalance	conductors
electric current	electric field	electrical potential	high resistance	insulators
kinetic	low resistance	metals	ohm	plastic
potential difference	resists	unbroken	volt	voltmeter

When objects rub together, charges may be transferred from one object to another, causing a ________________. However, if there is a continuous flow of charge, then an ________________ is produced. ***Current*** is defined as the amount of charge flowing past a point in a circuit per second. The unit used to measure current is the ________________. An instrument called an ________________ is used to measure current.

Charges flow easily through ***conductors***, such as ________________. Charge does not flow easily through ***insulators***, such as air, rubber or ________________. In other words, ________________ have low resistance compared to ________________. ***Resistance*** is how well something ________________ the flow of charge, therefore doing work by transforming electrical energy into some other form. The unit used to measure resistance is the ________________.

If there is an ________________ loop of conducting material, then current can flow, and this is called a ***closed circuit***. However, if the path is ________________, for example by an air gap, then current cannot flow, and this is an ***open circuit***. So if we use copper wires to connect components in a circuit, then we provide a ________________ path for current to flow through. However, to switch off a circuit, we can create an air gap that has ________________ and does not allow charges to flow across.

Charges in a circuit also require ________________ energy to transform into kinetic energy, in order to produce a current. In other words, electrical energy in a circuit is really the ________________ energy of moving charges. The electrical potential energy can be provided in the form of a battery or electrical powerpoint, which produces an ________________ between the positive and negative terminals. The electrical potential energy used per coulomb of charge between two points in a circuit is also known as ________________ or ***voltage***. The unit used to measure voltage is the ________________. An instrument called a ________________ is used to measure voltage.

Components in series and parallel

3. Describe how to draw components in series and parallel, by filling in the blanks.

ammeter	closed switch	different	first	light bulb	negative	parallel
positive	resistors	same	series	sides	splits up	voltmeter

Circuits are often drawn in rectangular paths, with components placed on the ________ of the rectangles, but never the corners. The current flows from the ________ terminal, represented by the long line on the battery, through the circuit, then to the ________ terminal, represented by the short line. In Circuit A, shown to the right, starting from the positive terminal, current passes through a wire, ________, wire, ________, wire, ________, then a final wire.

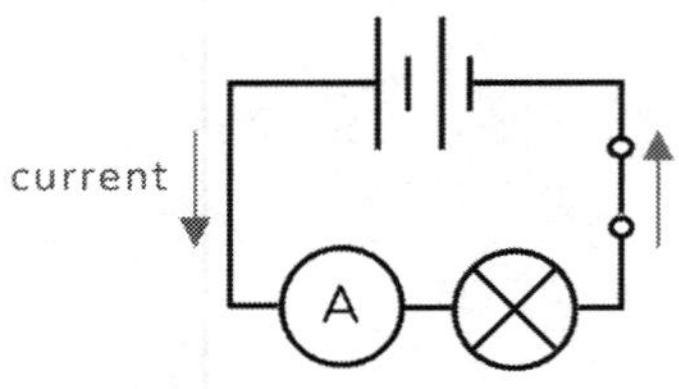

All components are in series

All the components in Circuit A are connected in ***series***, meaning that current passes through the ________ path to go through them.

An ammeter must always be connected in ________ with the component it is measuring the current through, in order to work.

In Circuit B, shown to the right, starting from the positive terminal, current passes through a wire, but then splits into two ________ paths. The first path consists of two ________ in series, joined by wires together then back to the battery. The second path has a ________, connected in parallel with the first resistor. ***Parallel*** means current ________ to pass through more than one path to reach the components.

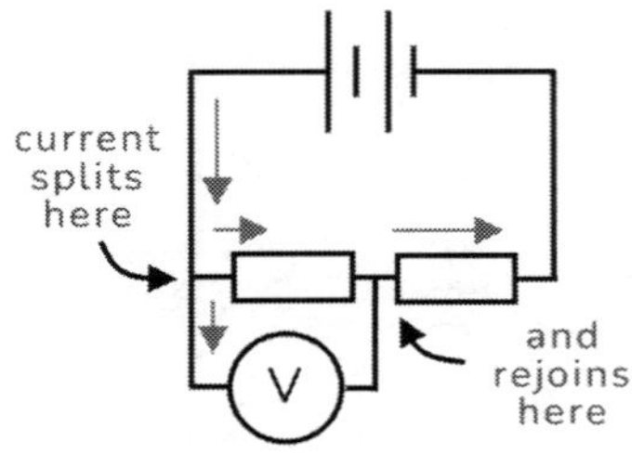

The voltmeter is in parallel with the first resistor

A voltmeter must always be connected in ________ between the points it is measuring the potential difference across. Therefore this voltmeter is only measuring the voltage across the ________ resistor and not the second.

4. Consider the circuit shown to the right.

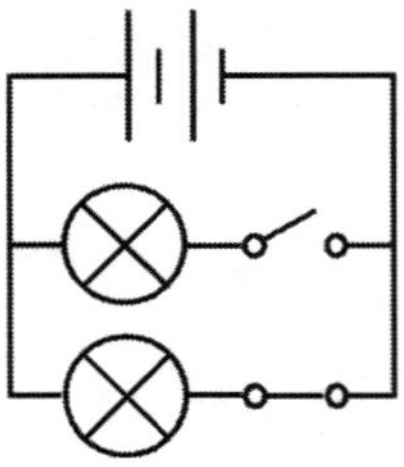

a) Describe the circuit, using the names of the components in them and whether they are arranged in series or parallel.

____

____

____

____

b) How many light bulbs are turned on in this circuit?

____

5. Consider the circuit shown to the right.

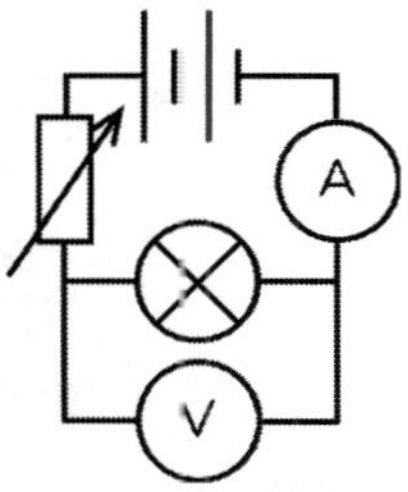

a) Describe the circuit, using the names of the components in them and whether they are arranged in series or parallel.

b) What is the voltmeter measuring?

6. Draw the following circuits.

a) A battery is connected to two light bulbs in parallel. A single open switch is in series with them both, which is able to turn both on and off at the same time.

b) A battery is connected to two resistors in series. A voltmeter measures the voltage across both of them.

c) A battery is connected to a closed switch and a resistor in series. A voltmeter measures the voltage across the switch and an ammeter measures the current through it.

d) A battery is connected to two light bulbs in parallel. Each light bulb is connected to a switch in series, so that each can be independently turned on and off.

12.3 Voltage and current

1. Describe voltage and current in a circuit, by filling in the blanks.

AC	ampere	anode	batteries	chemical potential	charge
cathode	coulomb	current	DC	electrical potential	enters
leaves	negative	opposite	positive	potential difference	safety
socket	time	voltage	work done		

Voltage, or ________________, is the electrical potential energy per unit charge and is measured in volts:

$$V = \frac{W}{q}$$

where: $V =$ ________________ across two points (V)

$W =$ work done or change in ________________ energy (J)

$q =$ ________________ (C)

This formula can also be written with ________________ as the subject:

$$W = qV$$

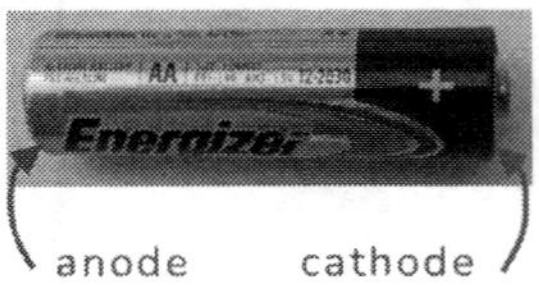

Inside a battery, chemical reactions transform ________________ energy into electrical potential energy, resulting in a positive charge at one terminal, called the ***cathode***, and a negative charge at the other terminal, called the ***anode***. An electric field exists from the ________________ to the ________________, in the same direction as the flow of positive charge.

Memory aid: Cathode has a "t" in it, which looks like a "+". Anode has an "n" for negative.

Current is the rate at which ________________ charge flows:

$$I = \frac{q}{t}$$

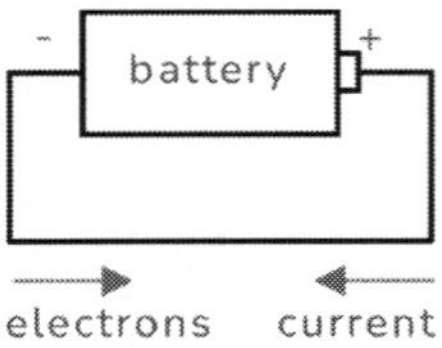

where: $I =$ ________________ (A)

$q =$ amount of charge passing a point (C)

$t =$ ________________ in which charge passes a point (s)

DC power symbols

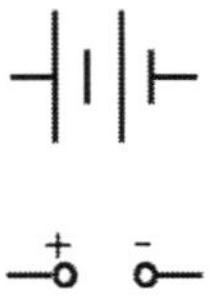

By definition, the ***cathode*** is the terminal the current ________________ a power source from, and the ***anode*** is the terminal the current ________________ the power source.

AC power symbol

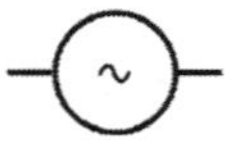

Historically, current was defined before anyone knew what kind of charges moved, but after J. J. Thomson discovered the electron in 1897, we now know that it is in fact the ________________ electrons that move in metals. We are still stuck with the same definition of current though, so current is still taken to flow from the positive to negative terminal when electrons flow the ________________ way. In other words, if one ________________ of electrons pass a point in a circuit per second, then we say that one ________________ of current flowed the other way.

When current only travels in one direction, it is called *direct current*, or ________________. For example, ________________ are a DC source.

Australian sockets use AC: active terminals switch between positive and negative polarity

The current from an electric ________________, on the other hand, switches direction periodically, and is called *alternating current*, or ________________. In Australia, the top two holes switch between being the positive and negative terminals 50 times per second, or at a frequency of 50 Hz, with a peak voltage of around 240 V. The third hole is earthed as a ________________ feature.

2. On the circuit diagram shown, use arrows to indicate the direction of current and direction of electron flow.

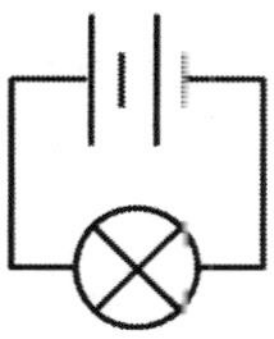

3. The picture shows a USB charger. The label states the following:

"Input: 100-240 V, AC 50-60 Hz, 0.18 A

Output: 5 V DC, 2 A, 10 W Max"

Contrast the input electricity supply with the output electricity supply.

Contrast means explain how things are different. Here, explain the key concepts in the label.

4. If a current of 1.00 A were to flow in a circuit, how many electrons would pass a point per second?

5. A mobile phone charger with an output of 2.0 A is used to charge a phone for 45 minutes. How many coulombs are delivered to the phone?

6. Turn off circuits before touching uninsulated wires! Otherwise, current can pass through your body to the ground. A direct current of 60 mA can make your muscles seize up and unable to relax, because electrocution disrupts your nervous system, which relies on electrical signals between your brain and muscles. If a 60 mA current passes through the body for 0.25 s, how many electrons is that?

7. One AA battery supplied a voltage of 1.5 V and current of 1.8 A for 35 minutes, before the voltage decreased as the battery started going flat.

a) How much charge passed through the battery, in coulombs, during this time?

b) What was the work done on this charge?

8. An LED light bulb connected to a 2.0 V supply for 5.0 seconds uses 0.16 J of energy.

a) How much charge in coulombs passed through the LED?

b) What current passed through the LED?

12.4 Resistance, ohmic and non-ohmic conductors

Resistance

1. Describe resistance, by filling in the blanks.

current	heat	high	higher	light
low	ratio	resistance	transform	voltage

The *resistance* of any component in an electric circuit is measured in ohms and is defined as the ________ of voltage applied across it to the current through it:

$$R = \frac{V}{I}$$

where: R = resistance (Ω)

V = ________ between two points (V)

I = ________ (A)

This equation tells us that the higher the voltage, the ________ the current, and that this ratio is the same if the ________ of the material stays the same. This equation also predicts that if the same voltage is applied across different materials, conductors with ________ resistance will have a higher current through them than insulators with ________ resistance.

Resistance in a circuit will ________ electrical energy into some other form. For example, a light bulb transforms electrical energy into ________. Metal wires transform electrical energy into ________.

Example: An electrician wants to find out the value of a resistor. They set up the circuit shown and obtain a reading of 12 V on the voltmeter and 36 mA on the ammeter. What is the value of the resistor?

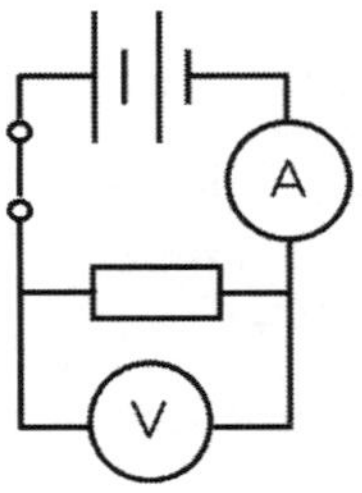

Answer:

$R = \frac{V}{I} = \frac{12}{36\times10^{-3}} = 330\ \Omega$

2. Compare voltage, current and resistance using the table.

Concept	Voltage	Current	Resistance
Definition in words			
Definition as equation			
Unit name and symbol			
Circuit symbol	Voltmeter	Ammeter	Resistor

3. An old type of light bulb has a tungsten filament (thin wire) in it, which gives off light and heat when current is passed through it. A current of 5.0 A may cause the filament to melt and the light bulb will break. It has a resistance of 2.0 Ω. What is the maximum voltage that will not break the light bulb?

____

____

4. The table below records the voltage applied to two different materials, silicon and wet wood, and the current through each of them.

Material	Voltage applied (V)	Current (mA)	Resistance (Ω)
Silicon	6.0	2.3	
Wet wood	12	3.0	

a) Complete the table by calculating the resistance of each material.

b) What variables should have been kept the same to ensure that a fair test was carried out in order to compare the resistance of the materials?

....

c) Explain which material appears to be the better conductor.

....

5. Copper is often used in wiring because it has a very low resistance and therefore current flows easily. One particular roll of copper wire has a resistance of 8.75 Ω per kilometre. If 50.0 m of this wire were connected to a car battery which applies 12.9 V to it, how much current would flow?

First work out the resistance of 50 m. Watch your units!

....

....

....

Ohmic and non-ohmic conductors

6. Compare ohmic and non-ohmic conductors, by filling in the blanks.

break	cannot	current	defines	depends	enormous
gradient	non-ohmic	Ohm's Law	ohmic	ohmic	one direction
proportional	resistance	reverses	same	voltages	

Ohmic conductor

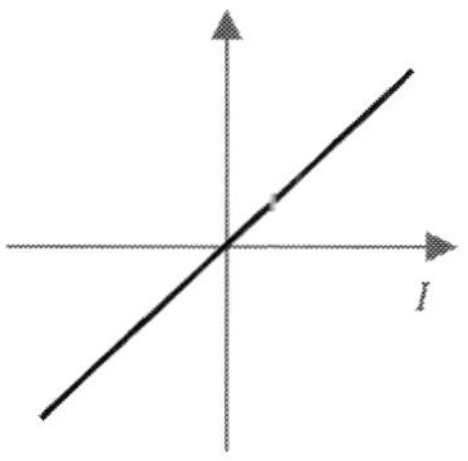

If a component stays the same resistance when different voltages are applied to it, then it is described as *ohmic*. This means that if you experimentally measure the current through a material when different voltages are applied, the graph of voltage on the y-axis and current on the x-axis will be a straight line, with voltage .... to current. The .... of the graph will be constant and equal to the resistance. In other words, ... conductors obey an experimental trend called Ohm's Law:

$$V = IR$$

where: R = constant resistance (Ω)

V = voltage between two points (V)

I = ... (A)

You can graph voltage and current on either axis. If voltage is on the x-axis, for an ohmic conductor, the gradient is: $\frac{1}{R}$

Ohm's Law looks like the same equation as the definition of ...., but here it is used to describe how the resistance of a component behaves if different .... are applied to it.

Resistors are designed to be ohmic – that is, they stay the .... resistance over a large range of applied voltages. If you switch the direction of the voltage but keep it the same magnitude, the resistance stays the same, meaning that current .... but stays the same magnitude.

Metal wires are ______________ to some extent – they stay a constant resistance if they can stay at the same temperature, but if they get hot, their resistance increases. For example, a filament light bulb, which contains a thin tungsten wire, is close to ohmic if only switched on for short periods of time, but does not obey ______________ if left on to get hot.

Filament light bulb

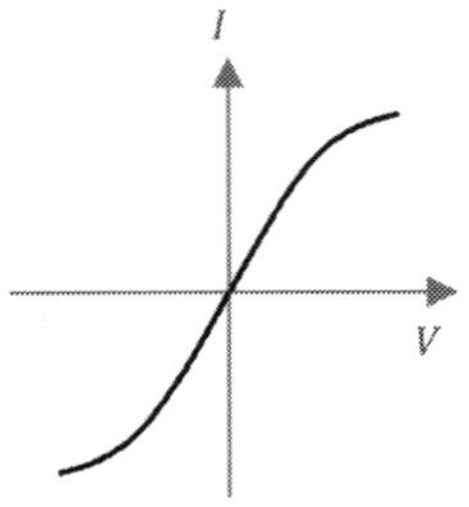

A light emitting diode (LED) is an example of a ***non-ohmic*** conductor – that is, its resistance ______________ on the applied voltage. LEDs will only light up and let current through in ______________. If the voltage is applied the wrong way, the resistance is ______________, so almost no current flows. If a large enough voltage is applied the wrong way, the LED will ______________.

LED

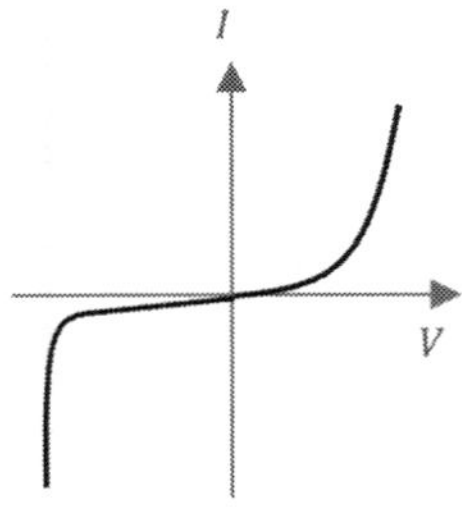

Because ______________ conductors do not have constant resistance, the gradient of a voltage-current graph ______________ be used to calculate the resistance. Instead, you need to use the equation that ______________ resistance, substituting the actual measurement of voltage and current to calculate resistance at that voltage.

7. The graph shows the current through an LED when different voltages were applied across it.

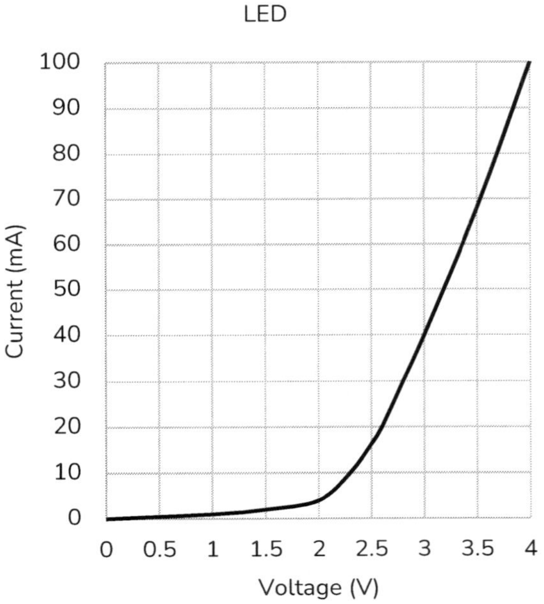

a) Estimate the resistance of the LED when 1.5 V was applied across it.

____

b) What is the resistance of the LED when 3.0 V was applied across it?

____

c) What is the resistance of the LED when 4.0 V was applied across it?

____

d) Analyse whether the LED is an ohmic conductor.

____

____

____

____

____

____

____

8. An experiment was done to investigate whether a resistor behaves like an ohmic conductor and determine its resistance, using the circuit shown.

a) Describe a method that could be used for this investigation to obtain the results shown in part (b). The first step is done for you.

1. Set up as shown in the circuit diagram.

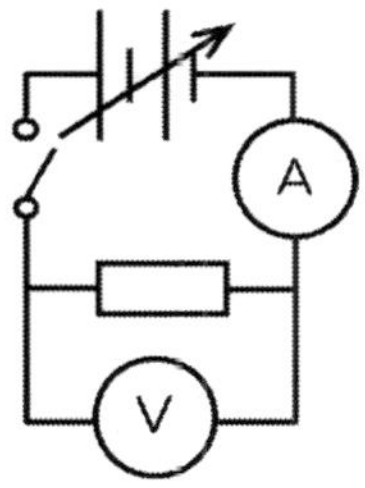

Always nclude steps to:
- improve validity, accuracy and reliability; and
- process data to answer the aim.

b) The following results were obtained. Graph the results.

Voltage (V)	Current (A)
2	0.4
4	0.8
6	1.2
8	1.6
10	2.0
12	2.4

V-I graph for a resistor

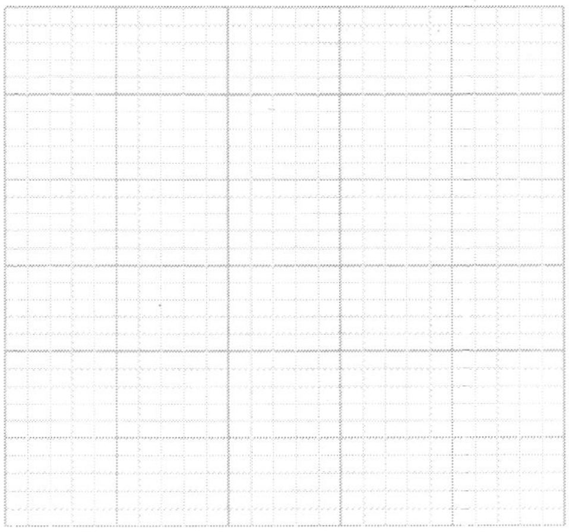

c) Use the graph to determine the resistance of the resistor.

12.5 Experimental skills – Resistance and resistivity

1. Describe the variables that affect the resistance of a conductor, by filling in the blanks.

charge	cross-sectional area	electrons	electrostatic	high
hotter	independent	kinetic energy	length	low
resistance	resistivity	useful	wasteful	work

The resistance of a straight conductor depends on the material, its length and cross-sectional area:

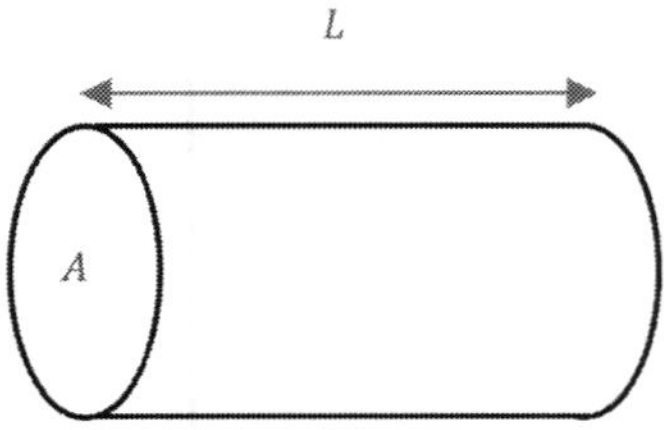

$$R = \frac{\rho L}{A}$$

where: $R =$ ________________ (Ω)

$\rho =$ resistivity (Ωm)

$L =$ ________________ of the conductor (m)

$A =$ ________________ of the conductor (m^2)

Resistivity is a property of the material and describes how well the material allows current to flow, ________________ of its size and shape. Since current is the rate at which charge moves past a point, resistivity therefore depends on how much ________________ is free to move in a material and what obstacles it encounters when it moves. In metals, the electrons in the outermost energy shell of the atom, known as ***valence electrons***, are the most free to move, since they are furthest from the positive nucleus, and therefore experience the least ________________ force from it. Metals therefore have ________________ resistivity. In non-metals, valence electrons are more strongly attracted to the nucleus, therefore less free to move, so they have ________________ resistivity compared to metals.

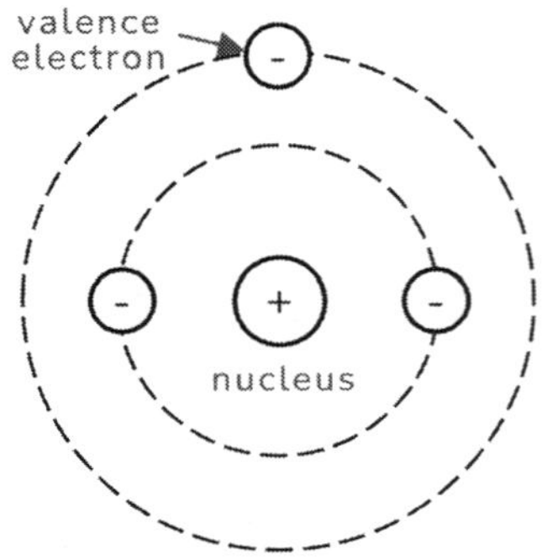

As ________________ move through a conductor, their movement is hindered by the other parts of the atoms, causing resistance. Collisions between the electrons and other parts of the atoms transform the ________________ of the electrons into heat, and this is called ***resistive heating***. Therefore, ________________ is done by resistance in circuits. Resistive heating is ________________ in a toaster, but ________________ in power lines that distribute electricity between power stations and homes.

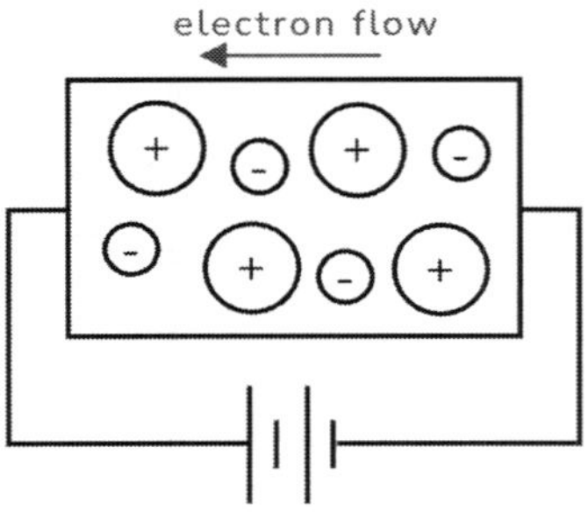

If a conductor gets ________________, the atoms in it will have more kinetic energy and are able to collide more with the flow of electrons. This makes it harder for electrons to flow, therefore increasing ________________.

2. A student observes that when the whole school goes out to lunch at the same time, they move faster through wide, short corridors compared to long, narrow corridors. Explain how this models resistance.

____

____

3. Explain why the unit for resistivity must be Ωm (ohms times metres).

____

____

4. The resistivities of some materials are given in the table.

a) Classify each as a conductor or insulator.

b) Which is the best conductor in the table?

c) Which is the best insulator?

Material	Resistivity (Ωm)
Carbon (graphite)	3.0×10^{-5}
Copper	1.7×10^{-8}
Glass	1.0×10^{9}
Sulfur	1.0×10^{17}
Tungsten	5.6×10^{-8}

5. An experiment is to be conducted to investigate effect of the length of a nichrome wire on the resistance of the wire. Nichrome is an alloy (mixture of metals) containing nickel and chromium, with a higher resistivity than most metals. You are provided with the following equipment for the experiment:

1 m nichrome wire	power pack	voltmeter	electrical tape
wires with alligator clips	metre ruler	ammeter	switch

a) Describe a safety risk involved in investigating this aim, and a precaution to minimise the risk.

b) Complete the method.

1. Set up as shown in the diagram.
2. Clip the two wires with alligator clips so that

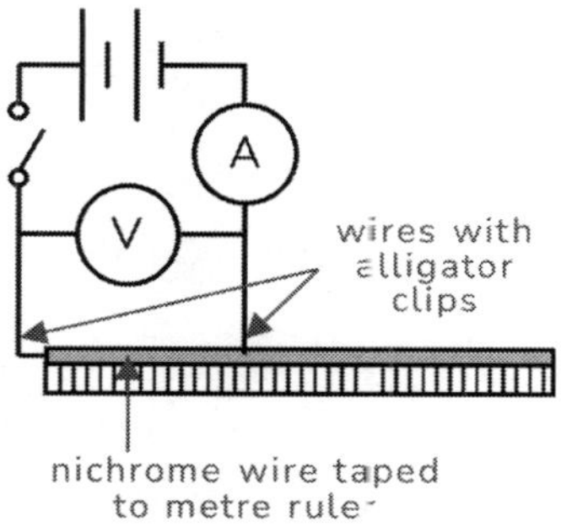

c) Outline a hypothesis for this experiment. Give a reason to support your hypothesis.

6. Sketch what you would expect the graph of the resistance of the wire vs length of the wire to look like. What does the gradient of the graph represent?

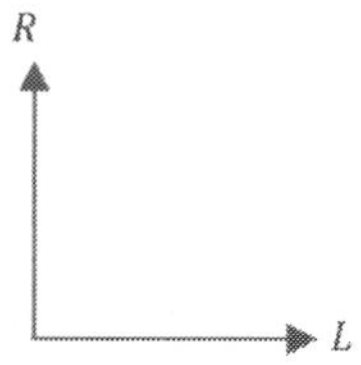

7. In another experiment, you are given different widths (called gauges) of nichrome wire and are required to investigate whether the cross-sectional area of the wire affects its resistance.

a) What would you need to keep the same to conduct a valid experiment?

b) What would you graph to determine a mathematical relationship between the cross-sectional area and resistance? What would the gradient of the graph represent?

12.6 Electrical energy and power

1. Explain how equations for electrical energy and power are derived, by filling in the blanks.

ammeter	ammeter	change in	charge	current
energy	increases	joules	power	time
transformed	voltage	voltmeter	watts	work done

Recall that work done on charges in a circuit, measured in ______________, is given by:

$$W = qV \quad ...(1)$$

where: W = work done or ______________ electrical potential energy (J)

q = ______________ (C)

V = ______________ between two points (V)

It is often impractical to measure the amount of charge in a circuit though, and much easier to measure current, using an ______________. Recall the definition of current is:

$$I = \frac{q}{t} \quad ...(2)$$

where: I = ______________ (A)

q = amount of charge passing a point (C)

t = ______________ in which charge passes a point (s)

Equation (2) can be written as:

$$q = It \quad ...(3)$$

Substituting equation (3) into equation (1) gives:

$$W = VIt \quad ...(4)$$

Therefore, to calculate how much ______________ a component uses in an electric circuit, you just need to measure the voltage across the component with a ______________, the current through it with an ______________ and the time for which the circuit was on.

Recall that power is the rate at which energy is ______________, measured in ______________:

$$P = \frac{W}{t} \quad ...(5)$$

where: P = ______________ (W)

W = ______________ (J)

t = time in which the work was done (s)

Substituting equation (4) into equation (5) gives:

$$P = VI$$

Therefore, the power from an electrical power supply ______________ with both voltage and current.

2. The label on the travel adapter shown states the following:

"Input: 100-240 V, AC 50-60 Hz, 0.35 A

Output: 5 V DC, 2.0 A"

Maximum power from the adapter is used to charge a phone for 30 minutes.

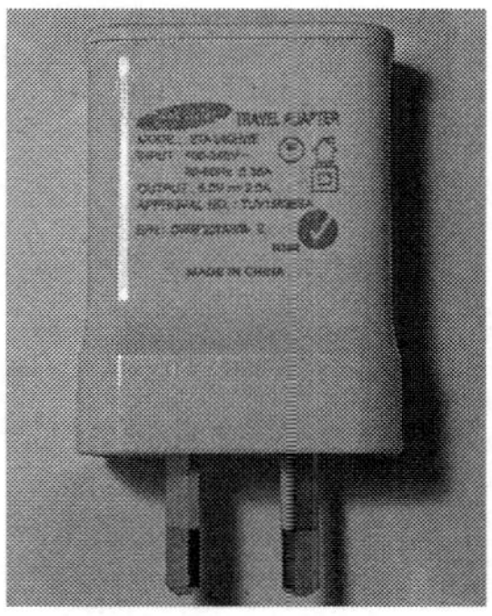

a) How much output power does the adapter provide?

b) How much energy does the phone gain, assuming the charger is 100% efficient?

100% efficient means no energy is converted to wasted forms, in this case heat.

3. Electric kettles use resistive heating from metal wire to heat water. A portable electric kettle that plugs into a car's 12 V electrical supply draws 120 W of power.

a) What is the current through the electric kettle?

b) It takes 83.8 kJ to heat a cup of water from 20.0° C to its boiling point of 100° C. How long will this kettle take to do this?

Recall k means kilo or $\times 10^3$.

4. Two variations on the equation $P = VI$ that are commonly used are $P = \frac{V^2}{R}$ and $P = I^2R$.

Recall the definition of resistance.

a) Show how both of these variations can be derived from $P = VI$.

b) $P = I^2R$ is useful for calculating how much resistive heating occurs, as it tells us the rate at which electrical energy is transformed into heat. A toaster heats bread using 1100 W supplied to a long, thin wire, with a resistance of 75 Ω. How much current runs through the toaster?

5. A school power pack contains a safety feature called a circuit breaker. This automatically switches off the power pack when the current exceeds 6.0 A, often making a "click" sound. A student connects a 5.0 m long wire, with a resistance of 0.070 Ω, to the terminals of the power pack, set to deliver 4.0 V DC. No other component is connected to the circuit.

a) Explain whether the circuit breaker is triggered.

b) Assuming the circuit stays on, how much power would be converted to heat in the wire?

6. A 20 Ω resistor and 100 Ω resistor are each connected to a separate battery supplying 12 V. Analyse whether this means the power supplied to each resistor is the same.

7. The kilowatt-hour (kWh) is the most common and convenient unit for energy used on electricity bills and energy ratings on appliances. One kilowatt-hour is the energy used by a 1 kW appliance in one hour.

a) What is 1.0 kWh in joules?

b) A 2.0 kW air conditioner is turned on for 5.0 hours. How much energy does it use, in kilowatt-hours and in joules?

8. Alkaline batteries come in many sizes, but actually all deliver around 1.5 V. The difference is the total amount of energy contained. The table shows typical values in watt-hours for some common sizes, which may vary between brands.

Size	Capacity (Wh)
AA	4.05
C	12
D	18

Each battery listed in the table is connected to a 5.0 Ω light bulb and ammeter, as shown in the circuit diagram. Assume the light bulb's resistance stays constant.

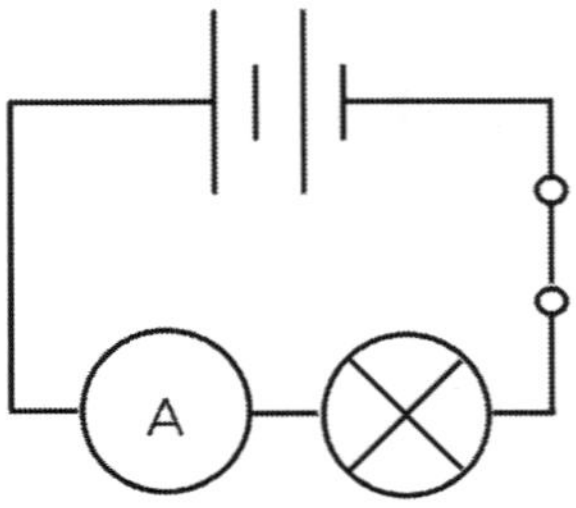

a) Determine what the ammeter reading would be for each battery.

b) If each battery was new when used in the circuit, and the light bulb was left on, calculate how many hours until the light bulb would dim and each battery would run out.

12.7 Series circuits

1. Describe current, voltage and resistance in a series circuit, by filling in the blanks.

added	ammeter	charge	current	divided	energy
equals	one	resistance	same	time	voltages

In series, current passes through only ________________ path to go through all components. The number of electrons entering a point in a circuit equals the number of electrons leaving it, as ________________ is conserved. So in the circuit shown, the charge q_1 passing through resistance R_1 ________________ the charge q_2 passing through resistance R_2:

$$q_1 = q_2 \quad ...(1)$$

Since current is given by $I = \frac{q}{t}$, in any given ________________ t the charge passing a point is $q = It$, so equation (1) also means:

$$I_1 t = I_2 t$$

Therefore the current from the battery I_T and the current through each light bulb respectively, I_1 and I_2, must be the ________________:

$$I_T = I_1 = I_2 \quad ...(2)$$

This means it would not matter where you put an ________________ in a series circuit, as you would measure the same current.

The change in electrical potential energy in the battery, W_T, equals the sum of the work done by the light bulbs W_1 and W_2, as ________________ is conserved:

$$W_T = W_1 + W_2$$

Since $W = qV$:

$$qV_T = qV_1 + qV_2$$

As charge is conserved, the voltage supplied to the circuit is equal to the sum of the ________________ across each component in series:

$$V_T = V_1 + V_2 \quad ...(3)$$

The resistance of a component is the voltage across it ________________ by the current through it: $R = \frac{V}{I}$ or $V = IR$. Expressing equation (3) in terms of current and ________________:

$$I_T R_T = I_1 R_1 + I_2 R_2 \quad ...(4)$$

From equation (2), the ________________ throughout the circuit is the same, so:

$$R_T = R_1 + R_2 \quad ...(5)$$

This means that the total resistance in a series circuit is equal to the resistances of the components in series ________________ together.

Resistors in series

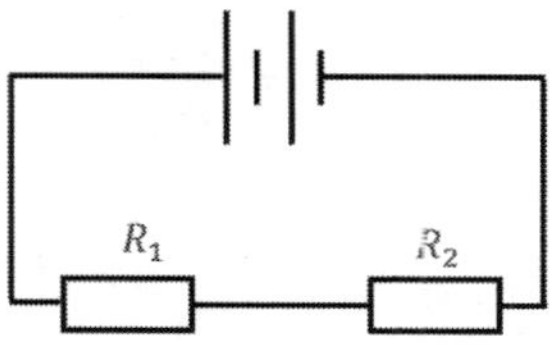

The ammeters all measure the same current

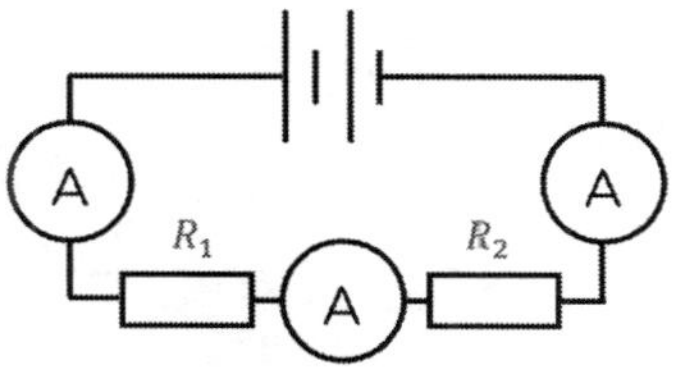

This is like skipping in a circle – everyone must skip at the same speed, otherwise they bump into each other.

$V_T = V_1 + V_2$

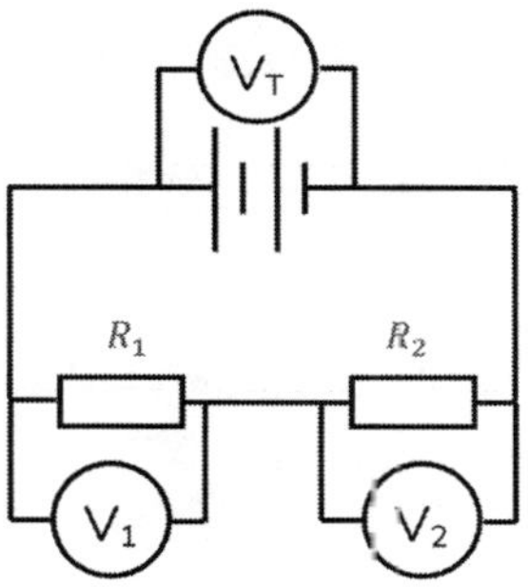

This:

... can be replaced by this:

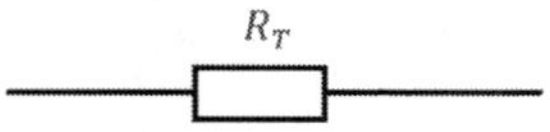

Example: A circuit was connected as shown in the diagram.

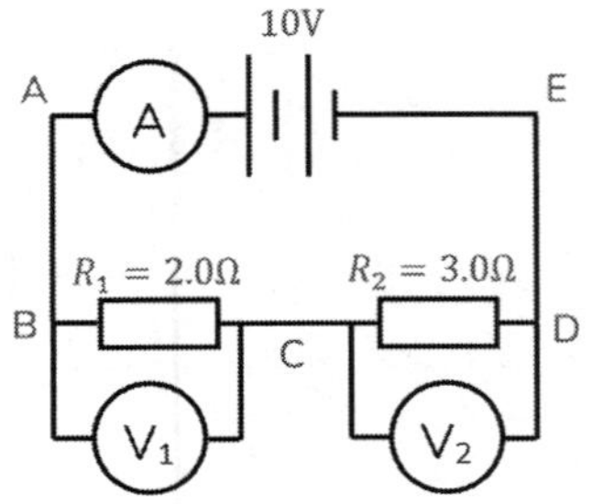

a) Calculate the total resistance of the resistors.
b) What would be the reading on the ammeter?
c) What would be the reading on each voltmeter?
d) Sketch a graph of the electrical potential vs position, from point A through the circuit components to point E.

In each calculation, the values of V, I and R must be between the same two points in the circuit.

In calculations we often ignore the resistance of wires and ammeters, since they are usually very low.

Answer:

a) $R_T = R_1 + R_2 = 2 + 3 = 5\ \Omega$

b) $V = IR$

$10 = I \times 5$

$I = 2\text{ A}$

c) $V_1 = IR_1 = 2 \times 2 = 4\text{ V}$

$V_2 = IR_2 = 2 \times 3 = 6\text{ V}$

d)

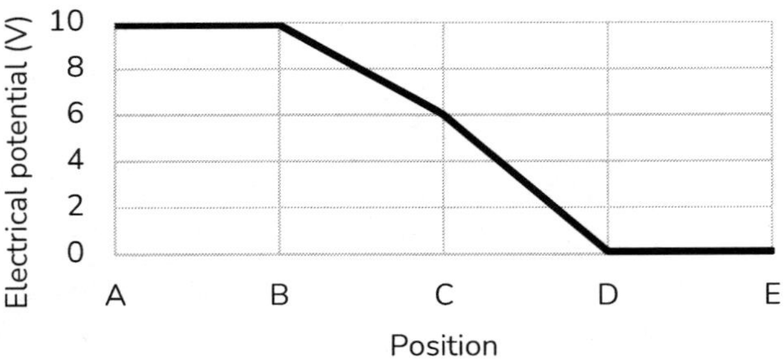

2. What is the total resistance shown?

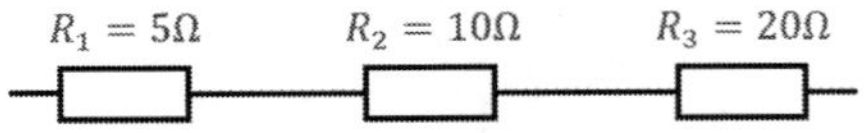

____

3. What resistor should be added in series to give a total resistance of 330 Ω?

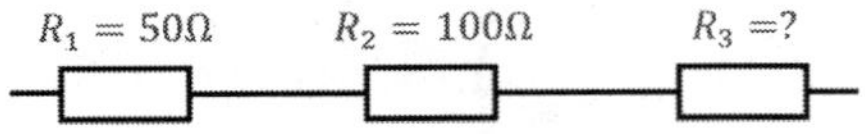

____

4. A circuit was connected as shown in the diagram.

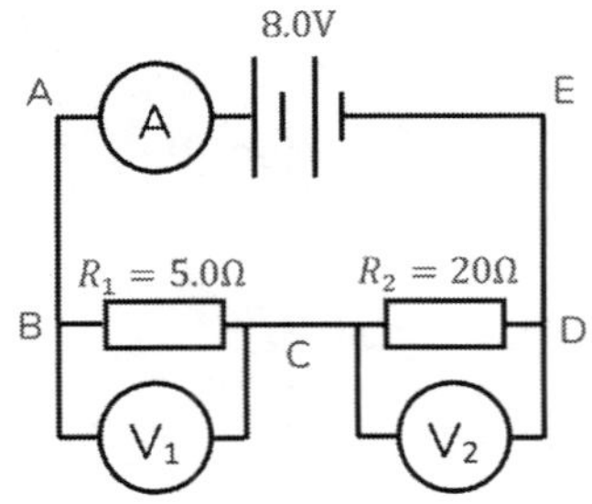

a) Calculate the total resistance of the resistors.

____

b) Calculate the reading through the ammeter and each voltmeter.

____

____

____

c) Explain what the sum of the readings through the voltmeters should equal.

____

____

d) Sketch a graph of the electrical potential vs position.

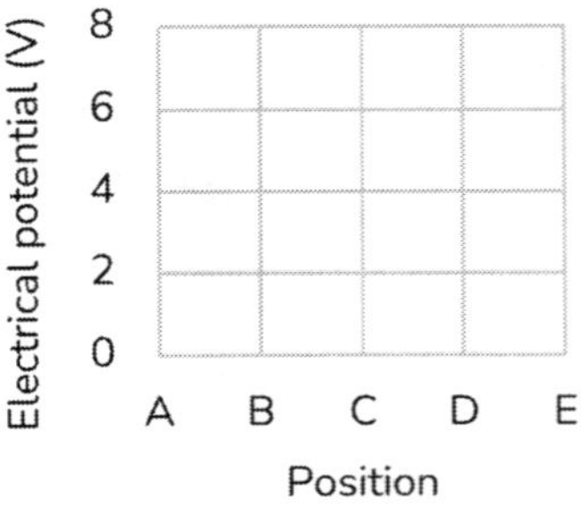

5. An experiment was carried out to investigate the effect of adding identical 5.0 Ω light bulbs in series.

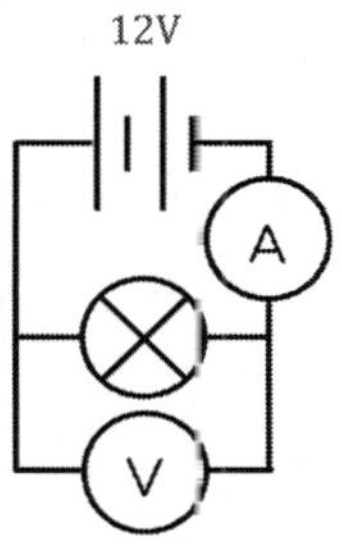

a) First, one light bulb was connected to a 12 V power supply and the voltage and current measured, as shown. Determine the voltage across the light bulb, current through it and power given off by the light bulb.

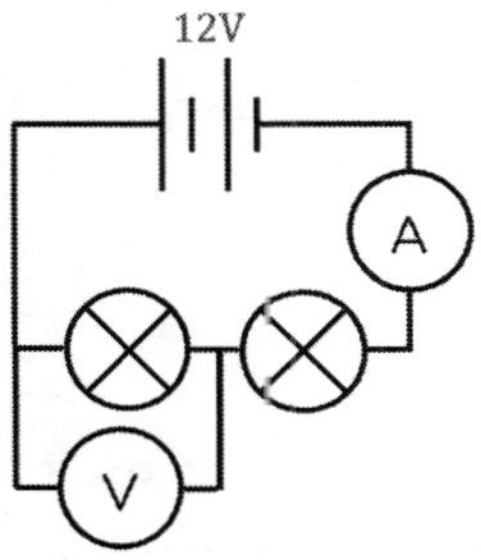

b) Next, a second identical light bulb was connected in series, as shown. Determine the voltage across the first light bulb, current through it and power given off by the first light bulb.

The order in which a question asks for values does not have to be the order in which you find them. It may be helpful to find another value first.

c) Compare the voltage, current and power calculated in part (a) and (b), as well as the brightness of each light bulb.

Brightness is related to power, since it is the rate at which electrical energy is converted to light.

d) Compare the total power drawn from the power supply in part (a) and (b).

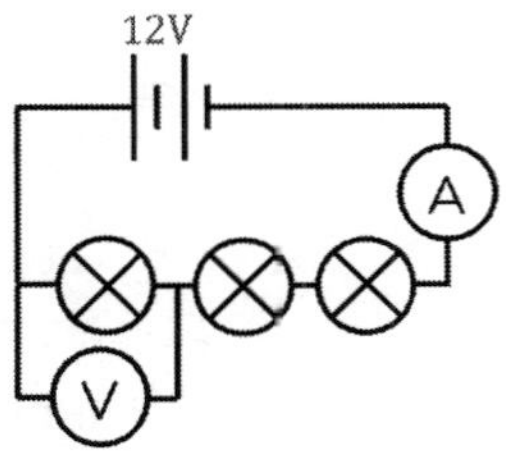

e) Now a third identical light bulb is added in series. Compare the voltage across the first light bulb, current through it and power given off by the first light bulb, to the values obtained with a single light bulb in part (a).

f) Outline three reasons why you would not want to connect the light bulbs in your house in series.

6. Two light bulbs of resistance $R_1 = 5.0\ \Omega$ and $R_2 = 7.5\ \Omega$ respectively are connected in series with a 10 V power supply, with R_1 closer to the positive terminal, as shown in Setup A.

Setup A

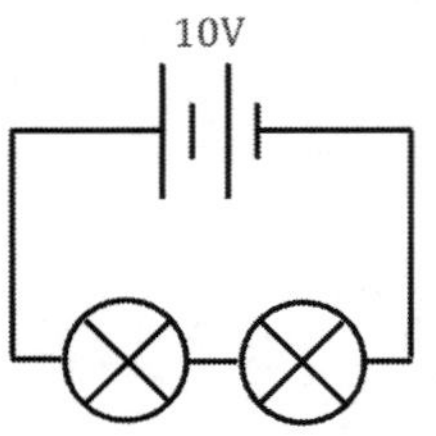

a) Calculate the power dissipated by each light bulb and compare their brightness.

Setup B

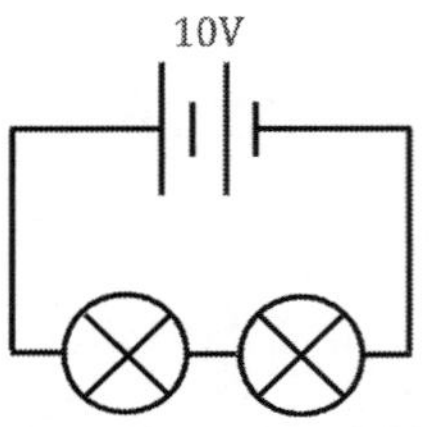

b) The position of the light bulbs is then switched, as shown in Setup B. Analyse whether this affects the brightness of each light bulb.

7. A red LED is recommended to operate with a voltage of 1.8 V across it, at which a current of 20 mA will flow through it.

a) What is the resistance of the LED when operating as recommended?

b) Lower than the recommended voltage of 1.8 V, the LED will not be bright enough. Higher than this voltage, the LED may break. The LED is connected to a 3.0 V power supply. What value of resistor must be added in series with the LED so that the LED operates as recommended?

8. One way to adjust the brightness of a light bulb is to use a variable resistor in series with it, as shown in the diagram. Explain what happens to the current, voltage and brightness of the light bulb if the resistance of the variable resistor increases.

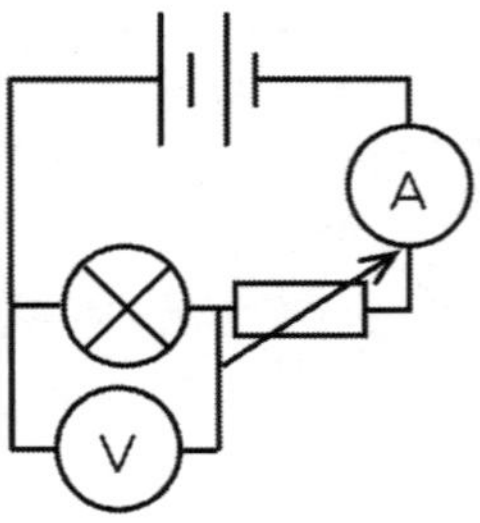

9. A 5.00 Ω light bulb is to be mounted at the front gate of a large property. An extension cable 45.0 m long connects the light bulb to a 12.0 V battery at the house, where it can be switched on and off. With such a long wire, we cannot ignore its resistance, which is 1.2 Ω in total.

a) Calculate the current through the light bulb.

b) Compare the power emitted by the light bulb using the extension cable, to the power emitted if the light bulb were to be directly attached to the battery without the extension cable.

c) How much power is lost via resistive heating in the extension cable?

12.8 Parallel circuits

1. Describe current, voltage and resistance in a parallel circuit, by filling in the blanks.

ammeters	battery	conserved	current	divided by
multiple	power	resistance	same	smaller
sum of	sum of	transformed	voltages	voltmeters

$I_T = I_1 + I_2$

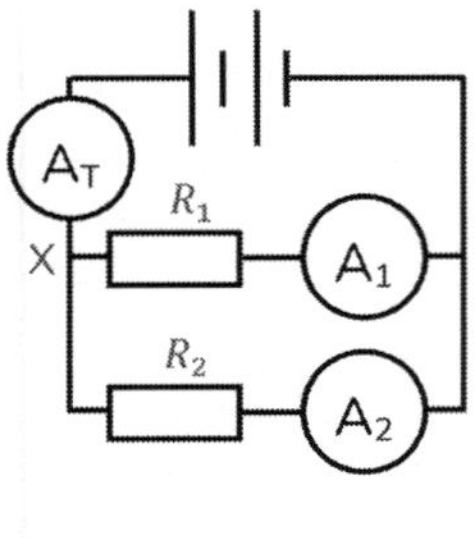

In *parallel*, current splits into ________ paths to go through components.

Charge is ________, so in the circuit diagram shown, the total charge q_T entering point X from the positive terminal must equal the ________ q_1 passing through resistance R_1 and q_2 passing through resistance R_2, giving:

$$q_T = q_1 + q_2 \quad ...(1)$$

Since ________ is given by $I = \frac{q}{t}$, if we divide both sides by the amount of time passed t, equation (1) gives:

$$I_T = I_1 + I_2 \quad ...(2)$$

This means the reading on ammeter A_T equals the sum of_the readings on ________ A_1 and A_2.

The voltmeters all measure the same voltage

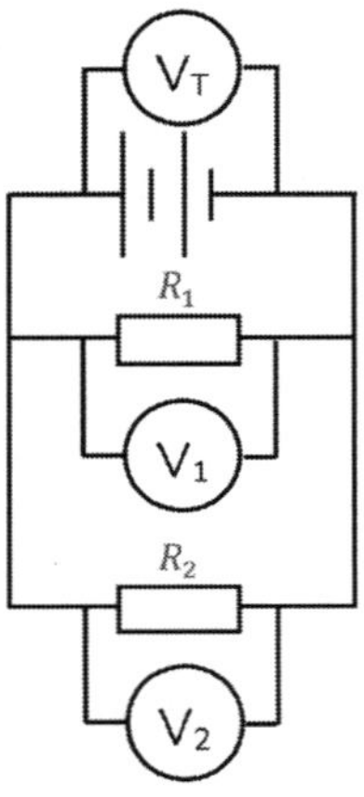

If we assume that wires have minimal resistance, all the electrical potential energy from the battery is ________ in the resistors. This means that the voltage, or electrical potential energy per unit charge, measured by each of the ________ is the same:

$$V_T = V_1 = V_2 \quad ...(3)$$

The total electrical power supplied by the ________ is:

$$P_T = I_T V_T$$

So multiplying equation (2) by V_T tells us what the total ________ is:

$$I_T V_T = I_1 V_T + I_2 V_T$$

From equation (3), we can see all the ________ are equal, so:

$$I_T V_T = I_1 V_1 + I_2 V_2$$

This means that the total power from the battery is equal to the ________ the power used by each resistor:

$$P_T = P_1 + P_2 \quad ...(4)$$

This:

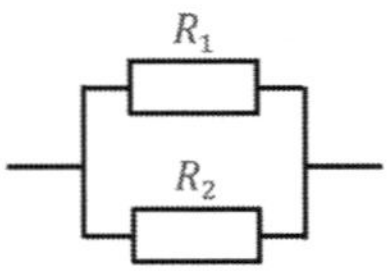

... can be replaced by this:

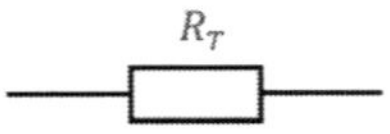

Resistance of a component is the voltage across a component ________ the current through it: $R = \frac{V}{I}$. Rearranging this gives $I = \frac{V}{R}$.

Expressing equation (2) in terms of voltage and ________ then gives:

$$\frac{V_T}{R_T} = \frac{V_1}{R_1} + \frac{V_2}{R_2}$$

Recall from equation (3) that the voltage across each resistor is the ________, so:

$$\frac{1}{R_T} = \frac{1}{R_1} + \frac{1}{R_2} \quad ...(5)$$

This means that the more resistors are added in parallel, the ________ the total resistance will be.

Example: A circuit was connected as shown in the diagram.

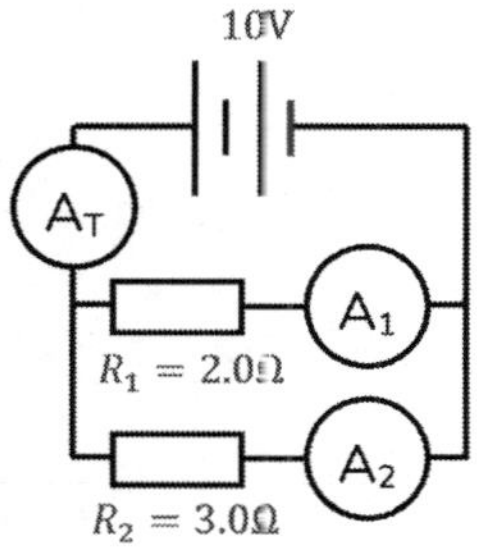

a) Calculate the total resistance of the resistors.

b) Explain whether adding a resistor in parallel increases or decreases the total resistance of the circuit.

c) What would be the voltage across each resistor?

d) What would be the reading on each ammeter?

Answer:

a) $\frac{1}{R_T} = \frac{1}{R_1} + \frac{1}{R_2} = \frac{1}{2} + \frac{1}{3}$

$R_T = 1.2\ \Omega$

b) Part (a) shows the total resistance is less than either of the individual resistors, so adding a resistor decreases the total resistance of a parallel circuit. The alternate pathway makes it easier for current to flow.

c) 10 V because there is no voltage drop between the battery and the resistors.

d)

$V_T = I_T R_T$	$V_1 = I_1 R_1$	$V_2 = I_2 R_2$
$10 = I_T \times 1.2$	$10 = I_1 \times 2$	$10 = I_2 \times 3$
$I_T = 8.3$ A	$I_1 = 5$ A	$I_2 = 3.3$ A

Notice the total current is the sum of the currents through each path.

2. What single resistor would be equivalent to the resistors shown?

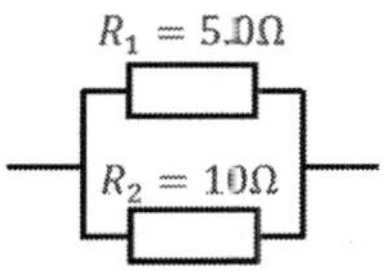

3. What single resistor would be equivalent to the resistors shown?

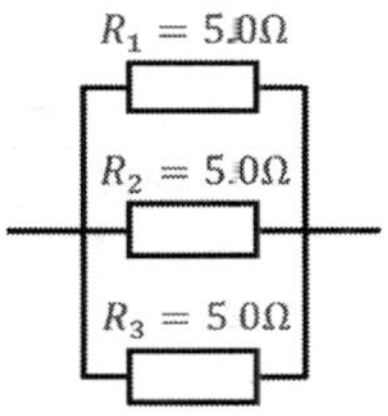

4. What resistor R_2 should be added in parallel to give a total resistance of 1.0 Ω?

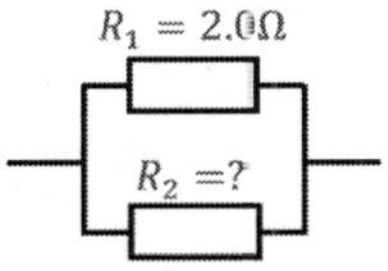

5. What resistor R_3 should be added in parallel to give a total resistance of 5.0 Ω?

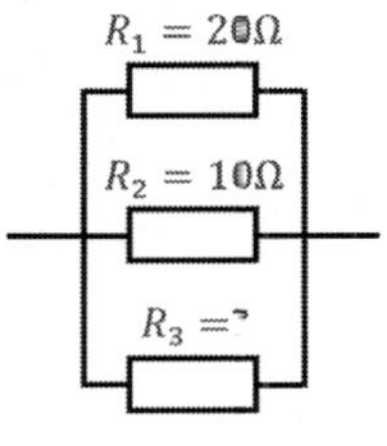

6. A circuit was connected as shown in the diagram.

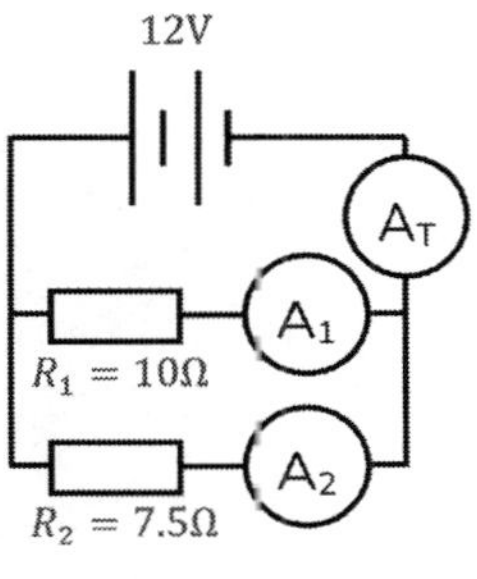

a) Calculate the total resistance of the circuit.

b) Calculate the reading through each ammeter.

7. An experiment was carried out to investigate the effect of adding identical 5.0 Ω light bulbs in parallel.

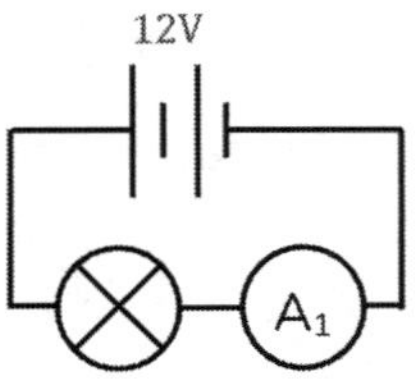

a) First, one light bulb was connected to a 12 V power supply and the current measured, as shown. Determine the current and power given off by the light bulb.

b) Next, a second identical light bulb was connected in parallel, as shown. Determine the current through each light bulb and total current from the battery. Compare these values to those obtained in part (a).

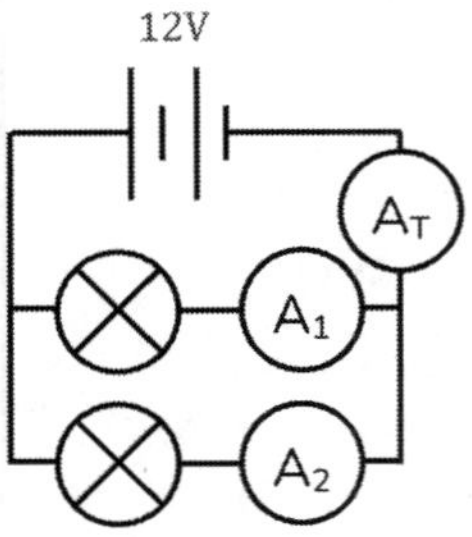

> You can find total current two ways:
> - finding total resistance first; or
> - adding the currents in each path.

c) Compare the power used by each light bulb and their brightness in part (a) and (b).

d) Compare the total power drawn from the power supply in part (a) and (b).

e) Outline three reasons why you would want to connect the light bulbs in your house in parallel rather than series.

8. The circuit shown contains two filament light bulbs. The current shown on ammeter A_1 is 2.5 A and the current shown on ammeter A_2 is 2.0 A.

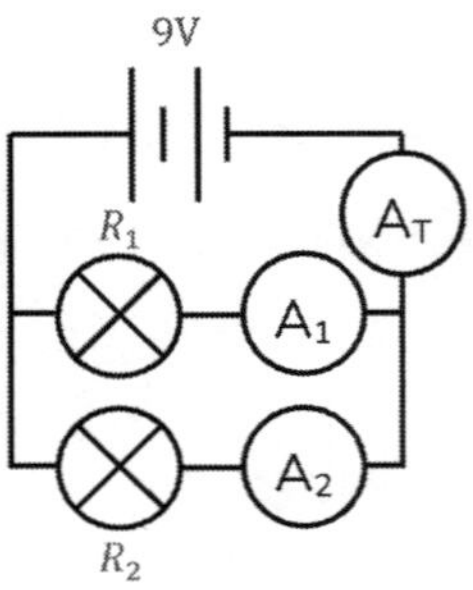

a) What is the reading on ammeter A_T?

b) How much power is used by each light bulb, and which one is brighter?

c) Calculate the resistances of each light bulb and the effective total resistance of the circuit.

9. Household circuits have a safety feature called a circuit breaker, which automatically switches off the circuit when the total current exceeds a certain amount. A household with a single 240 V DC power outlet has a circuit breaker which will switch off when 12 A is exceeded. Plugged in parallel are a 800 W refrigerator, 2000 W heater and 150 W electric blanket.

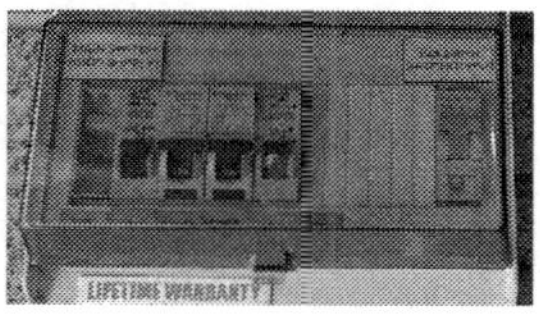

Circuit diagram

a) Draw a circuit diagram to represent this circuit. Use a resistor to represent each of the appliances and a switch to represent the circuit breaker.

b) Explain whether operating all of these appliances at the same time will trip the circuit breaker.

10. A short circuit happens when a path of lower resistance diverts current from a normal circuit, usually unintentionally. The circuit diagram shows one way you could short circuit a light bulb, connected to a 12 V school power pack. Assume the light bulb has a resistance of 5.0 Ω and the wire in parallel with it has a resistance of 0.025 Ω.

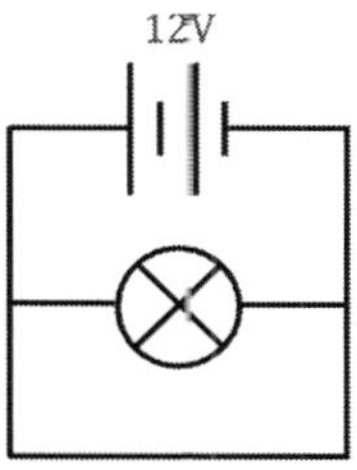

a) How much total current would be drawn from the power pack? Would this trip the circuit breaker built in to the power pack, which switches off if a current of 6.0 A is exceeded?

b) Explain why a short circuit can be dangerous if there were no circuit breaker.

12.9 Mixed circuit problems

Example: A circuit was connected as shown in the diagram.

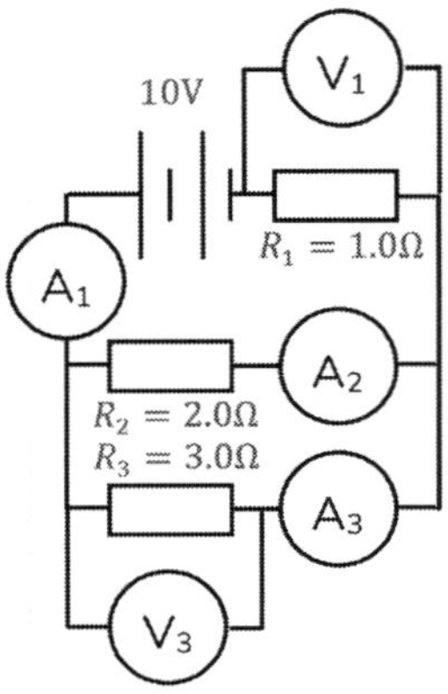

a) Describe the circuit in words.
b) Calculate the total resistance of the resistors.
c) Calculate the current and voltage through each meter shown.
d) What is the total power drawn from the power supply?

Answer:

a) Resistors R_2 and R_3 are in parallel with each other, and in series with resistor R_1 and a 10 V DC power source. Each resistor is in series with an ammeter. Resistors R_1 and R_3 are each in parallel with a voltmeter.

b) Let R_{23} be the equivalent resistance to R_2 and R_3:

$$\frac{1}{R_{23}} = \frac{1}{R_2} + \frac{1}{R_3} = \frac{1}{2} + \frac{1}{3}$$

$$R_{23} = 1.2\ \Omega$$

Let R_T be the total resistance of the circuit:

$$R_T = R_1 + R_{23} = 1 + 1.2 = 2.2\ \Omega$$

Find the total resistance of each parallel and series part separately.

Find values in whatever order you have enough data for.

c) A₁ measures the total current in the circuit:

$$V_T = I_1 R_T$$

$$10 = I_1 \times 2.2$$

$$I_1 = 4.5\ \text{A}$$

When using $V = IR$ remember each value must come from the same part of the circuit.

V₁ measures the voltage over R_1. The current through R_1 is the same as total current I_1:

$$V_1 = I_1 R_1 = 4.5 \times 1 = 4.5\ \text{V}$$

It may make it easier to keep track of data if you write your working around the circuit diagram next to the relevant part of the circuit.

V₃ measures the voltage over R_3, which is the same as over R_2. The total voltage drop along the circuit must be 10 V:

$$V_T = V_1 + V_3$$

$$V_3 = 10 - 4.5 = 5.5\ \text{V}$$

Currents through A₂ and A₃ are then:

$$I_2 = \frac{V_2}{R_2} = \frac{5.5}{2} = 2.8\ \text{A}$$

$$I_3 = \frac{V_3}{R_3} = \frac{5.5}{3} = 1.8\ \text{A}$$

d) $P = VI = 10 \times 4.5 = 45\ \text{W}$

1. What single resistor would be equivalent to the resistors shown?

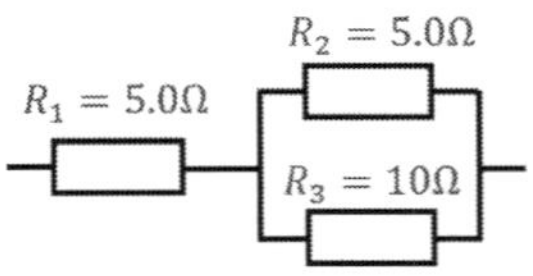

....

....

2. What single resistor would be equivalent to the resistors shown?

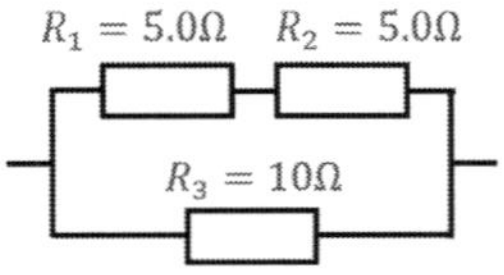

....

....

3. You have four 2 Ω resistors. Draw a circuit diagram to show how you would connect all of them together, to produce the equivalent resistances shown.

a) 5 Ω

b) 1.5 Ω

4. In the circuit shown, calculate:

a) the voltage across each resistor;

b) the current through each resistor;

c) the total current drawn from the power source;

d) the power used by each resistor; and

e) the total energy drawn from the power source in 5 minutes.

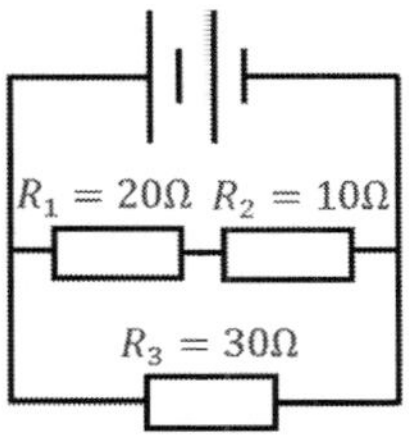

Find values in whatever order you have enough data for.

5. The reading on ammeter A_1 is 178 mA. Determine the readings on the other meters in the circuit.

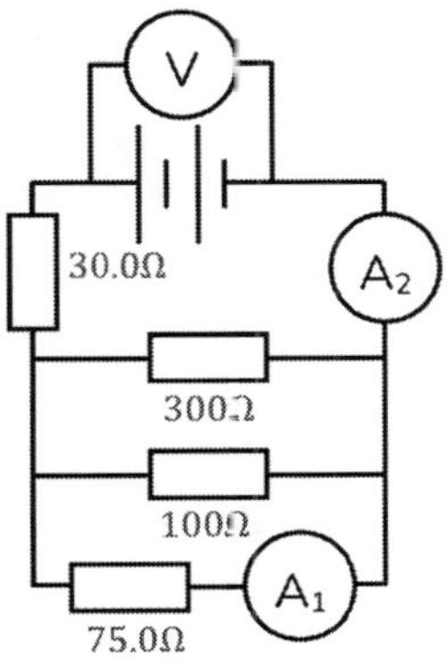

6. Three identical light bulbs, R_1, R_2 and R_3, each have a resistance of 5.0 Ω. Compare their brightness when they are connected in the circuit shown.

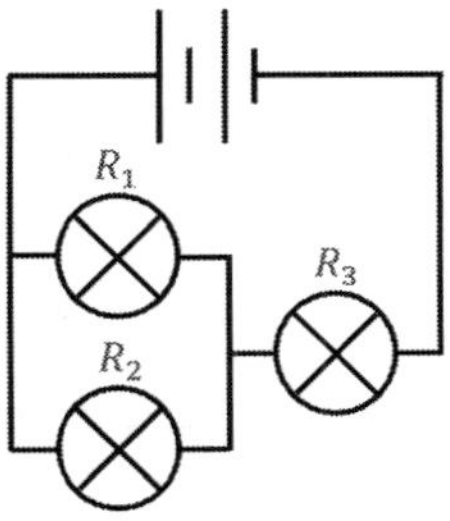

Don't just rote learn assumptions – always work through the logic.

7. The circuit diagram is for a simple heat mat, which consists of two 20 Ω heating elements, represented by resistances R_1 and R_2. By closing different combinations of the switches S_1, S_2, S_3 and S_4, the heat mat can be operated at three different settings – low heat, medium heat and high heat.

For each setting, identify which switches need to be closed, then calculate the total current through the circuit and the power used by the heat mat.

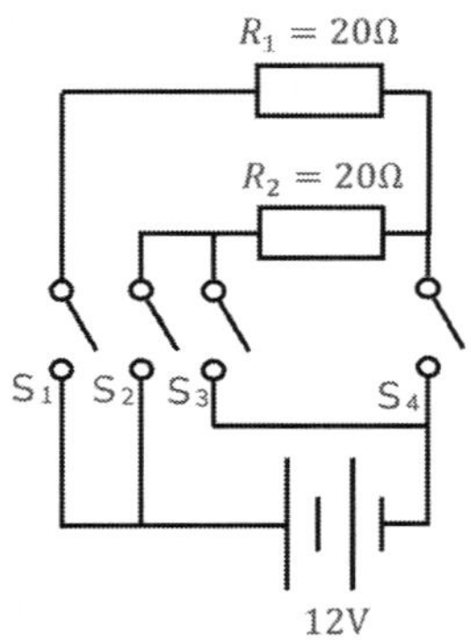

Put your finger on the positive terminal and physically trace the path of the current towards the negative terminal, to work out if it is a series or parallel circuit.

8. The circuit shown was connected. Each light bulb in the circuit has a resistance of 2.0 Ω.

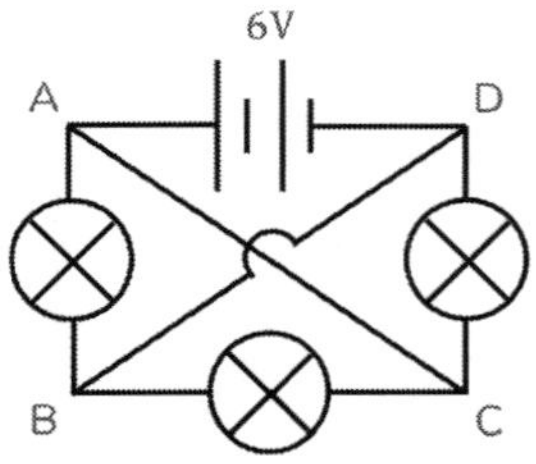

The curved bit of the wire in the middle indicates two overlapping wires that are unjoined.

a) Which points in the circuit are at the same electrical potential as the positive terminal of the battery?

b) Which points are at the same electrical potential as the negative terminal?

c) What happens if the wire connecting points A and C and the wire connecting points B and D are removed from the circuit? Compare the brightness of the light bulbs in the circuit and total power used before and after the removal of the wires.

Try drawing this circuit in a simpler way.

13.1 Magnetism and finding true north

Our modern day compass is still just a permanent magnet floating in liquid.
The end pointing north is the north pole, and the other end is the south pole.
N
S
Magnets do not just experience force from Earth's magnetic field. Like poles repel
S N N S
N S S N
and unlike poles attract.
S N S N
We picture invisible ***magnetic field lines*** to help us understand this mysterious force.
S N
The direction of the ***magnetic field*** is defined as the direction that a compass points.
N
The strength of the magnetic field is greater if field lines are closer together.
N
Permanent magnets are not the only things that have a magnetic field.
Electric currents also produce magnetic fields.
+
−
1820
So electricity and magnetism are closely related.
We call the interaction of electric and magnetic fields ***electromagnetism***.
Electric currents can therefore stop a compass pointing north.
But actually... a compass doesn't point exactly to true north!
True north, or the geographic north pole, is a point on the axis around which Earth rotates.
A compass will lead you close to it, but not quite.
MAGNETIC NORTH
GEOGRAPHIC NORTH
?
It may be a coincidence that compasses point close to true north.
N S
now
Ancient rocks show Earth's magnetic field has changed in the past.
past
Earth's magnetic field might be caused by liquid iron in Earth's core, which moves to form electric currents.
However, the origin of Earth's magnetic field remains a mystery today.

13.2 Magnetic fields around permanent magnets

1. Describe magnetic fields by filling in the blanks.

attract	closer	compass	cross	density	field	invisible
iron	magnetic field	model	north	north	pole	repel
resultant	south	south	stronger	tangent	touch	vector

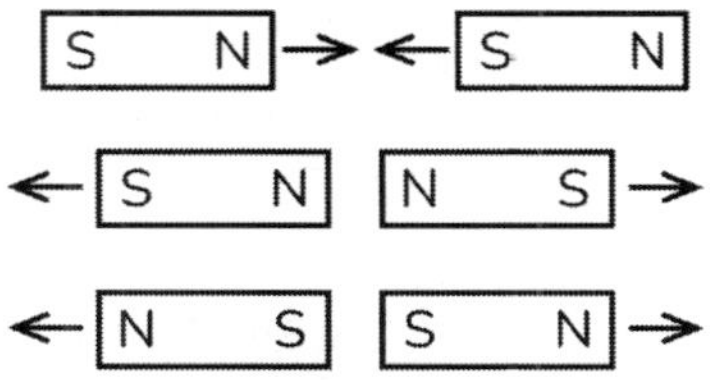

A ________________ is a region where an object experiences a non-contact force. A ________________ is a region that will exert a force on magnets and magnetic materials. For example, a fridge magnet will exert a magnetic force on objects containing ________________, such as a paper clip. The fridge magnet only needs to be close enough to the paper clip, but does not need to ________________ it.

Magnets have a north pole and a south ________________. If the north pole of one magnet is brought close to the south pole of another, they will ________________. If like poles of two magnets are brought close together, they will ________________.

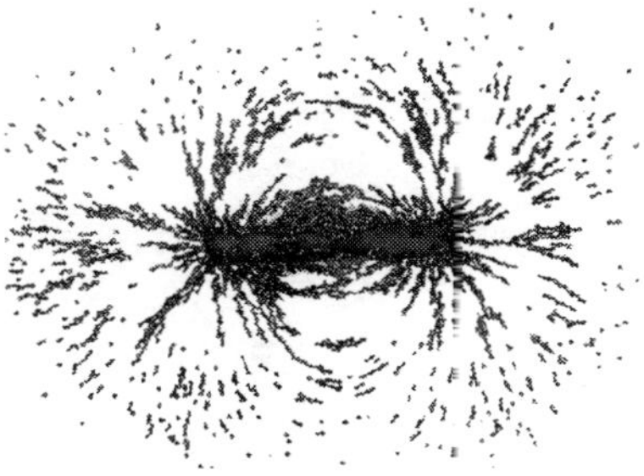

Magnetic fields are ________________, but sprinkling iron filings in the region will reveal their shape. We can ________________ the magnetic field by drawing field lines that follow this shape.

The strength of the magnetic field is represented by the ________________ of the field lines. In other words, if field lines are closer together, the magnetic field is ________________. Therefore, field lines around a bar magnet will be ________________ together near the poles than away from the poles.

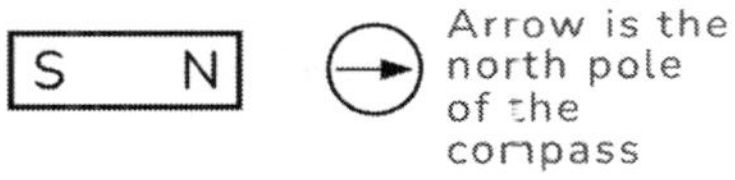

The direction of a magnetic field is defined as the direction that the north pole of a ________________ will point. Since a compass is actually a magnet, this means that the north pole of a compass will be repelled from the ________________ pole of a bar magnet, and attracted towards the ________________ pole of a bar magnet.

This means that the direction of magnetic field lines around a bar magnet go from its ________________ pole to its ________________ pole.

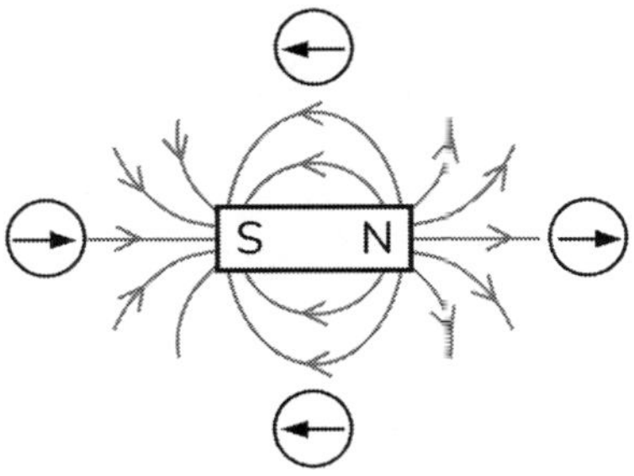

Since the direction of the field lines shows the direction a compass will point, this means the compass direction is given by the ________________ to the magnetic field line at each position.

It also means that magnetic field lines cannot ________________, because the compass will not point in two different directions.

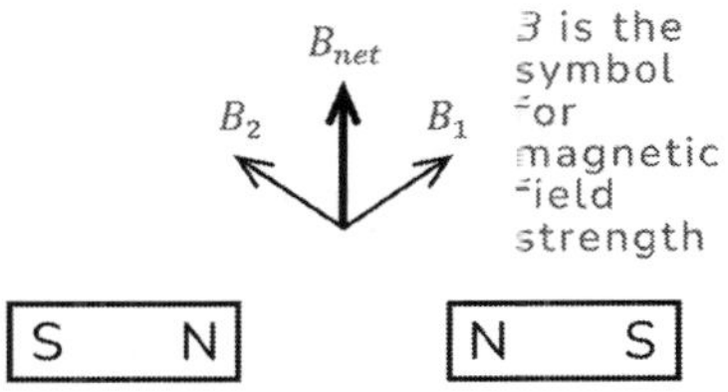

Because a magnetic field has both strength and direction, it is a ________________ quantity. So if there is a magnetic field exerted by two different magnets, the strength and direction of the magnetic field at one point is given by the ________________ of the magnetic fields from each magnet.

2. A bar magnet is shown to the right.

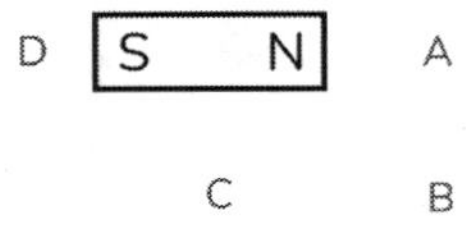

a) Draw an arrow at each point A, B, C, and D, showing which way a compass would point if it were placed at that point.

b) The magnetic field lines around the bar magnet are now shown. Describe the relationship between the arrows you drew in part (a) and the magnetic field lines shown.

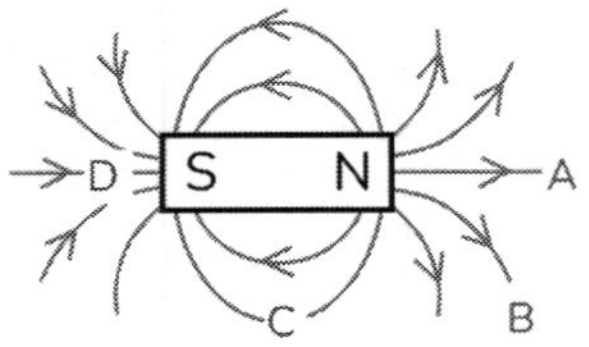

---

c) Explain which point, A, B, C, or D, has a stronger magnetic field.

---

d) The bar magnet is rotated by 90° as shown. Draw the magnetic field lines around it.

3. The magnetic field between pairs of bar magnets is shown. Identify whether X and Y are north or south poles on each.

a)

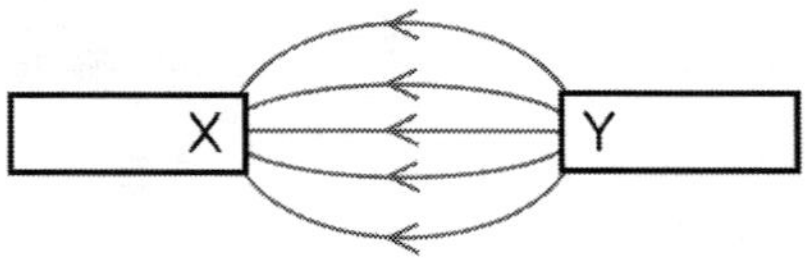

b)

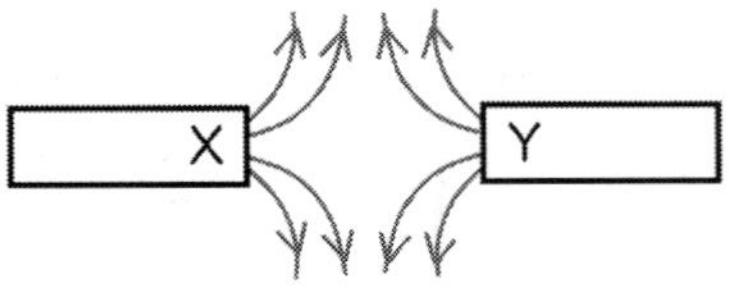

4. For each pair of bar magnets below, draw the magnetic field between the pair of bar magnets and compare the strength of the magnetic field at points X and Y.

a)

---

b)

---

5. Draw an arrow showing the direction of the magnetic field at point X. Show your working, by drawing the magnetic field from each individual magnet, then finding the resultant.

a)

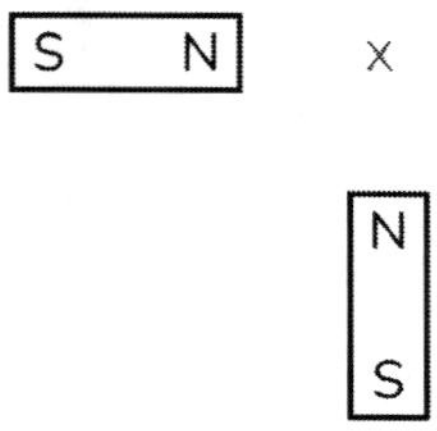

b)

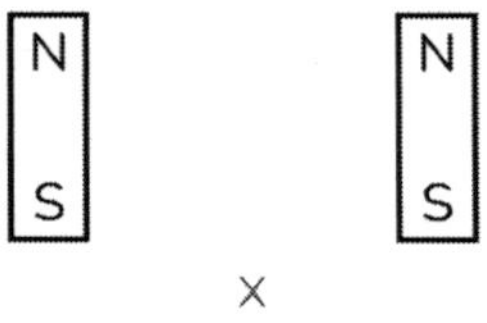

c)

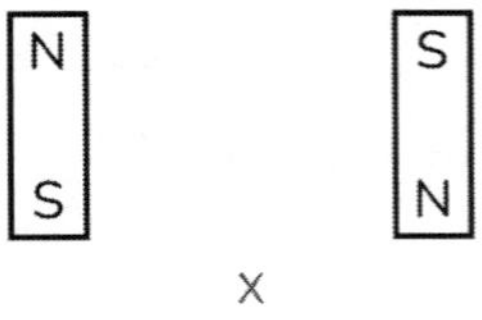

6. Magnets come in a variety of shapes and sizes, but they all have a north pole and south pole. To model the magnetic field using field lines, the same rules apply. Draw the magnetic field lines around the following magnets, focusing on the area near or between the poles.

a) Disc magnet:

Often used for holding things, for example fridge magnets.

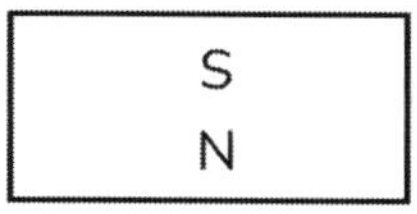

b) Horseshoe magnet:

Good for gripping heavy objects, looks good in cartoons.

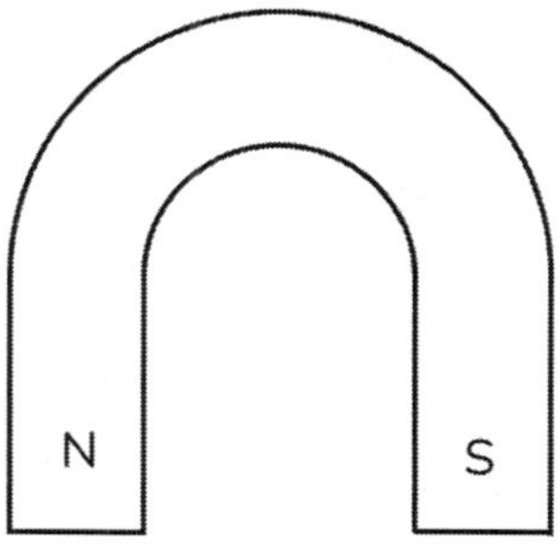

c) Curved magnets:

Used in motors, analog ammeters and voltmeters, so that electric current through a coil in the magnetic field makes the coil turn.

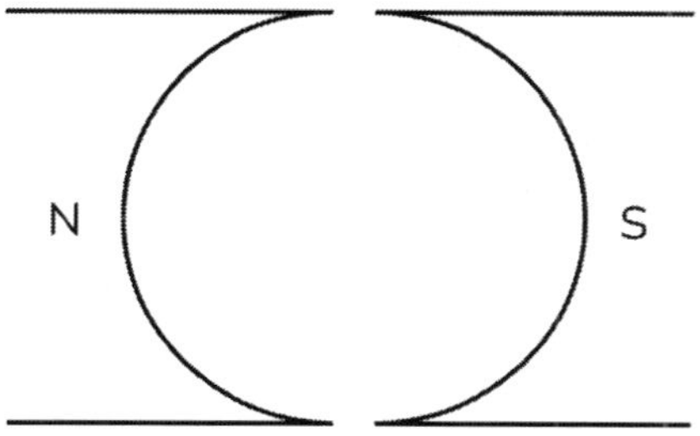

These are the poles of two curved magnets, which create a radial magnetic field between them. The other pole on each magnet is not shown.

d) Ring magnets:

Used in loudspeakers. Current through a coil around the central pole makes the coil vibrate in the magnetic field, which makes the speaker cone vibrate.

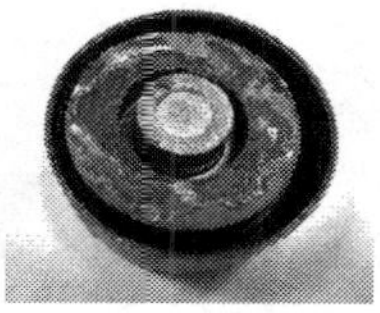

Top view:

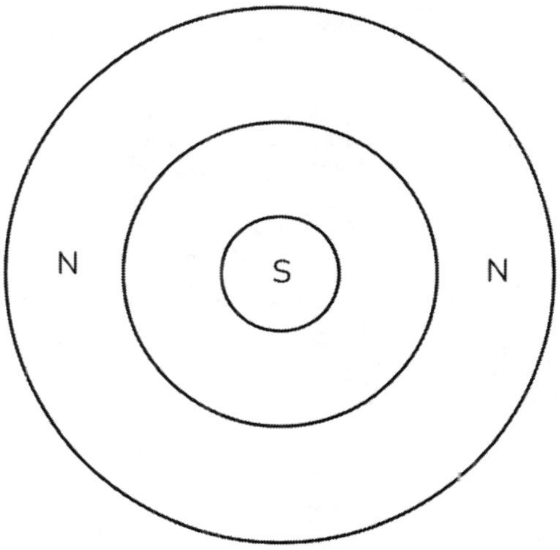

Side-on view:

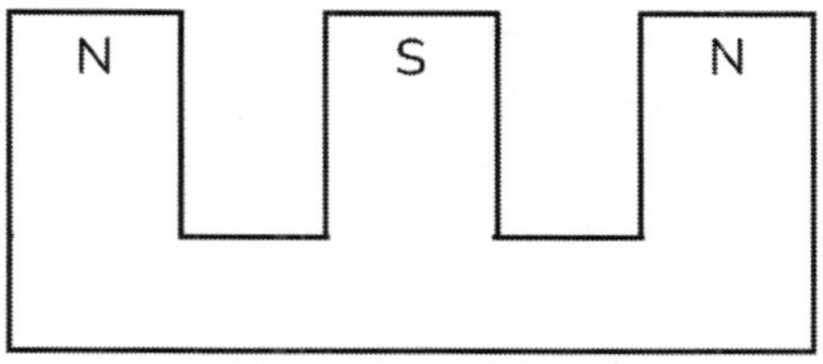

13.3 Magnetic fields around currents

Long, straight current-carrying conductors

1. Describe the magnetic field around straight conductors by filling in the blanks.

away	circle	clockwise	compass needle	cross
decreases	dot	Earth's	electromagnetism	electrons
fingers	further apart	left	positive	produce
right	straight conductor	thumb	two dimensional	

In 1820, Hans Christian Oersted showed that when a current flowed through a straight wire, it could make a ________________ deflect, so that it pointed at right angles to the wire. When the current was switched off, the compass went back to pointing in the direction of the ________________ magnetic field.

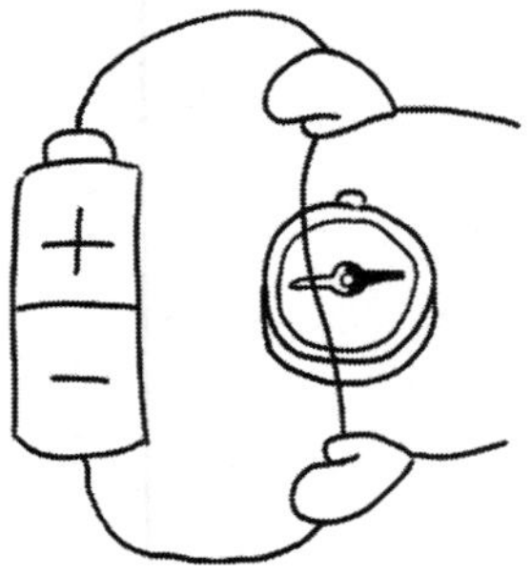

Right hand grip rule

This implies that an electric current can ________________ a magnetic field. It shows that electricity and magnetism are linked, opening up a new area of study in physics called ________________.

One way to remember the observed direction of the magnetic field around a ________________ is the "*right hand grip rule*". Point your right ________________ in the direction of the current. Then, the direction your ________________ would close if you were gripping something gives the direction of the magnetic field. This direction goes in a ________________ all the way around the current-carrying conductor. Recall that current direction is defined as the direction ________________ charge flows, which is the opposite direction to the flow of ________________.

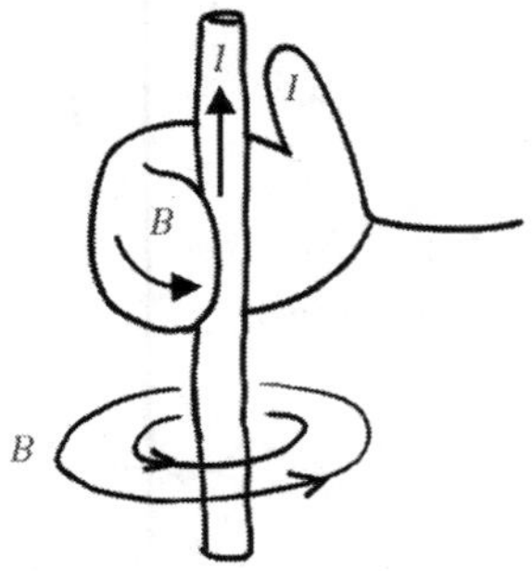

As fields extend in all three dimensions, we need new symbols to represent them on a ________________ page. We can represent the direction out of the page with a ________________ – imagine an arrow is coming towards you and what you see closest to you is the arrow tip. We can represent the direction that is into the page with a ________________ – imagine the arrow is going away from you and what you see closest to you is the tail feathers.

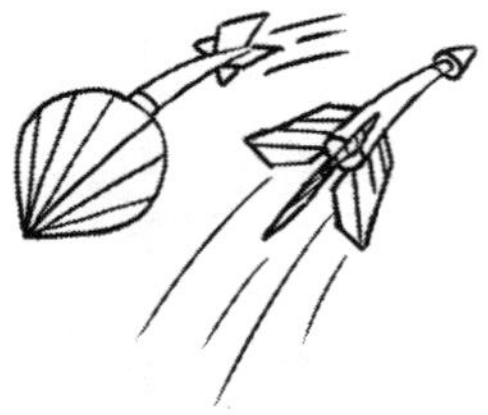

Therefore, to represent the magnetic field around a current going up the page through a straight conductor, we can use dots on the ________________ of it, to show the magnetic field is out of the page, and crosses on the ________________ of it, to show the magnetic field is into the page.

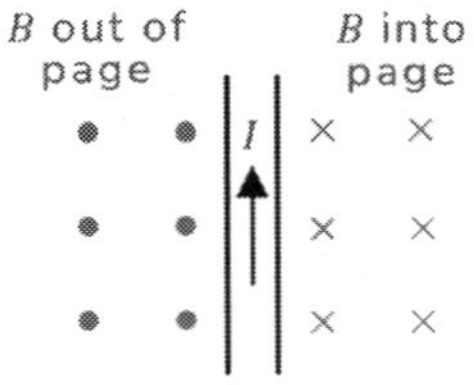

We can also use dots and crosses to represent current direction, so if the current in the conductor is travelling ________________ from you, into the page, you use a cross. The right hand grip rule would tell you that the magnetic field around this current is ________________.

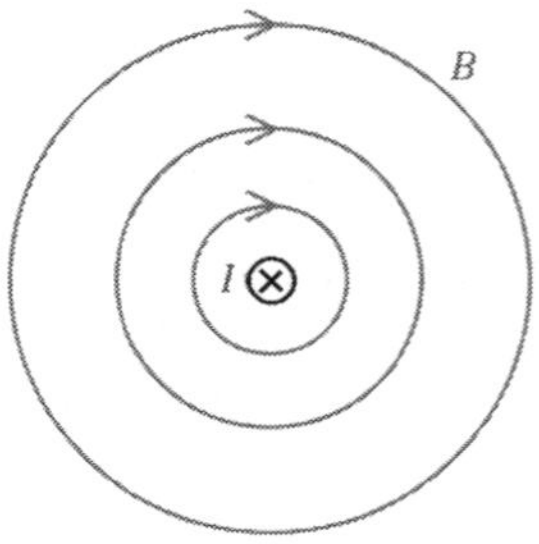

The strength of the magnetic field as you move away from the conductor ________________. Therefore, you need to draw the magnetic field lines ________________.

2. Draw the magnetic field around the following current-carrying conductors.

a)

b)

c)

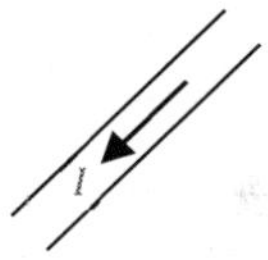

Use your right hand!

To represent fields in 2D on a page, you can draw dots, crosses or circles.

3. Two wires, A and B, carry a current of the same magnitude, as shown. In what direction is:

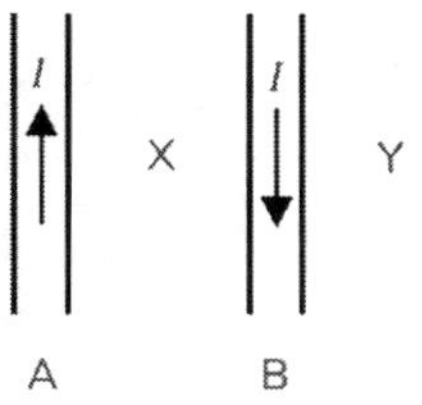

a) the magnetic field at point X due to wire A? __________________

b) the magnetic field at point X due to wire B? __________________

c) the resultant magnetic field at point X? __________________

d) the magnetic field at point Y due to wire A? __________________

e) the magnetic field at point Y due to wire B? __________________

f) the resultant magnetic field at point Y? __________________

Recall that magnetic field strength gets weaker as distance from the wire increases.

4. Circle the option that identifies the direction of the magnetic field at points X and Y in the following scenarios.

a) The two currents have the same magnitude. X is halfway between them.

X: left / right / up / down / into page / out of page / zero

Y: left / right / up / down / into page / out of page / zero

b) Y is the same distance from each side.

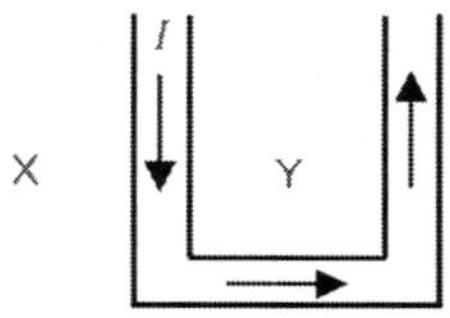

X: left / right / up / down / into page / out of page / zero

Y: left / right / up / down / into page / out of page / zero

Recall the resultant magnetic field is the vector sum of the magnetic fields from each object producing one.

If you feel like you're twisting your arm, try rotating the page instead.

5. The direction of the magnetic field at a point, near a straight current-carrying conductor, is either stated or shown by a compass. State the direction of the current.

a)

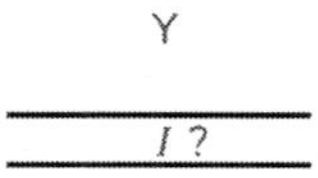

Magnetic field at Y is into page

Current direction: ______________

b)

Current direction: ______________

c)

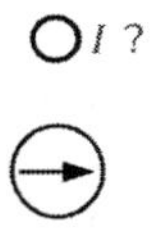

Current direction: ______________

6. Alternating current (AC) means current that switches direction periodically. A power source of alternating current is connected to a high resistance conductor, as shown. Explain what happens to the magnetic field at point X.

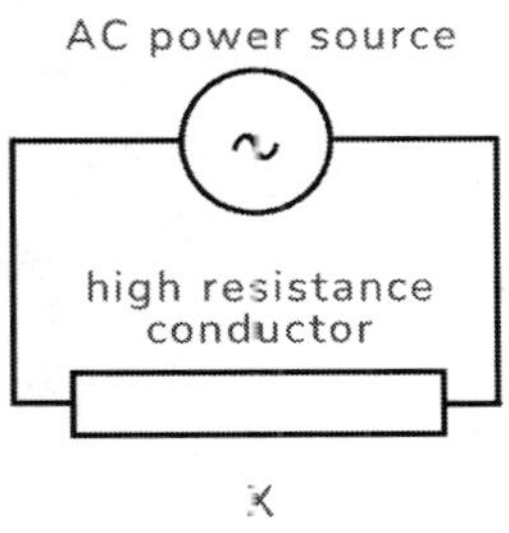

____

____

____

Current-carrying coils

7. Describe the magnetic field around a current-carrying coil by filling in the blanks.

bar magnet	compass	current	fingers
into the page	magnetic field	north	right hand coil
right hand grip	single coil	south	stronger
thumb	turns	vector sum	weaker

When current flows along a conductor that is not straight, it still produces a ________________. The magnetic field at a point will be the ________________ of the magnetic field from each bit of the conductor.

Right hand coil rule

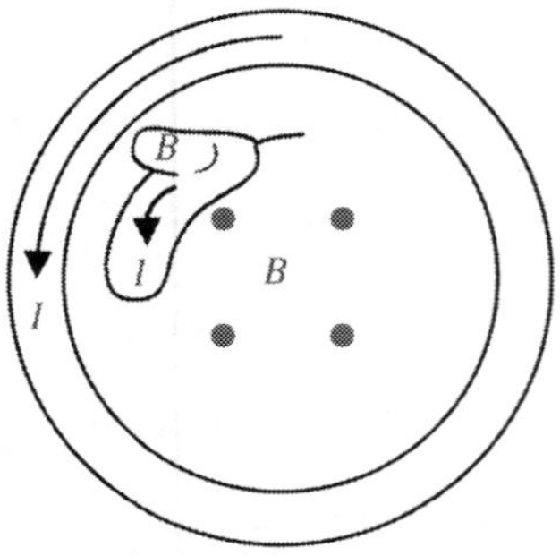

The magnetic field inside a circular coil with a current travelling clockwise in it will therefore be ________________. This is because if we use the "________________ rule" and stick our thumb on each point on the circular coil in the current direction, our ________________ always point into the page in the middle of the coil.

To work out the direction of the magnetic field in a coil faster, we can use the "***right hand coil rule***". This time, your fingers wrap in the direction of the current in the coil and your ________________ represents the direction of the magnetic field inside the coil.

A ***solenoid*** consists of a long wire looped around many turns around a cylinder. Each turn contributes a magnetic field inside the solenoid the same way a ________________ does. Therefore, the more turns, the ________________ the magnetic field inside the solenoid will be, with the direction given by the ________________ rule. Solenoids can often have hundreds of ________________.

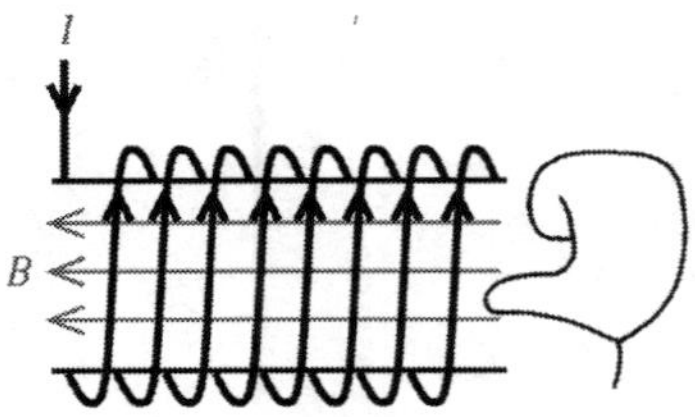

This solenoid has 8 turns

Outside the solenoid, there is still a magnetic field, which gets ________________ away from the ends of the solenoid. The shape of the magnetic field outside the solenoid is similar to the magnetic field around a ________________. Unlike a bar magnet, we can turn the magnetic field from the solenoid on and off, by switching the ________________ on and off. This is how an ***electromagnet*** works, and these are found in automated locking systems.

The magnetic field lines outside the solenoid go from the ________________ pole to the ________________ pole of the electromagnet. However, inside the solenoid, they go from the south pole to the north pole, since this is the direction that a ________________ inside the solenoid would point.

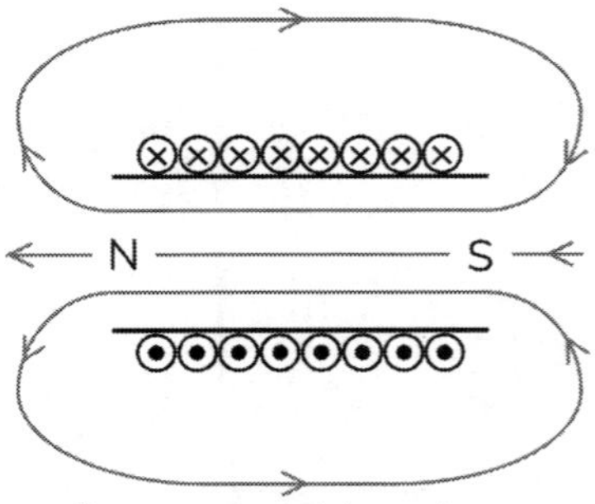

Same solenoid as above, cross section shown

8. Which way must the current travel to produce the magnetic field shown? Circle the correct answer.

a)

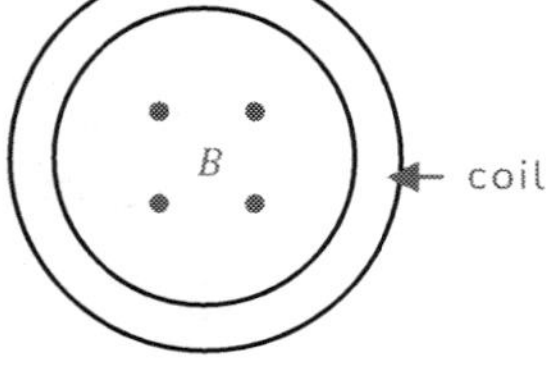

Clockwise / anti-clockwise

b)

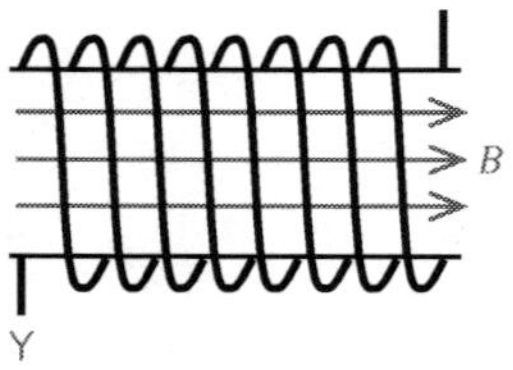

From X to Y / From Y to X

c)

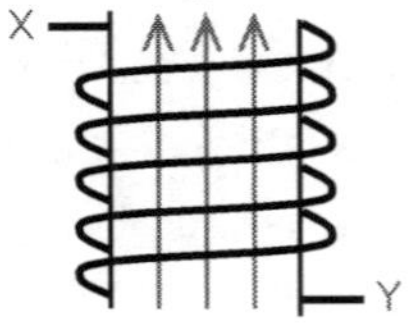

From X to Y / From Y to X

9. A magnetic field is produced by a current in the following scenarios.

a) Draw the magnetic field inside the coil.

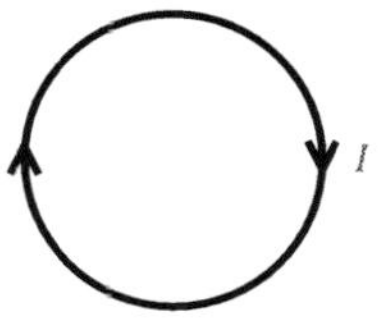

b) Draw the magnetic field in the solenoid and identify the poles.

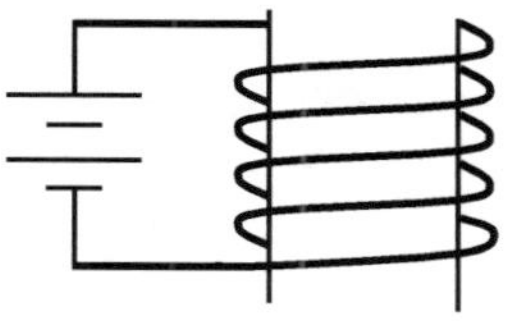

c) Draw the magnetic field between the solenoids and identify the poles.

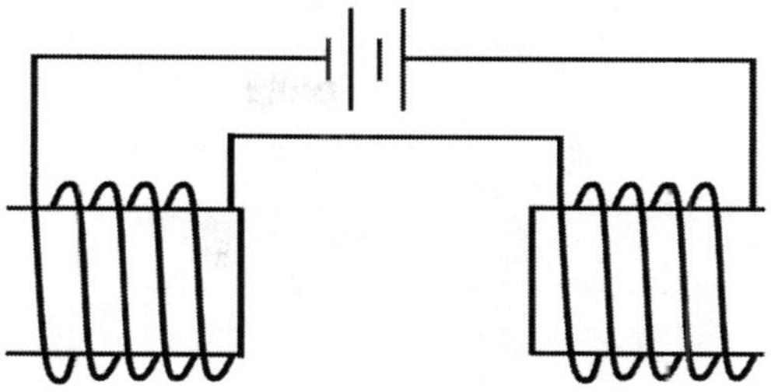

10. In the TV series Stranger Things, a curious group of kids notice that their compass points to a laboratory as they walk around it, instead of pointing north. Propose and explain what could be underground at the laboratory if this happened in real life.

....

....

11. A similar magnetic field is produced by two different objects, shown below. Add a drawing of each of the objects that could create the magnetic field.

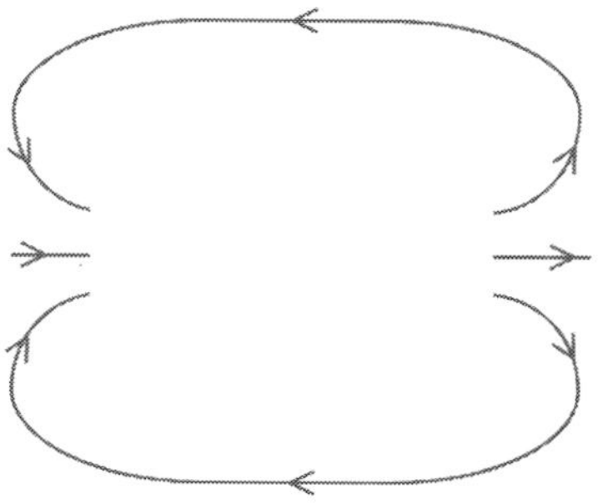

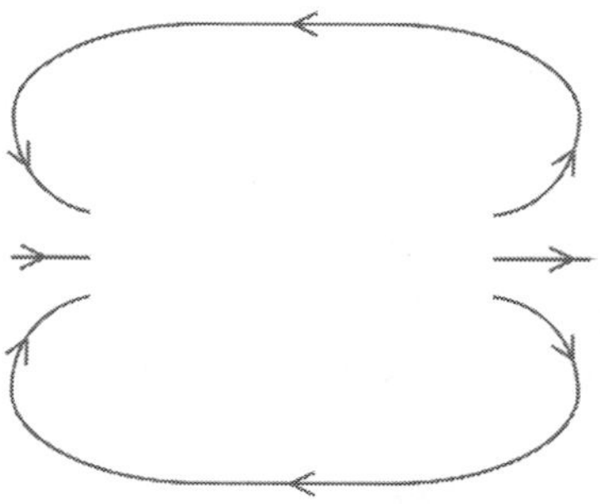

12. Explain why the magnetic field inside a solenoid is stronger than the magnetic field produced by a straight conductor, carrying the same current.

....

....

13. A bar magnet hangs from a rubber band, above a solenoid attached to a power source. Explain what happens to the magnet when the switch is closed.

....

....

....

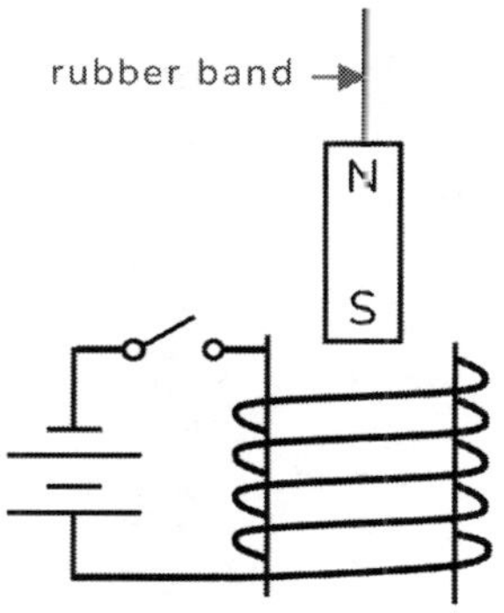

13.4 Magnetic field strength

Long, straight current-carrying conductors

1. Describe magnetic field strength around a straight conductor, by filling in the blanks.

constant	current	distance	force	iron	right hand grip
shape	small	vacuum	vector	vector sum	

Magnetic field strength is related to how much ______________ is exerted on magnetic objects by an external magnetic field, and is measured in ***tesla*** (T).

The magnetic field strength at a point due to a current-carrying conductor is given by the ______________ of the magnetic field strength contributed by each bit of the conductor. If you change the ______________ and size of the conductor, you change the magnetic field strength. For a long, straight current-carrying conductor, the magnetic field strength is given by:

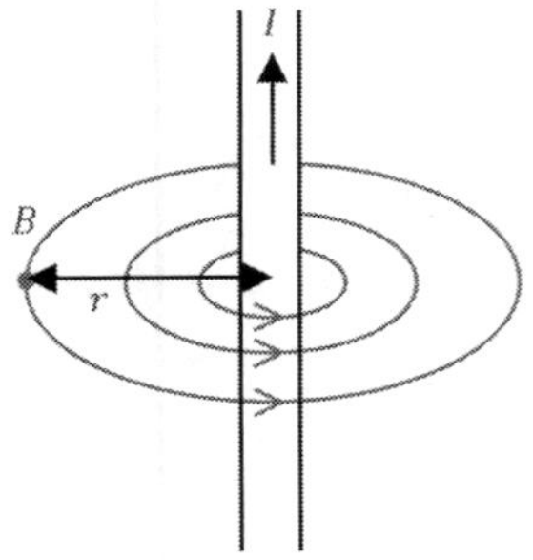

$$\vec{B} = \frac{\mu_0 I}{2\pi r}$$

where: $\vec{B}$ = magnetic field strength (T)

$\mu_0 = 4\pi \times 10^{-7}$ (N A^{-2}) (permeability of free space)

I = ______________ through the conductor (A)

r = perpendicular ______________ from the conductor (m)

μ_0 is a ______________ called the ***permeability of free space***. ***Magnetic permeability*** means how much a material affects the strength of a magnetic field passing through it. The permeability of free space is defined for a magnetic field passing through a ______________. For a magnetic material such as ______________, the permeability would be much higher.

This formula assumes that r is ______________ compared to conductor length.

Magnetic field strength is a ______________, so it has a magnitude and a direction, given by the ______________ rule for a straight conductor.

Free smartphone apps allow it to be used as a compass, to measure magnetic field direction, or a magnetometer, that measures magnetic field strength. The reason why smartphones have a magnetometer built in is so navigation using maps works.

Example: Two long wires are shown, one carrying a current of 1.5 A and the other 1.0 A. Determine the magnetic field at points X and Y.

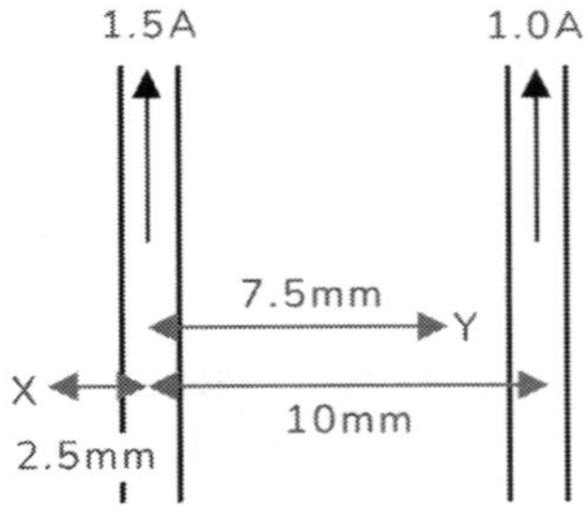

Answer:

At X from left wire: $\vec{B} = \frac{\mu_0 I}{2\pi r} = \frac{4\pi \times 10^{-7} \times 1.5}{2\pi \times 2.5 \times 10^{-3}} = 1.2 \times 10^{-4}$ T out of page

At X from right wire: $\vec{B} = \frac{\mu_0 I}{2\pi r} = \frac{4\pi \times 10^{-7} \times 1}{2\pi \times 12.5 \times 10^{-3}} = 1.6 \times 10^{-5}$ T out of page

Total at X: $\vec{B} = 1.2 \times 10^{-4} + 1.6 \times 10^{-5} = 1.36 \times 10^{-4}$ T out of page

At Y from left wire: $\vec{B} = \frac{\mu_0 I}{2\pi r} = \frac{4\pi \times 10^{-7} \times 1.5}{2\pi \times 7.5 \times 10^{-3}} = 4 \times 10^{-5}$ T into page

At Y from right wire: $\vec{B} = \frac{\mu_0 I}{2\pi r} = \frac{4\pi \times 10^{-7} \times 1}{2\pi \times 2.5 \times 10^{-3}} = 8 \times 10^{-5}$ T out of page

Total at Y: $\vec{B} = 8 \times 10^{-5} - 4 \times 10^{-5} = 4 \times 10^{-5}$ T out of page

2. The table below gives a rough idea of the size of some magnetic fields. Write each in scientific notation. Some of the table is done for you as an example.

Object	Magnetic field	Magnetic field in scientific notation
Human brain	1 pT	1×10^{-12} T
Toaster in use	500 nT	
Earth's magnetic field at equator	31 μT	
Fridge magnet	5 mT	
Neodymium magnet	1.25 T	1.25 T
Strongest continuous magnetic field produced in a laboratory	45.22 T	4.522×10^{1} T

3. The unit for magnetic field strength, tesla (T), is not an SI unit. Use the formula for magnetic field strength to deduce the equivalent unit for a tesla in SI units, comprising kg, A and s.

4. A long wire is shown, which carries a current of 2.5 A.

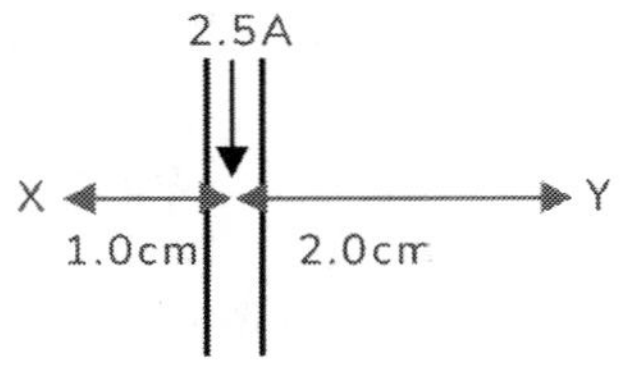

a) Determine the magnetic field at each of points X and Y.

b) Sketch a graph of the magnetic field strength vs distance from the wire.

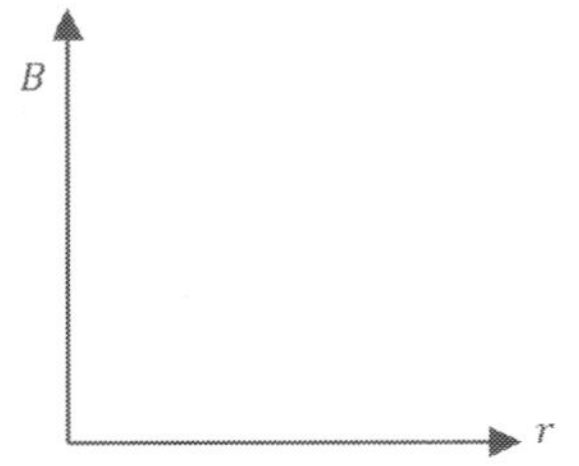

5. The cross sections of two long wires are shown, one carrying a current of 2.0 A and the other 3.0 A. Determine the magnetic field at points X and Y.

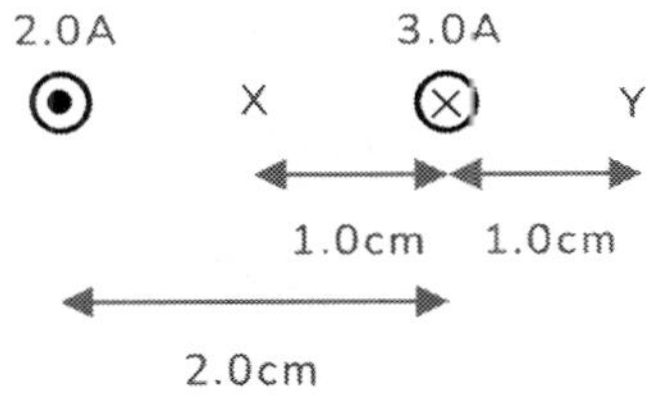

Draw vectors on the diagram to represent the magnetic field from each wire, to help with working out the resultant.

6. In a Sydney laboratory, the horizontal component of Earth's magnetic field is measured to be 50 μT and makes a compass sitting horizontally on a bench point north. A long, straight wire is positioned 1.5 cm above the compass, parallel to the bench top, running in a north-south direction. What current would be required to make this compass point north east, and which way must it flow through the wire?.

Vector diagram

____

____

____

____

Use pens as visual aids to represent the compass, magnetic fields and the wire.

Solenoids

7. Describe magnetic field strength due to a solenoid, by filling in the blanks. Use the diagrams to help you.

coil	current	length	length	magnetic field	permeability
proportional	solenoid	tightly	turns	turns	uniform

A coil with many turns is called a ______________. If a solenoid is much longer than its diameter, and the coils are very close together, the magnetic field inside the solenoid is ______________. Its direction is given by the right hand ______________ rule.

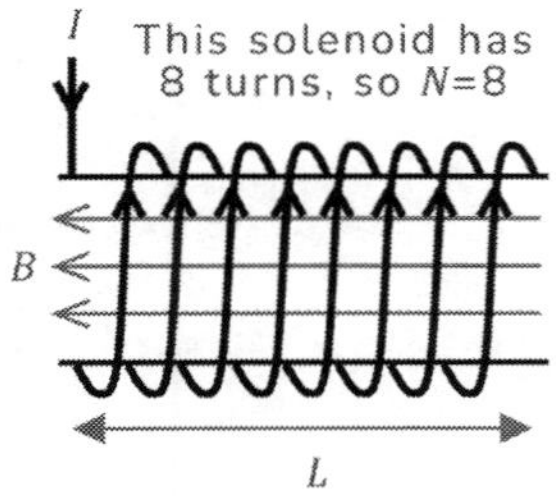

The magnetic field strength inside a solenoid is given by:

$$\vec{B} = \frac{\mu_0 N I}{L}$$

where: $\vec{B}$ = ______________ strength (T)

$\mu_0 = 4\pi \times 10^{-7}$ (N A^{-2}) (______________ of free space)

I = ______________ through the conductor (A)

N = number of ______________

L = ______________ of solenoid (m)

Same solenoid as above, cross section shown

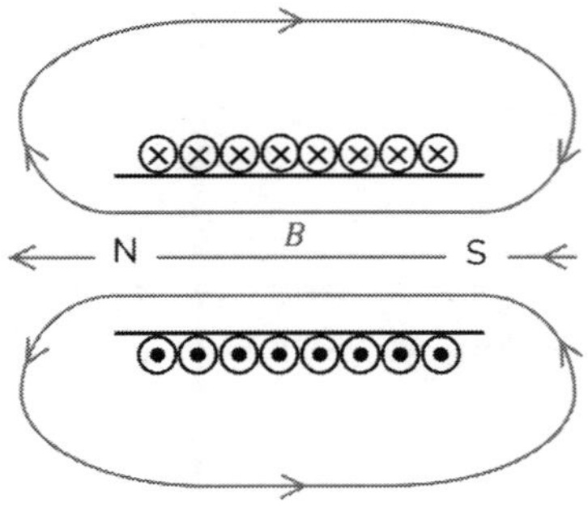

The formula tells us that the magnetic field strength is ______________ to the current, which means that the strength of an electromagnet made using a solenoid can be varied by changing the current.

The magnetic field strength can also be increased by increasing the number of ______________ or decreasing the ______________ of the solenoid. This means that packing the coils ______________ together, so there are more turns per length, will increase the magnetic field.

Example: A solenoid has a 13 cm coil length. When a current of 5.0 A flows through the solenoid, a magnetic field of 27 mT is produced inside the coil. How many turns does the solenoid have?

Answer: $\vec{B} = \frac{\mu_0 N I}{L}$

$0.027 = \frac{4\pi \times 10^{-7} \times N \times 5}{0.13}$

$N = 560$ turns

8. A small solenoid has a length of 6.00 cm and 2160 turns, with a resistance of 131 Ω. When connected to a 12.0 V power supply, what would be the magnetic field inside the solenoid?

9. The graph shows the magnetic field strength inside a solenoid as the current is varied. Calculate the number of turns per centimetre in the solenoid.

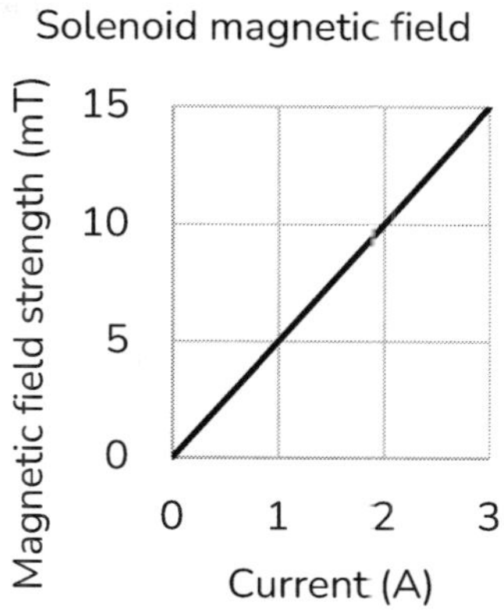

$\frac{N}{L}$ gives turns per metre

10. Explain why an effective solenoid needs to have a tightly wound coil.

11. Magnetic resonance imaging (MRI) uses magnetic fields to detect causes of injuries and tumours from cancer. MRI works by detecting hydrogen nuclei in us, which each have a small magnetic field. A patient undergoing MRI is put inside a large solenoid, which produces a strong magnetic field that makes the magnetic fields of the hydrogen nuclei align in the same direction. Also, a pair of coils, called RF coils, sit on each side of the large solenoid. A current through the RF coils is turned on for a short time, to add a second magnetic field. Once the current in the RF coils is turned off, the time taken for the hydrogen nuclei to re-align with the original magnetic field gives information about the patient's tissues.

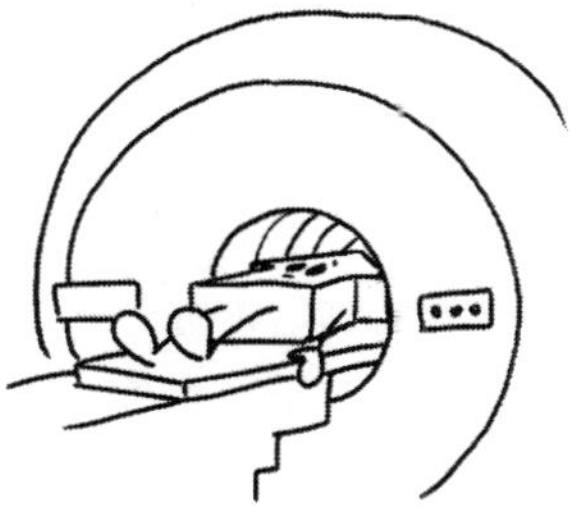

Side view

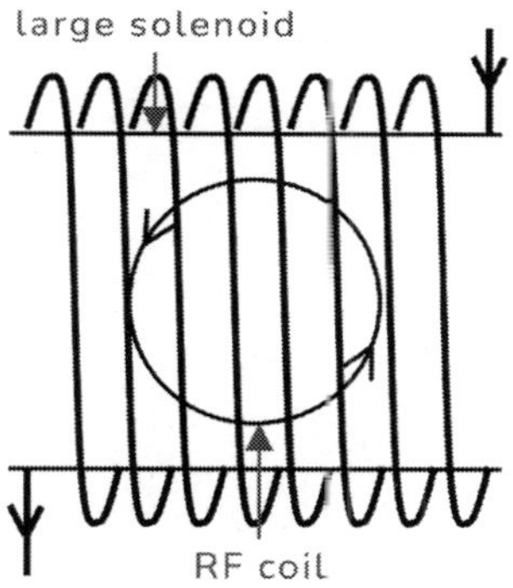

Top down view
cross sect on

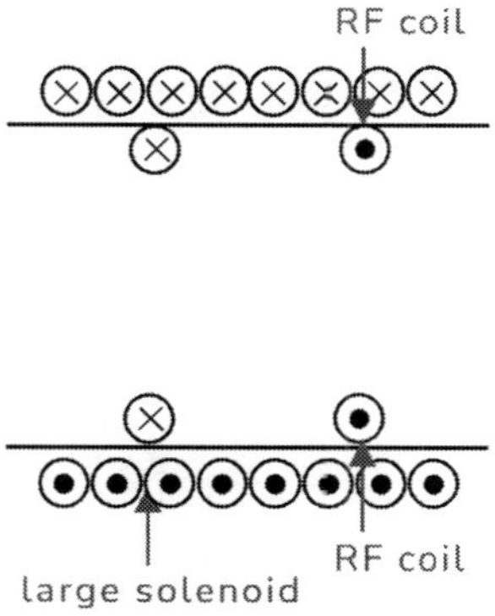

The diagrams show a simplified model of the large solenoid and RF coils, when they are both carrying a current.

a) Sketch a vector diagram showing the direction of the resultant magnetic field, in the top down view.

b) When the current in the RF coil is on, the resultant magnetic field direction changes by 18.0° from the magnetic field due to the large solenoid alone. The RF coil alone produces a magnetic field of 0.500 T. The large solenoid is 1.00 m long and has 9550 turns. What is the current in the large solenoid?

13.5 Ferromagnetism

1. Describe ferromagnetism, by filling in the blanks.

added to	agitated	align	cancel	different	direction
domains	ferromagnetic	iron	magnetic field	moving charges	nucleus
neodymium	unlike	not	negatively charged	oxygen	permanent
same	short time	strongly	no magnetic moment	weakly	

Atoms have ________ electrons that orbit around a ________ and spin on their axes. ________ create currents, and currents create magnetic fields. So a moving electron creates a tiny magnetic field, called a ***magnetic moment***, meaning it will rotate to ________ itself with an external magnetic field.

In this diagram, the magnetic moments are represented by the arrows going up.

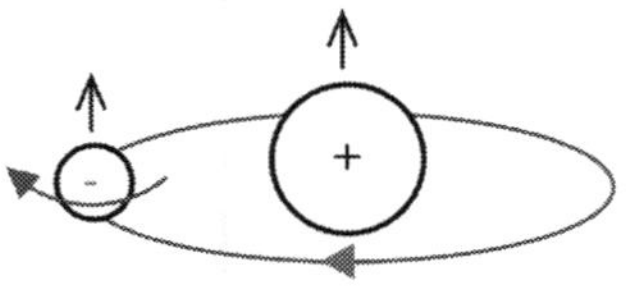

In many atoms, the magnetic moments of all the electrons cancel out, so overall the atom has ________. Elements with atoms like this, such as copper, are ________ attracted to magnets.

However, ***paramagnetic*** materials, such as aluminium and ________, have atoms with a permanent magnetic moment. The magnetic moments of all the atoms in the material are in random directions, so they ________ each other out, leaving no magnetic field overall. Magnetic moments in paramagnetic materials will align themselves in the ________ direction as an external magnetic field though. So paramagnetic materials are ________ attracted to magnets.

Paramagnetism: magnetic moments of atoms when there is no external magnetic field

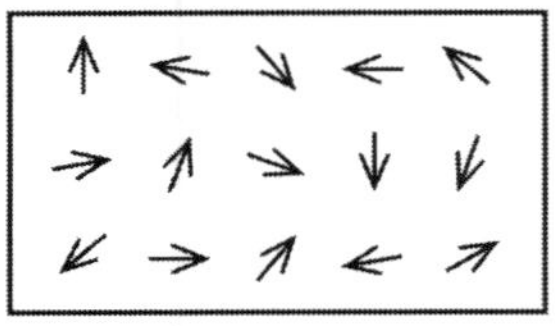

Paramagnetism in external magnetic field

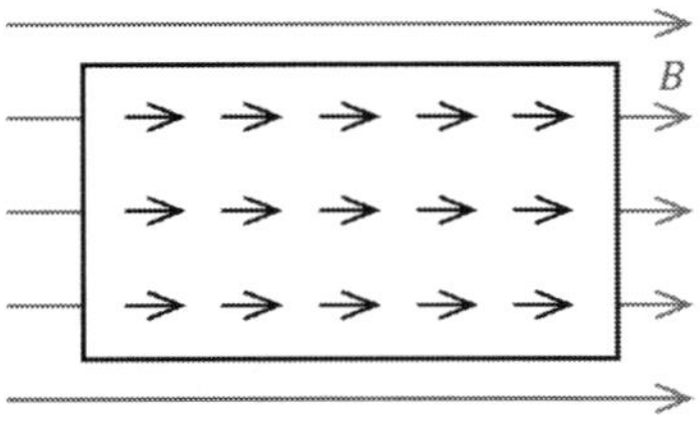

Ferromagnetic materials such as ________, nickel and cobalt, also have atoms with a permanent magnetic moment. ________ paramagnetic materials, in ferromagnetic materials, the atoms group into regions called ***domains***, in which neighbouring atoms have their magnetic moments aligned in the same direction. Normally, the domains' magnetic fields are in ________ random directions. However, the domains' magnetic fields will align themselves in the same direction as an external ________. This causes ferromagnetic materials to be ________ attracted to magnets. The total magnetic field is increased, because the magnetic field from the ferromagnetic material is ________ the external magnetic field.

Ferromagnetism: magnetic moments within each domain are aligned, even when there is no external magnetic field

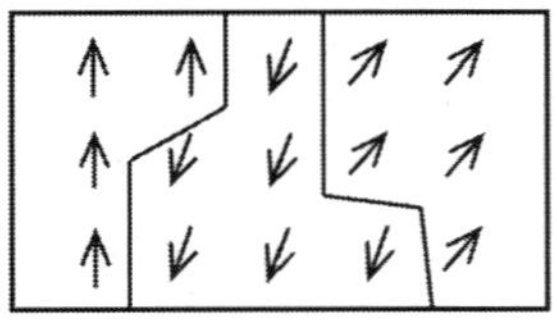

When the external magnetic field is removed, the domains retain the same alignment for a ________. This is why stroking an iron nail many times with the same pole of a magnet will temporarily magnetise the nail too, because the domains' magnetic fields are now in the same ________. The nail is demagnetised when the domains return to random directions by being ________, for example by hitting or heating the nail. This type of iron is called ***soft iron***, where the overall magnetic field is not ________.

Ferromagnetism in external magnetic field

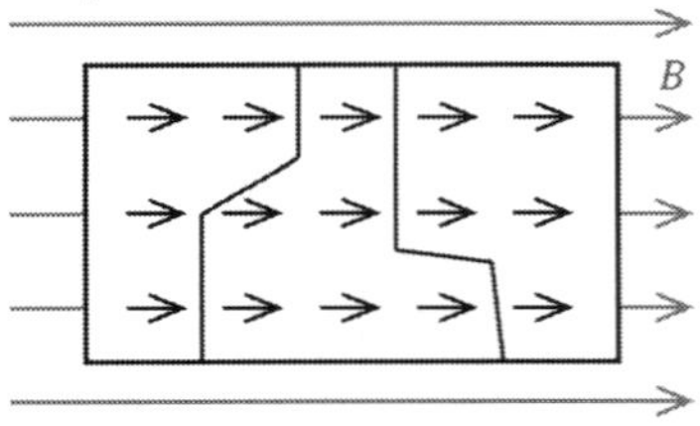

Permanent magnets are made of ***hard iron***, with added impurities that stop ________ returning to random alignments.

Alloys containing iron are often also ________. For example, strong neodymium magnets contain a mixture of the elements iron, ________ and boron.

2. The model shows an electron spinning on its axis and orbiting around a nucleus. Explain why this creates a magnetic field pointing upwards.

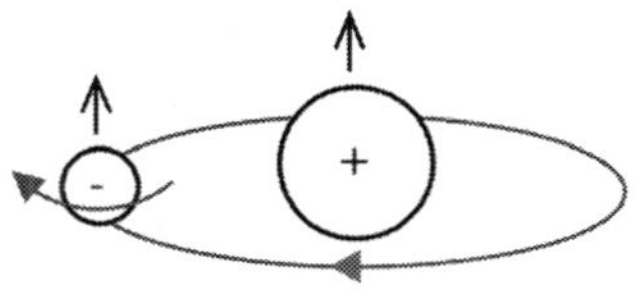

3. Compare paramagnetic and ferromagnetic materials.

4. Explain the difference between soft iron and hard iron.

5. Compare the behaviour of a piece of copper, a piece of soft iron and a compass, when a bar magnet is brought close to each.

Copper will

This is because

Soft iron will

This is because

A compass will

This is because

6. A pin made of iron is stroked from blunt to pointy end with the south pole of a bar magnet 50 times.

a) The pictures below show the domain walls in the pin and the magnet. Use arrows to represent the magnetic moments in the domains.

The unmagnetised pin

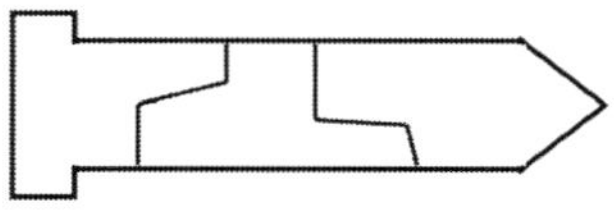

The bar magnet and pin after stroking

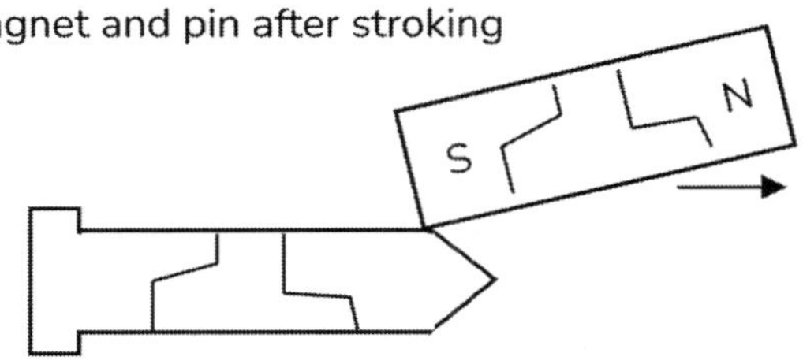

b) After stroking it with a magnet, the pin can now be used to pick up an unmagnetised pin. Explain why, with reference to the atoms inside the pin.

c) If the magnetised pin is threaded through a small piece of wax paper and floated in water, it will rotate to point in the same direction each time, if there are no magnets and electric circuits nearby. Explain why.

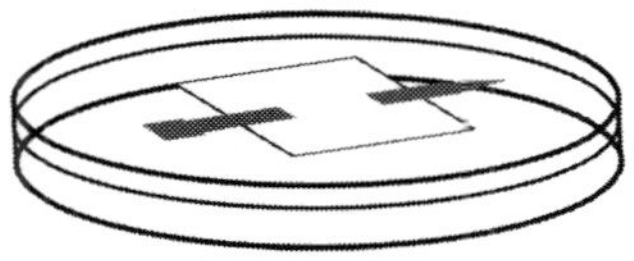

7. An electromagnet consists of a solenoid, or long coil of wire, wrapped around a soft iron core, as shown in the diagram.

Electromagnets are often used to pick up objects containing iron in real life, such as scrap metal, or in automated door locks, and may be connected in a circuit as shown.

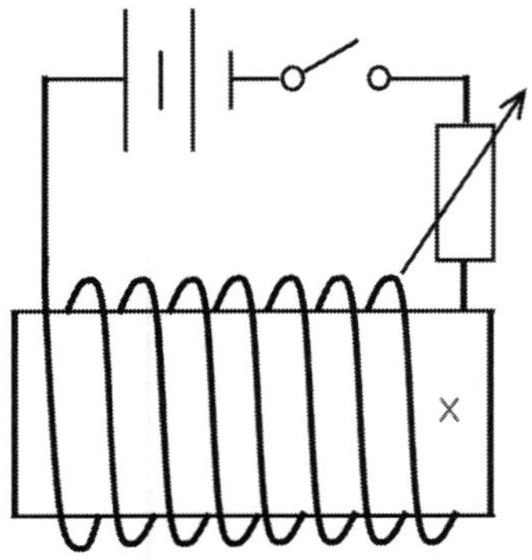

a) Identify whether X would be a north or south pole in the diagram, when the switch is closed.

Identify the direction of current in the coil first.

b) Explain the advantages of an electromagnet, compared to a bar magnet.

"Advantages" means more than one advantage must be explained.

The diagram gives hints about what can be changed in an electromagnet.

Refer to relevant equations in explanations.

c) The voltage of the power source is 120 V and the total resistance of the circuit is 50 Ω. The solenoid has 1000 turns and a length of 25 cm. The soft iron core is removed so there is just air in the core. Calculate the magnetic field inside the solenoid when the switch is on.

Which two equations are needed?

d) Justify the use of a soft iron core in the electromagnet, instead of an air core or a hard iron core.

Compare the magnetic fields in each case, to explain why it is an advantage to use a soft iron core.

8. True or false? Explain your reasoning.

a) All metals are magnetic.

b) If a bar magnet is cut in half, each piece will still have a north pole and south pole.

c) Soft iron is not as strong as hard iron, so it cannot be used for construction.

d) The copper used in the coil of an electromagnet is ferromagnetic.

9. A compass actually does not point north exactly... thinking this way is a simplification! Geographic north is defined as the direction to the North Pole, which is a point on Earth's axis of rotation. However, the Earth rotating has no direct link to creating a magnetic field around Earth. It therefore cannot explain why a compass should point to the North Pole.

The angle between a compass' direction and where geographic north lies can vary up to 30° depending on where you are on Earth. So hikers need to know how much to correct compass readings by in their location!

a) The diagram shows the shape of the field lines around Earth, in a simplified model of the Earth's magnetic field. In this model, the Earth behaves like there is a giant bar magnet in its core, as shown. Explain whether X is the north pole or south pole of this imaginary magnet.

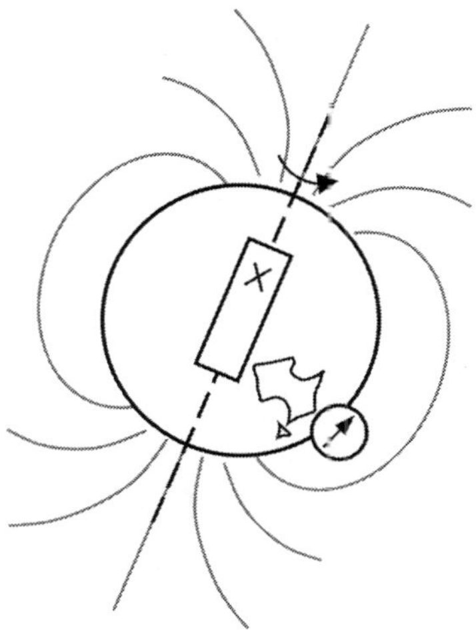

b) **Challenge:** The deeper into the Earth, the hotter it gets. This is why humans have only been able to dig down to 12.262 km underground, at the Kola Borehole in Russia. At high temperatures, ferromagnetism is destroyed. Explain why this happens, and relate this to why a simplified model of the Earth behaving like a bar magnet has limitations.

Simplified models can still be useful in science, but it is important to know the limitations, or problems, with them.

c) **Challenge:** Look at the field lines in the simplified model of Earth's magnetic field. Apart from a bar magnet, what else could produce a magnetic field of this shape? What would need to happen inside the Earth if your proposal was a better model to explain Earth's magnetic field?

The most accepted hypothesis at the moment for what causes Earth's magnetic field is called the dynamo theory. Look it up – what you have been learning about magnetic fields will help you understand it.

Part 4: Waves and thermodynamics

D	E	U	N	I	E	L	E	C	T	R	O	M	A	G	N	E	T	I	C	C
E	C	E	F	O	O	T	D	O	P	P	L	E	R	E	F	F	E	C	T	O
S	E	D	I	N	F	R	A	R	E	D	I	D	S	C	O	V	E	R	O	N
T	U	W	A	V	E	L	E	N	G	T	H	I	L	E	D	T	H	E	T	S
R	G	P	R	I	R	E	E	F	R	E	Q	U	E	N	C	Y	N	H	A	T
U	O	U	E	B	S	R	E	E	R	F	F	M	N	E	R	C	T	W	L	R
C	U	E	S	R	E	T	A	E	O	A	C	B	S	B	E	A	T	H	I	U
T	M	G	O	A	P	S	D	D	H	Y	C	U	M	L	S	E	V	C	N	C
I	K	I	N	T	I	O	R	N	I	L	A	T	E	N	T	H	E	A	T	T
V	L	O	A	I	T	U	S	S	T	A	N	D	I	N	G	W	A	V	E	I
E	P	S	N	O	R	N	T	I	A	D	T	R	U	V	H	T	Y	I	R	V
E	E	P	C	N	U	D	W	O	T	F	G	I	K	M	E	A	C	C	N	E
B	O	E	E	S	V	N	X	Z	W	I	H	N	O	E	A	I	O	O	A	L
N	C	C	D	I	S	P	E	R	S	I	O	N	L	N	T	B	N	N	L	G
U	O	I	L	O	N	G	I	T	U	D	I	N	A	L	R	C	V	D	R	H
O	M	F	T														E	U	E	N
X	P	I	Y														C	C	F	X
C	R	C	N														T	T	L	R
T	E	H	I	V	I	S	I	B	L	E	L	I	G	H	T	N	I	I	E	A
E	S	E	E	S	E	N	T	D	E	L	L	E	G	L	O	A	O	O	C	Y
M	S	A	N	C	Y	R	T	W	I	A	P	E	R	I	O	D	N	N	T	S
P	I	T	C	H	A	W	S	E	O	F	P	I	T	C	H	M	A	N	I	O
E	O	C	T	F	V	G	U	E	R	I	F	A	S	I	O	N	B	F	O	A
R	N	A	R	T	Y	A	L	O	S	F	U	R	R	L	W	R	U	D	N	M
A	H	P	A	O	Y	M	T	A	N	Q	E	C	A	R	T	A	G	D	H	P
T	V	A	N	T	W	M	R	E	E	K	U	R	U	C	I	E	V	D	V	L
U	D	C	S	V	U	A	A	E	U	R	R	A	E	T	T	L	L	E	N	I
R	K	I	V	D	L	R	V	B	R	J	E	L	R	N	R	C	N	N	E	T
E	U	T	E	H	R	A	I	E	F	A	F	F	J	E	C	O	M	E	C	U
P	V	Y	R	M	W	Y	O	R	D	E	D	B	R	E	L	E	U	R	U	D
P	H	A	S	E	R	E	L	C	R	I	T	I	C	A	L	A	N	G	L	E
R	E	W	E	R	A	R	E	F	A	C	T	I	O	N	C	N	W	Y	H	O
G	L	A	S	R	I	N	T	E	N	S	I	T	Y	C	A	T	E	P	S	L

14.1 Riding the waves

Instead the water moves up and down.
Yes, all waves involve vibrations around an *equilibrium position*,
where the particles would be if there was no wave.
The maximum displacement from the equilibrium position is called the *amplitude*.
That happens at the *crests* and *troughs*.
We call the distance between crests the *wavelength*.
So we are half a wavelength apart.
The time taken for one complete oscillation
is a *period*
and the number of oscillations per second is the *frequency*.
But I don't want to just oscillate... I want to ride to the beach!
OK, we'll need to swim to where the waves break.
Now the water is not behaving like a wave anymore, so it will transfer matter.
That wave looks big, and it's about to break!
Paddle! Match the wave speed!
Guess I was too slow.
What do you mean by wave speed, anyway?
Wave speed is the distance travelled by the wave per second.
In other words, it is how fast the crests move past a point.
t = 1 second

We can calculate it using the ***wave equation***: $v=f\lambda$
where:
v = wave speed (ms^{-1})
f = frequency (Hz)
λ = wavelength (m)
Here's another big crest
Go!
I use some gravitational potential energy to give me more kinetic energy, and...
Wheee!!!
Where do waves come from anyway?
Waves are caused by vibrations. Water waves are usually created by wind on the ocean.
The wavelengths get bigger out to sea!
Yes, water waves travel faster in deeper water.
So according to $v=f\lambda$, if wave speed increases, the wavelength increases!
The speed of a wave depends on the substance, or ***medium***, it travels through.
Changing water depth is a change in medium.
When it is so shallow that the ocean floor drags on the wave, the wave breaks.
Huh. The waves also bend so that they are parallel to the shore.
Good observation. That bending of waves as they change speed is called ***refraction***.

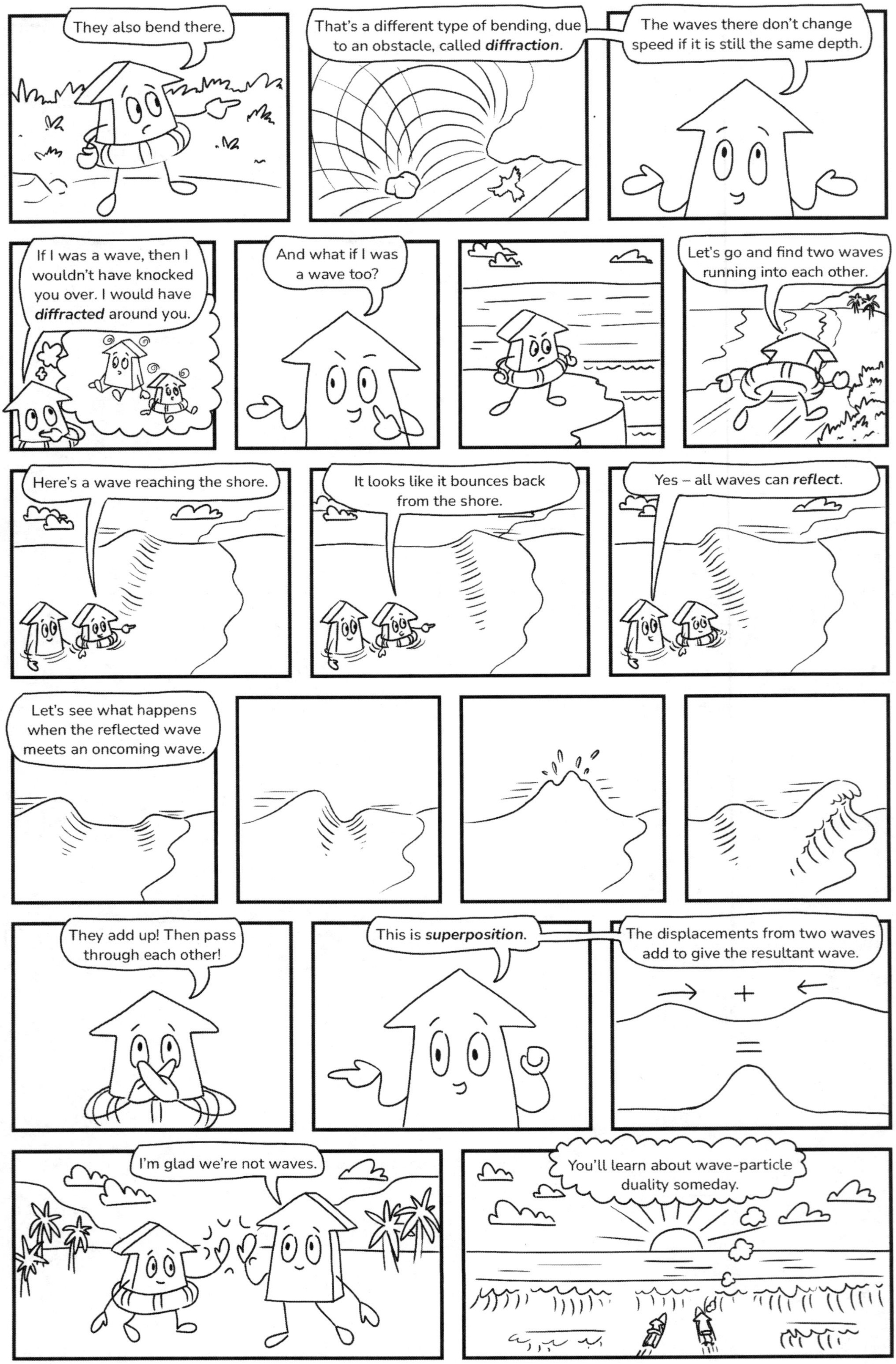
They also bend there.
That's a different type of bending, due to an obstacle, called ***diffraction***.
The waves there don't change speed if it is still the same depth.
If I was a wave, then I wouldn't have knocked you over. I would have ***diffracted*** around you.
And what if I was a wave too?
Let's go and find two waves running into each other.
Here's a wave reaching the shore.
It looks like it bounces back from the shore.
Yes – all waves can ***reflect***.
Let's see what happens when the reflected wave meets an oncoming wave.
They add up! Then pass through each other!
This is ***superposition***.
The displacements from two waves add to give the resultant wave.
+
=
I'm glad we're not waves.
You'll learn about wave-particle duality someday.

14.2 Describing waves

What is a wave?

1. Describe what is classified as a wave, by filling in the blanks.

air	energy	gases	light	matter	medium
no particles	particles	sound	vacuum	vibrating	

A ***wave*** is something that transfers ______________ but not ______________. For example, when someone talks, sound is transferred from their mouth to your ear, but ______________ is not. Heat and light are forms of energy that reach us from the Sun, but this process does not involve ______________ being moved from the Sun to Earth.

In waves, particles may still be moving, but they are ______________ on the spot.

Mechanical waves need a ______________ to travel through, which contains particles of matter. For example, ______________ is a mechanical wave and can travel through solids, liquids and ______________. However, sound cannot travel through a vacuum, which is a region of space with ______________.

Electromagnetic waves can travel through a ______________. For example, ______________ can travel through space, even though there are very few particles in it.

2. "A rock drops into water and makes a splash. As energy has been transferred from the rock to the water, the rock is a wave." Assess whether this statement is correct.

A wave is defined as ____.

Similar to a wave, the rock ____.

However, unlike a wave, the rock ____.

Therefore, ____.

3. When a rock is dropped in water, ripples spread out. When the ripples reach a leaf floating on the water, the leaf bobs up and down.

a) Explain why the ripples fit the definition of a wave.

____

____

____

____

b) Explain whether the ripples are mechanical or electromagnetic waves.

____

____

Wave features

4. Describe features used to describe waves, by filling in the blanks.

acceleration	along	amplitude	away	bunch up	compressions
consecutive	crests	equilibrium	highest	in phase	lowest
out of phase	perpendicular	rarefactions	same	sine wave	spread out
trough	vibrations	water waves	wavelength		

All waves involve ________, resulting in repeating patterns in the medium the wave travels through.

In *transverse waves*, these vibrations are ________ to the direction in which energy travels. Transverse waves often look like a ________. ________ behave approximately like transverse waves.

The ***amplitude*** is the furthest distance a particle vibrates away from its normal position, or ________ position, when there is no wave. In transverse waves, ***crests*** are the ________ points displaced by the amplitude and ***troughs*** are the ________ points.

Points on the wave which are doing the same thing at the ________ time are described as *in phase*. For example, all the points on the wave that are at a crest at the same time are ________. This means they have the same displacement, velocity and ________. All the points that are at a ________ are also in phase with each other. The distance between two ________ points in phase is called the ***wavelength***. ________ and troughs are half a wavelength out of phase with each other.

Transverse wave

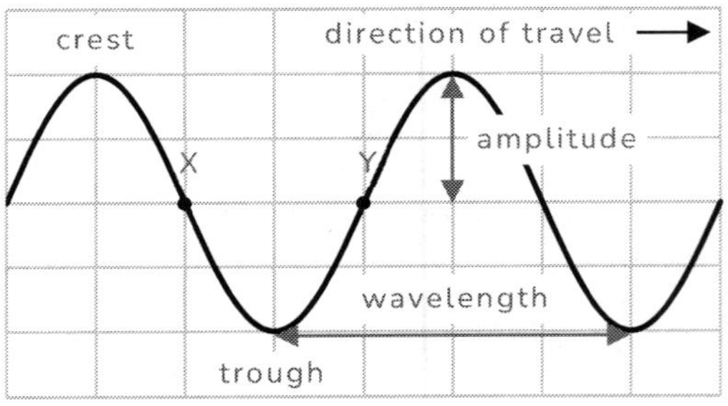

The same transverse wave a moment later

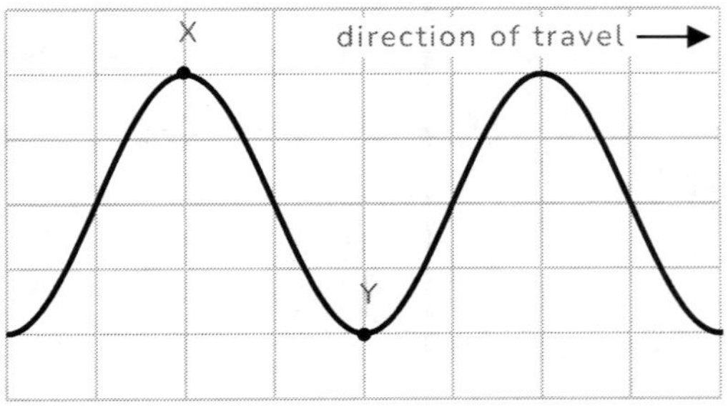

In *longitudinal waves*, vibrations are ________ the direction in which energy travels. Sound is a longitudinal wave, so when you hear a sound, it means the particles between the source of the sound and your ear are moving towards then ________ from your ear repeatedly. This causes ***compressions***, where the particles ________, and ***rarefactions***, where particles ________. In other words, ________ are high pressure regions, while ________ are low pressure.

Longitudinal wave

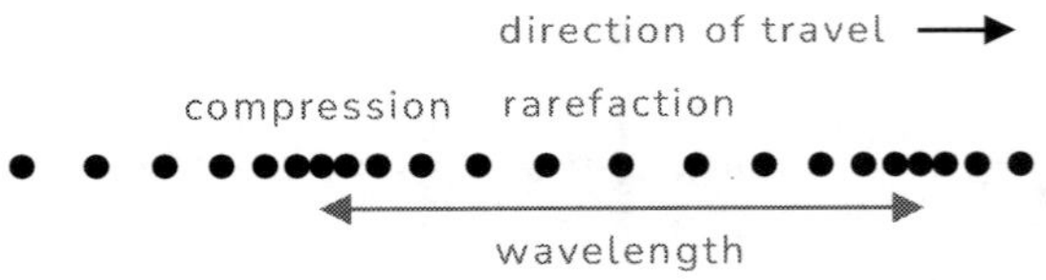

The particles in the middle of each compression will all be in phase, meaning that neighbouring compressions are all a ________ apart. However, particles in the middle of a compression are half a wavelength ________ with particles in the middle of a neighbouring rarefaction.

The furthest distance a particle vibrates away from its equilibrium position is still called the ________ in a longitudinal wave.

5. Identify whether a transverse or longitudinal wave is shown in each picture.

a)

b)

6. The wave in the diagram is travelling to the right.

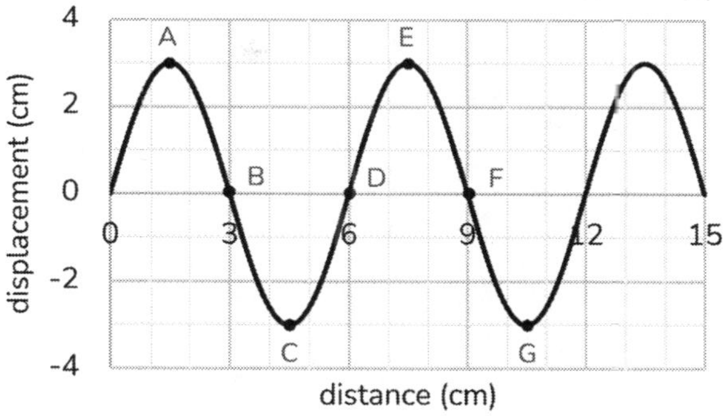

a) What is the amplitude? ____________

b) What is the wavelength? ____________

c) Identify all the points in phase with each other. ____________

d) Identify two points that are half a wavelength out of phase with point A. ____________

e) Explain whether this is a transverse or longitudinal wave.

____

7. Draw a transverse wave on the graph paper with an amplitude of 2 cm and wavelength of 4 cm.

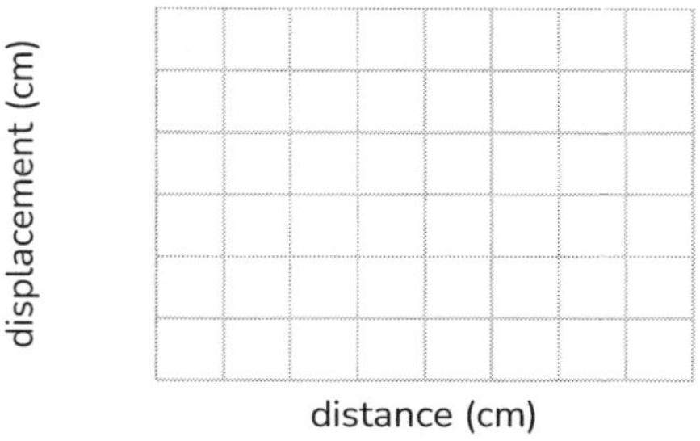

8. The first diagram shows particles in a medium in their equilibrium position, when there is no wave. The second diagram shows the position of the particles when a wave passes through it.

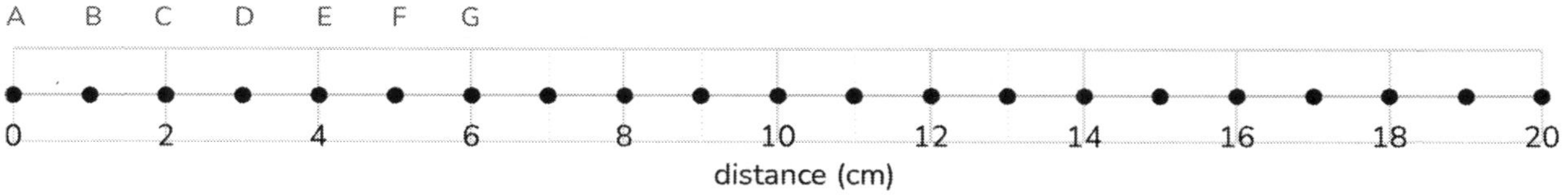

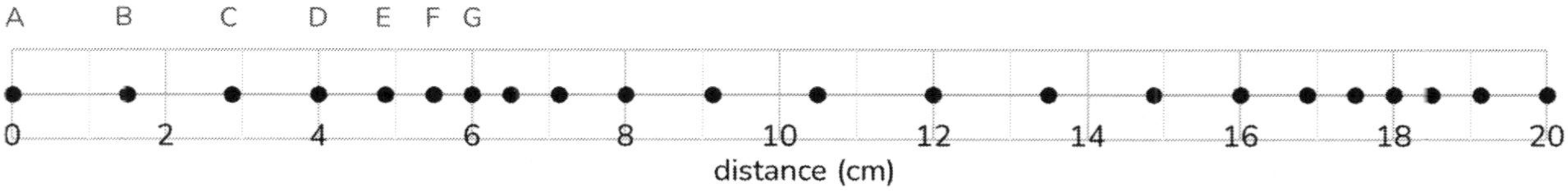

a) Which particle, from A to G, is in the middle of a compression? ____________

b) Which particle, from A to G, is in the middle of a rarefaction? ____________

c) What is the wavelength of the wave? ____________

d) Which particle, from A to G, is at its amplitude? ____________

e) On the diagram, identify a particle in phase with the answer in part (d). Explain your reasoning.

____

f) Identify a particle on the diagram that is half a wavelength out of phase with the particle identified in part (d). Compare their displacement from the equilibrium position.

____

14.3 Timing waves

1. Describe period and frequency, by filling in the blanks.

crest	day	frequency	frequency	hertz	number	passed
period	period	return	second	vibrations	wavelength	

When there is a recurring event, the ***period*** is the time between one event and another. For example, the period of Earth's rotation on its axis is one ________________.

The ***frequency*** of a recurring event is how often it occurs in a given time period. For example, the ________________ of a teenager's heart beat might be 60 beats per minute. This would be the same thing as 1 beat per ________________. Therefore the ________________ of this teenager's heart beat would be 1 second.

As waves involve recurring vibrations, the ***period*** is the time taken for a particle to undergo one complete vibration and ________________ to the same spot. For example, when a surfer bobs up and down due to ocean waves, the period is the time between one crest of a wave until the next ________________. Therefore in one period, one ________________ of ocean wave has passed the surfer.

The frequency of a wave is the number of ________________ per second. For example, the frequency of the ocean waves would be the ________________ of times the surfer goes down, then back up, per second. The frequency is also the number of wavelengths that has ________________ the surfer per second. The unit for frequency is the ________________ (Hz).

The equation for the relationship between period and frequency is:

$$T = \frac{1}{f} \text{ or } f = \frac{1}{T}$$

where: $T =$ ________________ (s)

$f =$ ________________ (Hz)

Example: When an orchestra is tuning, an oboe plays the note "A" for the other instruments to tune to. This note creates sound waves that make the air particles vibrate with a frequency of 440 Hz. How long would it take for an air particle to vibrate once when this sound passes through it?

Answer: $T = \frac{1}{f} = \frac{1}{440} = 0.00227 \text{ s} = 2.27 \text{ ms}$

2. An earthquake wave makes the ground shake up and down with a period of 5.0 s. What is the frequency of this earthquake wave?

____

3. A metronome keeps musicians in time when they are practising. If the metronome is set to tick at 120 beats per minute, what is its frequency in hertz, and its period?

____

____

4. A ceiling fan rotates with a frequency of 50 Hz. How long does it take to rotate once?

____

5. You watch a buoy on the ocean bob up and down. The time it takes for five wavelengths to pass the buoy is 20 s. What is the frequency of the ocean wave?

____

14.4 Wave speed

Wave equation

1. Describe the relationship between wave speed, frequency and wavelength, by filling in the blanks.

crest	distance	energy	frequency	frequency	period	period
second	time	wave speed	wave speed	wavelength	zero	

The speed of a wave, like the speed of anything, is the ________________ it moves per ________________ in the direction of travel. Specifically, wave speed refers to the speed of the ________________ transfer, and not the speed of any individual particle, which would have an average velocity of ________________ over one period.

One way in which wave speed can be calculated is to pick a part of the wave, for example the ________________, and measure how far it travels per second:

$$v = \frac{s}{t} \dots (1)$$

where: v = wave speed (m s^{-1})

s = distance travelled by the wave (m)

t = ________________ in which the distance is travelled (s)

The time taken for one wavelength pass a point is also the ________________. Substituting this into equation (1):

$$v = \frac{s}{t} = \frac{\lambda}{T} \dots (2)$$

where: v = ________________ (m s^{-1})

λ = wavelength (m) (Note: λ is a Greek symbol, pronounced "lambda")

T = ________________ (s)

Recall the ________________ of a wave is related to period by:

$$f = \frac{1}{T} \dots (3)$$

Substituting this into equation (2) gives the ***wave equation***:

$$v = f\lambda$$

where: v = ________________ (m s^{-1})

f = ________________ (Hz)

λ = ________________ (m)

Example: Earthquakes cause many types of waves through the Earth. The fastest type is the P wave, or primary wave, which is a longitudinal wave. P waves can have a range of frequencies. One P wave travels at 6.0 km s^{-1} in rock, producing vibrations of frequency 6.0 Hz. What is the wavelength of this wave?

Answer:

$v = f\lambda$

$6 \times 1000 = 6 \times \lambda$

$\lambda = 1000$ m

2. Another type of earthquake wave is the S wave, or secondary wave, a transverse wave. Like P waves, S waves can also have a range of frequencies.

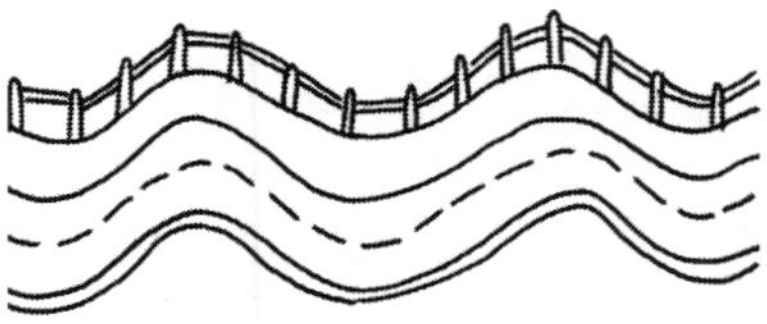

a) One S wave travels at 4.0 km s^{-1} in rock, producing vibrations of frequency 6.0 Hz. What is the wavelength of this wave?

Information about P waves is given in the example.

b) If an earthquake travelled through a plantation of regularly spaced trees, explain the difference that could be observed due to the P waves and S waves.

3. You are sitting in a stationary canoe on a river, when a motor boat passes by, causing waves which spread out towards the river banks. When the waves reach you, your canoe starts bobbing up and down. It takes 3.0 s to bob up and down 5 times. You estimate the crests of the waves are 3.5 m apart and the amplitude of the waves is 0.25 m. How fast are the waves moving?

4. The speed of a wave depends on the medium it travels through. Sound travels at different speeds in different air temperatures, which is why musicians must regularly tune their instruments, especially if the weather changes. A musical note of wavelength 78 cm is played. What frequency is this:

a) in air at 0°C, with a speed of sound of 331 m s^{-1}?

b) in air at 30°C, with a speed of sound of 349 m s^{-1}?

Medium and velocity

5. Explain why the medium a wave travels through affects its speed, by filling in the blanks.

elastic potential	equilibrium	faster	inertia
kinetic	medium	tension	transformations

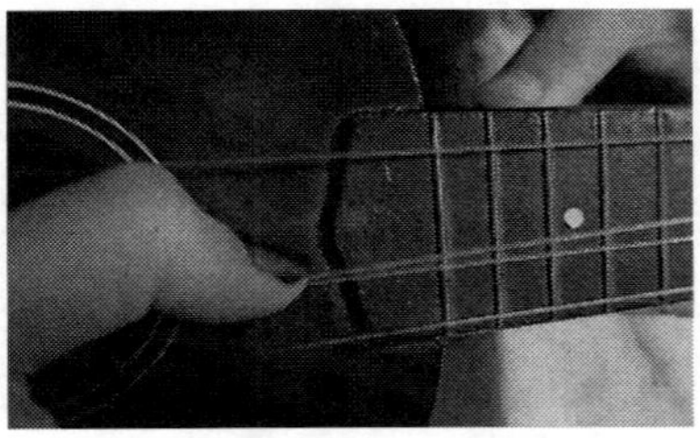

In mechanical waves, there must be some ***restoring force*** to return particles to their ________ position in order for a wave to travel. For example, when a ukelele string is plucked, ________ in the string pulls it back towards its equilibrium position. However, ________ keeps it moving past this point, so it vibrates. Increasing tension means the string can vibrate ________, increasing wave speed.

Mechanical waves involve continuous energy ________. For example, stretching a ukelele string gives it ________ energy, which transforms into ________ energy when released, which transforms back into elastic potential energy when it stretches out on the other side. The ________ affects how fast this energy transformation occurs.

6. Ash and Bee hold two ends of a spring, as shown. Bee holds one end still. Ash moves the other end up and down continuously to send waves through the spring.

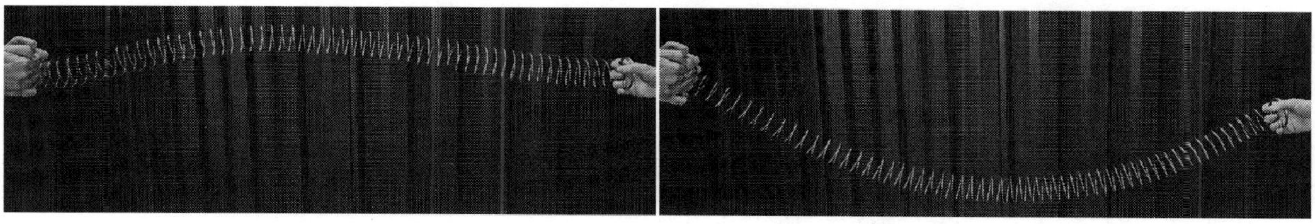

Explain what Ash needs to do in order to:

a) increase the amplitude of the wave.

....

b) increase the frequency of the wave.

....

c) increase the wavelength of the wave.

....

d) increase the speed of the wave.

....

7. Sound waves transfer energy through a medium through collisions between neighbouring particles. These collisions happen more frequently if the particles are close together.

Material	Speed of sound (m s^{-1})
Glass	5100
Sea water	1530
Air at 20°C	343

The table shows the speed of sound in different media.

a) Explain the difference in speed.

....

....

....

....

b) A sound of frequency 100 Hz is transmitted in each medium. What is the wavelength of sound in each medium? What happens to wavelength as the speed of sound in the medium changes?

....

....

....

....

8. Discover the science behind ocean waves, by filling in the blanks.

breaks	earthquakes	energy	gravity	medium	vibrate	wave

Ocean waves can be generated by wind or .... The waves travel along the surface of the ocean due to the restoring forces of surface tension and .... acting on the water molecules.

Changing the water depth changes the .... in which ocean waves travel. Generally, waves travel slower in shallower water as they come closer to shore, because water molecules drag along the bottom of the ocean as they .... Due to conservation of ...., as waves slow down, their amplitude increases. When the water is shallow enough to interrupt the water molecules' vibrations, the wave .... and it looks like the top of the wave forms a peak. A surfer on a breaking wave can move towards the shore instead of bobbing up and down, as it is not behaving like a .... anymore.

9. A ***wavefront*** is a line joining all adjacent points in phase. For example, points along a crest of an ocean wave are a wavefront, with the next wavefront on a crest one wavelength away.

The diagram shows wavefronts of a wave as it moves from the deep to the shallow area of a pool. A wave's frequency does not change as it moves into shallow water. If the crests in the deep area move at a speed of 4.4 m s^{-1}, what is the wave speed in the shallow area?

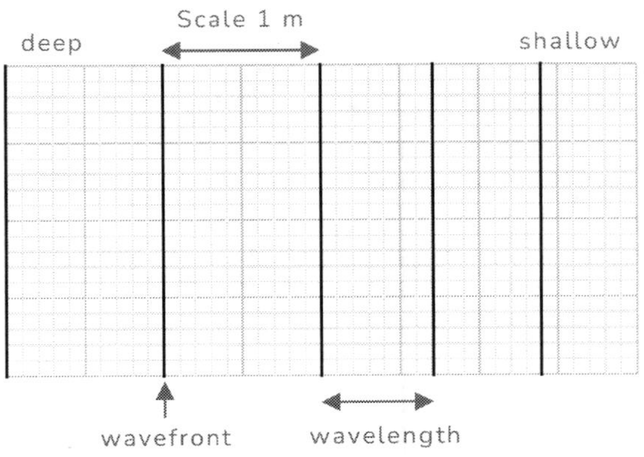

10. In a tank of water of uniform depth, waves are created by a small object dipping into the water at a constant period. The distance between wavefronts was measured. This was repeated using several different periods. The data was plotted on the graph.

Ripple tank with water

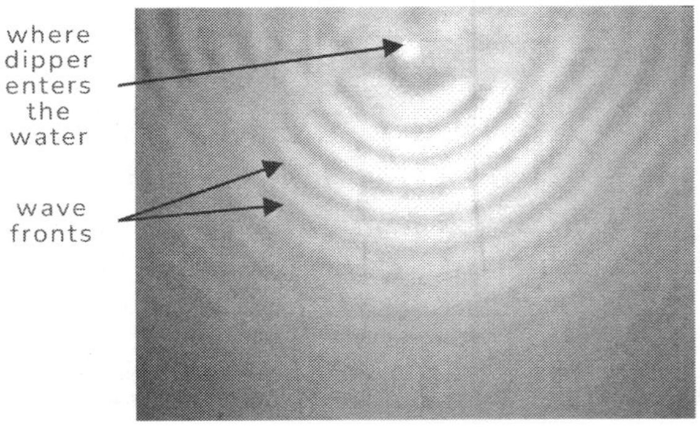

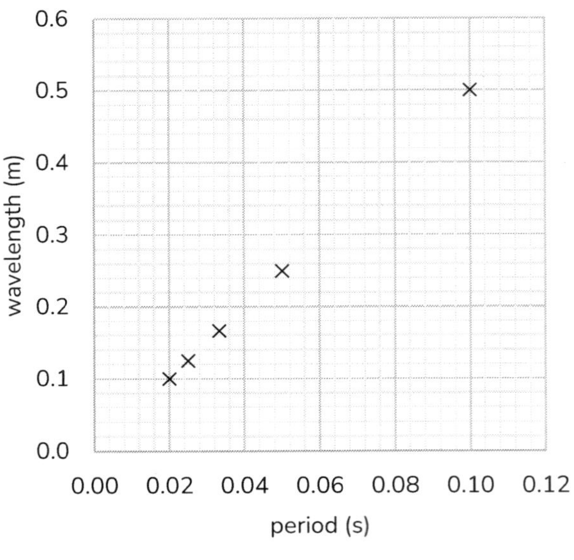

a) Describe the trend in the graph.

b) Explain what the gradient of the graph represents.

c) Calculate the speed of waves in the tank.

d) How could the velocity of waves in the tank be altered?

14.5 Graphing transverse waves

Displacement-distance graphs

Displacement-distance graphs show the displacement from equilibrium position of particles along the wave at a fixed time. For a transverse wave, the graph will look a bit like the shape of the wave.

Example: The graph shows a transverse wave.

a) What is the wavelength of this wave?
b) What is the amplitude of this wave?
c) Explain whether you can determine the period from just this graph.

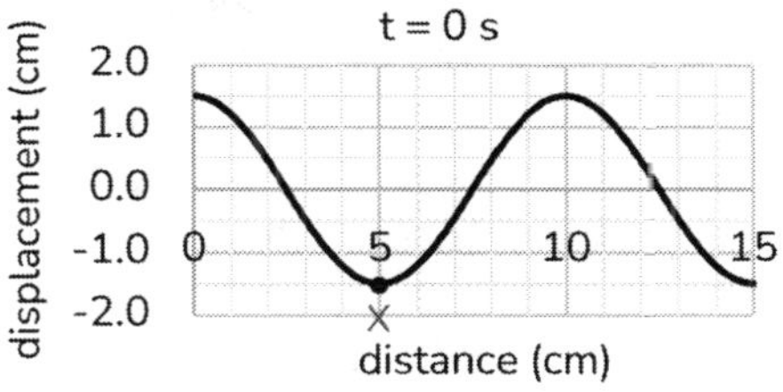

Answer:

a) 10 cm
b) 1.5 cm
c) No, because it has no information about time.

Example: The next two graphs also show the same wave as the previous example, but at later times.

a) Calculate the slowest wave speed that is consistent with these graphs.
b) What is the longest period of the wave that is consistent with these graphs?
c) Describe the motion of particle X between 0 to 0.4 s.

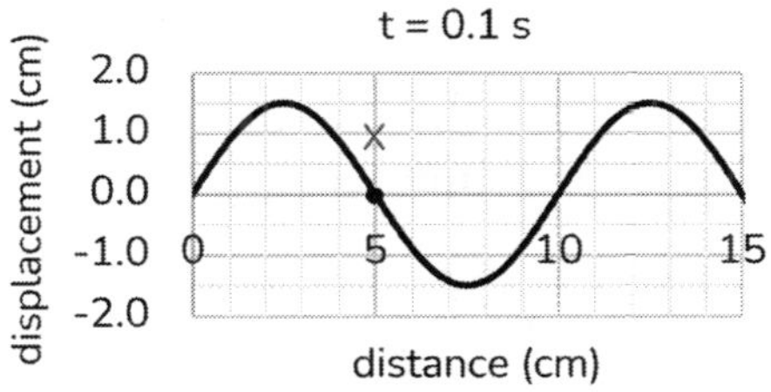

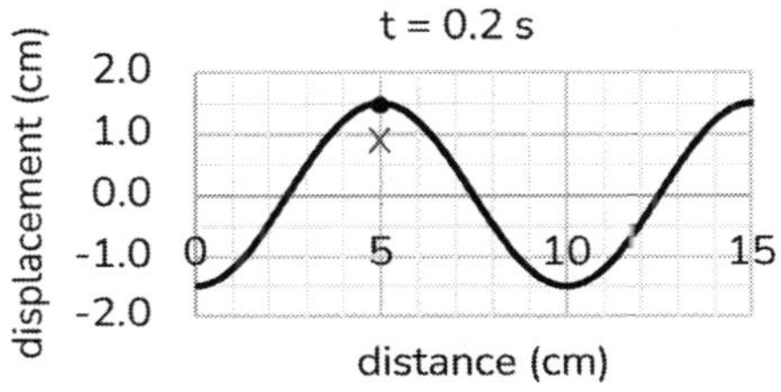

Answer:

a) The slowest speed corresponds with the smallest distance travelled by the wave from t = 0 s to t = 0.1 s, which is 2.5 cm to the right.

$v = \frac{s}{t} = \frac{2.5}{0.1} = 25 \text{ cm s}^{-1}$

(Other possibilities include travelling 12.5 cm right, or 7.5 cm left etc.)

b) $v = f\lambda = \frac{\lambda}{T}$

$25 = \frac{10}{T}$

$T = 0.4 \text{ s}$

c) It moves up to 1.5 cm, then back down, passing its equilibrium position to -1.5 cm, before returning to its equilibrium position at 0.4 s.

1. The graph represents the motion of a wave travelling at 1.0 m s^{-1} to the right.

   a) What is the wavelength of this wave? ____________

   b) What is the frequency of this wave?

   ____

   ____

   c) On the same axes, draw what the wave would look like in 1.5 s.

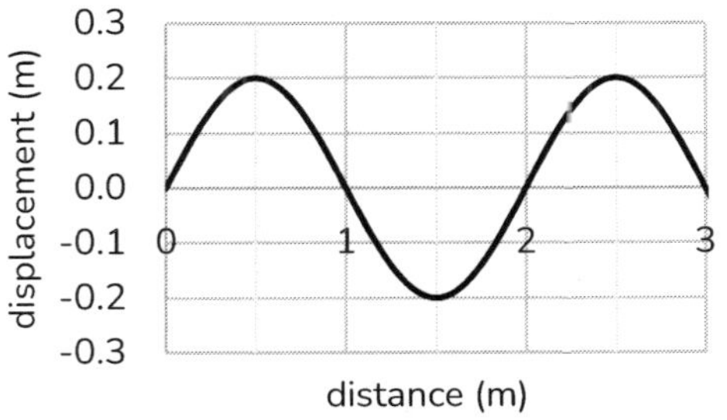

2. The graphs show a wave at two different times.

a) What is the wavelength of this wave? ______

b) Calculate the smallest speed of the wave that is consistent with these graphs.

______

______

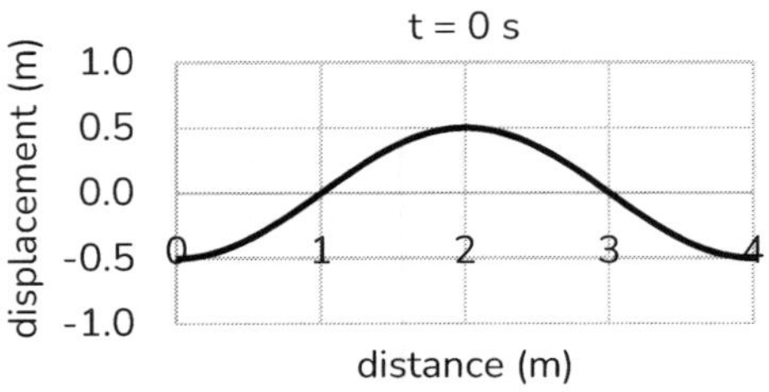

c) Explain what is the longest period of the wave that is consistent with these graphs.

______

______

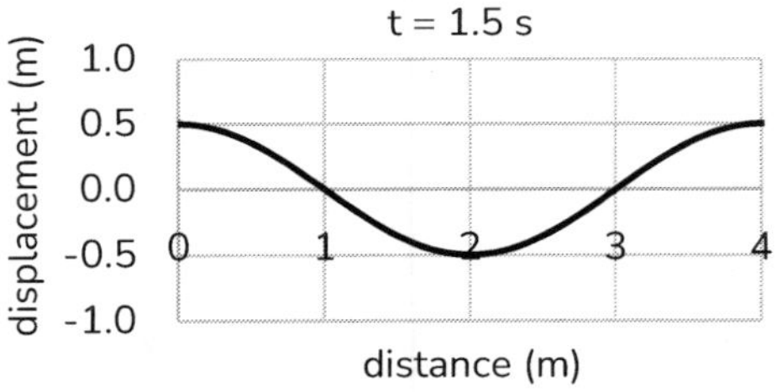

d) At what other times would the wave appear as shown in the second graph?

______

Displacement-time graphs

3. Describe displacement-time graphs, by filling in the blanks.

displacement	equilibrium	one particle	shifted	time	velocity	wave speed

A displacement-time graph for a wave tracks the position of ______ in the wave from its ______ position over time.

If you were to graph the ______ of a surfer bobbing up and down on ocean waves over ______, you would get a displacement-time graph. This would be the same graph for any other surfer bobbing up and down on the same waves, just ______ in time to account for when crests reach them.

As for any displacement-time graph, the gradient of this graph for a wave still represents the ______ of the object. This is the velocity of the individual particle, not the ______.

4. The graph describes the vertical motion of a surfer floating on the ocean as waves pass.

a) What is the amplitude of this wave? ______

b) Describe the surfer's position at 2 s.

______

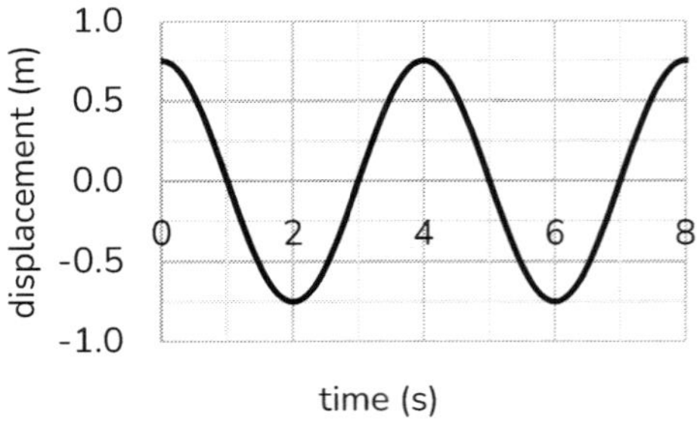

c) What is the period and frequency of this wave?

______

d) Explain why it is not possible to determine the wavelength and velocity of the wave.

______

______

e) When is the surfer at rest?

______

Recall what the gradient of a displacement-time graph represents. This does not change just because the graph describes a particle on a wave.

f) When is the surfer moving at maximum speed?

______

5. The first graph describes the vertical displacement of a boat, anchored in a harbour. The second graph describes the ocean wave that is making the boat move, at t = 0 s.

a) What is the frequency of the wave?

______

b) What is the wavelength of the wave?

______

c) What is the speed of the wave?

______

d) There are many boats anchored in the same harbour, experiencing the same wave. Where could there be another boat, with identical motion to this boat?

______

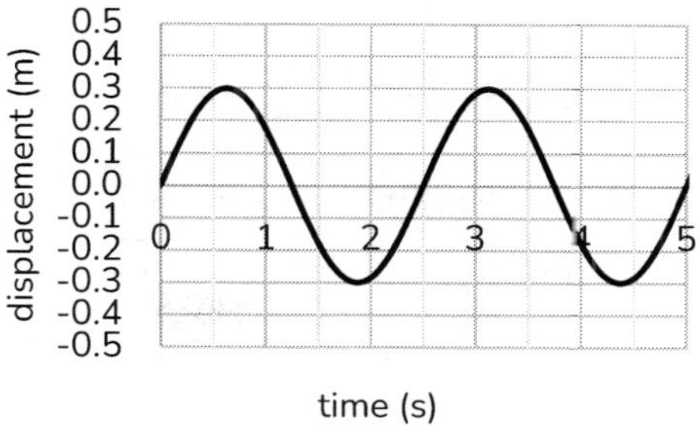

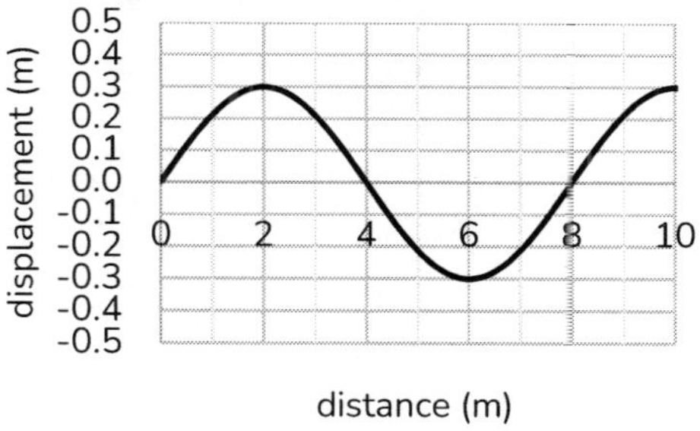

6. A buoy on a river bobs up and down 5 times in 10 s, with an amplitude of 10 cm. The speed of the wave that is making the buoy oscillate is 1.5 m s^{-1}. Draw on the axes provided:

a) a displacement-time graph for the buoy, over two periods, starting from its crest; and

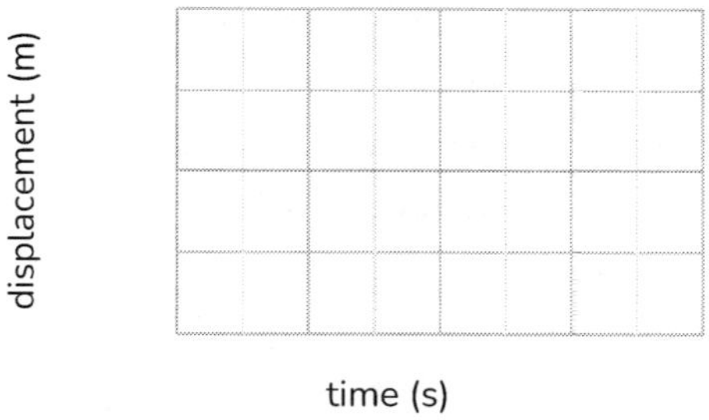

b) a displacement-distance graph for the wave, over two wavelengths, starting from a crest.

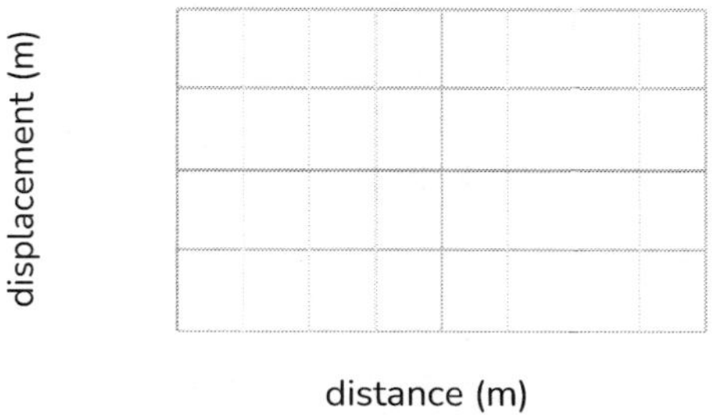

List the data provided and calculate the quantities needed first.

14.6 Sound waves

Making and observing sound waves

1. Describe sound waves, by filling in the blanks.

air particles	along	amplitude	compressions	energy	frequency	high
longitudinal	low	matter	medium	pitch	rarefactions	speed
tension	vibrate	volume	wave equation	wavelengths		

Sound is a type of wave, so it transfers energy without transferring ________________. When you hear a sound, energy is transferred to your ear, but ________________ are not, so you do not feel wind hitting your ear.

Sound is a ________________ wave. This means that it makes particles in the medium it travels through vibrate ________________ the direction of travel. This causes a higher density of particles in some areas, called ________________. In between are areas with a lower density of particles, called ________________. The compressions can also be described as areas of ________________ pressure, while the rarefactions are areas of ________________ pressure.

Higher frequency sound waves have a higher ________________. This means that the vocal cords of a soprano ________________ with a higher frequency than the vocal cords of a bass when they sing. Lower frequency sounds have a longer wavelength in the same medium, because they travel at the same ________________. This is how musical instruments can play at different pitches – a double bass will play deeper notes of a low frequency because the longer strings allow longer ________________ to vibrate on it than a violin with shorter strings.

As the amplitude of sound waves increases, the ________________ increases, so the sound is louder. For example, to make a louder sound on a guitar, you need to pluck the string harder, increasing the ________________ of the string's vibration. Increasing the amplitude therefore also increases the ________________ transferred by the wave.

When you tune a stringed instrument, for example a guitar, you tighten or loosen the string to change its ________________. This changes the ________________ that sound waves travel through, therefore changing the speed of waves. Higher tension means higher speed, therefore higher ________________ and pitch, according to the ________________ $v = f\lambda$.

2. When musicians in an orchestra tune their instruments, the oboe plays the note "A" with a frequency of 440 Hz. Assume the speed of sound in air is 340 m s^{-1}.

   a) What is wavelength of this note?

   ____

   ____

   ____

   b) Another note is played with a wavelength of 1.54 m. Compare the pitch and frequency of this note with the "A" used for tuning.

   ____

   ____

   ____

3. The range of frequencies that humans can hear is between 20 Hz and 20 kHz. Assume the speed of sound in air is 340 ms^{-1}.

a) What is the wavelength of the lowest pitch humans can hear?

b) What is the wavelength of the highest pitch humans can hear?

Graphing longitudinal waves

4. An oscilloscope is an instrument that graphs an electrical voltage as a function of time. It can be used to display heartbeat if it is connected to electrodes across the body. A microphone can also be connected to an oscilloscope to display a ***displacement-time graph*** of sound.

Sound reaching a microphone vibrates a thin diaphragm at the same frequency as the sound wave. The diaphragm is connected to a coil. The vibrations move the coil back and forth in a magnetic field, converting the kinetic energy of the coil into electrical energy. The voltage over time can be displayed by an oscilloscope. A bigger displacement by the sound wave will increase the voltage generated. The period of the signal is the same as the period of the sound wave.

A person sings a note into a microphone, producing the oscilloscope display shown.

a) What is the period and frequency of this note?

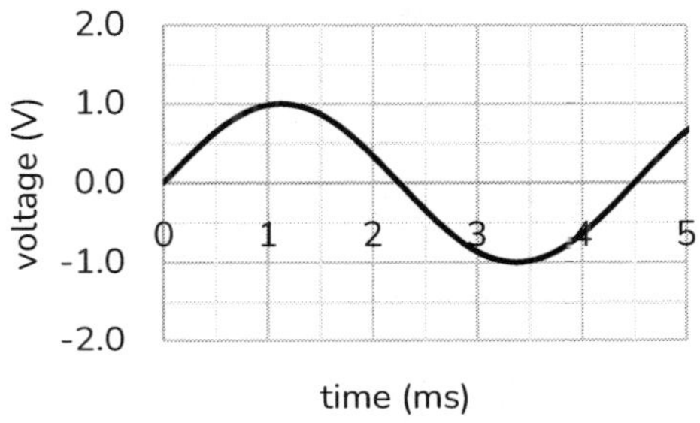

b) If the sound travels at 340 m s^{-1} in air, what is the wavelength in air of this note?

c) The person sings a different note into the microphone, producing the oscilloscope display shown. Compare how the two notes sound.

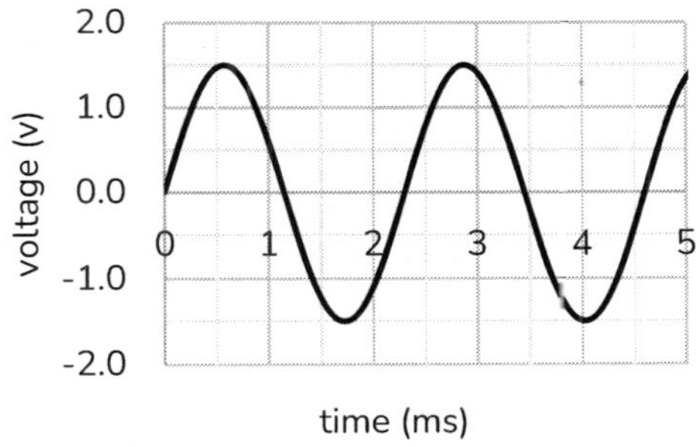

5. Longitudinal waves can also be described using ***displacement-distance graphs***.

The first diagram below shows the particles in a medium in their equilibrium position.

The second diagram shows the position of the particles when a wave travels through the medium, at time t = 0 s. A displacement-distance graph for the particles is also shown for this time. The distance on the x-axis is the equilibrium position of each particle. The y-axis shows the displacement of each particle from this equilibrium position. See that particles A, E and I are plotted as having a displacement of 0, as they are at equilibrium positions at t = 0 s. Particle C moved from its equilibrium position of 2.5 cm to 3.5 cm, a displacement of 1 cm to the right, which is taken as positive. Particle G moved from its equilibrium position of 7.5 cm to 6.5 cm, a displacement of 1 cm to the left.

The third diagram shows the position of the particles 0.5 s afterwards.

Equilibrium position

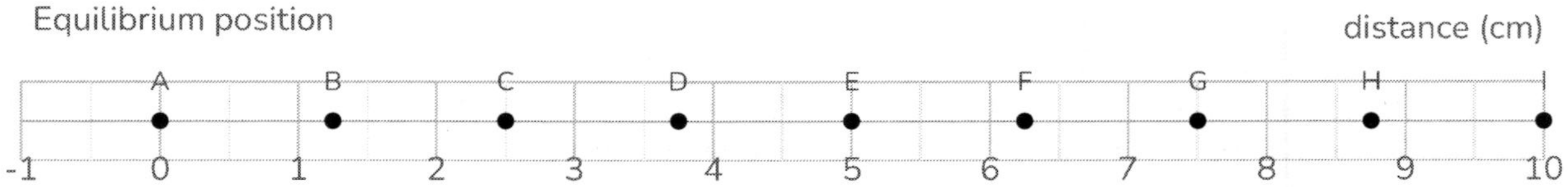

t=0 s

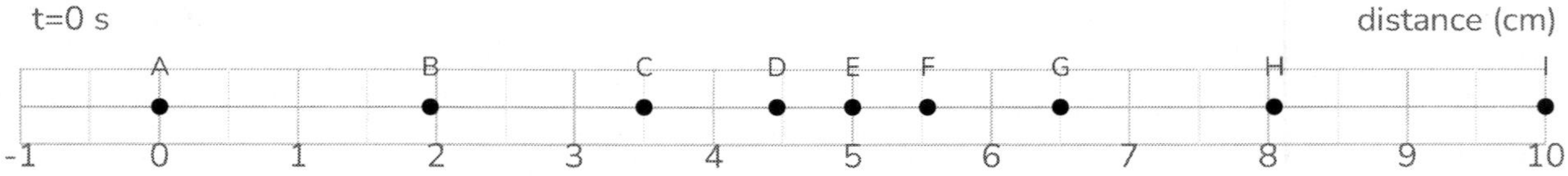

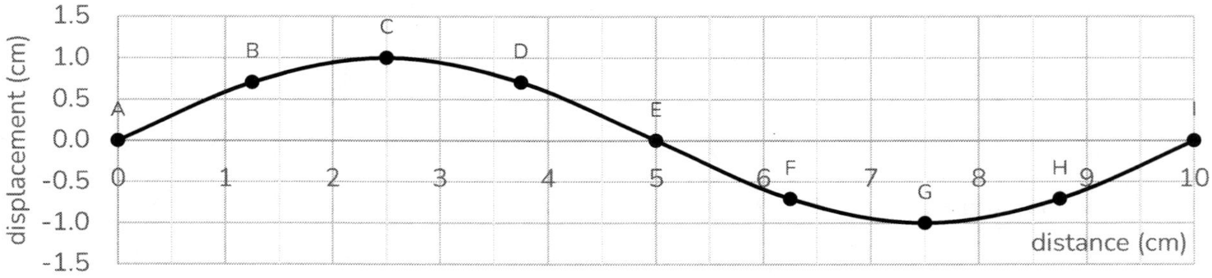

t=0.5 s

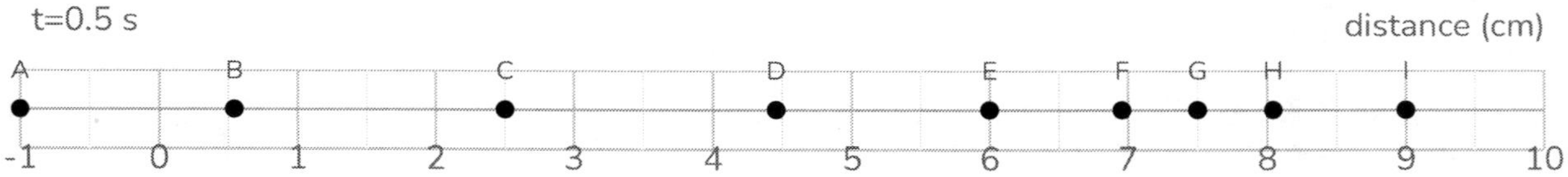

a) Plot the displacement-distance graph for the particles at 0.5 s on the same graph as for 0 s.

b) What is the wavelength of this wave? ______________

c) What is the amplitude of this wave? ______________

d) Which particle is in the middle of a compression at t=0.5 s? What is its displacement from its equilibrium position? ______________

e) Which particle is in the middle of a rarefaction at t=0.5 s? What is its displacement from its equilibrium position? ______________

f) Why is it difficult to use a displacement-time graph for a longitudinal wave to determine where the compressions are?

____

g) What distance to the right has the compression moved in 0.5 s? ______________

h) Determine the velocity of the wave.

____

6. Another way of representing a longitudinal wave is a ***pressure-distance graph***. Positive pressure denotes high pressure. Negative pressure denotes low pressure. No units are required for pressure at this level of physics – the most useful information to be gained from the graph is the wavelength.

The particles in the same wave as question 5 at t = 0 s are reproduced below, along with a corresponding pressure-distance graph.

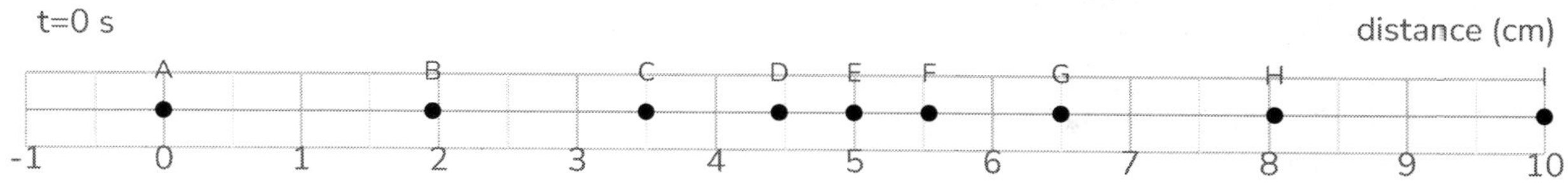

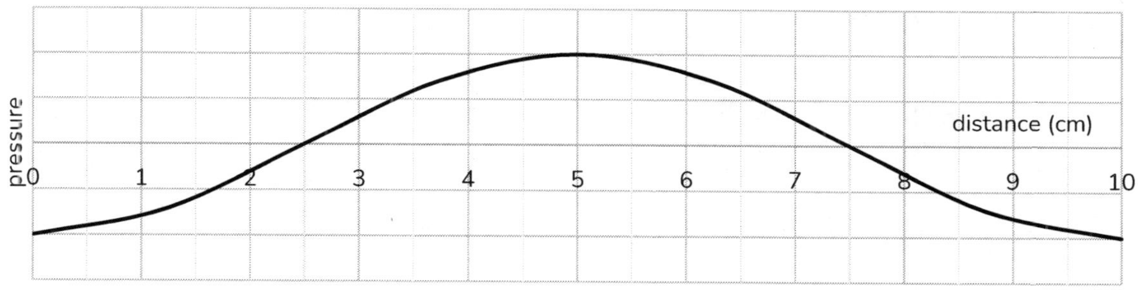

a) Compare the displacement-distance graph and pressure-distance graph for the sound wave, using the table below.

Feature	Displacement-distance graph	Pressure-distance graph
Describe what the y-axis represents		
What "crests" represent		
What "troughs" represent		
How it can be used to determine wavelength		

b) The particles in the wave in question 2 at t = 0.5 s are reproduced below. On the same graph as for t = 0 s above, graph the pressure-distance graph for the wave at t = 0.5 s.

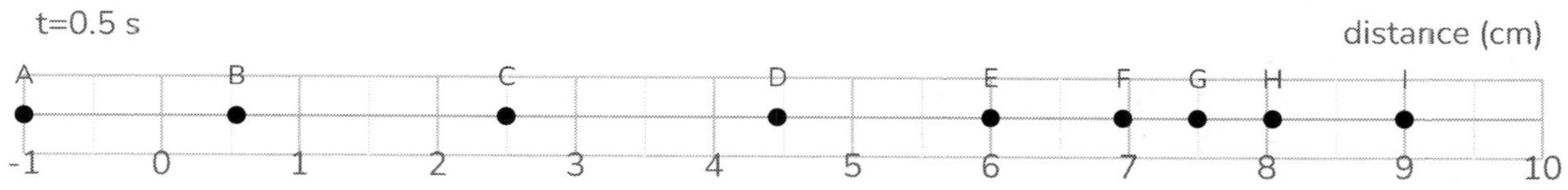

14.7 Light and electromagnetic waves

1. Describe electromagnetic waves, by filling in the blanks.

communications	currents	electrons	gamma	heat	light
magnetic	matter	medium	metal	nothing	resistance
spectrum	transverse	ultraviolet	vibrations	wavelength	wavelength

Sunlight reaches Earth, but ______________ is not transferred in the process, consistent with the definition of waves. Light is an example of an ***electromagnetic wave***, and the whole range of electromagnetic waves is called the electromagnetic ______________. Each type of electromagnetic wave has a different ______________ and frequency.

Electromagnetic spectrum

Wave	Wavelength
Gamma	< 0.01 nm
X-ray	0.01 – 10 nm
Ultraviolet	100 – 400 nm
Light	400 – 700 nm
Infrared	700 nm – 1 mm
Microwave	1 mm – 30 cm
Radio	> 30 cm

The Sun produces all types of electromagnetic waves. Those with the highest frequencies - ______________ rays, x-rays and ultraviolet rays - cause mutations in living cells, leading to cancer. Fortunately, the Earth's atmosphere blocks the gamma rays and x-rays and the ozone layer partially filters the ______________ rays. ______________ is the part of the electromagnetic spectrum that is visible to the human eye. Infrared is perceived by us as ______________. Microwaves and radio waves have many uses, including ______________.

Unlike mechanical waves, electromagnetic waves do not need a ______________ consisting of particles to travel through. However, like all waves, electromagnetic waves still involve ______________. These vibrations consist of changing electric fields and ______________ fields. The fields change in magnitude and direction in a pattern like a sine wave, that repeats every ______________. These changing fields are perpendicular to the direction the wave travels, so electromagnetic waves are ______________ waves.

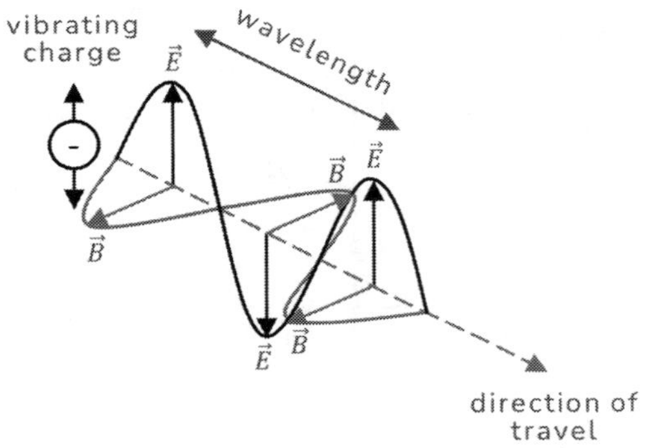

Electromagnetic waves can be created by changing electric fields, for example by turning an electric circuit on and off, which accelerates ______________ in the circuit.

Electromagnetic waves can be blocked using ______________ shielding, which has electrons that can move. The electric field makes electrons accelerate, turning electromagnetic energy into the kinetic energy of the electrons, then into heat due to the ______________ of the metal.

> Make radio waves by connecting one end of a wire to one terminal of a battery, then tapping the other terminal with the other end of the wire. This makes electrons in an antenna vibrate, so it can be received using an AM radio.

The electric fields are accompanied by magnetic fields, since electric ______________ produce magnetic fields.

All electromagnetic waves travel at the same speed in a vacuum, 3×10^8 m s^{-1}, which is also taken as its speed in air. ______________ we know of in the universe travels faster than this!

> Aluminium foil can block radio waves from reaching a mobile phone.

Use the information on the previous page to answer questions 2 to 8.

2. Which electromagnetic wave would travel the fastest through space from the Sun to the Earth?

3. The visible spectrum appears to us as the colours of the rainbow. Calculate the frequency of the following colours of light.

   Write the prefix for the wavelength in scientific notation.

   a) Red light of 650 nm

   b) Blue light of 475 nm

4. All radio stations are allocated a frequency of electromagnetic wave at which they are allowed to broadcast. The Triple J radio station uses the frequency of 105.7 MHz. Calculate the wavelength.

5. One of the electromagnetic waves used by an Australian mobile phone network transmits at 850 MHz.

   a) Calculate its wavelength.

   b) What type of electromagnetic wave is this?

6. A portable water purifier useful for camping emits radiation with a wavelength of 254 nm.

   a) Calculate its frequency.

   b) What type of electromagnetic wave is this?

7. Explain what wavelength and frequency means in relation to a light wave.

8. Compare sound and light.

Feature	Sound	Light
Is energy transferred?		
Is matter transferred?		
Can it travel through a vacuum?		
Type of wave		
Describe the vibrations involved		
Speed in air		

15.1 Inverse square law

1. Explain how light intensity is affected by distance from a light source, by filling in the blanks.

both sides	constant	dimmer	distances	electrostatic	energy
intensity	inverse square	lumens	radiated	second	SI unit
sound	source	sphere	subject	surface area	

The power passing through the hole is the same as the power reaching the wall

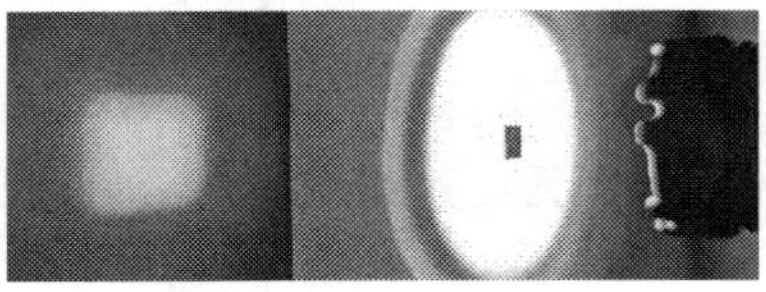

The further from a light bulb, the ________ the light. In other words, less power, or energy per ________, is received. However, due to the Law of Conservation of ________, the power emitted at the source must be equal to the total power that has ________ outwards in all directions. If we could track each bit of energy that is radiating outwards from the source, in the same amount of time, those bits of energy will have spread out over the area of a ________ centred on the light bulb.

Surface area of sphere $4\pi r^2$

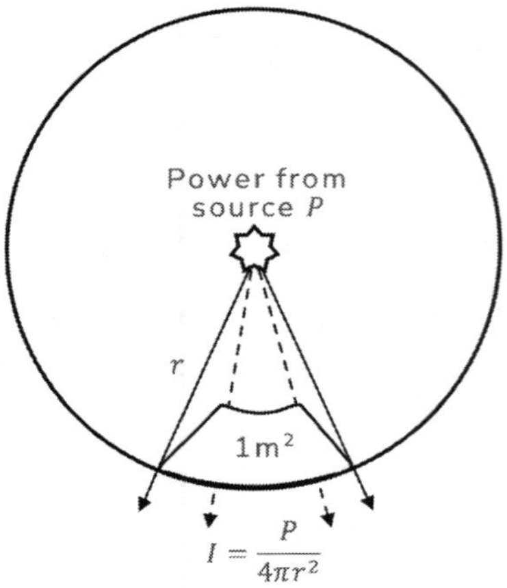

The ________ of a sphere is $4\pi r^2$ so the power is spread out across this area.

The amount of power per unit area, reaching a distance r away from the source of energy, is called the ________. So the intensity of light observed is given by:

$$I = \frac{P}{4\pi r^2}$$

where: I = intensity at a distance r away from the source (W m^{-2})

P = power emitted by the ________ (W)

r = distance from the source (m)

If the same light bulb is used, then $\frac{P}{4\pi}$ is a ________, so intensity is inversely proportional to the square of the distance from the source. This type of relationship is called an ________ relationship.

The inverse square relationship applies to all types of waves, including ________ waves. The inverse square relationship also describes many other phenomenon, including ________ and gravitational fields.

Comparing intensities at two different ________, r_1 and r_2, from the same source with power P, gives:

$$I_1 = \frac{P}{4\pi r_1^2} \text{ and } I_2 = \frac{P}{4\pi r_2^2}$$

Rearranging to make P the ________ then equating the formulae gives:

$$4\pi r_1^2 I_1 = 4\pi r_2^2 I_2$$
$$I_1 r_1^2 = I_2 r_2^2$$

Any unit for intensity or distance can be used in this equation, so long as the same unit is used on ________. The ________ for light intensity is the candela. When shopping for light bulbs, the brightness of the light bulb is often stated in ________ and the intensity of light received at a distance is measured in lux.

Mobile phone apps can be used to measure light intensity.

Example: A photographer uses a lamp to illuminate a subject. What will happen to the intensity of light on the subject if they double the distance between the lamp and the subject?

Answer: Let the original distance and final distance between the lamp and the subject be r_1 and r_2 respectively, so:

$r_2 = 2r_1$

When observing intensities from the same light source at different distances:

$I_1 r_1^2 = I_2 r_2^2 = I_2(2r_1)^2 = I_2 \times 4r_1^2$

$I_2 = \frac{1}{4} I_1$

The intensity will be reduced to one quarter, or 25% of the original intensity.

Example: Two similar styles of light bulbs, emitting 8.0 W and 12.0 W of power respectively, appear equally bright to an observer. The 8.0 W light bulb is 5.0 m away from the observer. What is the distance between the observer and the 12.0 W light bulb? Assume all the energy is emitted as light.

Answer: Let the intensity of light from the light bulbs be I_1 and I_2 respectively, with I_1 coming from the 8.0 W light bulb and I_2 coming from the 12.0 W light bulb.

The light bulbs appear the same brightness, so:

$I_1 = I_2$

$\frac{P_1}{4\pi r_1^2} = \frac{P_2}{4\pi r_2^2}$

$\frac{8}{5^2} = \frac{12}{r_2^2}$

$r_2 = 6.1$ m

2. Recall that $F_E = \frac{1}{4\pi\varepsilon_0}\frac{q_1 q_2}{r^2}$. Explain whether electric fields follow the inverse square law.

3. Mobile phones communicate by sending radio waves, which are received by antennas at base stations. The information is then transmitted through the phone network. A mobile phone is moved to three times the distance away from a base station. Show that the intensity of the radio wave received reduces to one ninth of the original intensity.

   Let $r_2 = 3r_1$

4. A photographer uses a lamp to illuminate the person they are photographing. They want to double the intensity of light from the lamp on their subject. Where do they need to move their lamp?

   Let $I_2 = 2I_1$

5. The stage lighting for a play uses four identical 650 W lights, hung from the roof 8.0 m above the stage. The same play is to be performed on a stage where the lights must be hung from a 12 m height. How many of the same type of stage light are needed to provide at least as much brightness as the previous stage?

What variable needs to be the same in both scenarios? Write an equation for each scenario, with this variable as the subject.

6. The intensity of light from a light bulb was measured at various distances away from the light bulb. The graph shows the processed data.

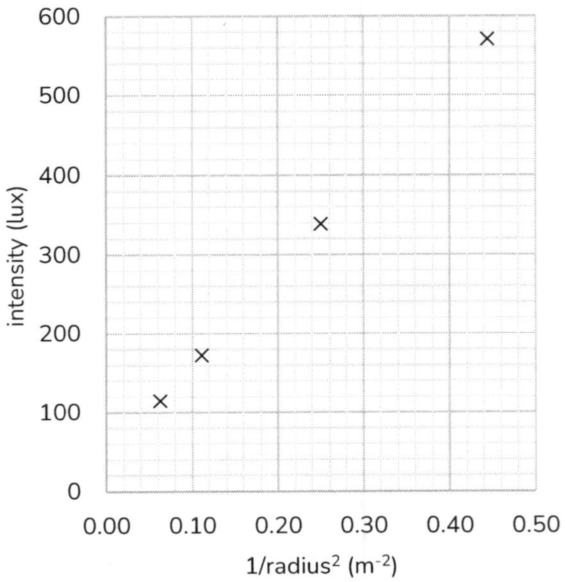

a) Justify the choice of x-axis, instead of graphing radius on it.

b) Propose an experimental reason why the data is not entirely consistent with the inverse square law.

Draw the line of best fit. What is the y-intercept?

c) Use the graph to estimate the intensity due to the light bulb alone, 1.0 m away.

Use the gradient.

7. An astronomical unit (AU) is the average distance between the Earth and Sun, which is 1.50 x 10^{11} km. The AU is a convenient measure for distances in our solar system. Our solar system's outermost planet, Neptune, orbits the Sun at an average distance of 30.0 AU. The intensity of sunlight on Earth is around 100000 lux. What is the percentage of sunlight on Neptune compared to Earth?

If a unit cancels out on both sides of an equation, there is no need to convert it to SI units.

8. Sirius is the brightest star in the night sky and appears 3.8 times brighter to us than Alpha Centauri. At the source, Sirius is 25 times more luminous than our Sun, whereas Alpha Centauri is only 1.5 times more luminous than our Sun. Alpha Centauri is 4.3 light years (ly) away from Earth and is in the closest star system to Earth. How far away is Sirius from Earth?

Alpha Centauri is one of the Pointers near the Southern Cross constellation

9. The Sun has a luminosity of 3.86 x 10^{26} W. Earth has an average distance in a year of 1.50 x 10^{11} m from the Sun.

a) How much solar power can be received per square metre on Earth?

Recall what power per unit area is the definition of.

b) What is the amount of energy that could be received per square metre on Earth over 9 hours of daylight, assuming the amount of solar power calculated in part (a) stays constant?

c) If a person uses 15 kWh of energy per day, and has 100% efficient solar panels, what area of solar panels would be needed to supply all of their energy needs? Estimate this using your calculation from part (b).

1 kWh is the amount of energy from 1 kW in 1 hour.

10. Sound waves also obey the inverse square law. The intensity of sound can also be measured in W m^{-2}. However, a more common unit is the decibel (dB). The threshold of human hearing, or an intensity of 1 x 10^{-12} W m^{-2}, is taken as 0 dB. There is still energy from sound waves below this – we just can't hear it. At this intensity, air particles in your ear are only vibrating with an amplitude of around 1 x 10^{-11} m!

The decibel scale is a logarithmic scale, meaning that instead of increasing in equal increments, the scale increases in powers of 10. This is more convenient for covering the huge range of intensities at which humans can hear. The table shows some examples of sounds and an estimate of their intensity:

Sound	Intensity (dB)	Intensity (W m^{-2})	Times bigger than 0 dB
Threshold of hearing	0	1 x 10^{-12}	$10^0 = 1$
Rustling leaves	10	1 x 10^{-11}	$10^1 = 10$
Whisper from 1 – 2 m	20	1 x 10^{-10}	$10^2 = 100$
Normal talking from 1 – 2 m	60	1 x 10^{-6}	$10^6 = 1000000$
Threshold of pain	120	1 x $10^0 = 1$	$10^{12} = 1000000000000$

Use the table to answer the questions below.

a) You can hear people whispering secrets to each other 1.0 m away from you, at a sound intensity level of 20 dB. What is the minimum distance they should move further away from you so the sound intensity level drops to 0 dB and you can no longer hear them?

b) Ash sits 1.0 m from a person talking and measures their sound intensity level to be 60 dB. Bee sits 100 m from the talking person. Explain whether the sound intensity level due to the talking would be within the threshold of Bee's hearing.

15.2 Reflection

1. Describe reflection, by filling in the blanks.

ball	crest	direction	in phase
light	media	out of phase	property
sound	trough		

Fixed end

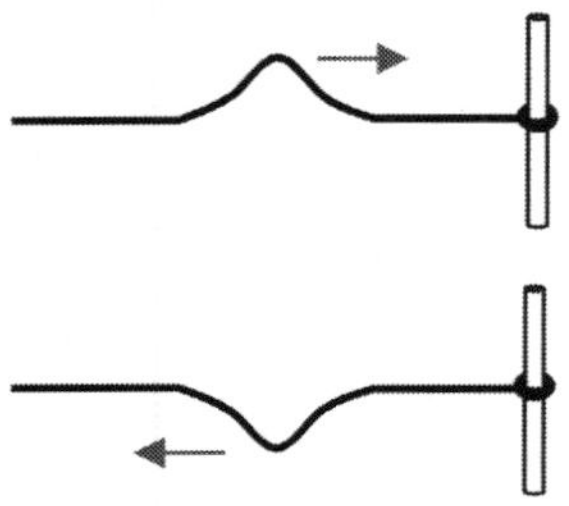

Reflection is the change in direction of a wave at the boundary between two ________________, so that the wave bounces back into the medium it started in. You see ________________ reflected from your face when you look in a mirror. An echo is a reflection of a ________________ wave. Reflection is a ________________ of all waves, but particles can also reflect. For example, a ________________ will bounce off hard ground.

Free end

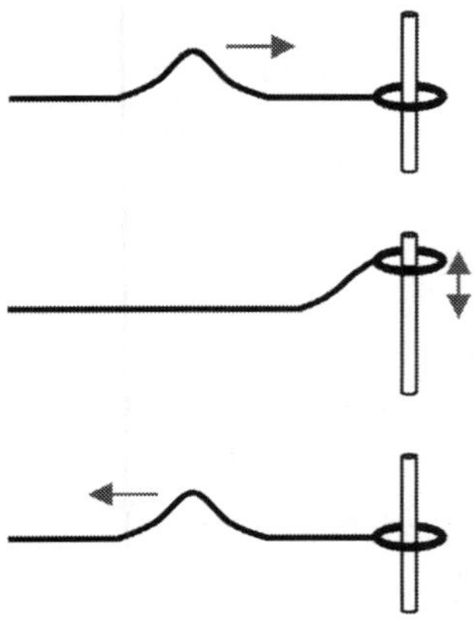

Reflection changes the ________________ of a wave, but not its speed, intensity, wavelength or frequency. Reflection can however change the phase of a wave. If you send a wave through a spring or rope so a crest reaches a fixed end, a ________________ will return, so the wave reflects half a wavelength ________________ back towards you.

However, if you send the same wave through the same spring towards an end that is free to move perpendicular to the direction of travel, then if a crest reaches the end, a ________________ will return. In other words, the wave will reflect ________________ back towards you.

2. Sound travels further than light in water, so sound is used for exploring underwater. A ship on the ocean surface emitting sound waves, or sonar, can use their reflections to measure distances to objects underwater. For example, using sonar data in the 1950s, Marie Tharp discovered the Mid-Atlantic Ridge, a volcanic mountain range under the North Atlantic Ocean. Her work disproved a belief at the time that the ocean floor was flat and provided evidence for plate tectonics. If the sound travelled in water at 1500 m s^{-1} and the echo of sonar transmitted vertically down to a section of the Mid-Atlantic Ridge returns back to the ocean surface after 3.3 s, how deep is the ocean floor at this point?

____

____

3. A pulse was sent through a slinky spring, stretched to a length of 5.0 m. The time taken for the pulse to travel one way, reflect off a fixed end and return back was 1.6 s.

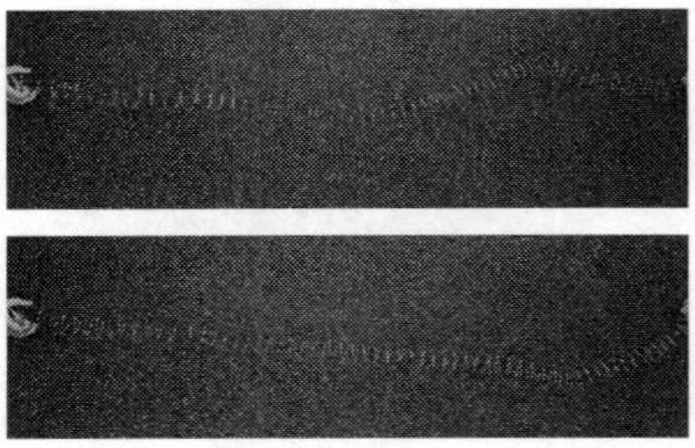

The same spring was then stretched to a length of 7.0 m and the same pulse was sent through the spring. The pulse returned after 1.9 s.

Explain the change in the speed of the pulse in the spring.

____

____

____

____

____

____

A pulse is a single vibration sent through a medium.

Always include relevant calculations.

Key words: medium, tension, restoring force.

4. Describe the Law of Reflection, by filling in the blanks.

boundary	equals	incident ray	normal	reflected ray

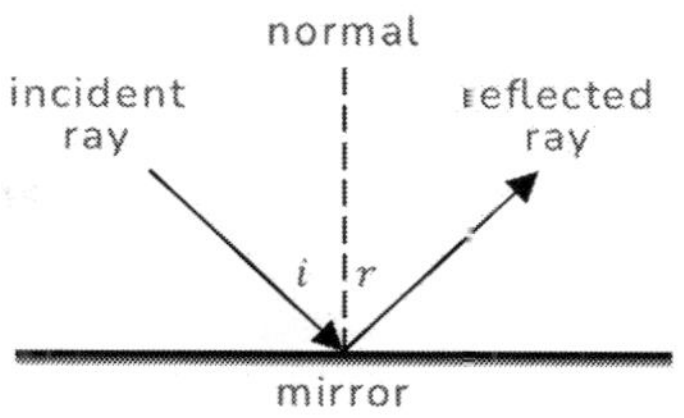

When an incident ray travels perpendicular to a reflective boundary, the ______________ is also perpendicular to the boundary. The line that is perpendicular to a boundary is called the ______________.

When an ______________ travels towards a boundary at an angle to the normal, it reflects back on the other side of the normal, so that the angle of incidence i ______________ the angle of reflection r:

$$i = r$$

This is called the Law of Reflection, which was first described by the Arab scientist, Alhazen. The angles i and r are measured between the normal and the ray, not between the ray and the ______________.

5. A light ray is incident on a mirror.

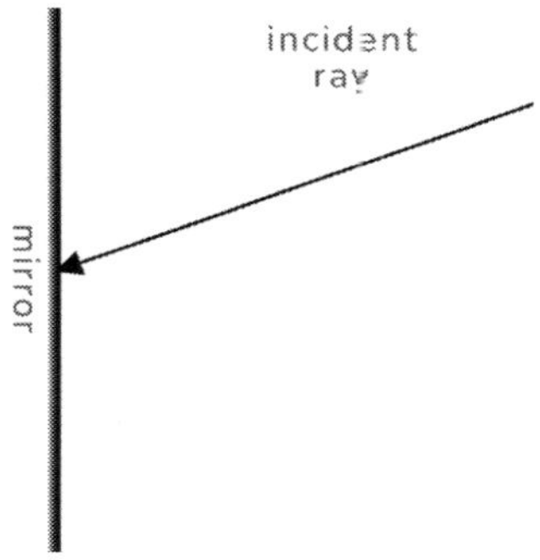

a) Use a protractor to measure the angle of incidence of the light ray.

____

b) Draw where the reflected ray will be, measuring angles using a protractor. Label the angle of incidence i, angle of reflection r and normal on the diagram.

6. The diagram shows the layout of a room where the walls are completely covered in mirrors, as seen from above.

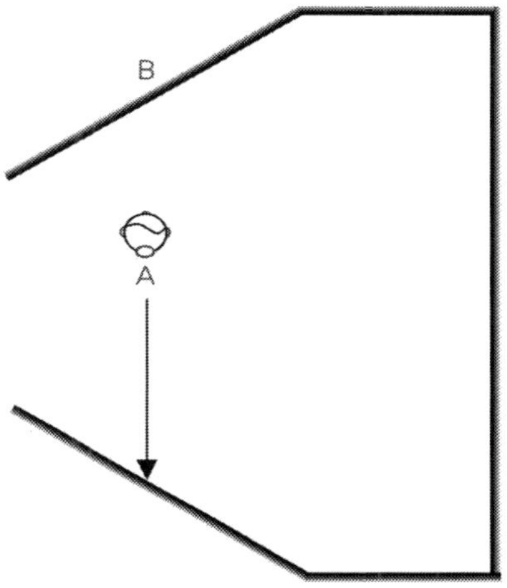

a) A ray of light is incident on one of the walls as shown. Draw where the ray of light will travel within the room.

b) What would you see if you stand at point A and look straight ahead at point B?

____

7. Explain how mirrors let us see a larger area, by filling in the blanks.

ground	halfway	incidence	light	reflected	same

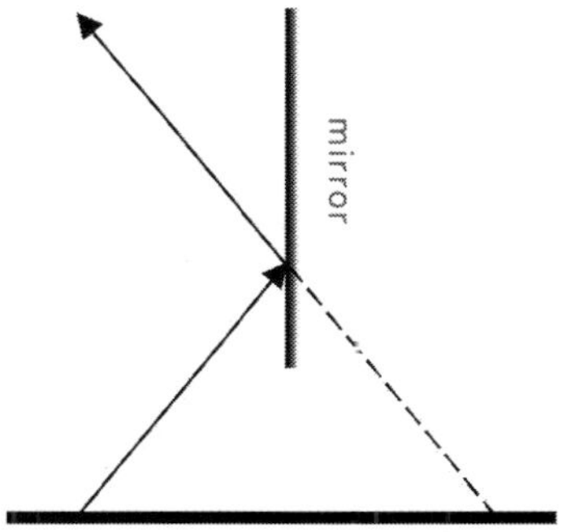

We only see if ______________ reaches our eye. To see an object, light must either be transmitted or ______________ by the object.

In a mirror, an object would appear to be in the ______________ direction as the light ray reaching our eye travelled. This means that we may be able to see our feet in a mirror, even if the mirror did not reach the ______________. The mirror just needs to reach low enough so that a light ray from our feet could be reflected by the mirror in such a way that the angle of ______________ equalled the angle of reflection. The point on the mirror at which a light ray from our feet would reflect towards our eye would be ______________ up between our feet and eyes, so the bottom of the mirror only needs to be as low as this to let us see our feet.

8. Describe reflection from curved surfaces, by filling in the blanks.

angles	concave	convex	corners	diverge	focus
incident	parabola	radio waves	reflect	tangent	

The Law of Reflection also applies to curved reflective surfaces, if angles are measured using a ______________ to the surface at the point an incident ray strikes it.

A curved surface that caves in towards the source of waves is described as ______________.

An example of a concave surface is a satellite dish, used to collect ______________ for communication or TV. The satellite dish is in the shape of a ______________, which in mathematics has the formula $y = ax^2$. The parabolic shape means that all rays that are ______________ on the dish parallel to its axis of symmetry will be reflected through a point, called the focal point or focus. The receiver for the radio waves is placed at the ______________, which is at the tip of the antenna, to collect a strong signal from all of the reflected rays.

Concave reflective surface - parabolic

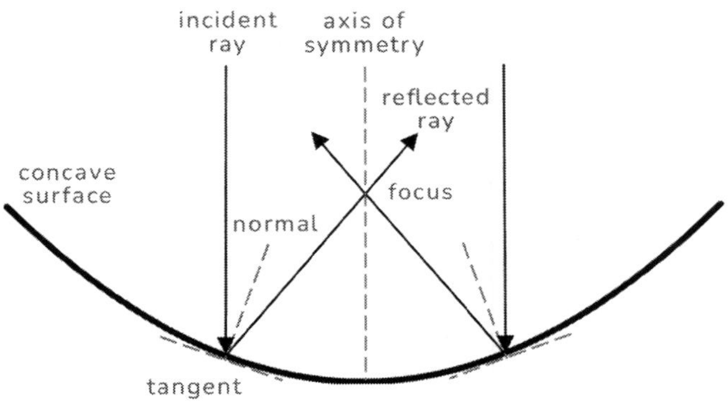

On the diagram, try drawing a different incident ray that is parallel to the axis of symmetry. Then draw its reflected ray. Does this also go through the focus?

A curved surface that bends away from the source of waves is described as ______________.

Convex mirrors do not ______________ waves towards a focus. Instead, incident rays that are parallel to the axis of symmetry will reflect so the rays ______________, meaning they spread and go in different directions.

Convex mirrors provide a wider field of view, because light from larger ______________ to the axis of symmetry can be reflected towards your eye. This is why convex mirrors are often at the end of narrow driveways, so drivers can see hazards around ______________.

Convex reflective surface

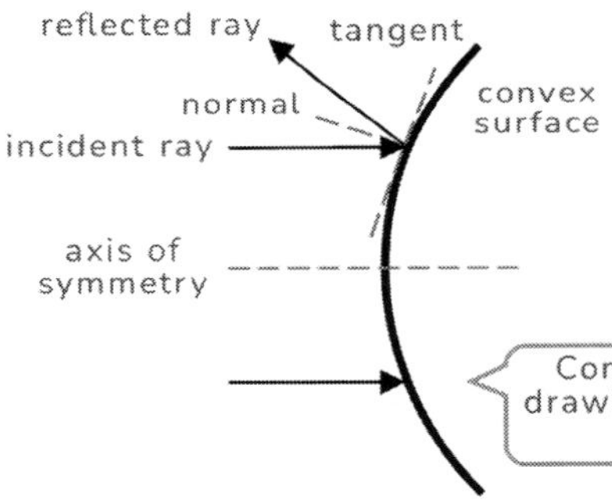

Complete the diagram by drawing the reflected ray for this incident ray.

9. Explain why car headlamps have a parabolic mirror behind them.

____

____

____

____

____

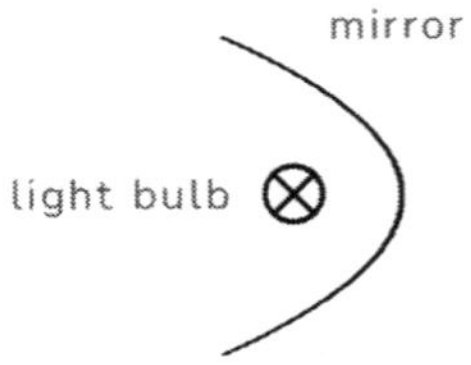

15.3 Superposition

1. Describe superposition, by filling in the blanks.

constructive	destructive	momentum	opposite
particles	unaffected	vector sum	velocity

Constructive interference of ocean waves

If two particles meet, such as two balls coming from opposite directions, they collide and change velocity, in a way that conserves ____________.

However, if two waves meet, they do not change ____________. Instead, they produce a resultant wave, so that the displacement of the medium at each point is given by the ____________ of the two waves meeting there. The waves themselves pass through each point ____________. This overlapping of waves is called ***superposition***. The production of the resultant wave is called ***interference***.

Interference is a property of waves, but not ____________.

____________ interference occurs when the resultant displacement of a particle due to two waves gives it a bigger displacement than it would have had due to either wave.

____________ interference occurs when the resultant displacement of the particle due to two waves becomes zero, because each separate wave would have displaced it by an equal magnitude in ____________ directions.

Example: A pulse is sent through a spring from each end, as shown.

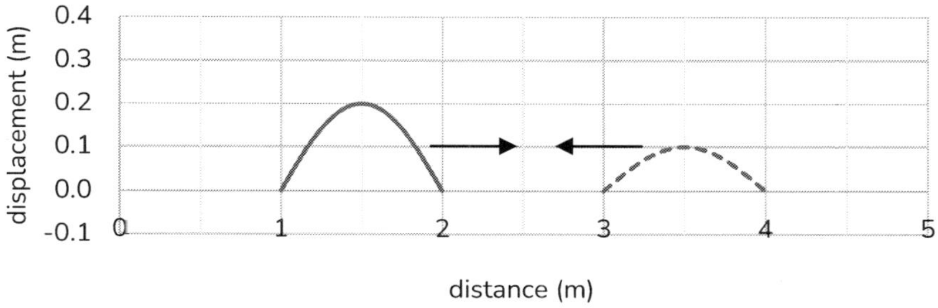

The pulses meet each other in the middle as shown.

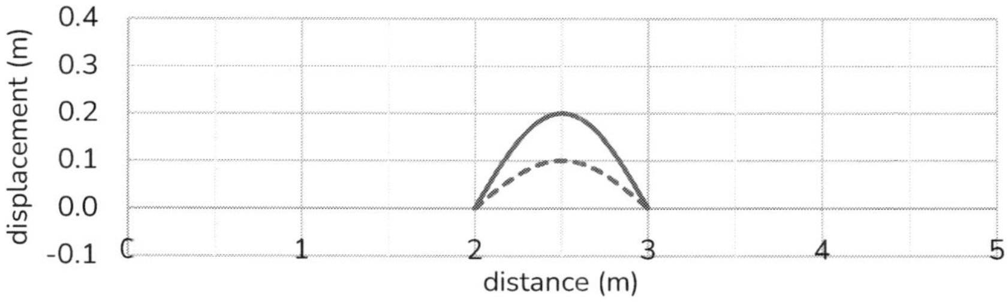

The resultant wave is now the spring actually looks when the pulses meet.

Draw the resultant wave at this instant and identify the type of interference.

Answer: This is constructive interference.

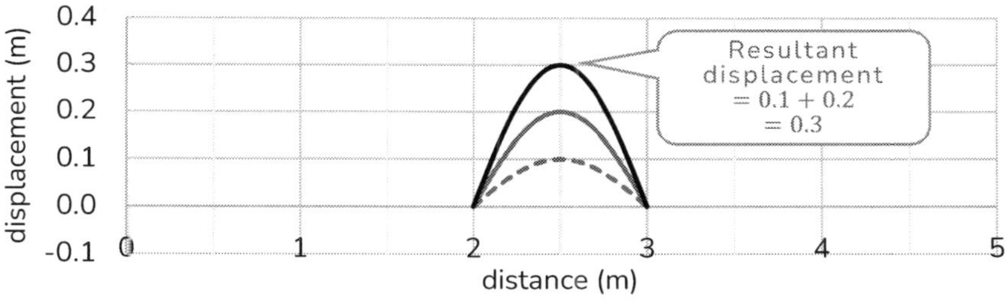

It's easiest to find the resultant displacement of the points at zero or maximum displacement first. Then sketch a curve that joins the points.

2. A pulse is sent through a slinky spring from each end, as shown.

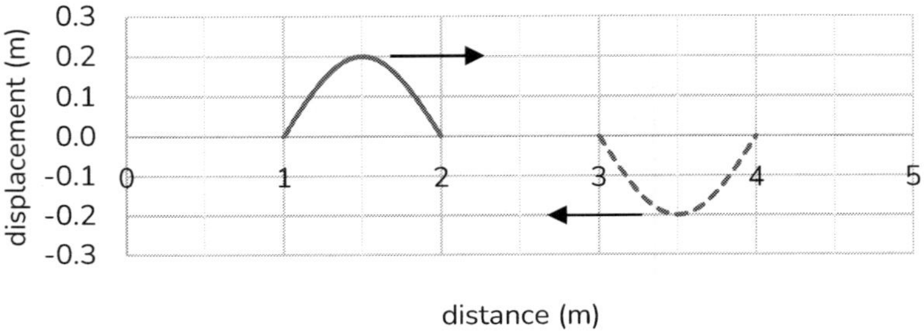

Explain what can be observed when the two pulses meet, as shown.

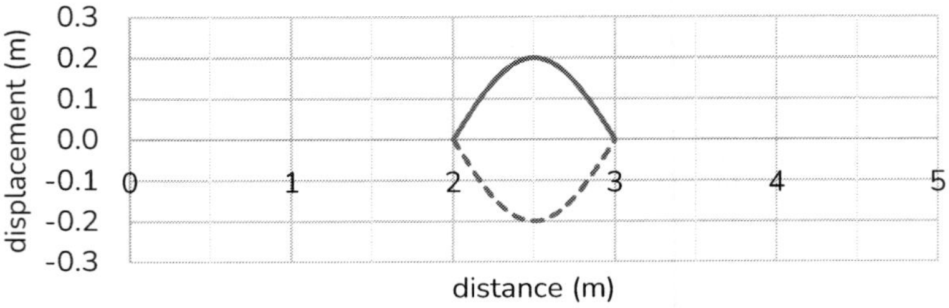

____

____

____

____

Example: Most transverse waves in nature have a shape like a sine wave. However, electronic signals can be produced so that the voltage-time graphs for them look like waves and can be analysed using similar techniques to waves. Square waves and pulses are often used in the digital world.

Two square pulses travel towards each other as shown.

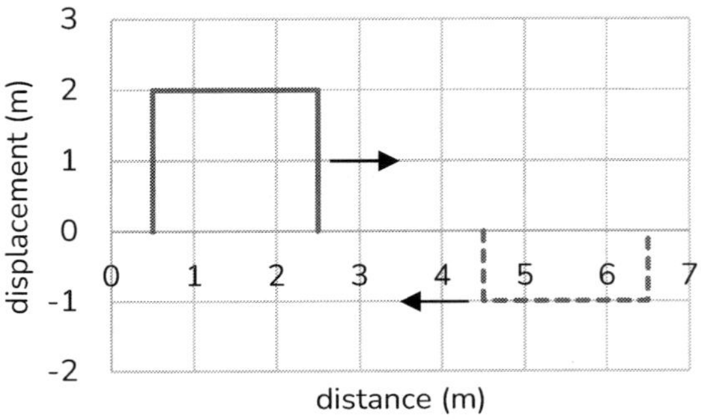

The pulses reach each other, as shown.

Draw the resultant pulse at this moment.

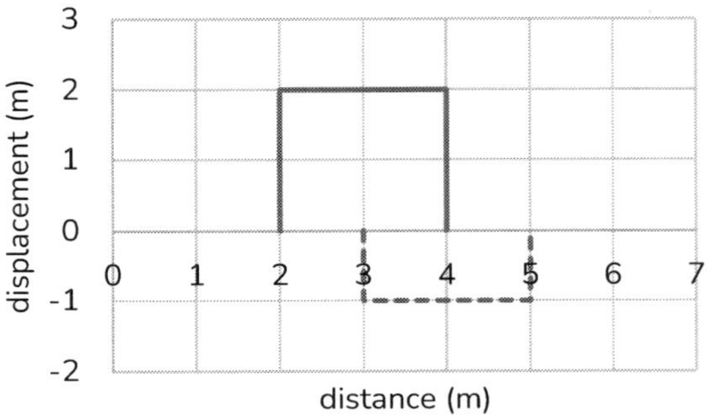

Answer: From 2 to 3 m, the only displacement is due to the first pulse.

From 3 to 4 m:

Displacement = +2-1 = 1 m

From 4 to 5 m, the only displacement is due to the second pulse.

3. Two pulses travel towards each other as shown.

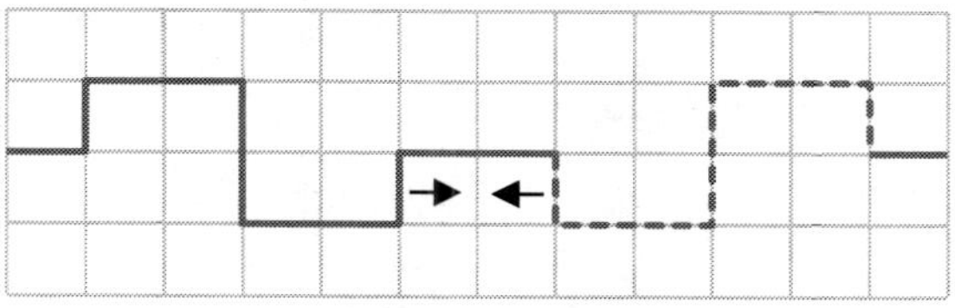

Explain, using sketches, how it is possible for the medium to experience:

a) destructive interference everywhere.

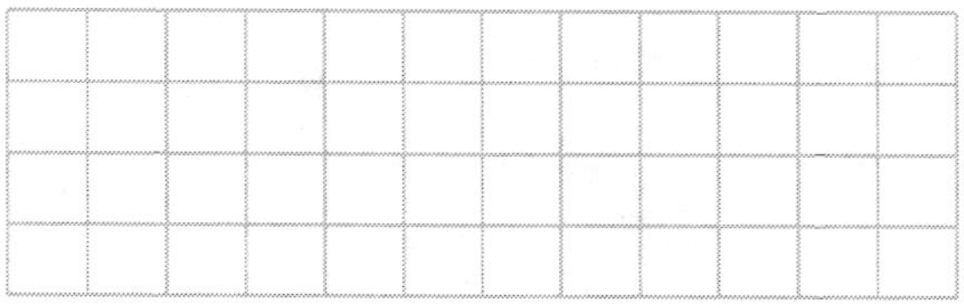

b) twice the amplitude compared to that which it would experience due to one pulse. (There are two possibilities.)

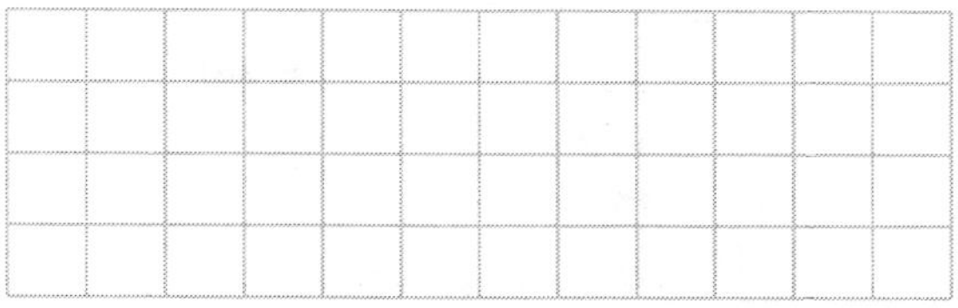

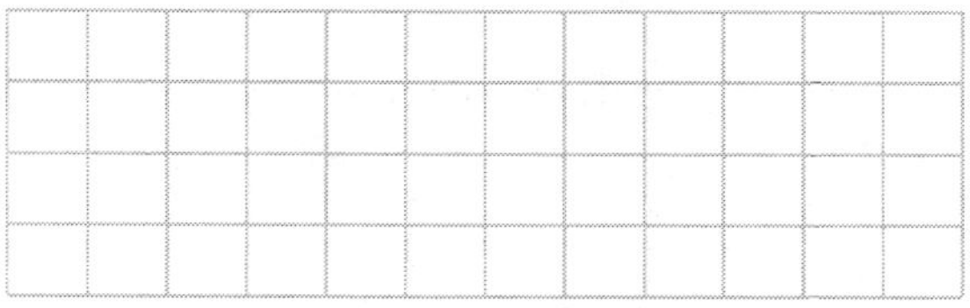

15.4 Standing waves on strings

Resonant frequencies

1. Explain how resonant frequencies produce standing waves, by filling in the gaps.

chaotic	energy	fixed end	frequency	half	move	out of phase
repeating	speed	superposition	tension	wave equation	wavelength	zero

Musical notes are due to resonant frequencies and the standing waves that they make.

Chaotic vibration

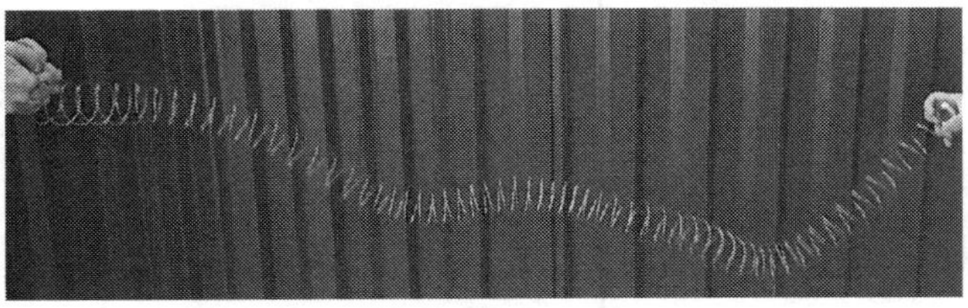

Fundamental frequency f_1

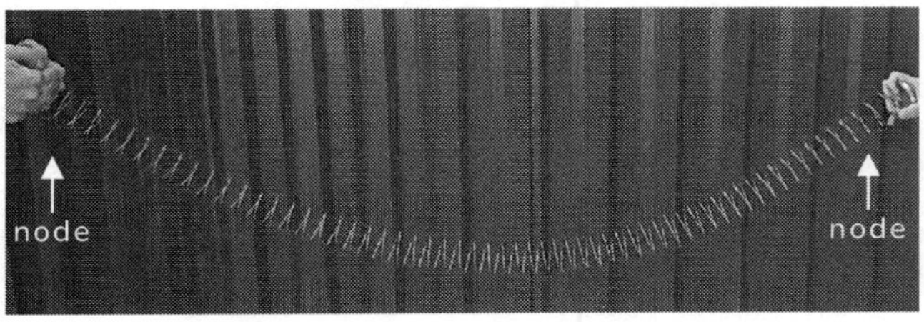

The second lowest resonant frequency f_2

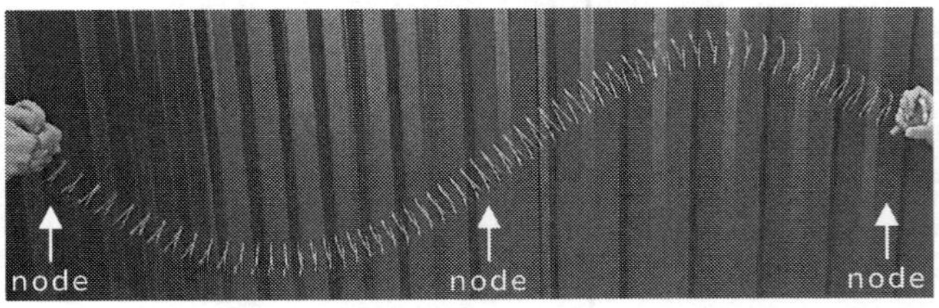

The third lowest resonant frequency f_3

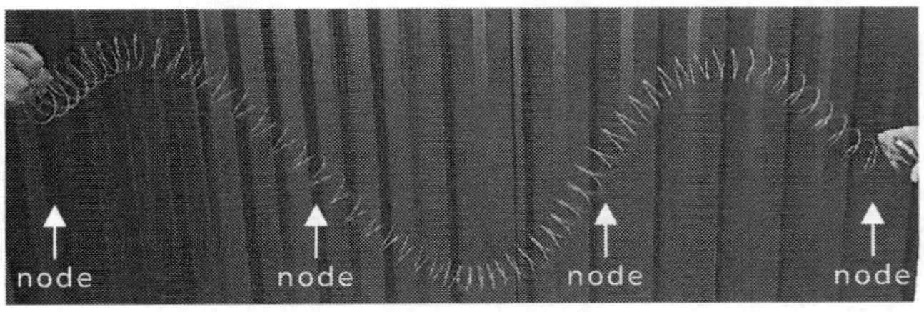

In reality, a node will be near your hand, since your hand needs to move a little to produce the vibrations.

If you hold the end of a string and vibrate it repeatedly to send a continuous wave through it, the wave would reflect half a wavelength __________ at the other end if it is a fixed end.

The resultant wave would be the __________ of the incident and reflected waves. For most frequencies of vibration, the resultant would look __________. It would feel difficult to keep vibrating your end up and down at the same frequency, since the reflected wave would return __________ to you, making your end vibrate with a different displacement to what you are giving the string.

However, for certain frequencies, called ***resonant frequencies***, a __________ pattern is produced, with the most basic pattern making the string move like a bit like skipping rope.

These frequencies need to produce wavelengths on the string which satisfy the boundary conditions of having a __________, called a node, at each end, since both ends are being held. A ***node*** is a point on the string where the total displacement due to the wave is __________. For example, the ends of guitar strings are nodes, as they do not __________ when the string is plucked.

The lowest resonant frequency, called the ***fundamental frequency***, f_1, will produce nodes at each end with __________ a wavelength between them.

The wavelength λ that a particular frequency produces on the string is given by the __________ of waves through the string, v, divided by the frequency f, according to the wave equation $v = f\lambda$. The speed is determined by the medium, so using the same string at the same __________ will mean that the waves travel through the string at the same speed. The longer the wavelength on the string, the lower the __________ that you must shake the end with.

In other words, if the distance between the ends is L, the __________ for this fundamental frequency will be $\lambda_1 = 2L$. Using the __________, the fundamental frequency will be equal to:

$$f_1 = \frac{v}{\lambda_1} = \frac{v}{2L}$$

A standing wave over one period

2. Describe what happens in a standing wave over one period, by filling in the blanks.

amplitude	constructive	constructive	crest	destructive	destructive	displacements
node	one	one quarter	resonant	superposition	transferred	trough

When an incident wave on a string reaches the other end as shown, so there is a crest in the middle, the reflected wave will also have a ______________ in the middle of the string, because it is reflected half a wavelength out of phase.

The resultant wave will be produced by the ______________ of the incident and reflected waves. The amplitude is therefore increased, with ______________ interference occurring.

At time $t = \frac{1}{4}T$, ______________ of a period later, the crest of the incident wave will have reached the other end. The reflected wave will therefore have a ______________ at the same end. The resultant wave at the other end still has zero displacement, producing a ______________. In fact, the resultant wave everywhere on the string experiences ______________ interference.

Another quarter of a period later, at time $t = \frac{1}{2}T$, there will be a trough in the middle of the string due to both the incident and reflected waves, so the resultant will therefore have a larger trough and exhibit ______________ interference.

Then another quarter of a period later, at time $t = \frac{3}{4}T$, the string will again experience ______________ interference.

______________ period later, the pattern repeats, with constructive interference producing a large crest in the middle. The middle of the string undergoes the largest ______________ compared to other points on the string, due to the fundamental frequency, and is called an ***antinode***.

Shaking the string at a ______________ frequency feels easy compared to other frequencies, as the resultant wave produces a node near your hand. The energy you put in will keep increasing the ______________ of vibration. However, there is no energy ______________ along the string overall, since the incident and reflected waves move energy in opposite directions. The waves produced by resonant frequencies are therefore called ***standing waves***.

$t = 0$

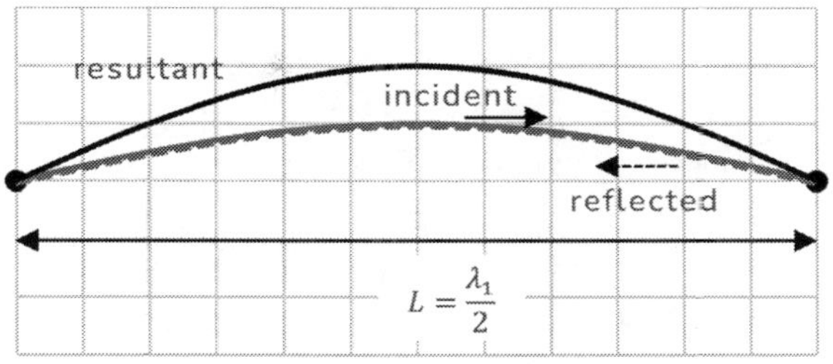

$t = \frac{1}{4}T$

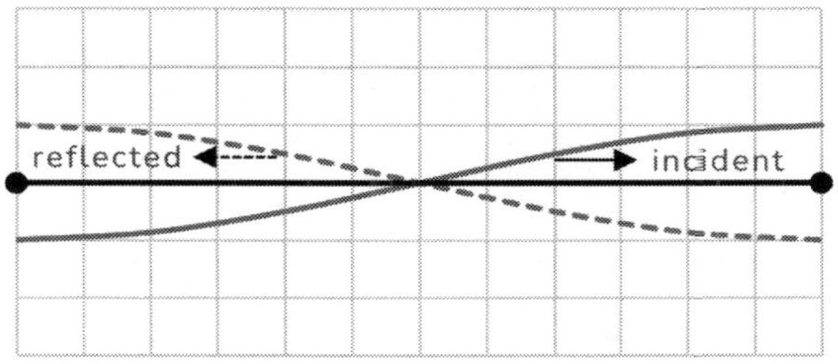

$t = \frac{1}{2}T$

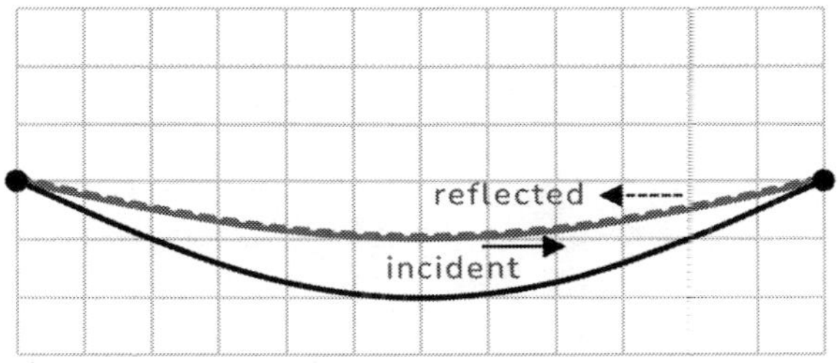

$t = \frac{3}{4}T$

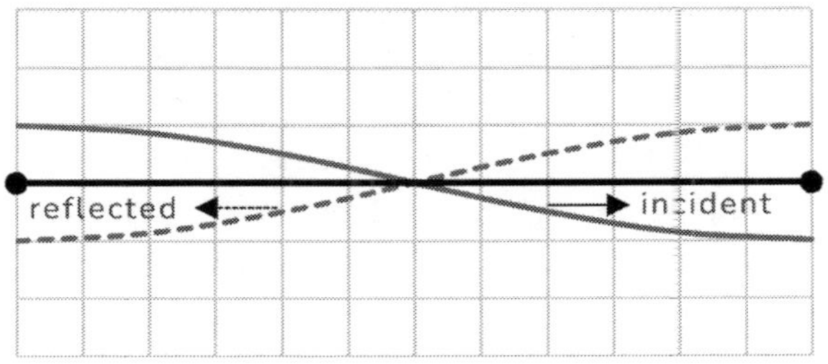

$t = T$

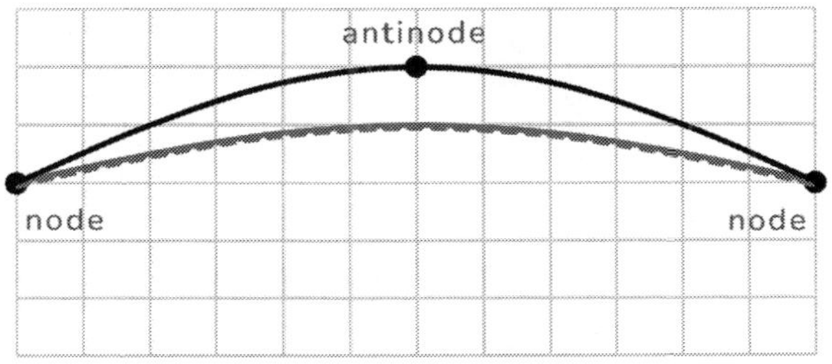

Higher resonant frequencies

3. Describe standing waves formed for the second lowest resonant frequency, by filling in the blanks.

double	half	harmonic	medium	node	pitch	standing

As well as the fundamental frequency f_1, there are higher resonant frequencies that will also produce ______________ waves. Collectively, these resonant frequencies are also known as harmonics. The second longest wavelength λ_2 of a standing wave on this string of length L will still have a ______________ at each end, but also a node in the middle, making the wavelength of the second harmonic $\lambda_2 = L$. The wavelength of the second harmonic is therefore ______________ the wavelength of the fundamental frequency.

Second harmonic

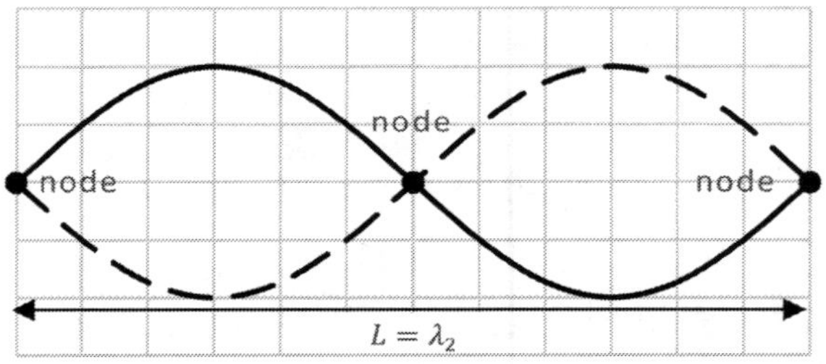

The velocity of the wave v is still the same, because the ______________ has not changed. Using the wave equation, the frequency of the second ______________ is:

$$f_2 = \frac{v}{\lambda_2} = \frac{v}{L}$$

The frequency of the second harmonic is therefore ______________ the fundamental frequency, $f_1 = \frac{v}{2L}$. When playing the second harmonic on a musical instrument, its ______________ will sound one octave higher in compared to the fundamental frequency.

4. The third harmonic on a string has the next longest wavelength, after the second harmonic.

   a) Draw the third harmonic. Nodes are given.

   b) Write an expression for its wavelength λ_3 in terms of L, the distance between the ends of the string.

   ____

   c) Write an expression for its frequency f_3 in terms of the speed of the wave v and string length L.

   ____

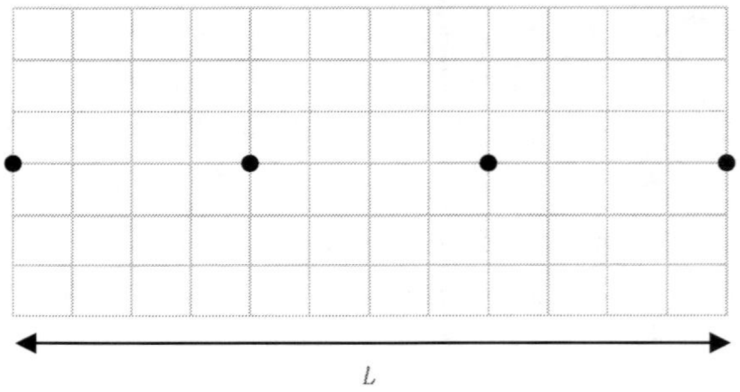

Compare the number of nodes for the fundamental, second harmonic and third harmonic. Notice a pattern?

5. The fourth harmonic on a string has the next longest wavelength, after the third harmonic.

   a) Draw the fourth harmonic. End nodes are given.

   b) Write an expression for its wavelength λ_4 in terms of L, the distance between the ends of the string.

   ____

   c) Write an expression for its frequency f_4 in terms of the speed of the wave v and string length L.

   ____

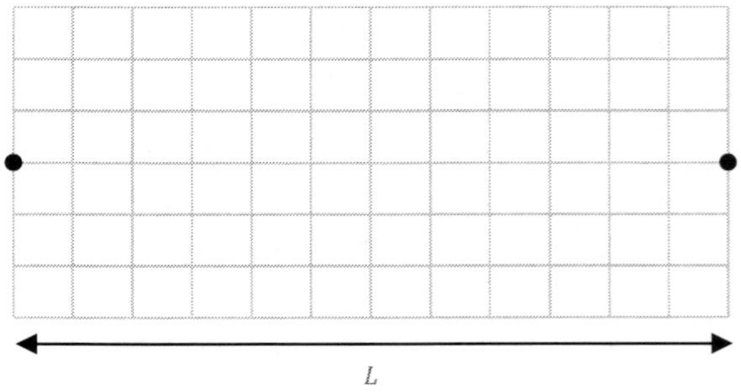

6. When describing a number pattern, a letter can be used to represent any integer – that is, a whole number. For example, the pattern of even numbers, 2, 4, 6, 8 can be represented as $2n$, where n is 1, 2, 3, 4, with the pattern also true for bigger integers. Odd numbers can be represented as $2n + 1$. This method is to represent a pattern is commonly used in both maths and science.

All the resonant frequencies f_n on a string with fixed ends also form a number pattern. If v is the wave speed and L is the distance between the ends of the string, which of the following formulae fits the pattern?

a) $f_n = \frac{nv}{L}$

b) $f_n = \frac{nv}{2L}$

c) $f_n = \frac{v}{nL}$

d) $f_n = \frac{2v}{nL}$

Substitute n=1 for the fundamental, n=2 for the second harmonic, then n=3 and 4 for the next harmonics into the formula. Which gives answers that match the frequencies from the previous questions?

7. An elastic band is stretched between two points 20.5 cm apart. When the elastic band is plucked, a note with a fundamental frequency of 310 Hz is recorded.

a) What is the speed of the wave on the elastic band?

What does the fundamental look like and what is its wavelength?

b) What is the wavelength and frequency of the second harmonic on this elastic band?

8. Explain why the pitch of a note produced by a string instrument changes if you reduce the length of the string that is able to vibrate.

Keywords: wavelength, fundamental. Use equations!

9. The standing waves for the fundamental frequency and second harmonic on a string are shown, with both having the same amplitude. Draw the resultant wave.

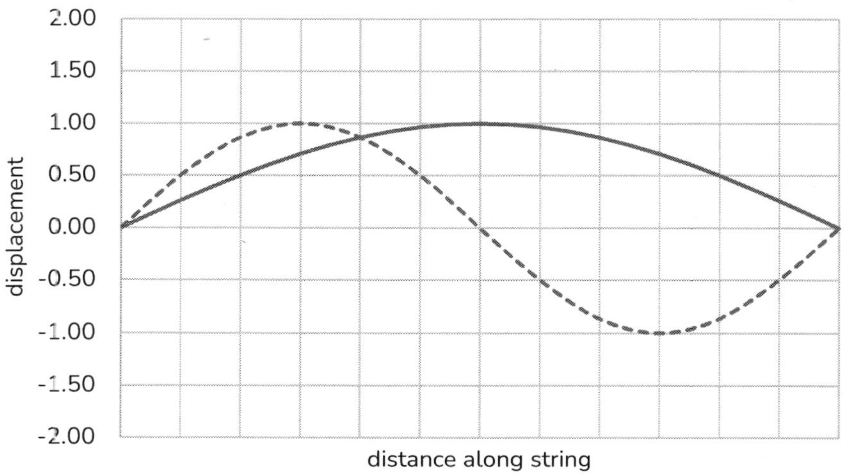

Recall the resultant displacement is the vector sum of the individual waves. The easiest points to add are those that:

- have zero displacement from one wave;
- opposite displacements from both waves, with a zero resultant; or
- are at the crest or trough of either wave.

Add more points if needed. Then join the resultant points with a smooth curve.

15.5 Standing waves in pipes

Assume the speed of sound in air, v, is 340 m s^{-1} for all calculations.

1. Describe how standing waves are formed in pipes, by filling in the blanks.

amplified	antinode	bottom	displacement	higher
longitudinal	medium	mouth	node	open
resonant	shorter	wavelength		

In wind instruments such as flutes, the ________________ through which waves travel is the air column. Unlike waves moving through strings, which are transverse, sound waves moving through air columns are ________________.

Standing waves are still able to form though, so the notes we hear from the musical instruments are the ________________ frequencies. In other words, sounds at the resonant frequencies are ________________ by the musical instrument. If we visualise the waves as ________________-distance graphs, with the distance representing the position along the air column, we can analyse standing waves in pipes in a similar way as standing waves in strings.

Boundary conditions can also be described using air pressure. At an open end, the equilibrium air pressure is the same as the air pressure outside the pipe. However, air particles press against a closed end, creating high pressure at the node.

At an open end of a pipe, air particles are free to move to the maximum displacement, so the boundary condition for open ends is an ________________. A flute is ________________ at both ends, so there would be an antinode at each end.

At a closed end, air particles are not free to move, so the boundary condition is a ________________. If you blow across the top of an empty bottle to produce a note, there would be a node at the ________________ of the bottle and an antinode at the ________________.

The fundamental frequency produces the standing wave with the longest ________________ that satisfies the boundary conditions of the pipe. All other resonant frequencies must also satisfy the boundary conditions, but will have ________________ wavelengths and ________________ frequencies than the fundamental frequency.

Displacement-distance graph for fundamental frequency on a closed pipe

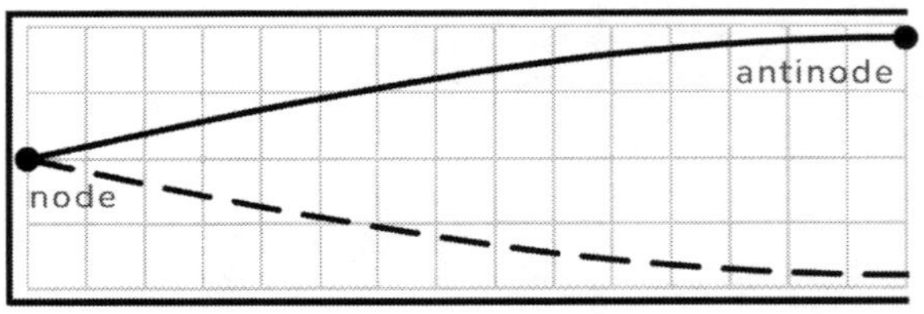

2. Musical instruments can play the sounds that form standing waves in them, but cannot play sounds that do not form standing waves.

   a) Explain what a standing wave is.

   ____

   ____

   ____

   b) If a note is played louder on a wind instrument, what happens to the standing wave formed in it?

   ____

Revise the information in previous worksheets.

3. A clarinet behaves like a pipe that is closed at one end, where the musician's mouth blows, and open at the other end.

One type of clarinet has a pipe length $L = 58$ cm.

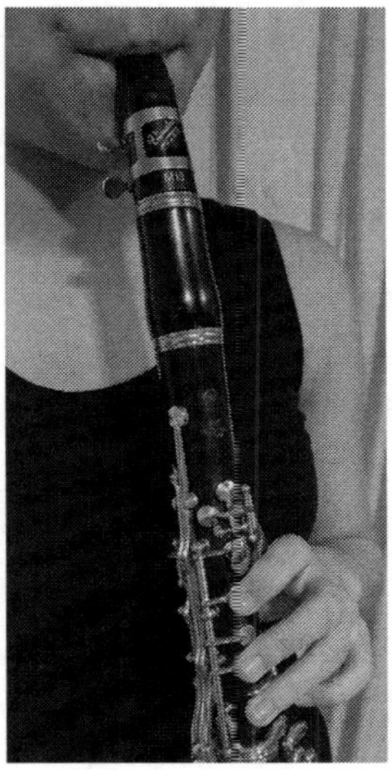

a) The diagram represents the standing wave formed by the fundamental frequency on the clarinet. Explain why its wavelength is $\lambda_1 = 4L$ and frequency is $f_1 = \frac{v}{4L}$. Calculate the frequency of the lowest pitch that this clarinet can play.

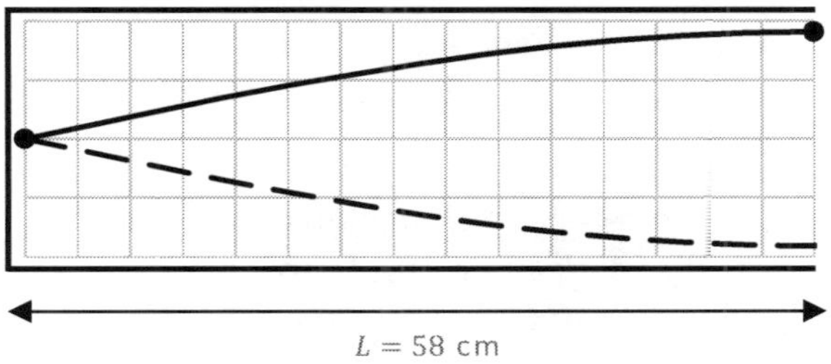

b) Draw the standing wave formed by the second harmonic for the clarinet. The nodes are given. Then write equations for its wavelength λ_2 and frequency f_2, in terms of L and v.

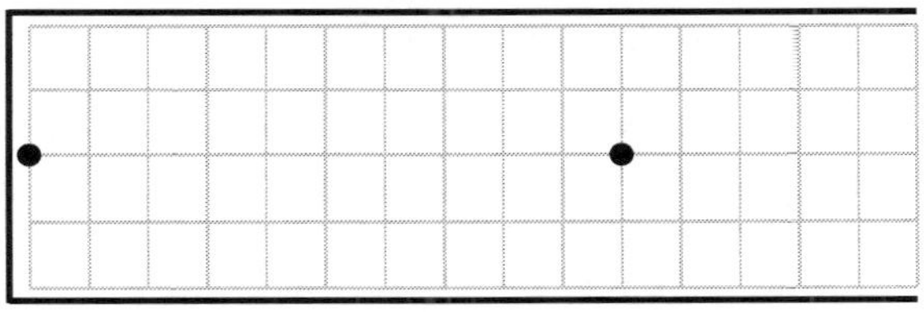

c) Draw the standing wave formed by the third harmonic for the clarinet. Then write equations for its wavelength λ_3 and frequency f_3, in terms of L and v.

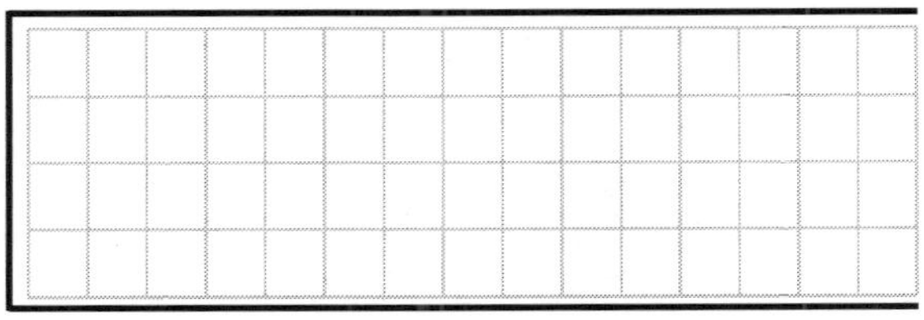

d) Which option below gives the formula for the resonant wavelengths and frequencies on a pipe that is closed at one end, where $n = 1, 2, 3$ and so on?

A	$\lambda_n = \frac{2L}{n}$	$f_n = \frac{nv}{2L}$
B	$\lambda_n = \frac{4L}{n}$	$f_n = \frac{nv}{4L}$
C	$\lambda_n = \frac{4L}{2n+1}$	$f_n = \frac{(2n+1)v}{4L}$
D	$\lambda_n = \frac{4L}{2n-1}$	$f_n = \frac{(2n-1)v}{4L}$

To draw the standing waves in parts (b) and (c), remember the boundary conditions. What is the pattern in the number of nodes?

Look at the pattern from parts (a), (b) and (c).

4. The human ear canal behaves like a closed pipe, with the ear drum at the closed end. The length of an adult's ear canal is often around 2.5 cm, whereas the length of a baby's ear canal is around 1.5 cm.

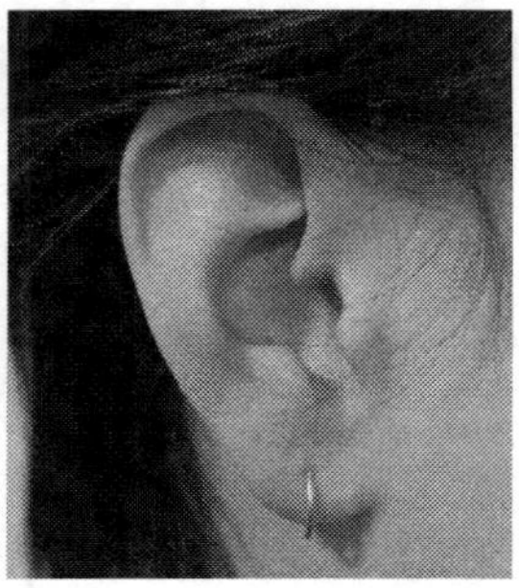

   a) Human hearing is most sensitive around the fundamental frequency of the ear canal. Calculate the frequency at which hearing is most sensitive for the adult.

   b) Explain what happens to the pitch at which hearing is most sensitive as a baby grows up.

Recall how to derive the equations and use these in your answers.

5. A didgeridoo behaves like an open pipe. The length of a certain didgeridoo is 1.2 m.

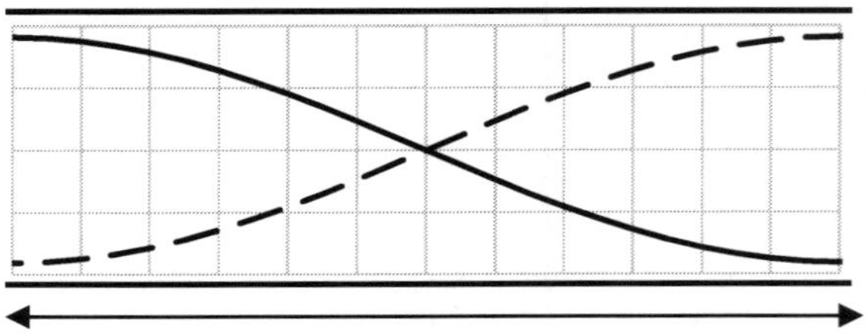

   a) The diagram represents the standing wave formed by the fundamental frequency on the didgeridoo. Write an equation for its wavelength λ_1 and frequency f_1, in terms of L and v. Calculate the frequency of the note with the lowest pitch that this didgeridoo can play.

   b) Draw the standing wave formed by the second harmonic for the didgeridoo. The nodes are given. Then write equations for its wavelength λ_2 and frequency f_2, in terms of L and v.

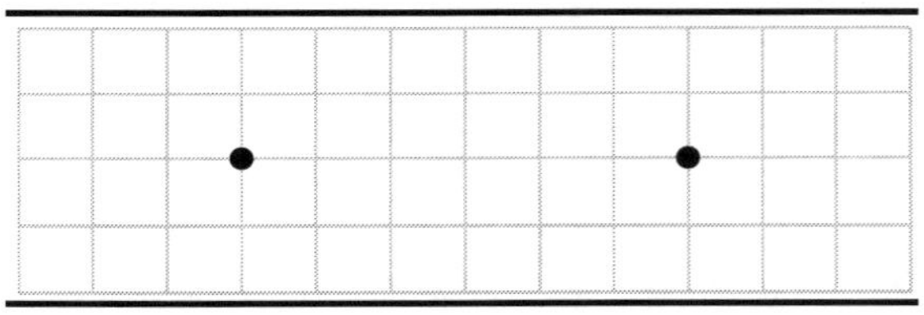

   c) Draw the standing wave formed by the third harmonic for the didgeridoo. Then write equations for its wavelength λ_3 and frequency f_3, in terms of L and v.

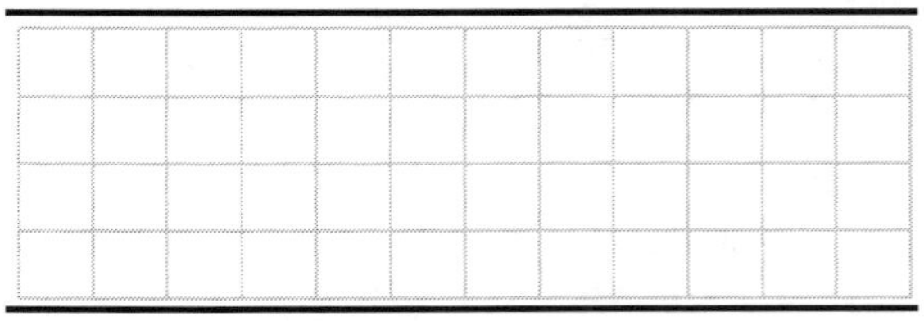

   d) Write equations for the resonant wavelengths λ_n and frequencies f_n on an open pipe, in terms of $n = 1, 2, 3$ and so on.

15.6 Interference from two sources

1. Describe what happens when there are two sources of waves, by filling in the blanks.

amplitude	constructive	destructive	equal	flat
in phase	integer	louder	meet	nanometre
one	out of phase	particles	path difference	rarefaction
reflection	silence	softer	sound	zero

Interference is a property of waves, but not ________. Interference occurs between a wave and its ________ returning towards the wave source. Interference also occurs when waves from different sources ________.

If two loudspeakers play the same single frequency in phase with each other, what can you hear as you walk around in front of them? Constructive interference occurs when the part of the wave reaching you from one loudspeaker is ________ with the wave reaching you from the other loudspeaker. For example, if a compression reaches you from both loudspeakers at the same time, the sound you hear will seem ________.

Destructive interference occurs when the parts of the waves from each loudspeaker reaching you are half a wavelength ________ with each other. For example, this happens if a compression reaches you from one loudspeaker and a ________ reaches you from the other at the same time. At these points, the sound you hear will seem ________. If you try this in a room where there are no echoes from the walls, at the points of destructive interference there would be ________.

The type of interference that occurs at a point depends on the difference in distance between that point and each source, also known as the *path difference*. The path difference is zero at all points an ________ distance from both sources. The waves reaching these points from both sources will always be in phase, so ________ interference occurs. This will also happens at points which are ________ wavelength further from one source than the other. In fact, constructive interference will happen at all points where the path difference is an ________ number of wavelengths, or $n\lambda$, where $n = 0, 1, 2$ and so on.

Displacement-distance graph from two sources are shown below.

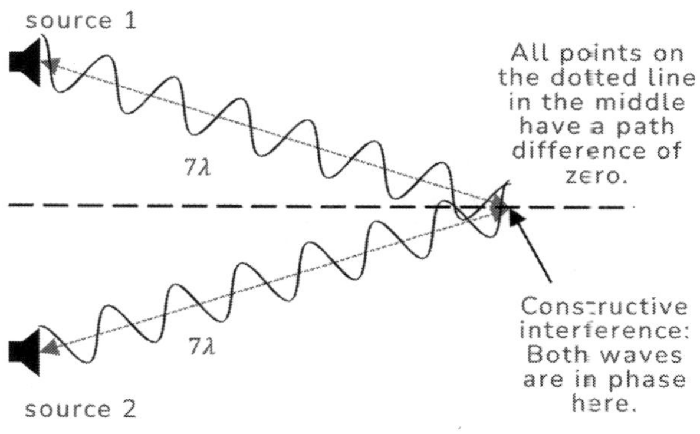

If the path difference is half a wavelength at a point, the waves reaching that point are out of phase, so ________ interference occurs. In fact, destructive interference will happen at all points where the ________ is $(n + \frac{1}{2})\lambda$, where $n = 0, 1, 2$ and so on.

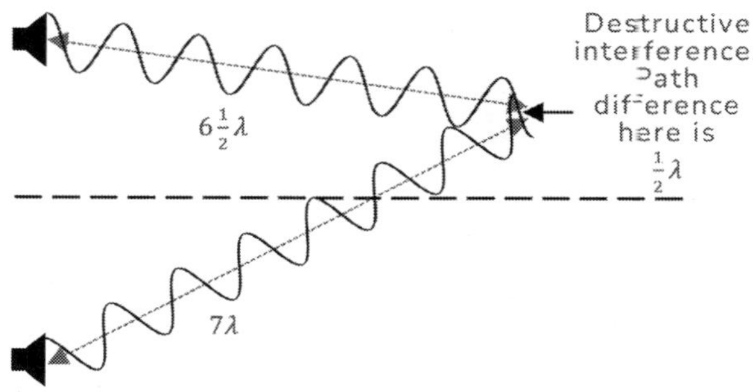

All other points will experience an amplitude somewhere in between the amplitude from constructive interference and ________.

Interference occurs with water waves too. Where constructive interference occurs, the water has a greater ________. Where destructive interference occurs, the water will look ________.

Evidence of light interference was only gathered in 1801, since the wavelength of visible light is on the ________ scale. This means that the path differences for light interference are tiny compared to ________.

2. Two loudspeakers are set up apart, as shown. The musical note middle C, with a frequency of 261 Hz, is played in phase through both speakers.

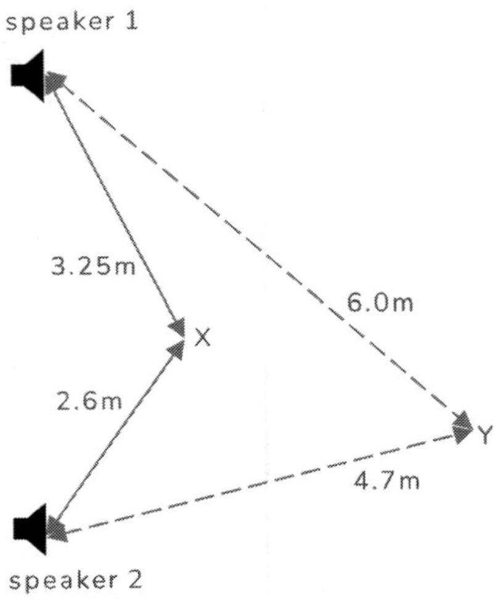

a) What is the wavelength of sound from the speakers? Assume the speed of sound is 340 m s^{-1}.

b) Explain what could be observed about sound at point X, which is 3.25 m from speaker 1 and 2.6 m from speaker 2.

Key words: path difference, wavelength, phase, interference, volume.

c) Explain what could be observed about sound at point Y, which is 6.0 m from speaker 1 and 4.7 m from speaker 2.

3. The same loudspeakers are set up apart, as shown.

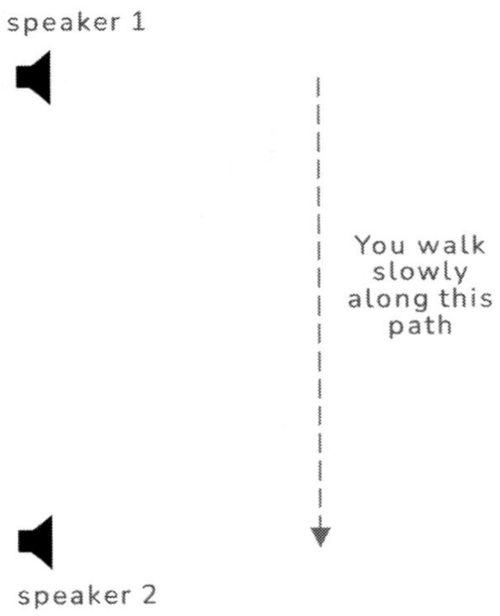

a) Again the musical note middle C, with a frequency of 261 Hz, is played in phase through both speakers. You observe the sound as you walk slowly along the path shown. Explain qualitatively what you observe.

b) Compare your observation in part (a) with what you observe if you repeat the experiment using a note with higher frequency.

15.7 Beats

1. Describe beats, by filling in the blanks.

average	difference	interfere	pitch	volume

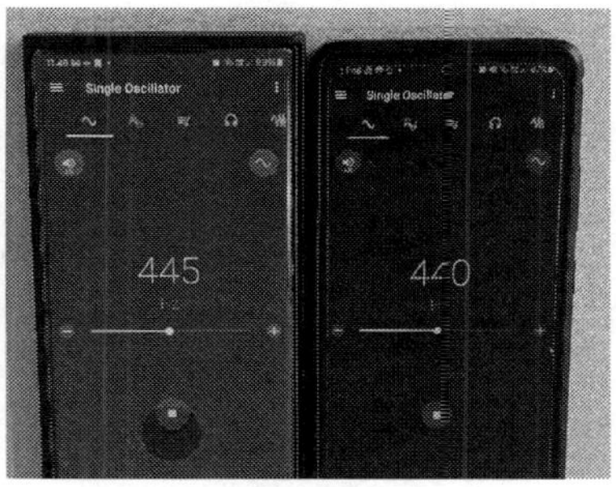

Beats are an interesting phenomenon that occurs when two sound waves of slightly different frequencies, f_1 and f_2, ________________ to produce a resultant wave. This resultant wave has an amplitude that varies with time, with the peaks occurring with a frequency f_{beat} that is the ________________ between f_1 and f_2:

$$f_{beat} = f_2 - f_1$$

The frequency of these peaks is called the ***beat frequency***.

The resultant wave also sounds like it has a frequency that is the ________________ of f_1 and f_2:

$$f_{resultant} = \frac{f_1 + f_2}{2}$$

The resultant wave sounds like it has a ________________ that is in the middle of f_1 and f_2 but with a ________________ that rises and falls rhythmically at the beat frequency.

Try this: Using a frequency generator app on two phones, choose an audible frequency to play on one and play another frequency with up to 5 Hz difference on the other. Listen to them both as they are held close together. The "woah, woah" sound you hear is beats. What happens when you change one of the frequencies a little bit?

Example: If sound waves of frequency 10 Hz and 8 Hz are superpositioned, what will be the frequency of the resultant wave and the beat frequency?

Answer: The resultant wave will have a frequency of $f_{resultant} = \frac{f_1+f_2}{2} = \frac{10+8}{2} = 9$ Hz.

Beat frequency will be $f_{beat} = f_2 - f_1 = 10 - 8 = 2$ Hz.

Sound wave with f_1 = 10 Hz

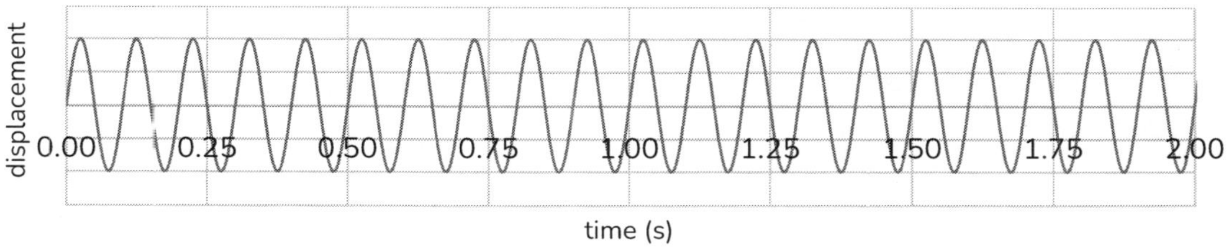

Sound wave with f_2 = 8 Hz

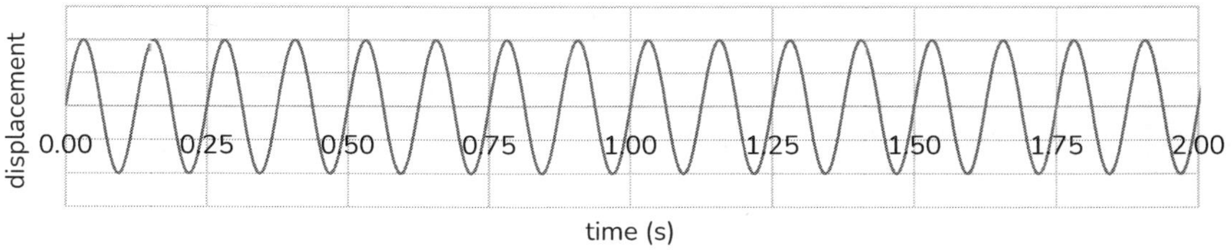

Resultant wave

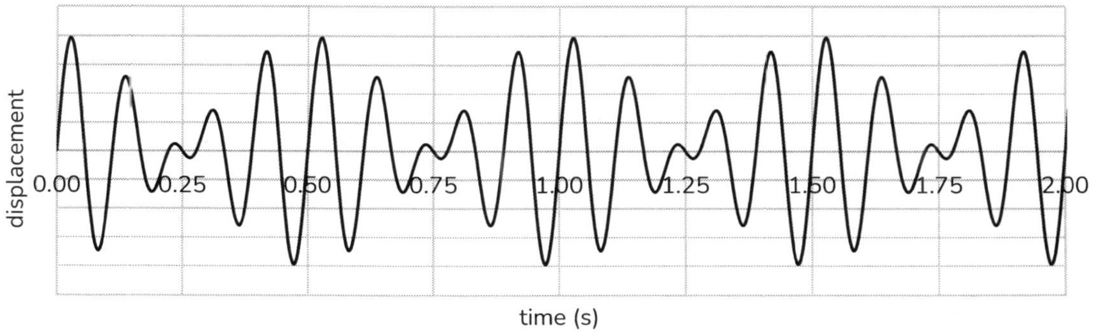

2. The displacement-time graph for two sound waves is shown below.

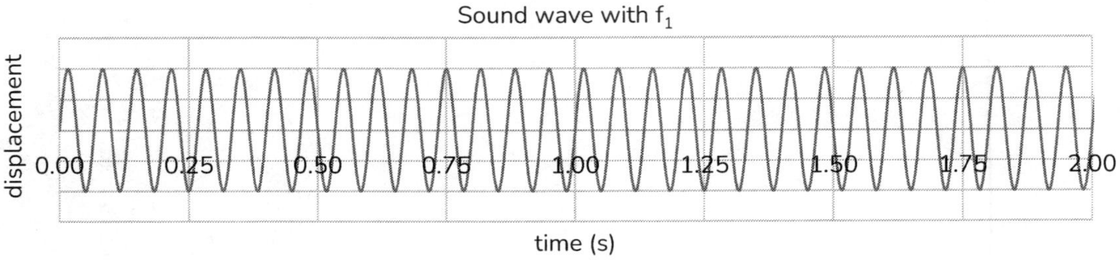

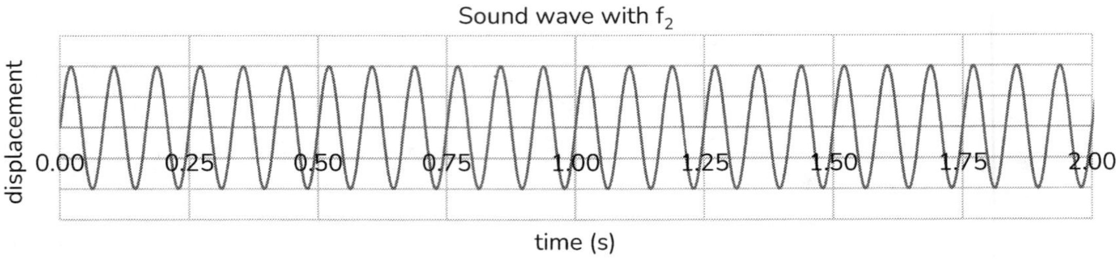

They are then played together. Determine:

a) the frequency of each sound wave f_1 and f_2;

From the first graph, there are ______ periods in ______ second, so ______

b) the frequency of the resultant wave; and

c) the beat frequency of the resultant wave.

Check if your answers match the displacement-time graph for the resultant wave shown below.

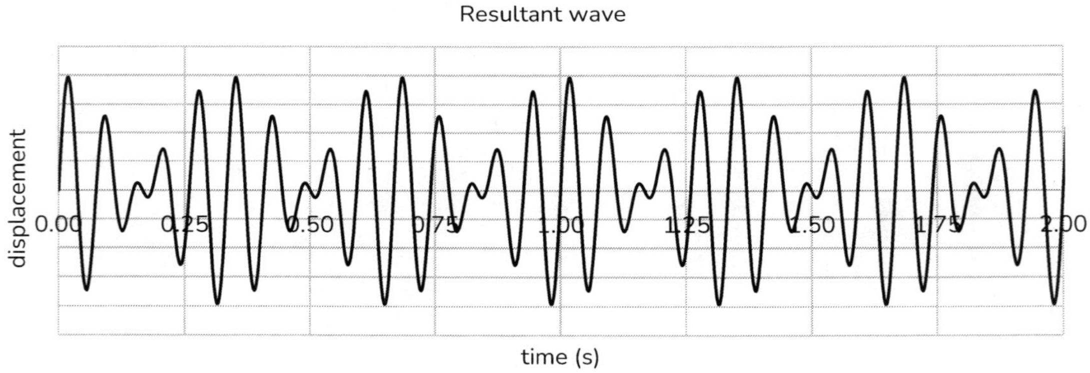

3. Before an orchestra plays, an oboe will play a note of frequency 440 Hz for the other musicians to tune to.

a) If another oboe plays a note of frequency 445 Hz, how many beats per second can be heard?

b) What happens to the beats as the second oboe is tuned so it gets closer to the frequency of the first?

15.8 Diffraction

1. Describe diffraction, by filling in the blanks.

absorbed	amplitude	aperture	bend
bigger	diffract	edges	energy
middle	particles	sheltered	smaller
spreading	superposition	wave	wavelength

Diffraction around a big obstacle

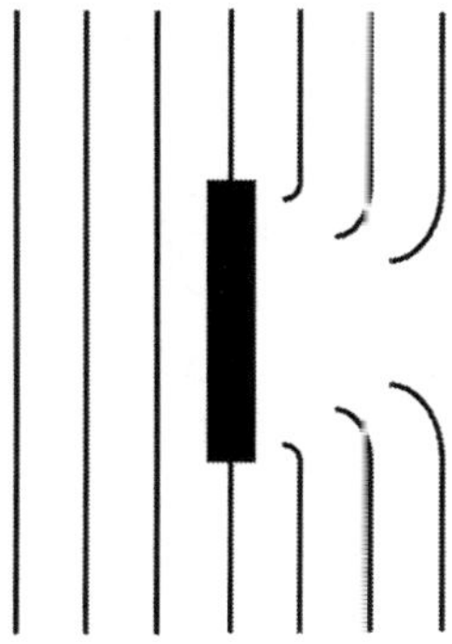

Diffraction around a small obstacle

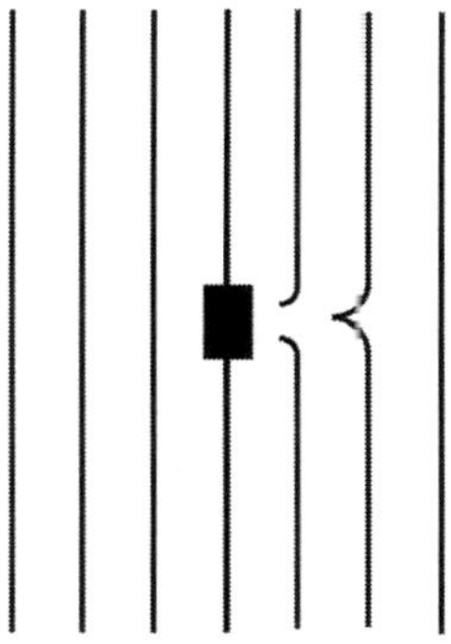

Diffraction through an aperture

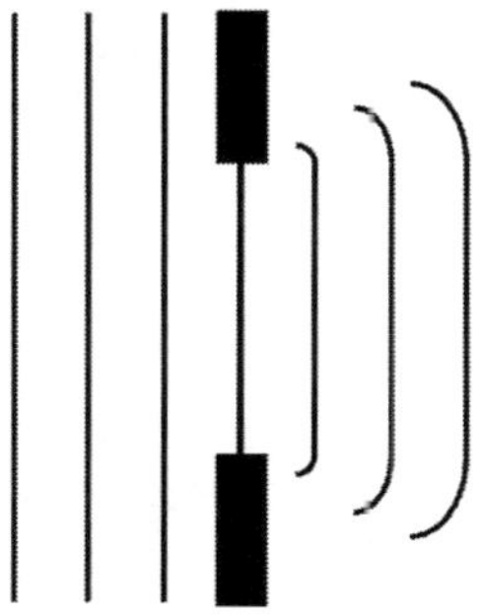

As well as ________________, another behaviour that waves exhibit and particles do not is diffraction. *Diffraction* is the bending of waves around obstacles or through a gap, also known as an aperture. If you stand around a corner from someone and shout, they can still hear you, because sound is a ________________ and will diffract around the corner. However, if you try to throw a ball at them, the ball will be blocked by the corner, because ________________ do not diffract.

Diffraction explains why ocean waves can still reach a beach, even if the beach is ________________ by a headland. The wavefronts ________________ around the headland, producing circular wavefronts as they spread out. The part of the wavefront hitting the edge of the obstacle acts a bit like a source of energy, with the energy ________________ out in all directions. However, some energy is lost from the wave as it is absorbed by the obstacle, so the ________________ of the wave is smaller as it diffracts around the obstacle. So the ocean wave bending around the headland will carry less ________________ than the wave before the headland.

How much diffraction occurs depends on the size of the wavelength and the obstacle or aperture. If an obstacle is small compared to the ________________, then the wave will diffract right around it. For example, if you stand in the ocean, the oncoming waves will ________________ around you, so the waves behind you do not look like they have been affected by your presence, because the distance between the crests of the ocean waves is likely to be ________________ than your width.

This is also why you cannot see viruses using a light microscope. Viruses are typically 20 nm to 400 nm in diameter, which is ________________ than the wavelengths of visible light, with a range of around 400 nm to 700 nm, so the light diffracts around the virus unaffected.

When wavefronts go through an aperture, diffraction is most noticeable when the wavelength is close to the size of the ________________. Diffraction still happens if the aperture is wider, but most of the wave will just pass through unaffected, so diffraction will only be observed close to the ________________ of the aperture and not the ________________. If the aperture is too small compared to the wavelength, most of the wave's energy will be ________________ by the edges of the aperture.

2. Reflection and diffraction can both lead to waves changing direction. Compare these two behaviours.

Feature	Reflection	Diffraction
Where does it occur?		
Describe the change in direction of the wave		
Does the wavelength change?		
Can particles exhibit this behaviour?		
Are observable effects affected by wavelength?		

For calculations, take the speed of sound as 340 m s^{-1} and the speed of light as 3×10^8 m s^{-1}.

3. If someone stands between you and a television, you can still hear the movie you were watching, but you cannot see the images. Explain why.

   Note that visible light has wavelengths of around 400 nm to 700 nm, whereas the range of human hearing is around 20 Hz to 20 kHz.

4. Obstacles can affect where radio stations can be received. A hill is between a radio transmission tower and a town. Analyse whether the town is more likely to receive signals from the radio tower transmitted at 702 kHz or 105.7 MHz.

5. Ash and Bee are talking on the phone during a thunderstorm, which they can both hear from their physical location. Ash says the thunder sounds high pitched, like a crack. However, Bee thinks the same thunderstorm sounds lower pitched, like a rumble. Explain how they can both be right.

15.9 Doppler effect

1. Describe the Doppler effect, by filling in the blanks.

apparent	away	decrease	frequency	ground	ground	high
increase	low	lower	observer	relative	source	towards

If you walk down a street and pass a house every 5 seconds, a car driving down the same street at five times your speed will pass a house every second – in other words, at a higher ________.

Similarly, if you and a source of sound move towards each other, then each wavelength reaches you faster, so frequency appears to increase and wavelength appears to ________. Consequently, you hear a more ________ pitched sound.

If the same source moves away from you, then each wavelength takes longer to pass you, so the wavelength of sound that you observe will ________. Consequently, the frequency you observe will decrease and you hear a more ________ pitched sound.

This is called the ***Doppler effect*** - the change in frequency of a wave observed when there is ________ motion between the source of the wave and the observer.

The Doppler effect explains why the engine noise of aeroplanes flying by appears to increase in pitch as they approach and decrease as they recede, if we observe the aeroplanes from the ________, but the passengers do not observe a change in pitch.

When a source and observer move towards each other, the apparent frequency detected by the ________ is given by:

$$f' = \frac{f(v_w + v_o)}{v_w - v_s}$$

where: f' = frequency detected by the observer (Hz)

f = frequency of the sound emitted at the ________ (Hz)

v_w = speed of the wave in the medium (m s^{-1})

v_o = speed of the observer relative to the ground (m s^{-1})

v_s = speed of the source relative to the ________ (m s^{-1})

So the faster the speed of the source v_s ________ from the observer, the higher the apparent frequency.

When a source and observer are moving away from each other, the ________ frequency detected by the observer is given by:

$$f' = \frac{f(v_w - v_o)}{v_w + v_s}$$

This formula looks like the formula used when they move towards each other, except that the plus and minus signs are swapped, so that the apparent frequency f' is ________ than the original frequency f.

So the faster the speed of the source v_s ________ from the observer, the lower the apparent frequency.

Both source and observer stationary

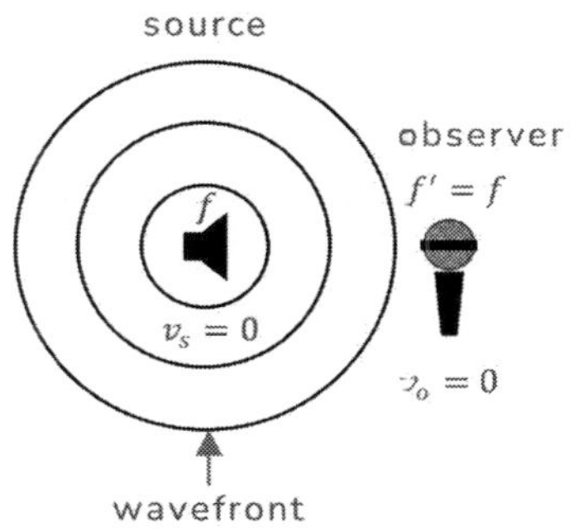

Source moving towards observer

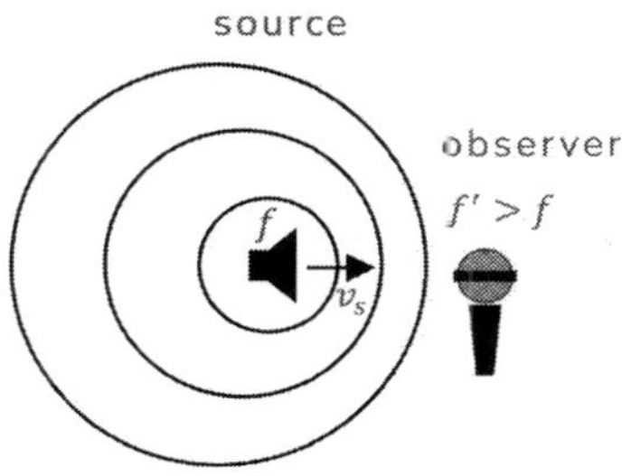

Source moving away from observer

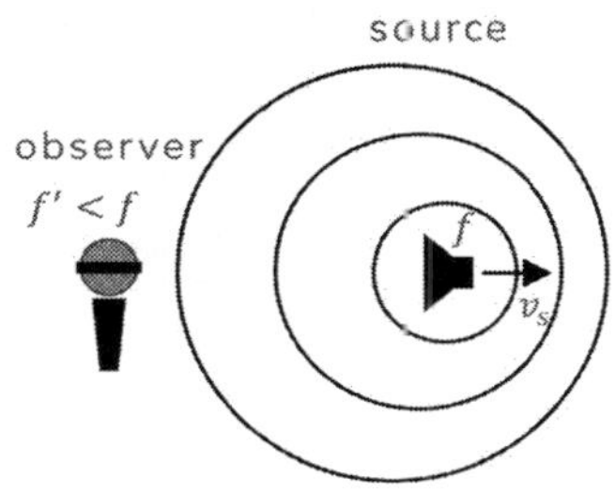

Source moving away faster

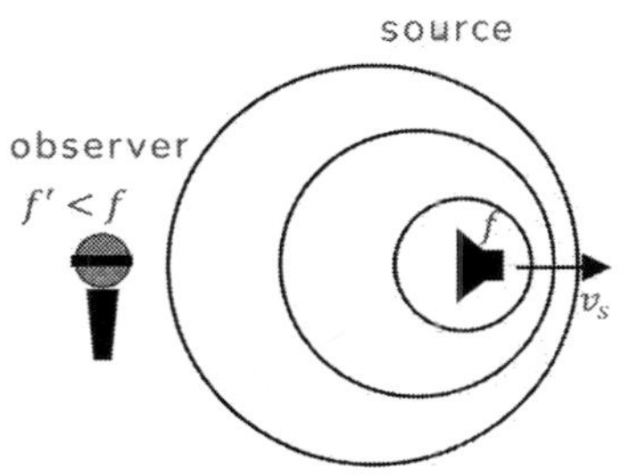

Example: A car is driven at 60 km h^{-1} when an ambulance approaches from the opposite direction. The car's driver first observes the ambulance's siren at a pitch of 1085 Hz. However, at the moment the ambulance is next to the car, the same siren is observed to be 950 Hz. How fast was the ambulance moving, in km h^{-1}?

Answer: The frequency observed as the ambulance is next to the car is the same as the frequency in the ambulance's frame of reference, since the sound is travelling at right angles to the ambulance's motion to reach the car, and therefore not affected by the Doppler effect at that moment. Therefore $f = 950$ Hz.

Using the equation for the source and observer moving towards each other, and converting speeds to m s^{-1}:

$$f' = \frac{f(v_w + v_o)}{v_w - v_s}$$

$$1085 = \frac{950\left(340 + \frac{60}{3.6}\right)}{340 - v_s}$$

$$v_s = 28 \text{ m s}^{-1} = 100 \text{ km h}^{-1}$$

Take the speed of sound in air as 340 m s^{-1} for all calculations.

2. Play a frequency on a mobile phone frequency generator app. Put the phone in a long sock. Spin the sock above your head while someone else observes the sound. A diagram of the path taken by the phone is shown. What would the observer hear? Label on the diagram the points in the phone's journey when these observations would occur.

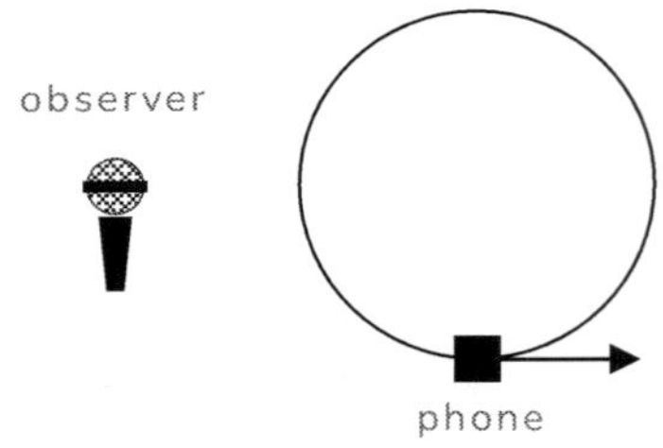

3. The formulae for the apparent frequency are often combined to describe both the source and observer approaching and receding from each other:

$$f' = \frac{f(v_w \pm v_o)}{v_w \pm v_s}$$

a) In the numerator (top of fraction), which sign, + or -, gives a larger apparent frequency f'?

b) In the denominator (bottom of fraction), which sign, + or -, gives a larger apparent frequency f'?

c) Explain what this combination of signs should be used for – the source and observer approaching each other or receding from each other?

d) Explain why treating v_o and v_s as velocities, rather than speeds, would lead to an incorrect calculation.

4. A mosquito's buzzing sound is due to the movement of its wings. One mosquito emits a frequency of 674 Hz when it hovers on the spot. If you sit still, what frequency will you hear if the mosquito flies:

a) away from you at 1.20 m s^{-1}.

b) towards you at 1.20 m s^{-1}.

5. Heated elements each emit a unique set of wavelengths of electromagnetic radiation, called a spectrum, that can be used to identify them. The electromagnetic radiation from stars and galaxies indicates that they all contain hydrogen. However, compared to the hydrogen spectrum observed in a laboratory on Earth, the hydrogen spectrum from most galaxies is shifted towards longer wavelengths. This effect is called the ***red shift***, and is like the Doppler effect for light.

Hydrogen spectrum on Earth

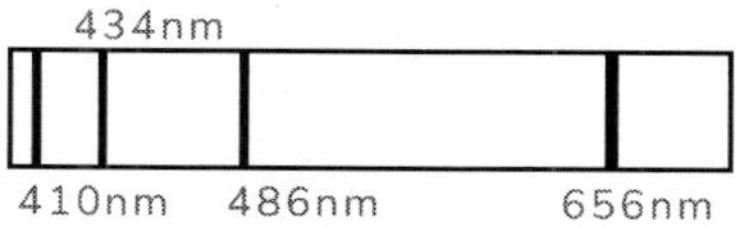

a) Explain what can be inferred about the motion of the galaxies from these observations.

b) The hydrogen spectrum for galaxies A, B and C, as observed from Earth, are shown. Compare the motion these galaxies.

Hydrogen spectrum from Galaxy A

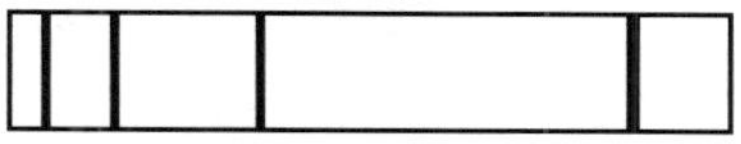

Hydrogen spectrum from Galaxy B

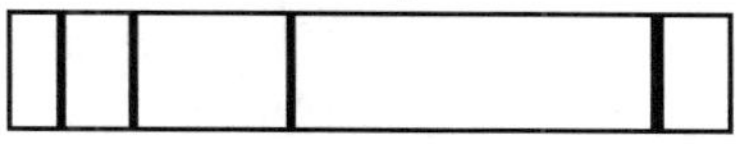

Hydrogen spectrum from Galaxy C

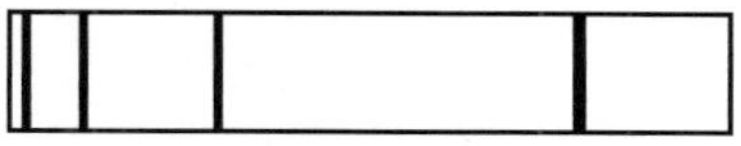

6. One application of the Doppler effect is to measure the speed of blood flow using ultrasound, which is sound of frequencies above the range that humans can hear. A device containing a speaker and microphone is pressed against a person's skin, to direct ultrasound of a known frequency into a blood vessel. The ultrasound is reflected off the moving blood, back towards the device.

One device emits ultrasound of frequency 1.5 MHz into a blood vessel, where blood flows at 12 cm s^{-1} towards the device. Sound waves travel at 1570 m s^{-1} in blood.

a) What is the apparent frequency of the wave from the blood's frame of reference, to the nearest Hertz?

b) The blood reflects the ultrasound back towards the device, at the same apparent frequency incident on it, calculated in part (a). However, the device observes the reflected ultrasound Doppler shifted a second time, since the blood is acting as a moving source. What is the frequency of the ultrasound that is detected by the device, to the nearest Hertz?

c) The emitted and reflected waves interfere to produce beats. Calculate the beat frequency detected by the device.

7. An alarm clock emits a frequency of 1610 Hz. A grumpy person drops the alarm clock out of an apartment window. How long would the alarm clock have been falling for when its apparent frequency is 1522 Hz, as observed by the stationary grumpy person?

15.10 Refraction

Refractive index

1. Describe refraction, by filling in the blanks.

air	away from	bend	bent	direction	equal to 1
faster	further	greater than	greater than 1	least time	less than
light ray	medium	medium	normal	ratio	slows down
speeds	towards	vacuum	zero		

A straw in a glass of water looks ________________ from most angles. The straw is not actually bent though – this phenomenon occurs because the ________________ reaching us from the straw bends as it travels from water to air.

The change in direction of travel of a wave as it moves from one ________________ to another is called ***refraction***. This bending occurs because waves travel at different ________________ in different media.

For example, light travels ________________ than anything we know in a vacuum, at 3 x 10^8 m s^{-1}. Light travels very close to this speed in ________________. However, as light enters other media, such as glass or water, it ________________. The ***refractive index*** of a medium gives the ________________ of the speed of light in a vacuum to the speed of light in that medium:

$$n = \frac{c}{v}$$

where: n = refractive index (this is a number with no units)

c = 3 x 10^8 m s^{-1} = speed of light in a ________________

v = speed of light in the ________________ (m s^{-1})

The refractive index of a vacuum is ________________, and air is usually taken to be the same. Any other medium will have a refractive index ________________, since light will travel slower in it.

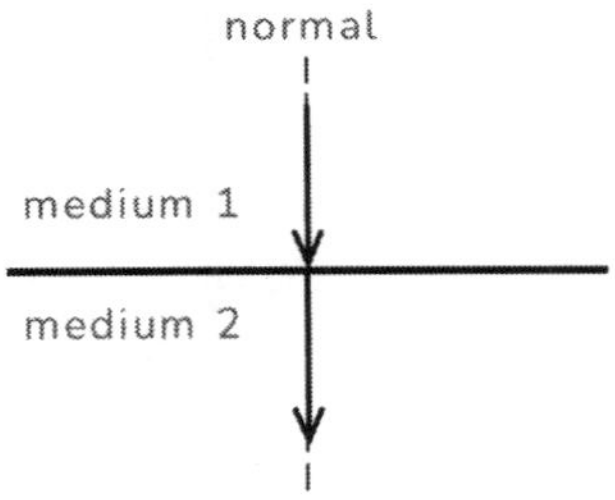

If light travels perpendicular to a boundary between two media, it would only change speed, but not ________________. In other words, a light ray travelling along the ________________ to a boundary would keep travelling through the second medium along the normal, but with a different speed. Another way of saying this is if the angle of incidence is zero (that is, the incident ray's angle to the normal is zero) then the angle of refraction is also ________________.

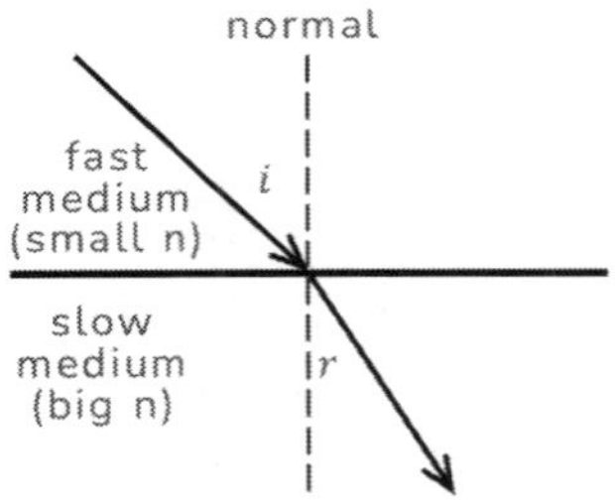

However, if the angle of incidence is greater than zero, then the change in speed will make the ray ________________. For example, a light ray travelling from air into water would slow down, as the refractive index of water is ________________ air. This makes light bend ________________ the normal, so the angle of incidence i is greater than the angle of refraction r. It is as if the light travels the path that would take it the least time to go from one point in the air to the water, rather than following a direct path which would cause it to travel ________________ in the slower medium.

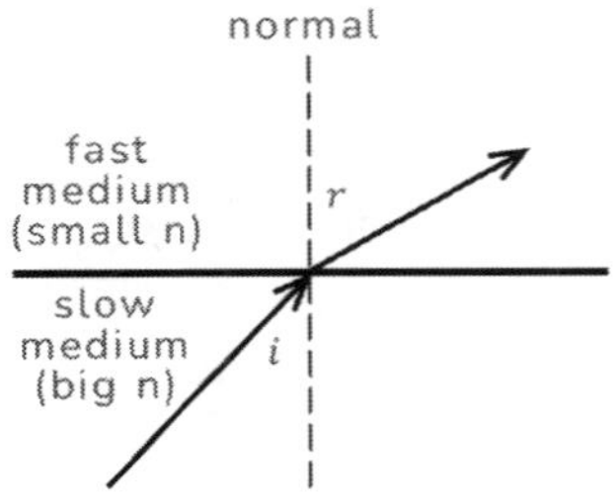

As light emerges from water into air, it speeds up, as the refractive index of air is ________________ water. This makes the light bend ________________ the normal, so the angle of incidence is less than the angle of refraction. Again, light follows the path that takes the ________________ between two points.

2. Some oil is spilled by an oil tanker on the ocean, leaving a layer of oil on the water. The refractive index of the oil is 1.43 and the refractive index of the water is 1.33.

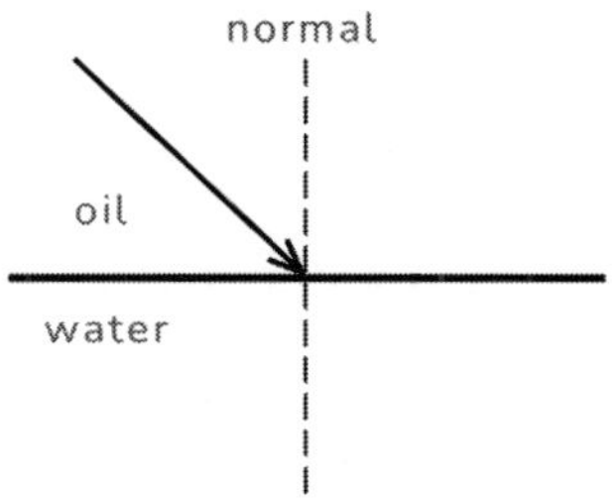

a) Calculate the speed of light in each medium.

b) A light ray travels through the oil towards the water, as shown. Sketch the path the light ray takes through the water. Label the angle of incidence, normal and angle of refraction on the diagram.

3. Refractive indices can be used in forensics to identify materials, such as glass samples. An unknown piece of glass is placed against a known type of glass with refractive index 1.50, and a laser is shone through both samples, as shown. Explain what can be inferred from this result.

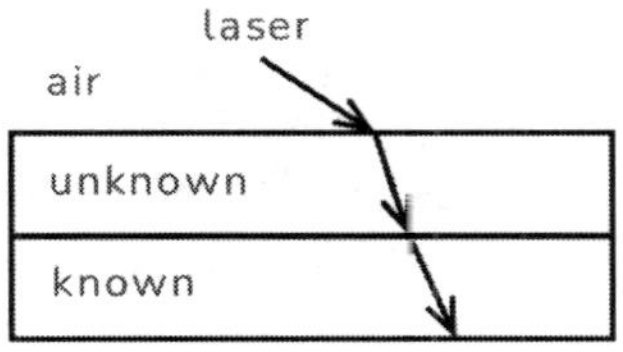

4. When a fish in a glass fish tank is viewed from outside the tank, it is difficult to judge its actual position and size. This is because the light we observe from the fish must travel through the water, glass, then air, experiencing refraction at each boundary.

Material	n
Air	1.00
Water at 20°C	1.33
Glass	1.52

a) The refractive indices of the materials is given. Determine the speed of light in each material.

b) A light ray that originally reflected off the fish reaches our eye outside the fish tank, as shown. Because our brain interprets light to travel in straight lines, the fish appears to be at position X. Sketch the path of the light ray through the water and glass, and explain where the fish would actually be.

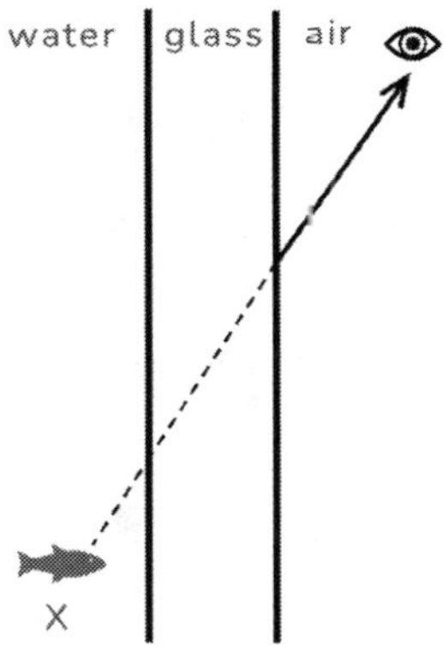

Wavelength and refractive index

5. Describe the relationship between wave speeds, wavelengths and refractive indices, by filling in the blanks.

frequency	refractive index	two media	type of wave
vacuum	wave equation	wave speeds	wavelengths

The ______________ of any medium light travels through is the speed of light in a vacuum c divided by its speed in the medium v:

$$n = \frac{c}{v}$$

This equation can be rearranged to make the speed of light in a ______________ the subject:

$$c = nv \quad ...(1)$$

If light travels between two media, with refractive indices n_1 and n_2 and ______________ of v_1 and v_2 respectively, then substituting these into equation (1) gives a relationship between these speeds and refractive indices:

$$n_1 v_1 = n_2 v_2$$

$$\frac{v_1}{v_2} = \frac{n_2}{n_1} \quad ...(2)$$

Substituting the ______________ $v = f\lambda$ for each medium gives:

$$\frac{f_1 \lambda_1}{f_2 \lambda_2} = \frac{n_2}{n_1} \quad ...(3)$$

However, the ______________ of a wave does not change as it passes from one medium to another, so:

$$f_1 = f_2$$

Substituting this into equation (3) and cancelling the frequencies gives a relationship between the ______________ in the media and refractive indices:

$$\frac{\lambda_1}{\lambda_2} = \frac{n_2}{n_1} \quad ...(4)$$

Combining equations (2) and (4) together, the ratios of velocities, wavelengths and refractive indices when a wave travels between ______________ is given by:

$$\frac{v_1}{v_2} = \frac{\lambda_1}{\lambda_2} = \frac{n_2}{n_1}$$

This formula can also be used to describe waves other than light which travel between media, but the value of the refractive indices will depend on both the ______________ and the media.

6. Water waves experience a change in medium if the water depth changes. However, their frequency stays the same. The diagram shows the wavefronts of water waves as they travel from deeper to shallower water.

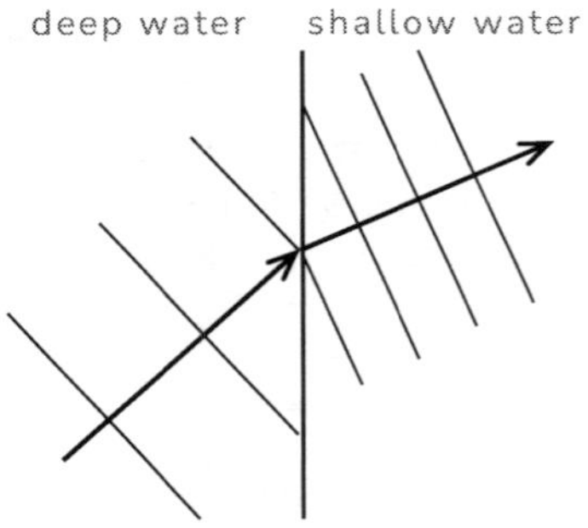

a) Explain why this causes a change in their wavelength and direction of travel.

____

____

____

____

b) A boat travelling in shallow water creates waves. The diagram represents the wavefronts as they move from shallow water to deep water. Explain the changes in the wave.

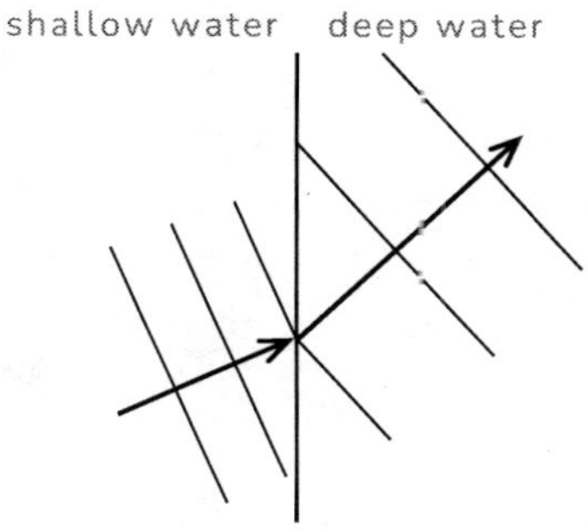

7. Light with a wavelength of 365 nm in water travels from the water through a submarine window, made of acrylic glass. The refractive index of water is 1.33 and the refractive index of acrylic is 1.53.

a) What is the speed of light in acrylic?

b) What is the wavelength of this light in acrylic?

8. Light reaching our eye first passes through the cornea, a transparent and protective covering at the front of the eye with a refractive index of 1.376. Between the cornea and the lens is a fluid called the aqueous humour, which has a refractive index of 1.336.

Compare the speeds, frequencies and wavelengths of light as light travels in the cornea and aqueous humour.

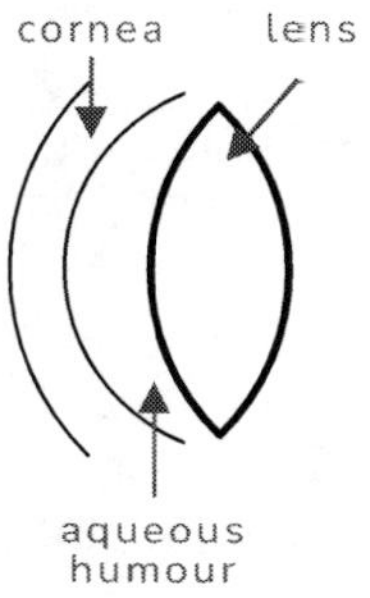

15.11 Angle of refraction

When a wave travels from a medium with refractive index n_1 to a second medium with refractive index n_2, the angle of incidence i and angle of refraction r are related to the speeds, wavelengths and refractive indices as follows:

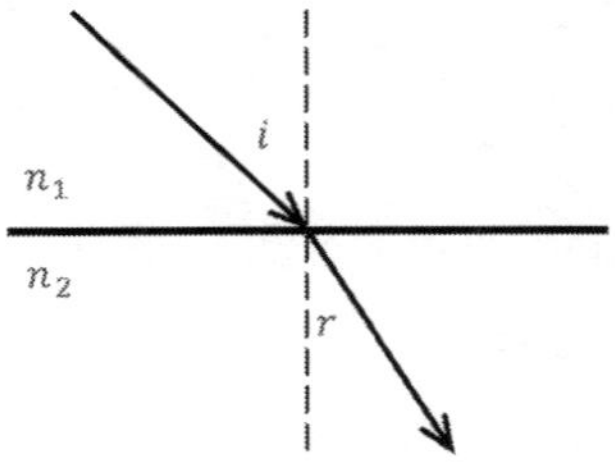

$$\frac{\sin i}{\sin r} = \frac{v_1}{v_2} = \frac{\lambda_1}{\lambda_2} = \frac{n_2}{n_1}$$

This relationship is also known as Snell's Law.

Example: A beam of light is shone from water to air at an angle of incidence of 30° to the surface of the water. The refractive index of water is 1.33. Determine the angle of the beam of light as it emerges into air and draw a diagram to represent the path taken by the light.

Answer: The angle of refraction is given by:

$$\frac{\sin i}{\sin r} = \frac{n_2}{n_1}$$

$$\frac{\sin 30°}{\sin r} = \frac{1}{1.33}$$

$r = 42°$

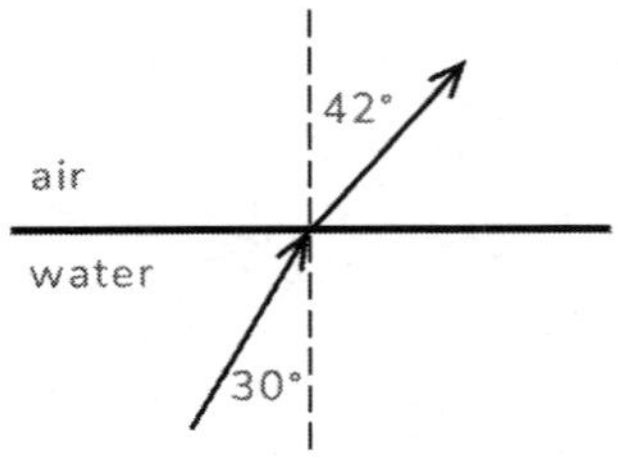

1. Refraction is used in the food industry to measure the concentration of solutions.

   The graph shows the relationship between the concentration of sugar dissolved in water at 20°C and its refractive index, for light of wavelength 589 nm.

   The same wavelength of light used to test a sugar solution of unknown concentration. An angle of incidence of 75° in air produced an angle of refraction of 44° in the solution. What is the concentration of the sugar solution?

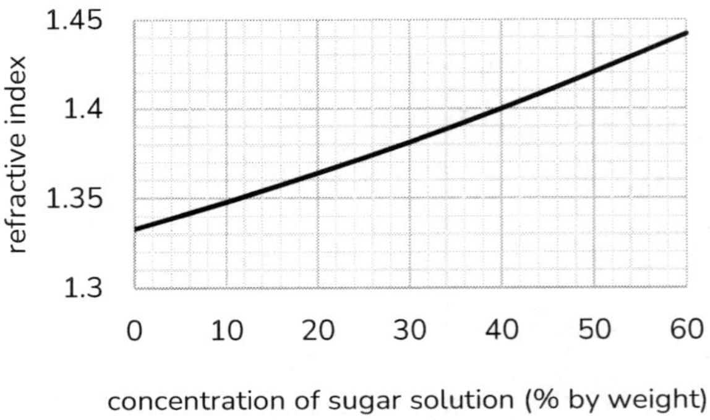

____

____

____

____

2. Mirrors are commonly made of a reflective surface, covered with glass. One mirror has a layer of glass 5.0 mm thick, with a refractive index of 1.5. Light of wavelength 520 nm is incident at 30° to the glass, as shown.

   Determine the path of the ray and sketch it on the diagram. Label the angles that the light makes to the glass and reflective surface, as well as the distance between the points where it enters and leaves the glass.

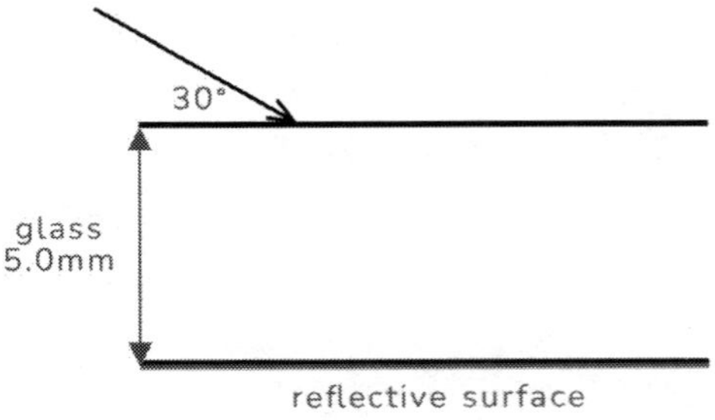

____

____

____

3. In an experiment to determine the refractive index of an unknown type of plastic, light with a wavelength of 615 nm was shone from air into a block of the plastic.

The table lists the angles of incidence used and resulting angles of refraction.

i (°)	r (°)	$\sin i$	$\sin r$
15	10		
30	20		
45	28		
60	36		
75	40		

a) Process the data to calculate $\sin i$ and $\sin r$ in the columns provided.

b) Graph $\sin i$ on the x-axis and $\sin r$ on the y-axis. Explain why this is a useful choice of axes.

c) Use the line of best fit to calculate the refractive index of the plastic. Explain why the line of best fit should be used, rather than an individual data point.

d) What is the speed of light in the plastic?

15.12 Lenses

1. Describe lenses, by filling in the blanks.

concave	converge	convex	diverge	tangent

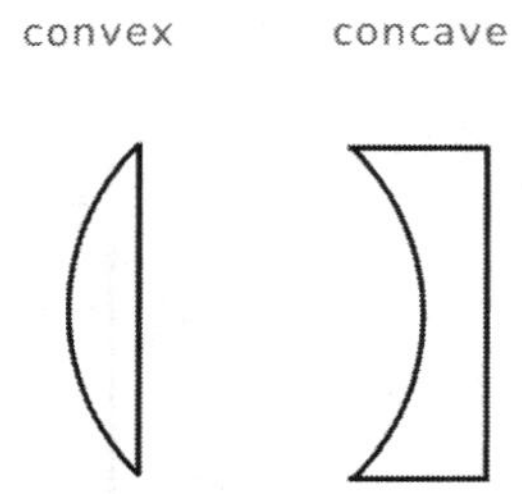

Lenses have curved surfaces which can refract waves in order to make them _______________ (come together) or _______________ (spread out).

A surface on a lens that caves in towards the source of waves is described as _______________, while a surface that curves away is _______________.

At any point on the curved surface, the normal is perpendicular to the _______________ to the curved surface.

Convex lenses

2. Describe convex lenses, by filling in the blanks.

concentrating	converge	convex	focus
principal axis	slows down	speed up	sphere

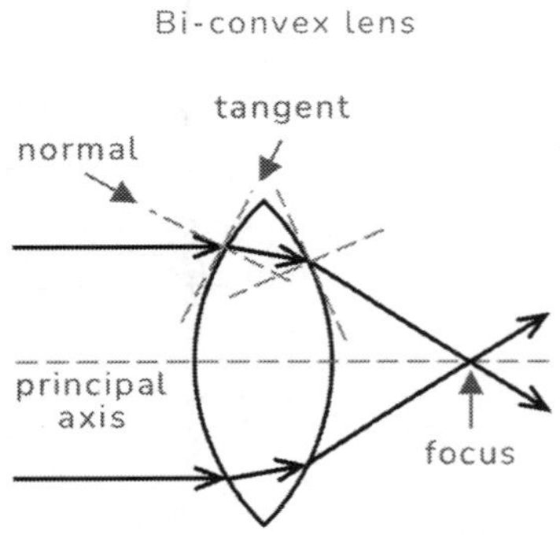

A magnifying glass is an example of a _______________ lens. In fact, it is a bi-convex lens, with both sides being parts of a _______________. An imaginary line through the middle of the lens, perpendicular to its surface is called the _______________. Light rays that are parallel to the principal axis will bend towards the normal, because light _______________. This bending makes the rays _______________ in the glass. When they reach the boundary to travel back into air, the rays _______________ and bend away from the normal. Because the glass is curved, this bending also makes the rays converge, so the rays meet at a _______________.

Sunlight passing through a magnifying glass can be directed to the focus, _______________ enough heat there to burn paper!

3. An object is between a convex lens and its focus. Light reflects off the object in all directions, but we only need to consider three rays to predict how the object appears through the lens.

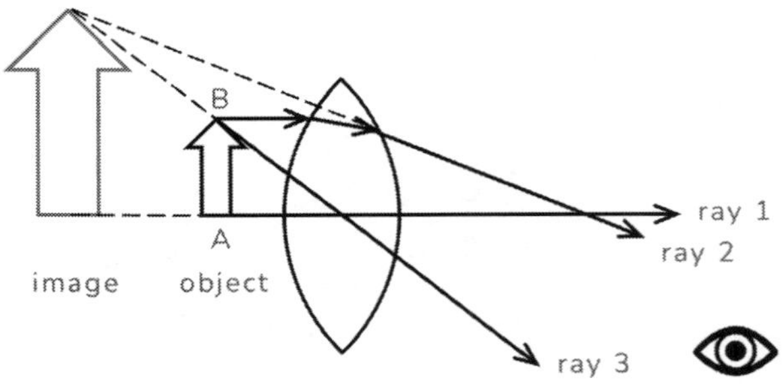

   a) Point A, the bottom of the object, is positioned on the principal axis. Explain why ray 1 from Point A travels through the lens without bending.

   ____

   ____

   b) Ray 2 reflects parallel to the principal axis from Point B, the top of the object. Name the point on the other side of the lens that this ray will travel through. _______________

c) Ray 3 travels from Point B through the centre of the lens. Convex lenses are shaped so that rays travelling towards their centre bend on entry at the same angle that they exit, so it is as if these rays pass through in a straight line. Extending Ray 2 and 3 to where the lines they travel along intersect gives where we will perceive a clear image of Point B to be. Note the rays do not actually travel to this point – it is our perception of light travelling in straight lines that makes us observe the image there. Compare the appearance of the image to the object.

4. Identify a similarity between the effect that a concave mirror and a convex lens has on light.

Concave lenses

5. Torches often have a concave lens – that is, it curves towards the light bulb on the inside. In fact, sometimes it can be a bi-concave lens and look indented from both sides. Light rays that are parallel to the principal axis are incident on the lens. Explain the effect of the lens on the torch light, with the aid of the diagram shown.

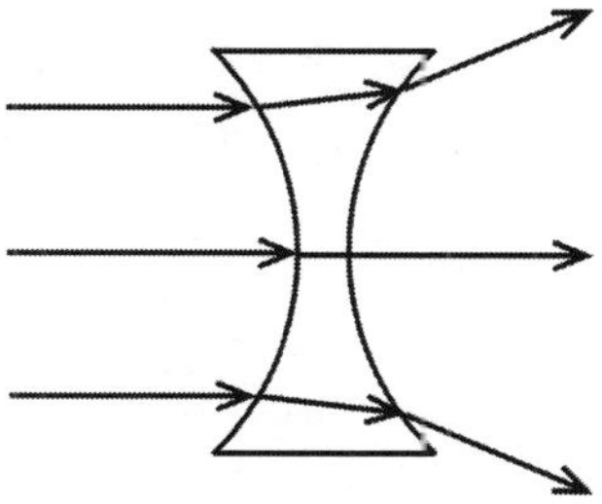

Key words: rays, civerge, refractive index, wider.

6. Explain the role of a concave lens in a peephole, with the aid of the diagram shown.

Peephole from plane cockpit

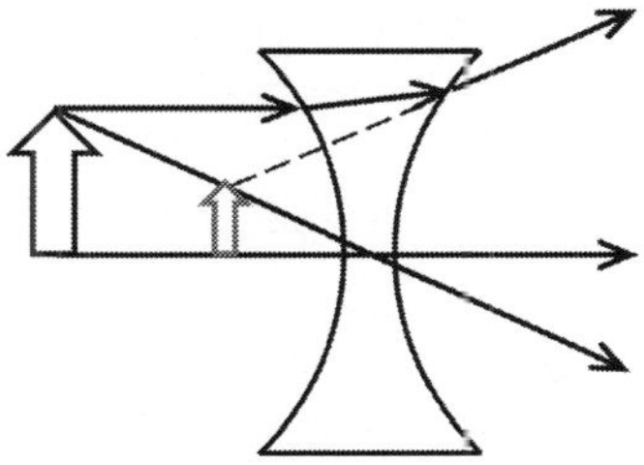

15.13 Dispersion

1. Describe dispersion, by filling in the blanks.

A prism can be used to disperse white light into the visible spectrum

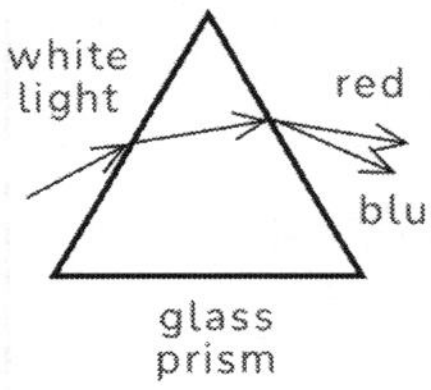

angles	blue	electromagnetic	higher	incidence
medium	phenomenon	rainbow	red	refraction
slower	Snell's Law	sound waves	vacuum	

What our eyes and brains interpret as white light is actually made of the visible wavelengths of the ______________ spectrum. However, refraction can separate white light into a ______________, so that we can see the different wavelengths in it. ______________ and violet light have wavelengths near the 400 nm end of the visible light spectrum, while ______________ light has wavelengths near the 700 nm end.

The rainbow occurs because refractive index depends on both the wavelength of the wave and the ______________ it is travelling through. This means that different wavelengths of light experience slightly different speeds in a medium other than a ______________. So if a ray of white light travels from a medium with refractive index n_1 to another medium with refractive index n_2, the different wavelengths of the ray start with the same angle of ______________ i but refract at different angles of ______________ r. The relationship is given by ______________:

$$\frac{\sin i}{\sin r} = \frac{v_1}{v_2} = \frac{\lambda_1}{\lambda_2} = \frac{n_2}{n_1}$$

When a ray of white light undergoes dispersion through glass, it is observed that blue light has a larger angle of refraction than red light, therefore blue light must have a ______________ speed in glass than red light, and a ______________ refractive index.

"Blue bends best. Red refracts rotten."

This separation of different wavelengths into different ______________ is called *dispersion*. Dispersion is a ______________ that can occur for all waves, but is most easily observed with light. There is not much dispersion of ______________ in air, but it becomes more significant for sound waves in solids.

2. Dispersion and diffraction are both phenomena that are affected by wavelength of the wave. Compare these two types of wave behaviour.

Feature	Dispersion	Diffraction
Where it occurs		
Does wave direction change?		
Does wave speed change?		
Does the wavelength change?		
Contrast the effect on shorter and longer wavelengths		

3. The graph shows experimental measurements of the refractive index of water at 20°C, as the wavelength changes in the visible spectrum.

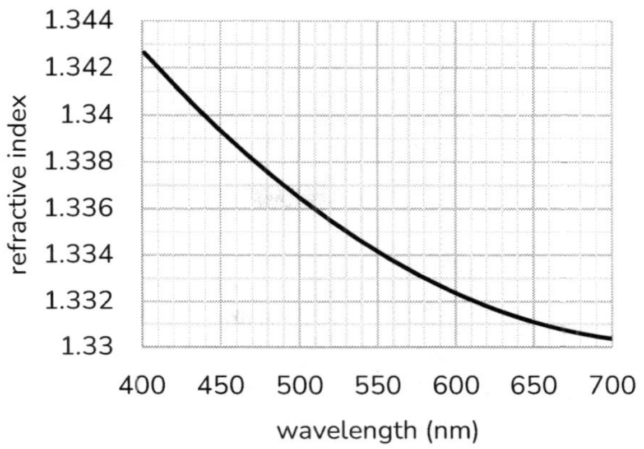

Consider blue light of wavelength 450 nm in air and red light with a wavelength of 670 nm in air.

a) Describe the trend shown in the graph and compare the refractive indices for blue and red light in water at 20°C.

b) Compare the speeds in water of these blue and red wavelengths.

c) Explain what happens to the wavelengths of the blue light and red light as they travel through water.

d) The refraction of a ray of the blue and red light together is observed as it travels from air to water, when the angle of incidence is at 25° and then at 85°. Explain which angle of incidence causes the most noticeable dispersion of light, with supporting calculations.

e) Sketch a diagram showing the path of the blue and red light in water, as a light ray containing both blue and red light enters water from air at an angle of incidence of 85°.

15.14 Total internal reflection

1. Describe total internal reflection, by filling in the blanks.

angle of incidence	equal to	greater	higher	less
lower	normal	reflect	refracting	

Total internal reflection occurs when a wave travelling from a slower medium towards a faster medium is reflected from the boundary, instead of ________________ as it leaves the slower medium. In other words, the ray must travel from a medium with a ________________ refractive index towards the boundary of a medium with a ________________ refractive index. If this ray exits the slower medium, the angle of refraction will be ________________ than the angle of incidence, and it will bend away from the ________________.

Refraction

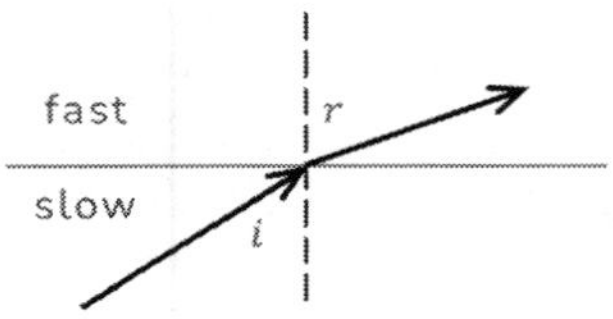

Critical angle

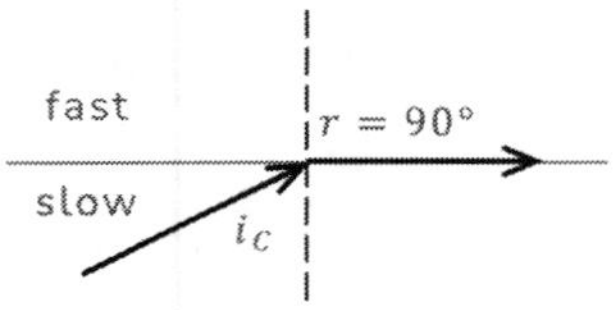

Total internal reflection

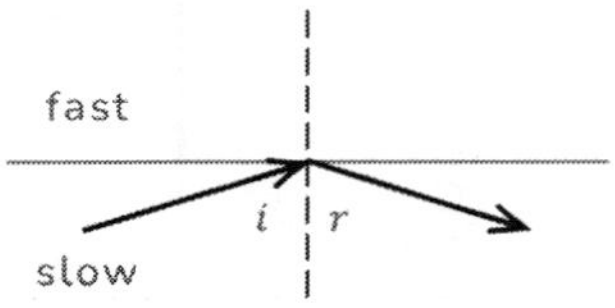

However, the refracted ray can only have an angle of refraction ________________ than 90° and still be in the faster medium. The ________________ that gives an angle of refraction of 90° is called the ***critical angle***, i_C:

$$\frac{\sin i_C}{\sin 90°} = \frac{n_2}{n_1}$$

$$\sin i_C = \frac{n_2}{n_1}$$

If the angle of incidence is bigger than i_C, the ray will no longer exit to the faster medium, but will instead ________________ from the boundary. The Law of Reflection still applies to total internal reflection, so the angle of reflection will be ________________ the angle of incidence.

2. Have you ever noticed that at night, sounds from a distance can sound louder? This is especially true across a body of water, such as a lake. Explain why, by filling in the blanks.

cooler	critical angle	direct line	kinetic energy	reflection
refract	refractive index	rises	slower	warmer

During the day, the sun warms the ground, so at night, heat ________________ from the ground through convection. This makes the air ________________ as the height above the ground increases. Sound travels faster in warm air, since the air particles have more ________________ compared to cold air and collide faster. By contrast, the air just above a lake stays ________________, since water does not heat up as fast as the ground during the day. Therefore sound travels ________________ in the cool air just above the lake.

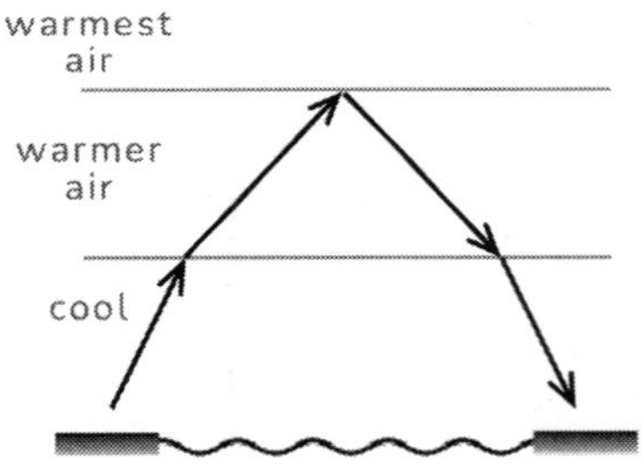

As height above the ground increases at night, the temperature increases and the ________________ of sound decreases. Therefore sound travelling from the ground towards the warm air above will ________________, bending the path of the sound downwards. If sound travels upwards so that the angle of incidence to a layer of warm air is greater than the ________________, the sound will undergo total internal ________________ and be directed back towards the ground. This means that if you hear a source of sound across a lake, the sound can travel to you both in a ________________ as well as along a second route via refraction and reflection, amplifying the sound.

3. Explain why total internal reflection cannot occur for a wave travelling from a medium with a lower refractive index to another with a higher refractive index.

____

____

Example: A diver inspecting an aquarium shines a laser from inside a fish tank full of water out towards the window of the fish tank, at an angle of incidence of 52°. The water has a refractive index of 1.33 and the acrylic that the window is made of has a refractive index of 1.49.

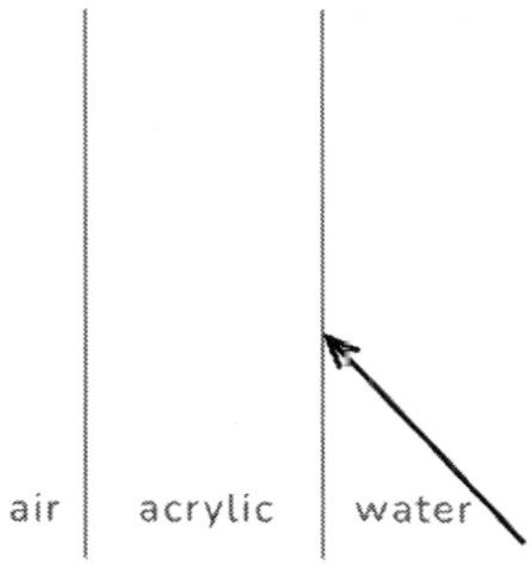

Explain the path of the laser at the boundary between:

a) the water and the acrylic window; and
b) the acrylic window and the air outside.

Answer:

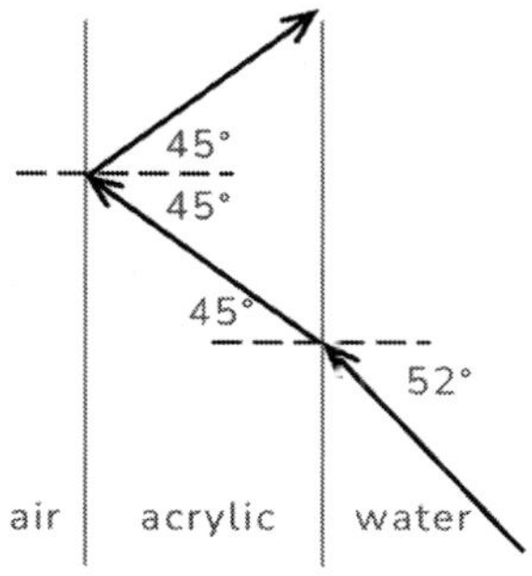

a) At the water-acrylic boundary, the light bends towards the normal, as it is moves from low to high refractive index. The angle of refraction is:

$$\frac{\sin i}{\sin r} = \frac{n_2}{n_1}$$

$$\frac{\sin 52°}{\sin r} = \frac{1.49}{1.33}$$

$$r = 45°$$

b) From part (a), the light reaches the acrylic-air boundary at an angle of incidence of 45°. It moves towards air, which has a lower refractive index than acrylic, so there is a critical angle:

$$\sin i_C = \frac{n_2}{n_1} = \frac{1}{1.49}$$

$$i_C = 42°$$

Since the angle of incidence is greater than the critical angle, the light will totally internally reflect at an angle of reflection of 45°.

4. If the water in an ocean is clear, you may be able to see fish underwater. However, if the fish looks up, it would only see light from above the water in a circle of light, because light from outside refracts as it enters the water. Outside this circle, the fish would only see light from the ocean floor undergoing total internal reflection.

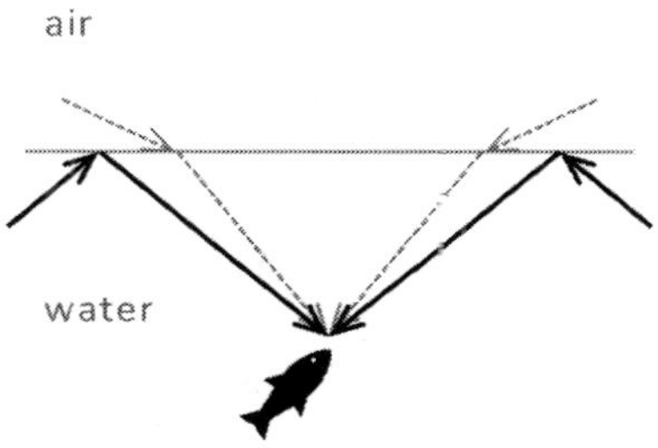

a) The refractive index of water at 20°C is 1.33. What is the critical angle for light travelling between water and air?

____

____

b) If a fish is 1.0 m below the surface of the water, what is the radius of the circle of light it could see on the surface?

____

____

5. Optical fibres are used in communication to transmit information via electromagnetic waves, usually infrared or visible light. An optical fibre consists of a glass core, surrounded by a cladding with a slightly lower refractive index. The glass core has a refractive index of 1.50.

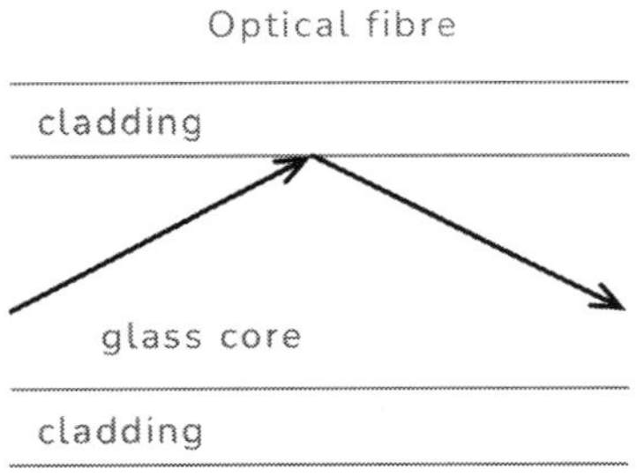

a) Calculate the speed of light in the glass core.

b) How long it would take for light to travel 1000 km in the glass core, if the wave were able to travel in a straight line through it?

Light from a phone torch can be transmitted via the optical fibre in an audio cable.

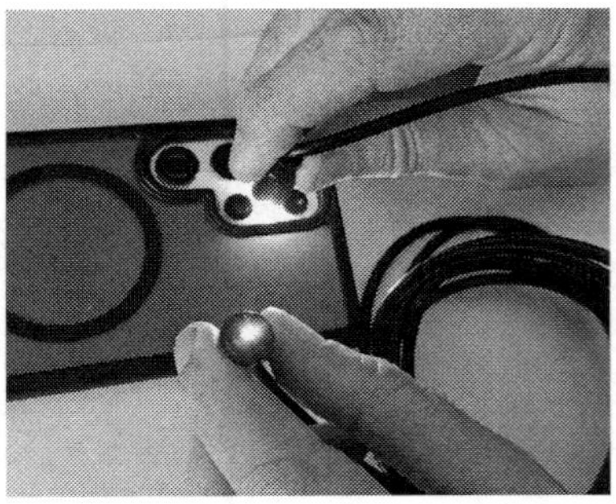

c) Compare the critical angle at which light would totally internally reflect from the boundary of the glass core, with and without cladding. Assume the cladding has a refractive index of 1.49 and the optical fibre is surrounded by air.

d) Discuss the effect of the cladding on the path that light can take in the optical fibre and the time taken for light to travel from one end of the optical fibre to the other.

The cladding doesn't just provide protection for the glass core – it is there for another important reason.

6. Diamonds sparkle, even though they do not emit light. This is due to diamond having a very high refractive index of 2.42, as well as the jeweller who cut the diamond knowing some geometry.

A typical cut has a flat surface on top, through which light enters, and a pointed base which is set into jewellery.

Consider a light ray entering through the flat surface on top, as shown, for the different angles of the base. Each cut shown is symmetrical.

Determine and sketch the path that the light ray will take in each case, and explain which angle would make the diamond appear the most sparkly.

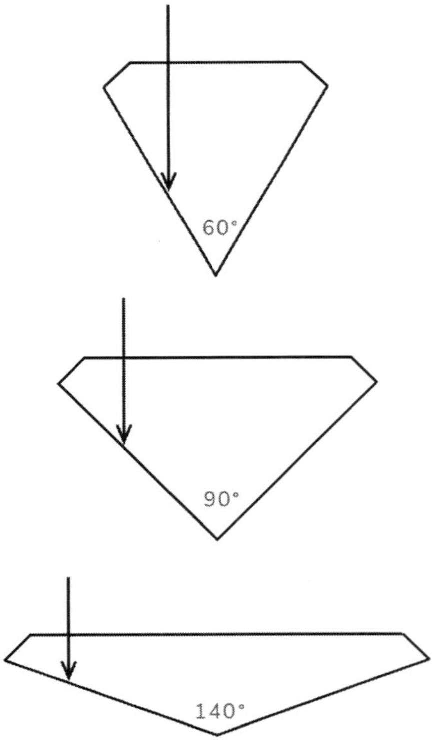

The angles of a triangle add up to 180°.
Cointerior angles on parallel lines add up to 180°.

16.1 Thermodynamics and perfect potatoes

The amount of heat in kilojoules needed to raise one kilogram of a substance's temperature by one degree is called ***specific heat capacity.***
The equation for this is:
$Q = mC\Delta T$
where:
Q = amount of energy (kJ)
m = mass of substance (kg)
C = specific heat capacity of substance (kJ kg^{-1} K^{-1})
ΔT = increase in temperature (K or °C)
So with less water, the same heat gives a bigger temperature increase.
How much heat do we need to heat water anyway?
I originally had 5 litres of water.
That's 5 kg of water. The density of water is 1 g/mL.
1g
1mL=1cm³
How convenient!
Well, the gram was originally defined using the density of water.
The specific heat capacity of water is 4.18 kJ kg^{-1} K^{-1}.
So to raise 5 kg of water from room temperature at 25°C to 100°C needs 5 x 75 x 4.18 kJ of heat.
Yes. Water actually has the highest specific heat capacity of any liquid we know.
And it's finally boiling! There are big bubbles rising!
That's heat rising due to ***convection***.
As the water particles gain kinetic energy, they move further apart, so the water becomes less dense and floats upwards.
Then as the water cools at the surface, it becomes more dense and sinks.
hotter water rises
cooler water sinks
Time for potatoes!
Oh no, now the water is not boiling anymore!
Well, since the potatoes were at room temperature, heat will transfer from the hot water to the potatoes until they are at the same temperature, or ***thermal equilibrium***.
hot
cold
Initial
Final
It's getting hotter again, but this time it depends on the specific heat capacity of the potatoes as well as water.

Meanwhile, how about we microwave the other half of the potatoes?
How does the microwave heat food? It's not convection or conduction, is it?
No, it's through **radiation**. That's the third way heat can be transferred.
It's actually electromagnetic radiation, isn't it?
Yes, heat transfer via radiation does not involve particles moving.
But once radiation hits particles, it can transfer energy to it.
The part of the electromagnetic spectrum that our nerve cells perceive as heat is called ***infrared***.
Hey, these potatoes have gone a bit soft!
How long do you think it will take for them to get crispy on the outside?
Ah... it needs a much hotter temperature to do that.
But this water isn't getting hotter...
Now that it's boiling, the temperature will stay constant.
100° C
The heat added now overcomes the electrostatic attraction between the water molecules, as the water evaporates. This is called the ***latent heat of evaporation***.
Instead of kinetic energy, heat is now transformed into potential energy.
liquid water
gaseous water
Is that why oil is used for cooking? Because it gets hotter?
Yes, the boiling point of cooking oils is around 300°C.
Actually, they start smoking before then, so we'll use 200°C.
I'll use the top rack to heat up this bit of oil.
Good thinking, it's the hottest part of the oven due to convection.
It will only take 5 minutes to heat the oil.

Oil must have a smaller specific heat capacity than water.
Yes. Plus I only used two tablespoons of oil per kilogram of potatoes, so it heats up fast.
Smells so good when the rosemary and garlic go in the oil!
We drain the potatoes now
and add them to the oil.
0
10
50
40
30
DING!
Give the potatoes a stir to turn them over.
That tray was hot! But why is the oven door so cool?
Take a closer look at it.
There's two panes of glass with a gap between them!
That air gap prevents conduction
because the air particles are too far apart to collide!
0
10
50
40
30
DING!
Perfect!
Let's celebrate!

16.2 Heat transfer

Heat can be transferred from hotter to cooler regions in three ways – convection, conduction and radiation. Heat is a form of energy, so it has no mass.

Convection

1. Explain why firefighters recommend that you "get down low and go go go" to escape a fire, by filling in the blanks.

coolest	electrostatic	gases	gravitational	hotter	kinetic energy
less dense	more dense	sideways	solids	upwards	

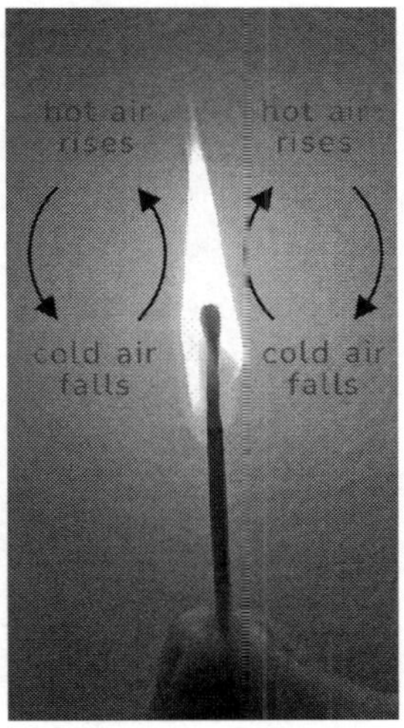

When air is heated, the air particles have more ______________ to overcome electrostatic attraction to each other, so they move further apart. As the hot air expands, it becomes ______________ and floats upwards. This means that the air closer to the ceiling is ______________ than the air closer to the ground. The current of rising hot air pushes particles of smoke ______________, which can be toxic to inhale.

This current of hot air also pushes air already near the ceiling ______________. As hot air moves away from the heat source, it cools, so the particles lose kinetic energy and are less able to overcome the ______________ force of attraction between each other. The cooler air is ______________ than hotter air and sinks due to ______________ force. This means that the ground is often the ______________ place in a burning room.

This process of hotter substances rising and cooler substances sinking is called ***convection***, and occurs in substances where particles are free to move, such as liquids and ______________, but not in substances where particles are strongly bonded to others, such as ______________.

2. Scientific studies of the effect of clothing on the heat gained by people in a hot desert show that loose robes keep people cooler than tight clothing or being half naked. Explain this conclusion.

____

____

Conduction

3. Explain how a double glazed window works, by filling in the blanks.

cold	collisions	conductor	gas	hot
insulator	kinetic energy	particles	solid	transfer

Double glazed window

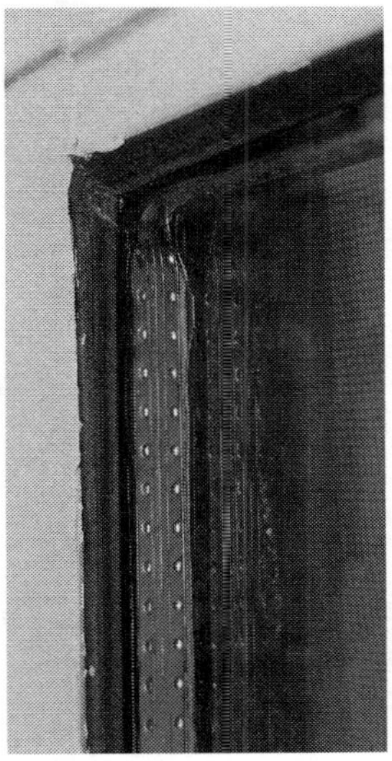

On a hot day, heat from the Sun makes particles in the outermost pane of glass vibrate with more ______________. Particles in glass can collide with neighbouring ______________ and transfer some of their kinetic energy to them. These particles in turn then collide with more particles to ______________ kinetic energy. This process is called ***conduction*** - the transfer of heat through a substance from the hotter parts to the colder parts, as a result of ______________ between neighbouring particles.

Conduction happens more easily in solids than gases, because particles in a ______________ are closer together and are more likely to collide and transfer kinetic energy. By contrast, the ______________ layer in a double glazed window will not easily conduct heat, because gas particles are more spread out. In other words, the glass is a better ______________ of heat than gas, which is a better ______________. This means double glazed windows will keep a house cool on ______________ days. They will also work on ______________ days to keep heat from inside escaping to the outside.

4. Diamond is an excellent conductor of heat, in fact better than metals. However, it is a poor conductor of electricity. Distinguish between what is required in a material to conduct electricity, as opposed to heat.

Radiation

5. Explain what heat as radiation is, by filling in the blanks.

electromagnetic	invisible	longer	particles	shorter	Sun	vacuum

The processes of convection and conduction both involve the kinetic energy of ________ in transferring heat. However, heat travels to Earth from the ________ without particles present, via ________ radiation. Heat in this form is also known as infrared. Infrared is ________ to our eyes, but can be detected by specialised human nerve cells called thermoreceptors in our skin.

The wavelength of infrared ranges from around 700 nm to 1 mm, which is ________ than the wavelength of visible light at 400 nm to 700 nm, but ________ than the wavelength of microwaves. Infrared travels at the same speed as light in a ________, 3×10^8 m s^{-1}.

6. The wavelength of infrared ranges from around 700 nm to 1 mm. Calculate its frequency range.

7. A light bulb was connected to a power pack. The light emitted was analysed using a spectrometer, which uses diffraction to separate the wavelengths of light produced and measure their intensity. The graphs show the results when 6 V and 12 V were supplied to the same light.

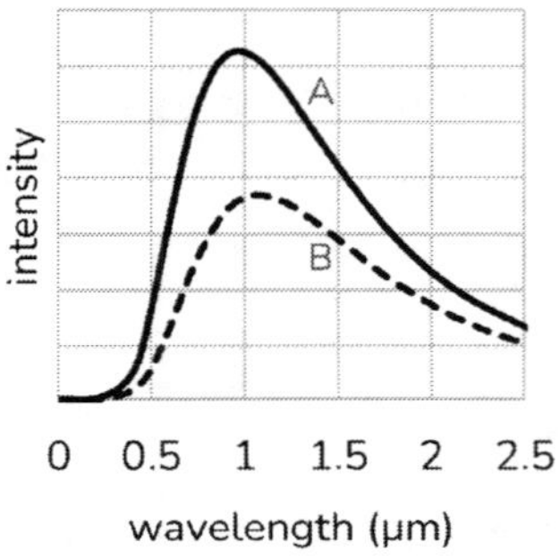

a) Explain which graph, A or B, was produced by the 12 V setting.

If there are relevant equations or physics principles, use them in your explanation.

b) Propose whether the light bulb is energy efficient.

Energy efficient appliances transform most of their energy into useful forms.

c) Compare the observations that could be made of the light bulb on the 6 V and 12 V settings, based on information from the graph.

What kinds of energy are being produced by the light bulb? Some conversion may be needed.

16.3 Skills – Conduction

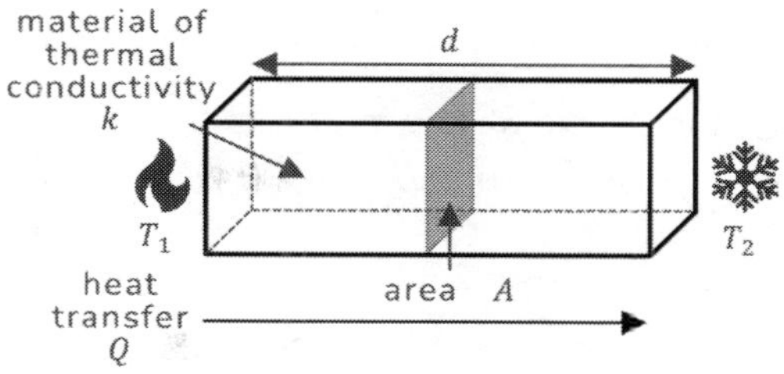

An architect is asked by a client to design a room with a large glass window, in order to provide a nice view.

They find the following equation on a physics website that describes the rate of heat transfer through a material:

$$\frac{Q}{t} = \frac{kA\Delta T}{d}$$

where: $\frac{Q}{t}$ = rate at which heat is transferred from one end to the other (W)

$\Delta T = T_1 - T_2$ = temperature difference between ends of conductor (K or °C)

A = cross-sectional area of conductor (m^2)

d = length of conductor (m)

k = thermal conductivity of the material

1. Derive the unit for the thermal conductivity k of a material, using the units for the other variables in the equation.

2. The client wants the window to be 4.0 m wide and 3.0 m high. The cheapest window glass has a thickness of 3.0 mm and a thermal conductivity of 0.96 W m^{-1} K^{-1}. On a cold day, it can be 10 °C outside. The client wants to keep the temperature of the room at a constant 25 °C, using electric heating.

   a) At what rate must the room be heated in order to maintain this temperature? Assume that heat in the room is only lost through the window and the temperature outside stays at a constant 10 °C.

   b) How much energy is consumed in heating the room this way in one hour?

3. The client is not happy about the amount of electrical energy used to heat the room. The architect suggests that they can change the electrical power consumed if they change the temperature that the room is heated to.

   a) Circle the correct words to describe the relationship:

   The power is **proportional to / inversely proportional to** the difference in temperature between the two sides of the window.

   b) Due to the Law of Conservation of Energy, the power consumed to maintain the room temperature must equal the rate of heat loss through the window, assuming that this is the only way heat escapes the room. Sketch a graph of the rate at which heat is lost vs temperature difference between the inside and outside, using the axes provided.

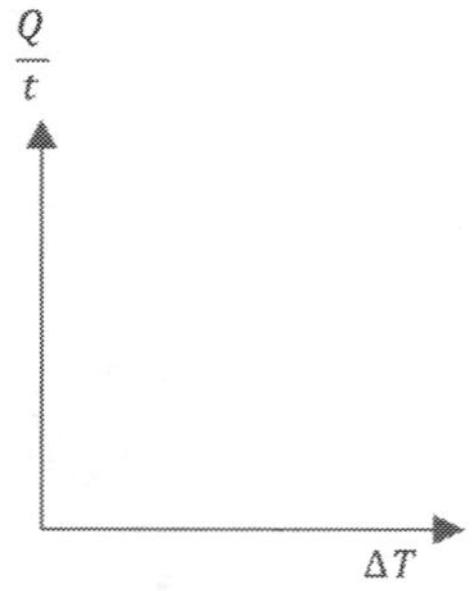

   c) If the client wants to use half the power needed to heat the room to 25 °C when it is 10 °C outside, what temperature should the room be heated to instead?

4. The client prefers to keep the indoor temperature at 25 °C and wants the architect to explore other ways of reducing heat loss through the window. The architect suggests changing the area of the window.

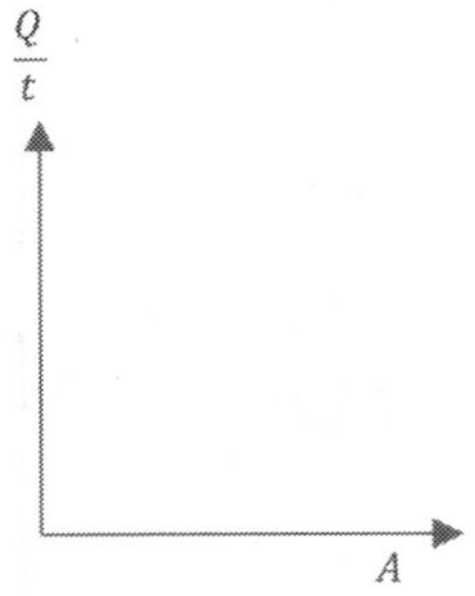

a) Describe the relationship between the rate at which heat is lost and the area of the window, referring to proportionality.

b) Sketch a graph of the rate at which heat is lost vs area of the window, using the axes provided.

c) Analyse what the gradient of this graph represents.

Recall linear regressions.

When analysing, use relevant calculations if data is provided.

5. The client does not want to change the area of the window either as the view would be limited. Therefore, the architect explores using glass with a different thickness in order to reduce heat loss.

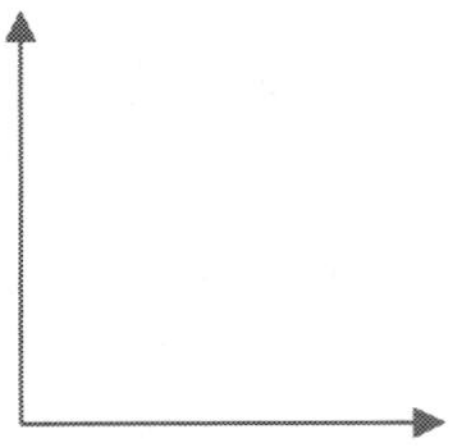

a) Describe the relationship between the rate of heat loss and the thickness of the glass, referring to proportionality.

b) Sketch a graph of the rate at which heat is lost vs thickness of the glass, using the axes provided.

c) Explain how you would use a different graph to quantitatively analyse the mathematical relationship between the rate of heat loss and the thickness of the glass. Sketch this graph using the axes provided.

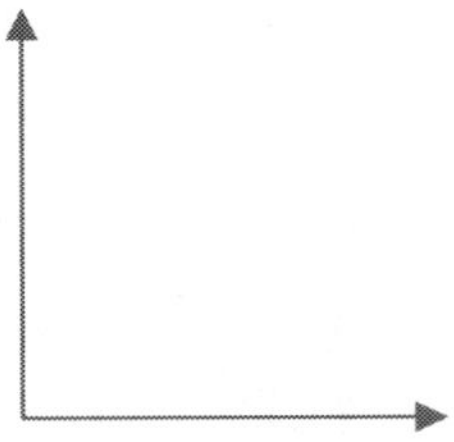

6. The client likes the proposal to change the thickness of the glass, and asks the architect if there are additional ways to reduce the rate of heat loss, which have not yet been explored. Discuss another variable that can be changed to reduce the rate of heat loss, with the aid of a graph.

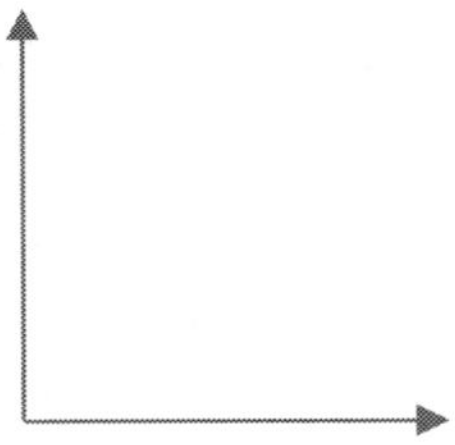

16.4 Temperature and heat

Temperature

1. Compare heat and temperature, by filling in the blanks.

average	Celsius	conduction	convection	joules
kinetic energy	moving	one	zero	

Heat can be transferred in matter via collisions of neighbouring particles to transfer kinetic energy, in a process known as ________, as well as a process involving hotter, less dense matter rising and cooler, more dense matter sinking, known as ________. The ***heat*** contained in a substance is therefore the total ________ of its particles. As heat is a form of energy, it is measured in ________.

The kinetic energy of each particle in a substance varies, so ***temperature*** is a measure of the ________ kinetic energy of particles in a substance. Two common units for temperature are degrees ________ (°C) and Kelvin (K). The Kelvin scale is often used in science, and the lowest temperature on it, 0 K, is called absolute ________. Theoretically, 0 K is the temperature at which all particles in a substance stop ________ and therefore have no kinetic energy. The Celsius scale is converted to Kelvin by:

$$K = °C + 273.15$$

Therefore, an increase in one degree Celsius is the same as an increase in ________ Kelvin.

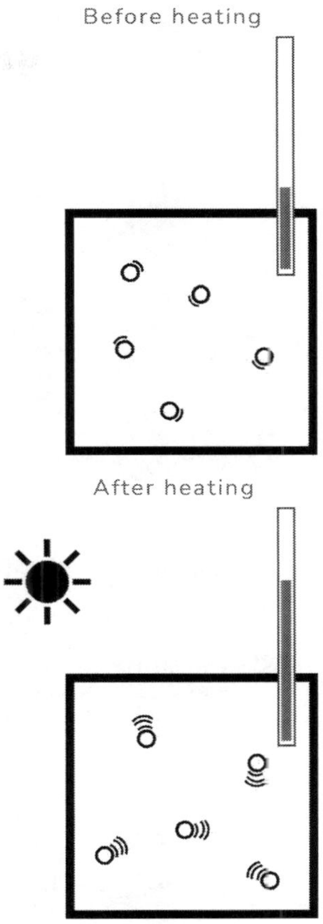

2. What are these temperatures in Kelvin?
   a) Freezing point of water 0 °C ________
   b) Room temperature of 25 °C ________

3. What are these temperatures in degrees Celsius?
   a) Absolute zero ________
   b) 373.15 K ________

Specific heat capacity

4. Describe what is meant by specific heat capacity, by filling in the blanks.

absorb	energy	hot	mass	same	sand	temperature	water

On a hot day, the ocean may feel pleasantly cool, but the sand on the beach can feel painfully ________. Although the water and sand are subjected to the ________ sunlight and surrounding air temperature, the water can absorb heat better than sand, without its temperature being raised as much. This ability to ________ heat energy is called *heat capacity*. The ***specific heat capacity*** is the heat energy required to raise the temperature of one gram of a substance by 1°C or 1K. So the specific heat capacity of ________ is much higher than ________, since the same amount of heat causes a bigger temperature increase in sand. The amount of energy (Q) needed to raise the temperature of a substance is given by:

$$Q = mC\Delta T$$

where: Q = amount of ________ (J)

m = ________ of substance (g)

C = specific heat capacity of substance (J g^{-1} K^{-1})

ΔT = increase in ________ (K or °C)

Example: On a sunny day at midday, the solar power falling on Earth's surface per square metre is around 1000 W m^{-2}. Compare the temperature reached by 1000 kg of water to 1000 kg of sand, if both were heated by the solar power falling on one square metre over one hour, and both had an initial temperature of 20°C.

Water has a specific heat capacity of 4.18 J g^{-1} K^{-1} and sand has a specific heat capacity of 0.830 J g^{-1} K^{-1}.

Answer: Amount of solar energy on 1 square metre in 1 hour: $E = Pt = 1000 \times 60 \times 60 = 3.6 \times 10^6$ J

Final temperature of water is given by: $Q = mC\Delta T$

$3.6 \times 10^6 = 1000 \times 10^3 \times 4.18 \times (T_f - 20)$

$T_f = 20.9°\text{C}$

Final temperature of sand is given by: $Q = mC\Delta T$

$3.6 \times 10^6 = 1000 \times 10^3 \times 0.83 \times (T_f - 20)$

$T_f = 24.3°\text{C}$

Sand gets hotter when subjected to the same amount of heat, due to its lower specific heat capacity.

Take water's specific heat capacity as 4.18 J g^{-1} K^{-1} for all calculations.

5. 1.0 kg of water at 90°C is cooled down to 65°C. Discuss whether the working out below correctly calculates the amount of energy lost:

   $Q = mC\Delta T = 1 \times 4.18 \times (65 - 100) = -146.3$ kJ

   Comment on the values substituted, units and negative sign.

6. A cup of water of volume 250 mL is heated from 25°C to 80°C in 5.0 minutes.

   If 2.0 L of water, which is initially at 25°C, is given the same amount of heat, what temperature would it reach?

   The density of water is 1.0 g mL^{-1}.

7. An experiment was conducted to investigate the effect of heating different liquids on their change in temperature.

   a) What needs to be controlled in this experiment so that it is a fair test?

   b) Account for the difference in temperature change, if a fair test was conducted.

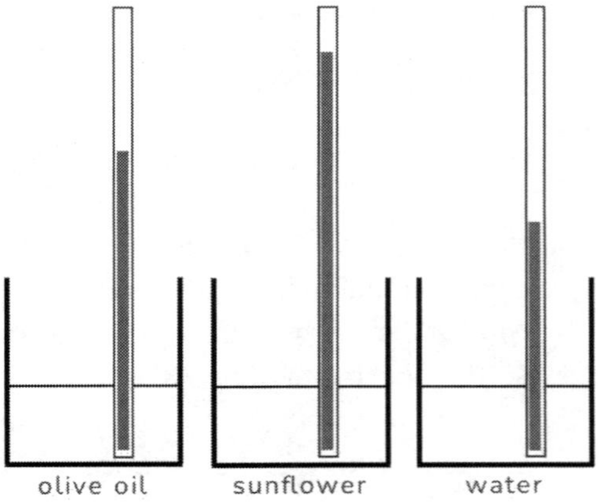

8. A portable kettle uses a voltage of 12 V and a current of 10 A to heat 800 mL of water. How many minutes will it take for this water to reach boiling point, if it was initially at 25°?

The boiling point of water is 100°C.

9. In the 1800s, James Joule conducted an experiment in which a large falling mass was used to spin a turbine inside an insulated container of water. This raised the temperature of the water, demonstrating that motion could transfer heat.

In a similar experiment, it was observed that the temperature of 250 g of water increased by 0.50°C when a turbine in it was rotated by a falling 20 kg mass. How far did the mass need to fall?

10. Over half the human body is made of water. The liquid with the highest known specific heat capacity is water. Explain how the specific heat of water impacts human body temperature when the temperature of our surroundings changes.

16.5 Thermal equilibrium

1. Describe how substances come to be in thermal equilibrium, by filling in the blanks.

collisions	conservation	heat energy	net	same

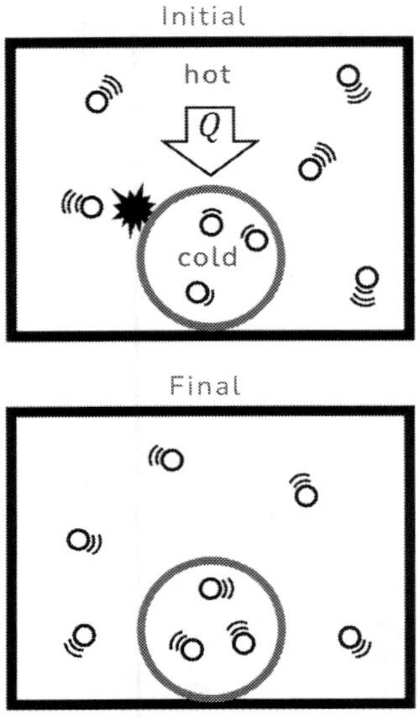

When substances at different temperatures come into contact, __________ is transferred from regions of high temperature to low temperature. This occurs via __________ between the particles in the substances. The amount of heat lost by the hotter substance must equal the amount of heat gained by the cooler substance, according to the law of __________ of energy. When both substances reach the __________ temperature, thermal equilibrium is reached.

At thermal equilibrium, collisions still occur, and individual particles may still transfer kinetic energy, but there is no __________ heat transfer between the two substances.

Example: Quenching is a process in metalwork, when a hot metal is submerged in water to cool it quickly. 1.0 kg of steel is heated until it is red hot, at a temperature of 460 °C. It is then submerged in 20 kg of water at 20 °C. The specific heat capacity of steel is 0.42 kJ kg^{-1} K^{-1} and the specific heat capacity of water is 4.18 J g^{-1} K^{-1}. What will be the temperature of the water once the water and steel reach thermal equilibrium?

Answer:

Heat gain by water = heat loss by steel

$Q_W = -Q_S$

$m_W C_W \Delta T_W = -m_S C_S \Delta T_S$

$20 \times 4.18 \times (T_f - 20) = -1 \times 0.42 \times (T_f - 460)$

$83.6T_f - 1672 = -0.42T_f + 193.2$

$T_f = 22\ °C$

Take water's specific heat capacity as 4.18 J g^{-1} K^{-1} for all calculations.

2. 30 g of milk at a temperature of 4.0 °C is added to a cup of tea containing 220 g of water at 95 °C. The specific heat capacity of milk is 3.93 J g^{-1} K^{-1}. Assume the specific heat capacity of tea is the same as water, 4.18 J g^{-1} K^{-1}.

a) What is the final temperature of the tea after the milk is added?

Heat gain by milk = heat loss by water

$Q_m = -Q_w$

b) Explain why the temperature of the water decreased, using the particle model.

3. To make jewellery, gold is often heated to shape it, then quenched in water to cool it. 10 g of gold is heated to 550 °C, then submerged in 10 g of water at 25 °C. When the gold and water reach thermal equilibrium, the temperature of the water is 40.72 °C.

a) Calculate the specific heat capacity of gold.

b) Explain why the final temperature of the water is not the average of the starting temperatures of the gold and water.

4. One way that a sustainably designed home can stay warm in winter is by using concrete flooring near sunny windows, so that radiation heats the concrete during the day. At night, the concrete transfers heat to the air in the home.

In one room, the concrete floor has been heated by the sun to 20 °C. The temperature of the air in the room suddenly drops to 10 °C, due to a door being left open. The door is then closed again.

a) The room has a square concrete floor with a side length of 5.0 m. The concrete floor is 0.10 m thick. The density of concrete is 2400 kg m^{-3}. What is the mass of the concrete floor?

$$density = \frac{mass}{volume}$$

b) The same room has a ceiling height of 4.0 m. Air has a density of 1.246 kg m^{-3} at 10 °C. What is the mass of the air in the room?

c) The specific heat capacities of concrete and air are 1.17 kJ kg^{-1} K^{-1} and 1.006 kJ kg^{-1} K^{-1} respectively. If the initial temperatures of the concrete floor and air are 20.0 °C and 10.0 °C respectively, what temperature will the air be when they are in thermal equilibrium? Assume that the room is well insulated, so no heat is transferred into or out of the room.

d) Explain how the concrete floor could be effective in keeping the room cool on a hot day.

16.6 Latent heat

When a substance is heated, it can undergo several changes of state, also known as phase changes. The table below compares the different states of matter and the changes to energy and temperature as the substance is heated.

Phase or phase change	solid 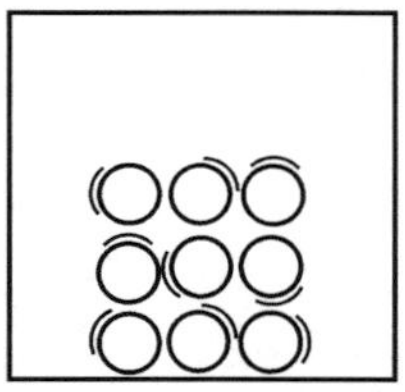	melting ⇨	liquid 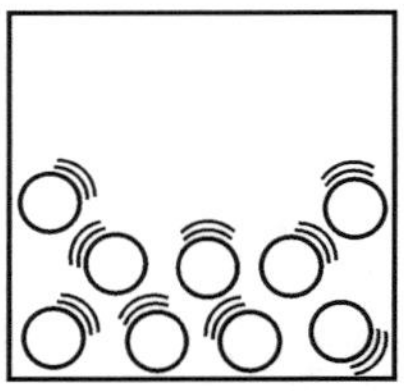	evaporation ⇨	gas 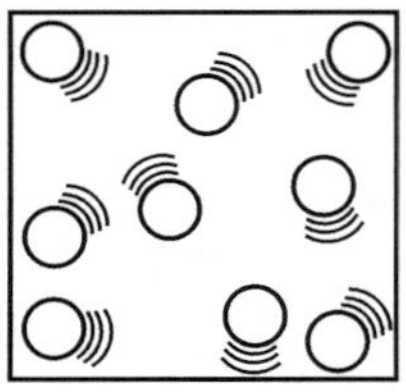
Energy gained by particles	kinetic $Q = mC_s\Delta T$	potential $Q = mL_f$	kinetic $Q = mC_l\Delta T$	potential $Q = mL_v$	kinetic $Q = mC_g\Delta T$
Temperature change	increases	unchanged	increases	unchanged	increases

1. Describe the energy changes that occur when water is heated, by filling in the blanks.

conservation	electrostatic	evaporate	freeze	gaseous
hidden	kinetic energy	liquid	mass	particles
potential energy	same	solid	state	

When water at room temperature is heated, the kinetic energy of its ____________ increases. This means the temperature increases, because temperature is a measure of the average ____________ of the particles.

When water reaches its boiling point at 100 °C, all the liquid water turns into water vapour, its ____________ state. However, the temperature of the water no longer increases and stays at 100 °C, until the phase change is complete, meaning the average kinetic energy of the particles stays the ____________. The heat transferred to the water does not just disappear though, as this would violate the law of ____________ of energy. Instead, it causes the phase change from liquid to gas, by overcoming the ____________ attraction between the water particles and increasing the electrical ____________ between them. Because the heat does not cause a temperature change, it is called ***latent heat*** – latent means ____________.

The ***latent heat of fusion*** L_f is the amount of energy that must be absorbed by a substance for it to melt, turning it from ____________ to liquid. It is also the same amount of energy a liquid substance must lose to its surroundings in order to ____________.

The ***latent heat of vaporisation*** L_v is the amount of energy that must be absorbed by a substance for it to boil or ____________, turning it from liquid to gas. It is also the same amount of energy a gaseous substance must lose to its surroundings in order to condense, turning it from gas to ____________.

The value of latent heat is experimentally determined using the formula:

$$Q = mL$$

where: Q = amount of energy required to change the ____________ of a substance (J)

m = ____________ of the substance (g)

L = specific latent heat of fusion or vaporisation (J g^{-1})

2. The graph shows the temperature of 1.0 kg of ice as heat energy was added to it and it changed state into water, then water vapour.

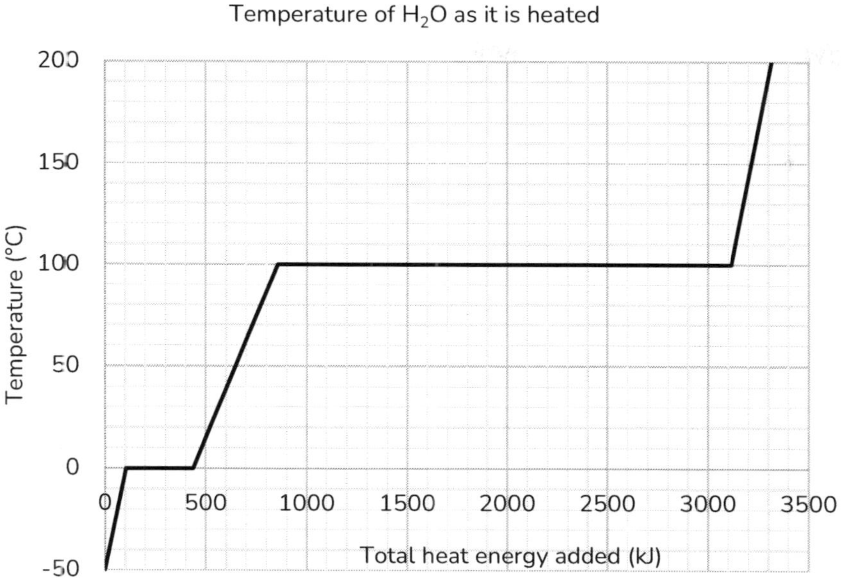

1 kg of water takes up 1L.

Identify the states of matter and phase changes in the graph.

a) Explain why the temperature stays constant for a while at 0 °C and 100 °C, even though heat is added.

Key words: electrical potential energy, electrostatic attraction, water particles, phase change.

b) Determine the latent heat of fusion, using the graph.

Recall the definitions. Then read the x-axis.

c) Determine the latent heat of vaporisation, using the graph.

d) How much energy was used to raise the temperature of the water from 0 °C to 100 °C?

e) Determine the specific heat capacity of water using the graph.

What does the gradient represent?

f) Compare the specific heat capacity of water to that of ice and water vapour.

3. Paraffin wax is often used in candles, and is a mixture which contains compounds called hydrocarbons, which have hydrogen and carbon in them.

The graph shows a heating curve for the paraffin wax in a candle, as 15 g of wax is heated.

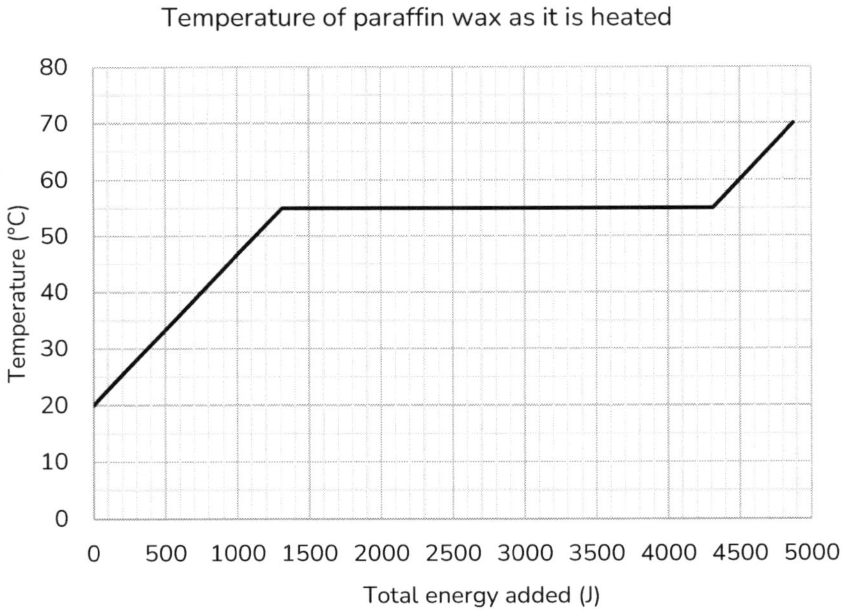

a) What is the melting point of this paraffin wax?

b) Determine the latent heat of fusion of this paraffin wax.

c) Explain the effect of the latent heat of fusion on how long candle lasts.

d) Determine the specific heat capacity of solid paraffin wax.

e) Determine the specific heat capacity of liquid paraffin wax.

Part 1: Motion and forces in 1D - Answers

1.2 Describing motion

1. Comparing terms

qualitative quantitative	Both describe observations but quantitative observations are numerical.
scalar vector	Both have magnitude but vectors also have direction.
distance displacement	Both describe position but distance is a scalar and displacement is a vector. Distance travelled refers to the actual path taken between two points, displacement refers to the change in position.
speed velocity	Both describe rate of change in position but speed is a scalar and velocity is a vector. Speed is calculated using distance and velocity using displacement.
instantaneous speed average speed	Both describe how fast something is travelling, but instantaneous speed is the exact speed at an instant. Average speed is the total distance travelled divided by time.
velocity acceleration	Both describe a rate of change. Velocity is change in displacement divided by time. Acceleration is change in velocity divided by time.

2. fast, heavy
3. acceleration, velocity
4. a) displacement
   b) distance
   c) acceleration
   d) velocity
   e) speed
   f) velocity
5. Speed:
   v = speed (m s^{-1})
   Δd = distance travelled (m)
   Δt = time (s)
   Velocity:
   $\vec{v}$ = velocity (m s^{-1})
   $\Delta\vec{s}$ = displacement (m)
   Δt = time (s)
6.

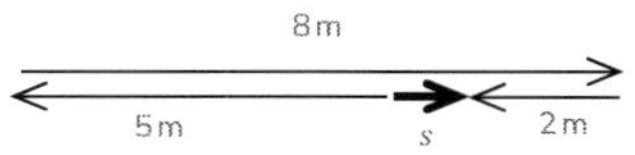

   a) $5 + 8 + 2 = 15$ m
   b) $+5 - 8 - 2 = -1$ m
   Displacement is 1 m to the right.
7. a) $6.8 + 2.7 = 9.5$ m
   b) $s = \sqrt{6.8^2 + 2.7^2} = 7.3$ km

$$\tan\theta = \frac{2.7}{6.8}$$

   $\theta = 22°$ so direction is S22°W

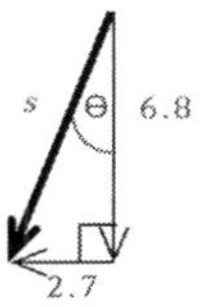

8. a) $123 + 123 = 246$ m
   b) $s = \sqrt{123^2 + 123^2} = 174$ m

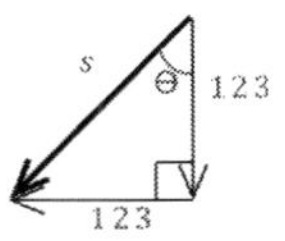

9. One possibility is that the snake travelled 5 units down, 5 units right then 5 units up.

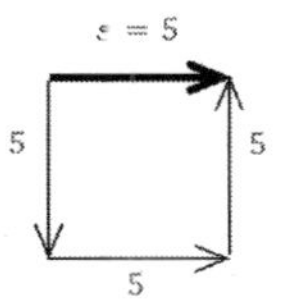

10. It is incorrect. They have calculated their distance travelled. The total displacement would be zero, since they return home, back to the starting point.
11. a) Zero
    b) $2\pi r = 2 \times \pi \times 6371 = 40000$ km
12. Do not assume that the Earth is flat and recognise that it is roughly spherical. You could be at the south pole and travel in a triangular shape along the sphere. You could also be anywhere along a latitude that is 11.59 km south from the north pole. After travelling 10 km north, you would travel a complete circle in a westerly direction with a circumference of 10 km and a radius of $\frac{10}{2\pi} = 1.59$ km, before returning to where you started.

1.3 Speed and velocity

SI units

1.

0.035 m	millimetre
0.297 m	centimetre
286000 m	kilometre

2.

0.010 s	millisecond
360 s	minute
18000 s	hour

3. $340 \text{ m s}^{-1} = 340 \times 3.6 = 1224$ km h^{-1}
4. Mach 9.6 = 9.6×340 m s^{-1}
   $= 3264$ m s^{-1}
   $= 3264 \times 3.6$
   $= 12000$ km h^{-1}
5. a) $v = \frac{\Delta d}{\Delta t} = \frac{100}{9.8} = 10.4$ m s^{-1} = 37.6 km h^{-1}
   b) 10.4 m s^{-1} = 10.4×3.6 km h^{-1} = 37.6 km h^{-1}
   This is slower than the car.

Rearranging equations

6. $v = \frac{\Delta d}{\Delta t}$
   $1.4 = \frac{\Delta d}{10 \times 60}$
   $\Delta d = 1.4 \times 600 = 840\ m$
7. The time you take to cross the street is given by:
   $v = \frac{\Delta d}{\Delta t}$
   $1.4 = \frac{7.5}{\Delta t}$
   $\Delta t = \frac{7.5}{1.4} = 5.36$ s
   How far the car travels in this time is given by:
   $v = \frac{\Delta d}{\Delta t}$
   $\frac{50}{3.6} = \frac{\Delta d}{5.36}$
   $\Delta d = 74.4$ m

Speed or velocity? Average or instantaneous?

8. The speedometer of a car shows you your speed at any one moment in time. This is your instantaneous speed. The odometer shows the total distance travelled. Dividing the odometer reading by the total time taken to travel it will give your car's average speed.
   The instantaneous speed can vary within this time. For example, if you are driving at 50 km h^{-1} then brake for a

red light, your instantaneous speed started off as 50 km h^{-1} and ended as 0 km h^{-1} but your average speed will be between these. However, if you drove at a uniform speed so there is no acceleration, then your average speed and instantaneous speed will be the same.

9. Velocity differs from speed because it also has direction. If a child rides a merry-go-round that spins at a uniform speed, the velocity is not the same, because the child changes direction. You can work out the child's speed by measuring the time it takes for the child to spin once around and return to the starting position. You also need to measure the distance the child sits from the centre of the merry-go-round, r. The speed throughout will be the circumference, $2\pi r$, divided by the time.
   However, since the child returned to the starting point, there is no displacement, so the average velocity will be zero. The instantaneous velocity at any point is the same magnitude as speed though, but the direction is at a tangent to the circular path travelled.

10. a) They assumed a uniform velocity of 60 km h^{-1} for the journey, which gives the time of 11.5 min. However, this does not consider that the road may change direction, or the instantaneous speed of the car must reduce for obstacles.
    b) $v = \frac{\Delta d}{\Delta t} = \frac{11.5}{30 \div 60} = 23$ km h^{-1}

11. $v = \frac{\Delta d}{\Delta t} = \frac{2 \times \pi \times 384000}{27.3 \times 24} = 3680$ km h^{-1}

12. a) $v = \frac{\Delta d}{\Delta t}$
    $2.4 = \frac{28+15}{\Delta t}$
    $\Delta t = \frac{43}{2.4} = 18$ s
    b) Since the player jogs at a steady 2.4 m s^{-1} throughout, this is also their instantaneous speed, average speed. and magnitude of instantaneous velocity. However, the direction of instantaneous velocity changes as the player goes around the corner.
    Calculating average velocity:
    Displacement: $\Delta d = \sqrt{28^2 + 15^2} = 31.8$ m
    Average velocity: $v = \frac{\Delta d}{\Delta t} = \frac{31.8}{43} = 1.8$ m s^{-1}
    Angle: $\tan\theta = \frac{15}{28}$
    $\theta = 28°$
    So average velocity is 1.8 m s^{-1} in a direction of 28° to the original jogging direction, which is smaller in magnitude to the average speed.

1.4 Motion diagrams for constant velocity

1. a) 8 units × 1 m = 8 m
   b) $T = \frac{1}{f} = \frac{1}{10} = 0.1$ s
   The number of periods shown in the diagram is 4.
   Total time = 0.1 × 4 = 0.4 s
   c) $v = \frac{s}{t} = \frac{8}{0.4} = 20$ ms^{-1} to the right
   d) The dots are equally spaced apart, meaning that the distance travelled in each time interval was the same, so the velocity is the same.

2. $s = 4.8 \times 0.5 = 2.4$ m
   $t = 3T = 3 \times \frac{1}{f} = 3 \times \frac{1}{30} = 0.1$ s
   $v = \frac{s}{t} = \frac{2.4}{0.1} = 24$ m s^{-1} to the right

3. a) $t = \frac{1\ min}{120\ steps} = \frac{60\ s}{120} = 0.5$ s
   b) In time 0.5 s, the displacement of the runner is their stride length:
   $v = \frac{s}{t}$
   $2.3 = \frac{s}{0.5}$
   $s = 1.15$ m
   c) Distance on scale $= \frac{1.15}{0.5} = 2.3$ grid units
   So footprints would be at 4.6, 6.9, 9.2 and 11.5 grid units.

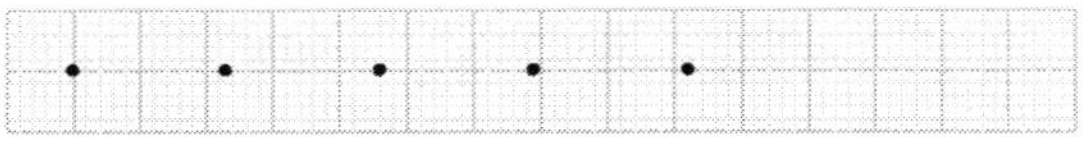

4. a) $v = \frac{s}{t} = \frac{12}{3 \times 0.1} = 40$ cm s^{-1} to the right
   b) $v = \frac{s}{t}$
   $40\ \text{cm s}^{-1} = \frac{s}{\frac{1}{24\ \text{Hz}}} = s \times 24$ s
   $s = 1.666 \ldots$ cm $= 1.7$ cm
   So dots should be at 1.7, 3.3, 5.0, 6.7, 8.3, 10.0 units.

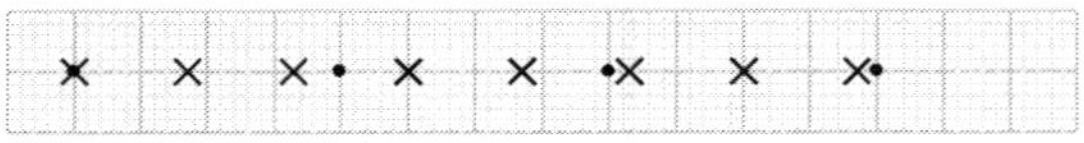

   c) 24 frames per second would give a smoother, more detailed video as there are more images. However, it would take the animator longer to make the video, as more photographs need to be taken of the car.

2.2 Uniform acceleration

Understanding acceleration

1. Acceleration is the rate of change in velocity. $\vec{a} = \frac{\Delta\vec{v}}{\Delta t}$
2. Uniform acceleration means acceleration stays the same, because the net force is unchanged.
3. a)

time (s)	velocity (m s^{-1})	Δv (m s^{-1})
0	19.6	0
1	9.8	- 9.8
2	0	- 9.8
3	- 9.8	- 9.8
4	- 19.6	- 9.8

Acceleration is uniform.
The object could be launched upwards, slow down, then fall due to gravity.

b)

time (s)	velocity (m s^{-1})	Δv (m s^{-1})
0	3	-
1	1.5	-1.5
2	0	-1.5
3	-1.5	-1.5
4	-3.0	-1.5

Acceleration is uniform.
The object could be sliding up a ramp, slow down, then slide back down due to gravity.

c)

time (s)	velocity (m s^{-1})	Δv (m s^{-1})
0	0	-
1	0.4	+ 0.4
2	1.0	+ 0.6
3	0.4	- 0.6
4	0	- 0.4

Acceleration is not uniform.
The object could swinging back and forth, speeding up as it approaches the middle, then slowing down as it reaches its amplitude.

SUVAT equations

4. $\vec{s}$ = displacement (m)
   $\vec{u}$ = initial velocity (m s^{-1})
   $\vec{v}$ = final velocity (m s^{-1})
   $\vec{a}$ = acceleration (m s^{-2})
   t= time (s)
5. a) Yes – acceleration due to gravity is 9.8 m s^{-2}
   b) No – the ground exerts an upward force depending on how much the ball is compressed
   c) No – constantly changes direction
   d) Yes – there is constant deceleration
   e) Yes – acceleration is provided only by gravity
   f) No - constantly changes direction
6. $\vec{s}$, $\vec{v}$, t
7. If positive is down:
   a) $\vec{v}^2 = \vec{u}^2 + 2\vec{a}\vec{s}$
   $\vec{v}^2 = 0^2 + 2 \times 9.8 \times 7.0$
   $\vec{v} = 11.7$ m s^{-1} down
   b) $\sqrt{2 \times 9.8 \times 7.0} \times 3.6$ km h^{-1} = 42.2 km h^{-1}
   The coin moves faster than the car.
8. If positive is up:
   a) $\vec{v}^2 = \vec{u}^2 + 2\vec{a}\vec{s}$
   $0 = \vec{u}^2 + 2 \times -9.8 \times 1.22$
   $\vec{u} = 4.89$ m s^{-1} up
   b) $\vec{v}^2 = \vec{u}^2 + 2\vec{a}\vec{s}$
   $0 = 4.89^2 + 2 \times -1.625 \times \vec{s}$
   $\vec{s} = 7.36$ m up
9. If positive is up:
   a) $\vec{v}^2 = \vec{u}^2 + 2\vec{a}\vec{s}$
   $0 = \vec{u}^2 + 2 \times -9.8 \times 12.0$
   $\vec{u} = 15.3$ m s^{-1} up
   b) $\vec{s} = \vec{u}t + \frac{1}{2}\vec{a}t^2$
   $0 = \sqrt{235.2} \times t + \frac{1}{2} \times (-9.8) \times t^2$
   $t(\sqrt{235.2} - 4.9t) = 0$
   $t = 0$ or 3.13 s
   The apparatus spends 3.12s in the air.
10. If positive is up:
    a) $\vec{v}^2 = \vec{u}^2 + 2\vec{a}\vec{s}$
    $0 = \vec{u}^2 - 2 \times -9.8 \times 6.73$
    $\vec{u} = \sqrt{131.908}\ m\ s^{-1}$ up
    $\vec{v} = \vec{u} + \vec{a}t$
    $\vec{0} = \sqrt{131.908} + (-9.8)t$
    $t = 1.17$ s
    b) If positive is up and the origin is the maximum height:
    $\vec{s} = \vec{u}t - \frac{1}{2}\vec{a}t^2$
    $-6.73 = 0 \times t + \frac{1}{2} \times (-9.8) \times t^2$
    $t = 1.17$ s
    This time is the same as part (a).
11. Let up be positive. Let u_0 be the launch velocity and acceleration due to gravity be $-g$.
    Time taken to go up is given by:
    $\vec{v} = \vec{u} + \vec{a}t$
    $0 = u_0 - gt$
    $t = \frac{u_0}{g}$
    The distance up s is the same as the distance down, given by:
    $\vec{v}^2 = \vec{u}^2 + 2\vec{a}\vec{s}$
    $0 = u_0^2 - 2gs$
    $s = \frac{u_0^2}{2g}$
    Time taken to go down is given by:
    $\vec{s} = \vec{u}t + \frac{1}{2}\vec{a}t^2$
    $\frac{u_0^2}{2g} = 0 - \frac{1}{2}gt^2$
    $t = \frac{u_0}{g}$
    which is the same time taken to go up.
12. a)

Time (s)	Displacement (m)	Velocity (m s^{-1})
0.5	1.495	0.54
0.7	1.407	-1.42
1.2	-0.528	-6.32

    b)

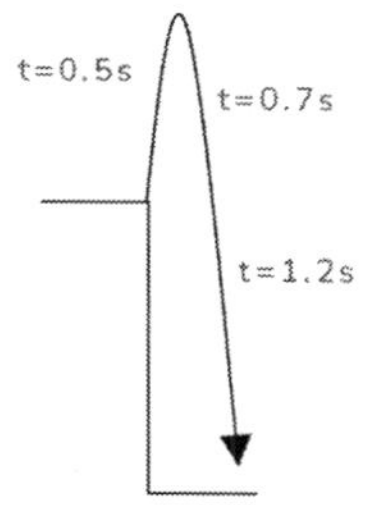

    c) Let up be positive and the origin the diving board.
    $\vec{s} = \vec{u}t + \frac{1}{2}\vec{a}t^2$
    $-3 = 5.44t + \frac{1}{2} \times (-9.8) \times t^2$
    $4.9t^2 - 5.44t - 3 = 0$
    $t = \frac{5.44 \pm \sqrt{5.44^2 - 4 \times 4.9 \times (-3)}}{2 \times 4.9}$
    $t = 1.51$ or -0.40 s
    Ignore the negative answer as this makes no real life sense. Time taken to reach the water is 1.51 s.
13. Let forward be positive.
    a) The car travels at a constant 50 km h^{-1} for 1.5 s before it brakes. Distance travelled is given by:
    $\vec{u} = \frac{\vec{s}}{t}$
    $\frac{50}{3.6} = \frac{s}{1.5}$
    $s = 20.8$ m
    Distance during braking is given by:
    $\vec{v}^2 = \vec{u}^2 + 2\vec{a}\vec{s}$
    $0 = (\frac{50}{3.6})^2 + 2 \times -7.0 \times s$
    $s = 13.8$ m
    Total distance = 20.8+13.8 = 34.6 m
    b) The car travels at a constant 60 km h^{-1} for 1.5 s before it brakes. Distance travelled is given by:
    $\vec{u} = \frac{\vec{s}}{t}$
    $\frac{60}{3.6} = \frac{s}{1.5}$
    $s = 25$ m
    Distance during braking is given by:
    $\vec{v}^2 = \vec{u}^2 + 2\vec{a}\vec{s}$
    $0 = (\frac{60}{3.6})^2 + 2 \times -7.0 \times s$
    $s = 19.8$ m
    Total distance = 25+19.8 = 44.8 m.
    This is an extra 10.2m compared to 50 km h^{-1}.
14. a) Using $s = ut + \frac{1}{2}at^2$ gives $a = \frac{2s}{t^2}$.
    b) Missing values are both 10.
    c) Yes
15. a) 150, 133, 125
    b) No

16. a)

time (s)	speed (m s^{-1})	displacement (m)
0	0	0
1.0	9.8	4.9
2.0	19.6	19.6
3.0	29.4	44.1
4.0	39.2	78.4

b) $v = \frac{s}{t} = \frac{78.4}{4.0} = 19.6$ m s^{-1}

c) At 2.0 s

17. Let the final time be t_f. Average speed is given by:

$$v_{av} = \frac{s}{t} = \frac{ut_f + \frac{1}{2}at_f^2}{t_f} = u + \frac{1}{2}at_f$$

Instantaneous speed at half time, $\frac{t_f}{2}$, is given by:

$$v = u + at = u + \frac{at_f}{2}$$

This is the same value as the average speed.

18. $v_{av} = \frac{s}{t}$

Displacement s is given by:

$$s = ut + \frac{1}{2}at^2 = ut + \frac{1}{2}\left(\frac{v-u}{t}\right)t^2 = ut + \frac{1}{2}(v-u)t$$

$$2s = 2ut + vt - ut = ut + vt$$

$$\frac{s}{t} = \frac{u+v}{2} = v_{av}$$

2.3 Visualising acceleration

1. a) The motion diagram shows the position of the object at a rate of 10 images per second. We can find the time interval between positions by dividing 1 by 10. We can find the displacement during each time interval by multiplying the number of grid units travelled by 0.1 m. The displacement of the object during each equal time interval is increasing. This means the object's average velocity must be increasing.
As acceleration is the change in velocity over time, this means the object's acceleration must be downwards.

b) $T = \frac{1}{f} = 0.1$ s

c) $t_1 = 0.1$ s, $t_2 = 0.2$ s, $t_3 = 0.3$ s, $t_4 = 0.4$ s

d)

t (s)	s (m)	v_{av} (m s^{-1})	v (m s^{-1})
0.1	0.05	0.50	0.98
0.2	0.20	1.00	1.96
0.3	0.44	1.47	2.94
0.4	0.78	1.95	3.92

e) Goes up by around 0.5 m s^{-1} each 0.1 s.

f) Yes, taking into account rounding and errors from the limit of reading of the grid.

g) $s = ut + \frac{1}{2}at^2$

$0.78 = 0 + \frac{1}{2}a \times 0.4^2$

$a = 9.8$ m s^{-2} downwards

h) See table in (d).

i) The average velocity during a time period is half the instantaneous velocity at the end of the time period.

j) $s = ut + \frac{1}{2}at^2 = 0 + \frac{1}{2} \times 9.8 \times 0.5^2$

$= 1.23$ m

2. a) False
b) True
c) True
d) True
e) True
f) True

3. a) The racecar was braking, as the displacement in each time interval is decreasing.

b) For the first to the last dot:

$s = ut + \frac{1}{2}at^2$

$38.6 = 31.3 \times 2.0 + \frac{1}{2} \times a \times 2.0^2$

$a = -12$ m s^{-2}

Using this in $s = ut + \frac{1}{2}at^2$ to calculate the other positions gives 14.2 m, 25.3 m, and 33.4 m, which matches the diagram.

4. 5 frames per second means dots appear at time intervals of 0.2 s. Positions in real life are given by:

$$s = ut + \frac{1}{2}at^2 = 0 + \frac{1}{2} \times 6t^2 = 3t^2$$

At 0.2 s, 0.4 s and 0.6 s, these positions are at 0.12 m, 0.48 m and 1.08 m respectively.

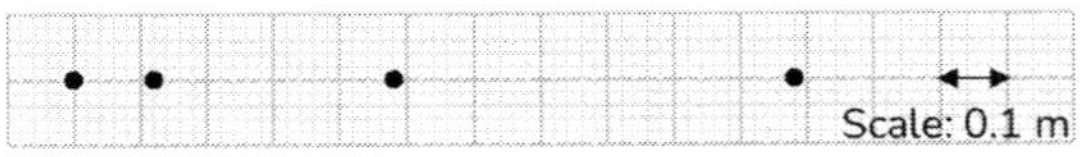

3.2 Experimental skills

1.

Term	Definition
variable	a factor that can change in an experiment
independent variable	the variable that is intentionally changed to investigate its effect
dependent variable	the variable that is observed or measured as a result of a change
controlled variable	variables that are kept the same throughout an experiment
fair test	an experiment where only one variable is changed
aim	a statement of the independent and dependent variable investigated
hypothesis	an educated prediction of the results of an aim
conclusion	a statement summarising the results of an investigation
validity	a correct test of the aim
accuracy	how close measurements are to the true value
reliability	whether repeat trials give similar results

2. a) Distance travelled
   b) Time taken
   c) Ball starts from rest, 10 degree angle, same wooden ramp, same tennis ball
   d) To investigate the effect of the distance travelled by a ball down a ramp on the time taken.
   e) As the distance travelled by the ball increases, the time taken will increase.
   f) Use of measuring tape and stopwatch
   g) Video the time taken to eliminate human reaction reaction time. Increase the length of the wooden ramp to increase the time being measured, to reduce the impact of the limit of reading on the measuring tape and timing device.
   h) Repeat steps 1 to 2 for two more trials.
   i) Add more distances to step 4 eg. 2.0 m, 2.5 m, 3.0 m.
3. a)

Distance (m)	Average time (s)	Absolute error	Percentage error
0.5	0.79	± 0.1	± 13%
1.0	1.1	± 0.1	± 9%
1.5	1.3	± 0.02	± 2%

   b) As the distance travelled by the ball increased, the time taken increased.
4. See the table for question 3.
5. a)

distance (m)	absolute error (m)	percentage error
0.5	± 0.0005	± 0.1%
1.0	± 0.0005	± 0.05
1.5	± 0.0005	± 0.03

   b)

time (s)	absolute error (s)	percentage error
0.79	± 0.005	± 0.6%
1.08	± 0.005	± 0.5%
1.33	± 0.005	± 0.4%

   c) Limits of reading in this experiment are not significant concerns because the percentage error due to them is very small, less than 1%.
6. a) For 1.0 s: $percentage\ error = \pm\frac{|0.25|}{1.0} \times 100\% = \pm 25\%$
   For 5.0 s: $percentage\ error = \pm\frac{|0.25|}{5.0} \times 100\% = \pm 5\%$
   The larger the time, the smaller the percentage error.
   b) The reliability of the results increased as the distance travelled increased. This could be due to the time taken increasing, which means the percentage error due to human reaction time decreases.
7. a) Random
   b) Systematic
   c) Random
   d) Random

4.1 Motion graphs

Reading line graphs

1. A line graph should be used when both axes show numerical data that is continuous. Line graphs are advantageous for showing numerical trends and predicting results.
2. The line of best fit shows the average trend, so it does not necessarily go through any of the data points.
3. 16 m s^{-1}
4. 25 m s^{-1}
5. 4 m s^{-1}
6. Pressing the accelerator
7. As the time increased, the speed increased.
8. The trend is valid. Most data points are close to the line of best fit. Variations might be from the driver not pressing the accelerator uniformly, changes in gear or road conditions.

Gradient of a graph

9. $a = \frac{\Delta v}{\Delta t} = \frac{27-9}{5-1} = 4.5$ m s^{-2}
10. Acceleration is constant because the gradient is constant. In other words, the graph is a straight line.
11. a) $a = \frac{\Delta v}{\Delta t} = \frac{3-17}{4-0} = -3.5$ m s^{-2}
    b) Braking
    c) 0
12. a) Distance
    b) The gradient represents speed:
    $\text{gradient} = \frac{\text{distance}}{\text{time}} = \text{speed} = 15$ m s^{-1}
    They are skating at a constant speed of 15 m s^{-1}.

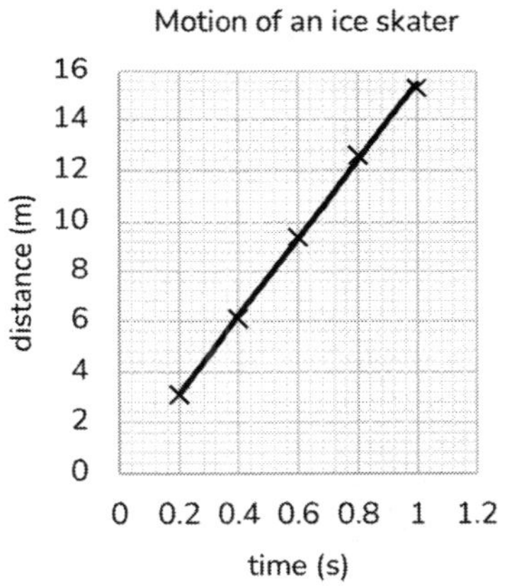

Gradient on curved graphs

13. When the line of best fit is curved, the gradient changes throughout. However, even if a velocity-time graph is curved, the gradient still represents acceleration. The acceleration will not be uniform though. Likewise, for a curved displacement-time graph, the gradient still represents velocity, as both its gradient and velocity are given by change in displacement over time. Velocity will not be uniform, meaning the object graphed is accelerating.
    The gradient at any one point can still be found by drawing a tangent to that point, which is a line that just touches the curve at one point. The gradient of the line at that point can then be calculated using the "rise over run" of the tangent.
    For example, in the displacement-time graph shown, at $t = 0$ the gradient of the tangent is zero, so the velocity of the ball is zero.

At $t = 0.5\ s$ the gradient of the tangent is:

gradient = velocity

$= \frac{1.4-0.72}{0.54-0.4} = \underline{4.9}$ m s^{-1}.

Down is taken as positive in the example, as the displacement increases as ball drops. Therefore a positive gradient, where the graph curves up the page, means the velocity is downwards.

14. a)

Time (s)	Velocity (m s^{-1})
0	0
0.2	1.96
0.4	3.92
0.5	4.9

b)

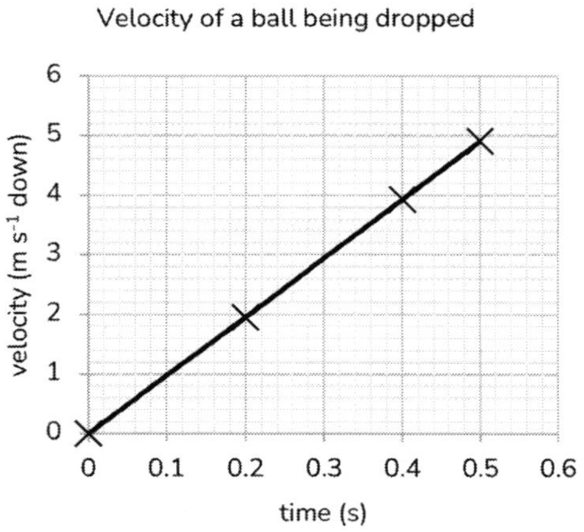

c) gradient = acceleration = 9.8 m s^{-2} downwards
This is acceleration due to gravity, since gravity is what makes a ball fall.

Area of graphs

15. In geometry, the area of many shapes is related to height and width. The area between a line graph and the x-axis is also calculated using the height, given by the change in the y-axis, and width, given by the change in the x-axis. Let's revisit the example from question 1. Speed v is on the y-axis and time t is on the x-axis, so the area is related to speed v multiplied by time t. From the definition of speed:

$$v = \frac{s}{t}$$
$$vt = s$$

So the area between the line graph and the x-axis on a speed-time graph gives distance travelled.

Looking closely at this particular graph, the area is in the shape of a trapezium, made of a rectangle and triangle. The area of the rectangle is given by:

$$hw = 9 \text{ m s}^{-1} \times (5-1) \text{ s} = 36 \text{ m}$$

This is also the distance that the car would have travelled if it kept going between 1 and 5 seconds at its initial speed u of 9 m s^{-1}:

$$s = ut = 9 \text{ m s}^{-1} \times (5-1) \text{ s} = 36 \text{ m}$$

The area of the triangle is given by:

$$\frac{1}{2}hw = \frac{1}{2}(v-u)t \ldots (1)$$

Rearranging the definition of acceleration, $a = \frac{v-u}{t}$, gives $v - u = at$. Substituting into equation (1):

$$\text{area of triangle} = \frac{1}{2}hw = \frac{1}{2}(v-u)t = \frac{1}{2}at^2$$

Comparing this to $s = ut + \frac{1}{2}at^2$ shows the area of the triangle therefore represents the extra distance the car travelled because it accelerated, rather than continuing at its initial velocity u:

$$\frac{1}{2}hw = \frac{1}{2}(v-u)t = \frac{1}{2}(27-9) \text{ m s}^{-1} \times (5-1) \text{ s} = 36 \text{ m}$$

So the total distance travelled by the car from 1 to 5 seconds was 36 + 36 = 72 m.

16. a) Travelling at a constant speed of 14 m s^{-1}
b) $v = \frac{s}{t}$
$s = vt = 14 \times 3 = 42$ m
c) The speed reduced to rest with an acceleration of:
$a = \frac{0-14}{2} = -7$ m s^{-2}
d) distance = area under graph $= \frac{1}{2} \times 2 \times 14 = 14$ m or
$s = ut + \frac{1}{2}at^2 = 14 \times 2 + \frac{1}{2} \times -7 \times 2^2 = 14$ m

17. a) Around 3.3 s
b) Car A has an acceleration, given by the gradient, of:
$a = \frac{\Delta v}{\Delta t} = \frac{20-0}{5-0} = 4$ m s^{-2}
Car B has an acceleration, given by the gradient, of:
$a = \frac{\Delta v}{\Delta t} = \frac{30-0}{5-2} = 10$ m s^{-2}
which is greater than A.
c) Distance travelled by Car A, given by the area, is:
$s = \frac{1}{2} \times 5 \times 20 = 50$ m
Distance travelled by Car B, given by the area, is:
$s = \frac{1}{2} \times 5 \times 30 = 45$ m
Car A is in front by 5 m.

18. a) Since area $= a \times \Delta t$ and $a = \frac{\Delta v}{\Delta t}$
therefore area $= \Delta v$
b) $\Delta v = \text{area} = 12 \times 100 + \frac{1}{2} \times (29-12) \times 100$
$= 2050$ m s^{-1}
$v_{final} = v_{initial} + \Delta v = 0 + 2050 = 2050$ m s^{-1}
c) No because acceleration is not constant.

19. Summary

speed	nothing	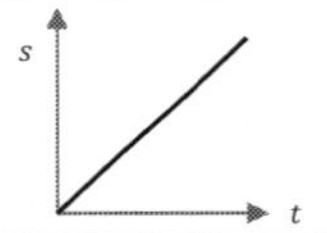
acceleration	distance travelled	
nothing	change in speed	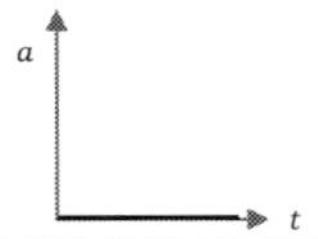

4.2 Motion graphs with vectors

1. a) $v_1^2 = u^2 + 2as_1$
$0 = u^2 + 2 \times -9.8 \times s$
$u^2 = 39.2$
$u = 6.26$ m s^{-1} up
b) $v_1 = u + at_1$
$0 = 6.26 + (-9.8)t_1$
$t_1 = 0.639$ s
c) $s = ut + \frac{1}{2}at^2$
$-4 = 6.26t + \frac{1}{2} \times (-9.8)t^2$
$t = \frac{6.26 \pm \sqrt{6.26^2 - 4 \times 4.9 \times (-4)}}{2 \times 4.9} = 1.75$ s
d) $v^2 = u^2 + 2as = 39.2 + 2 \times (-9.8) \times (-4) = 117.6$
$v = -10.84$ m s^{-1}
(answer must be negative as ball moves down)

Acceleration-time graph

2. $-9.8 \text{ m s}^{-2} \times 1.75 \text{ s} = 17.15 \text{ m s}^{-1}$
3. The area between the graph and the x-axis is at. The equation $v = u + at$ gives $at = v - u = \Delta v$ when rearranged.
4. The ball was moving up but ends up moving down.
5. $\Delta v = v - u$
   $v = \Delta v + u = -17.15 + 6.26 = 10.89 \text{ m s}^{-1}$
   This is the same as 1(d), allowing for rounding.

Velocity-time graph

6. The gradient of any velocity-time graph is found by dividing the rise, or change in velocity, by the run, or change in time, so it is equal to the acceleration. Since acceleration is constant, the velocity-time graph will have a constant gradient and look like a straight line. In this case, the gradient of the velocity-time graph will equal acceleration due to gravity of -9.8 m s^{-2}. The gradient is negative because we set our frame of reference to take upwards as positive.
   The line graph will start at the initial velocity of +6.26 m s^{-1} and end at the final velocity of -10.8 m s^{-1}. The line graph will cross the x-axis when velocity equals zero, which occurs at the peak of the ball's path.
7. gradient $= \frac{-10.8-6.4}{1.75} = 9.8 \text{ m s}^{-2}$
   The gradient represents acceleration due to gravity.
8. a) The statement has mistaken the graph for a displacement-time graph, reading zero velocity as zero displacement.
   b) The statement confuses negative velocity for decreasing velocity. Negative velocity here just means the ball is travelling downward. A larger negative number means a larger velocity downward.
9. a) The area of the triangle above the x-axis is:
   area $= \frac{1}{2} \times 0.65 \times 6.2 = 2.0 \text{ m}$
   This is the same as the distance the ball travels upwards, to its maximum height.
   b) Area of triangle below x-axis is:
   area $= \frac{1}{2} \times (1.75 - 0.65) \times 10.8 = 6.0 \text{ m}$
   This is the same as the distance the ball travels downwards, from its maximum height. Subtracting the two areas gives the final displacement of the ball from its starting position:
   $s = 2.0 - 6.0 = -4.0 \text{ m}$
   c) Adding the two areas gives the total distance travelled by the ball:
   $2.0 + 6.0 = 8.0 \text{ m}$

Displacement-time graph

10. The displacement-time graph shows the position of the ball over time. Therefore we can start by plotting the positions at times already known, being the initial position, peak of +2 m and final position of -4 m. The line graph will cross the x-axis where the ball returns to the origin.
    The previous velocity-time graph helps us work out shape of the line connecting these points, because velocity is the gradient of the displacement-time graph. Since the velocity changed throughout the journey, the gradient of the displacement-time graph will not be constant.
    As the ball goes up, the gradient of the displacement-time graph is positive, because velocity is positive. However, as it slows on the way up, the gradient decreases, so the line flattens.
    As the ball falls down, the gradient of the displacement-time graph is negative, because velocity is negative. However, as it speeds up as it falls, the negative gradient increases.
11. The gradient is given by the rise, representing change in displacement, divided by the run, representing change in time. Velocity is also change in displacement over time.
12. The gradient shows the ball's velocity is zero, because the tangent is a horizontal line.
13. The gradient represents the final velocity:
    gradient $= \frac{-4 - (-1.4)}{1.75 - 1.25} = 10.8 \text{ m s}^{-1}$

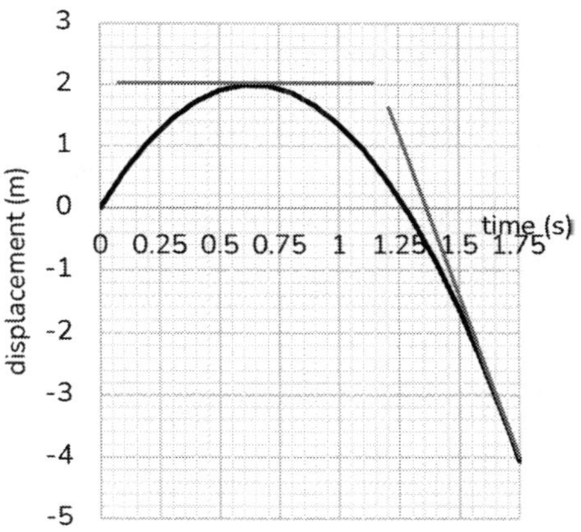

More questions

14.

Driving north at a constant velocity	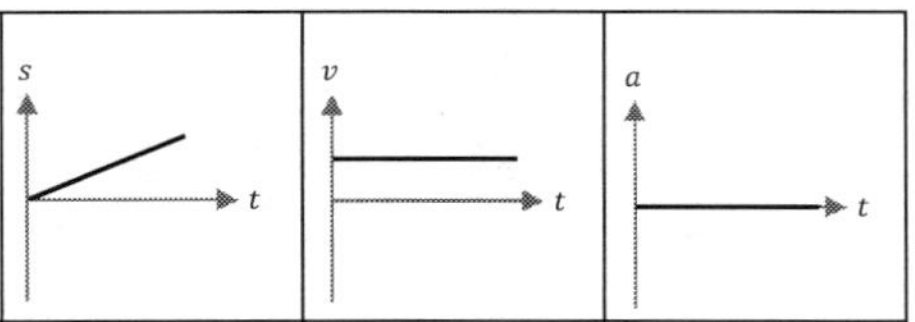
Driving south at a constant velocity	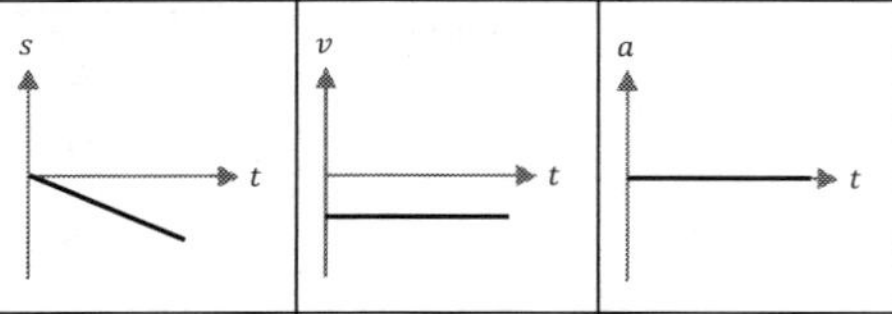
Constant acceleration from rest, moving northbound	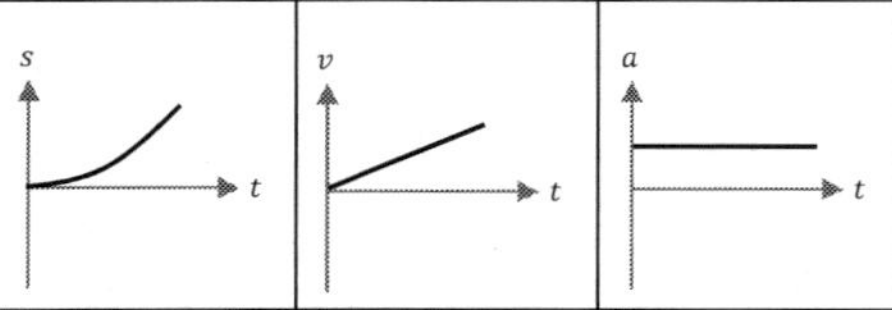
Constant acceleration from rest, moving southbound	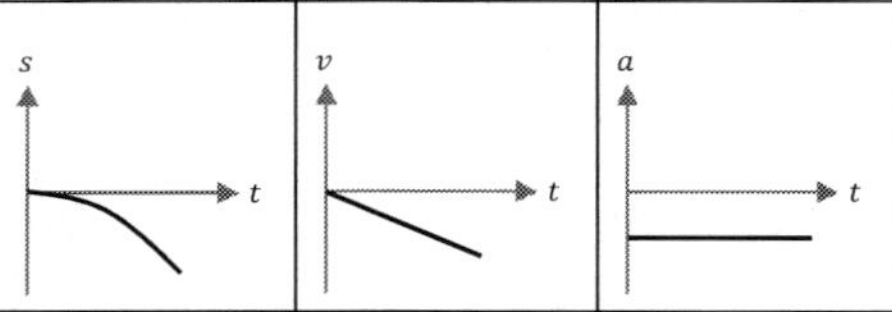
Driving north, braking with constant deceleration to stop	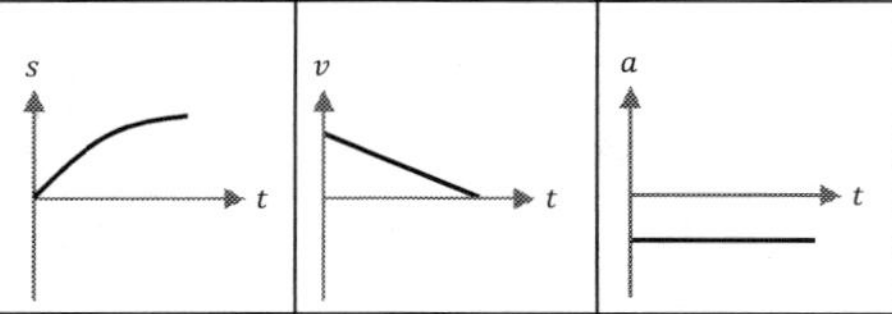
Driving south, braking with constant deceleration to stop	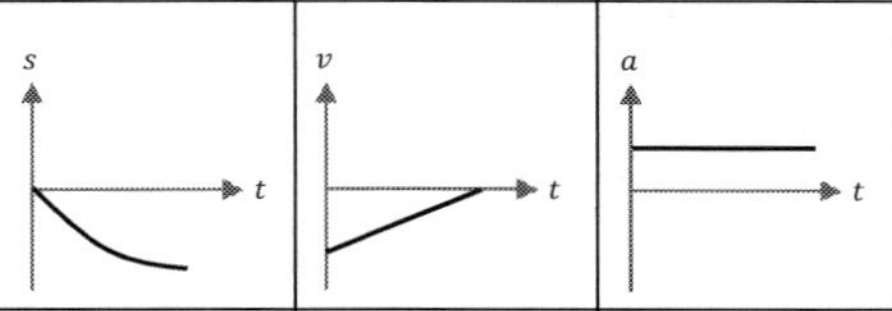
Driving north, braking with constant deceleration to U-turn, then accelerating at the same rate to go south	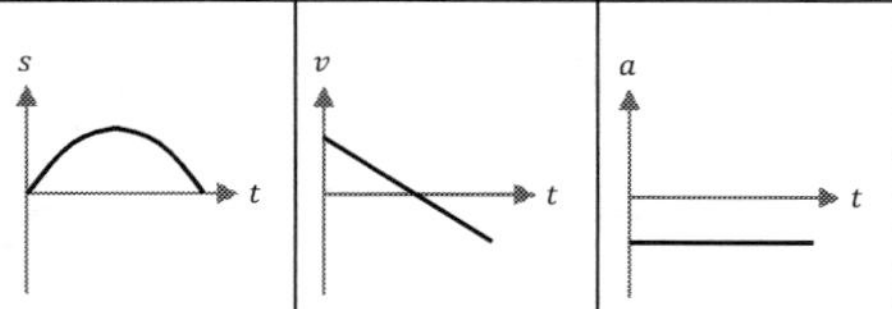

15. a) Down is positive, as the graph has a positive gradient, and the gradient represents acceleration due to gravity, which is down.

b) The astronaut stops momentarily at the peak after going up, before coming down.

c) Acceleration is given by the gradient:
$a = \frac{4-(-4)}{5} = 1.6 \text{ m s}^{-2}$

d) Distance is given by the sum of the areas between the graph and the x-axis:
$\text{distance} = \frac{1}{2} \times 2.5 \times 4 + \frac{1}{2} \times 2.5 \times 4 = 10 \text{ m}$

e) Displacement is given by the positive areas minus the negative areas. These are equal to each other, so displacement is zero and the astronaut is back on the ground.

16. a) The pendulum starts displaced by 0.2 m on one side, passing through a central point to swing to an equal distance of 0.2 m on the other side, before swinging back to its starting position.

b)

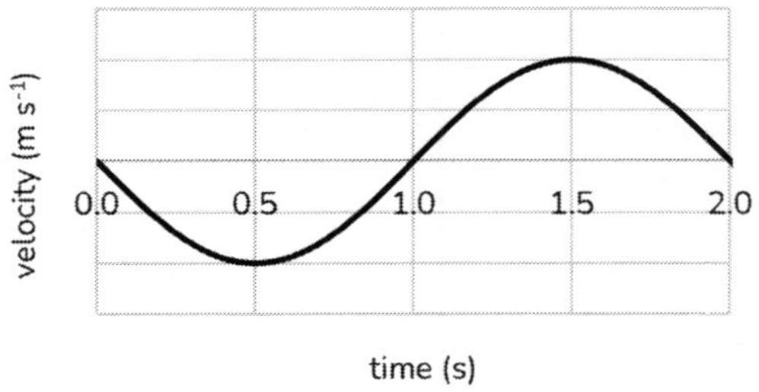

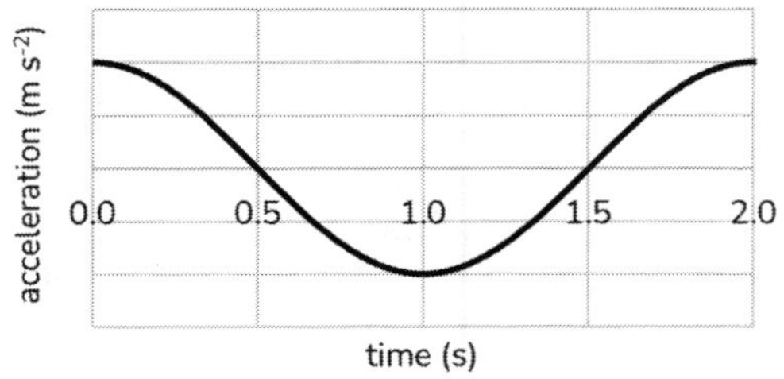

5.2 Introduction to forces

1. A force is a push or a pull, measured in newtons. Whenever there is a net force of zero on an object, the forces on the object are balanced, so the object's velocity does not change. Acceleration is caused by an unbalanced force. Forces acting on an object must either be due to something in direct contact with it or act via a field.

2.

Contact forces	Field forces
air resistance: force on an object as it moves through air buoyancy: force exerted by a fluid on objects floating on it friction: force between surfaces moving across each other normal force: a pushing force from a surface, acting perpendicularly to it tension: a pulling force that stretches objects	electrostatic force: force between charges gravitational force: attraction between any objects with mass magnetic force: force exerted by magnets on magnetic materials

3. a) Normal force is exerted by the floor of the elevator. Gravitational force is exerted by the Earth, and stays the same throughout the elevator journey.

b)

moving up at constant velocity	moving up but slowing down as it reaches a higher floor	accelerating downwards as it returns to the ground floor
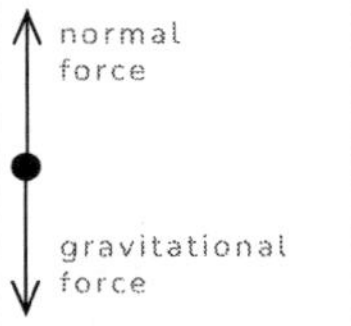		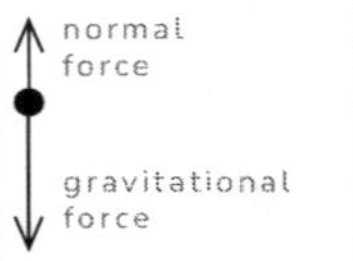

4.

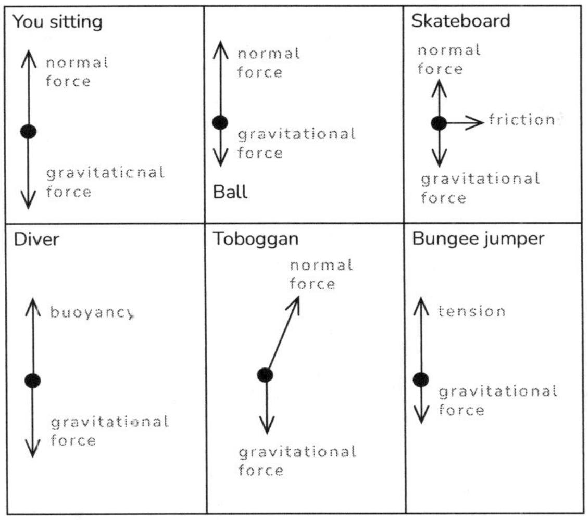

5. When you take a step forward, you must push your foot down and backward into the ground. Friction between your foot and the ground opposes your foot moving backwards and pushes you forward. If it were slippery, there would be less friction to stop your foot moving backward, so your foot would keep moving backward, due to inertia. This may result in you falling.
As you push your foot down into the ground, this causes an equal and opposite reaction force from the ground, pushing you up. This force from the ground on pushing on you is called the normal force, and it is the same magnitude as the force of you on the Earth. However, it causes a much bigger acceleration on you, since according to $F = ma$, the same force on a smaller mass causes a bigger acceleration.

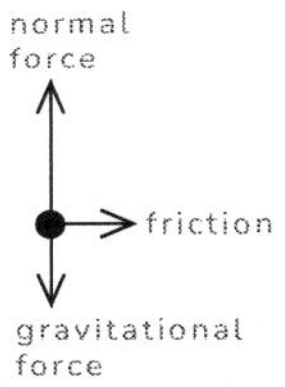

6. When a rocket is launched, gases produced from combustion of the fuel inside it are pushed downward and out. The gases exert an equal and opposite reaction force upward on the rocket, called thrust. When the thrust exceeds the gravitational force on the rocket, the net force will be upward and it will start to accelerate. The engine will exert a constant thrust on the rocket, but as fuel is used up, the total mass of the rocket decreases, leading to an increase in acceleration, according to $F = ma$.

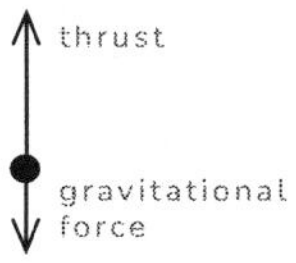

7. A moving car will keep moving at constant velocity in a straight line, unless acted upon by a net force, according to the Law of Inertia. To turn, there must be a force to change the direction of the car. This force is supplied by friction between the road and the tyres, pushing the car in the direction of the turn. This force is the reaction force to the car pushing the road outwards.

8. a) You pushing the water back creates a reaction force from the water pushing you forwards.
   b) Pushing harder increases force on the water and the reaction force of the water pushing on you increases by the same magnitude. Increasing force increases acceleration, according to $F = ma$.
   c) You keep moving forward according to the Law of Inertia, until water resistance slows you down.
9. $F = ma = 40 \times 1.5 = 60$ N forwards
10. a) 200 N, as the reaction force is equal to the action in the opposite direction.
   b) Ash: $a = \frac{F}{m} = \frac{200}{50} = 4$ m s^-2^
   Bee: $a = \frac{F}{m} = \frac{200}{70} = 2.9$ m s^{-2}
11. a) False – Inertia will keep an object moving at constant velocity.
   b) False – This decreases its acceleration due to a force, but has no effect on what that external force is.
   c) False – Balanced forces means net force on one object is zero, while action-reaction pairs involve two objects.
   d) True – this is an action-reaction pair.
   e) False – Normal force is in whatever direction is perpendicular to the surface, and its magnitude is the same as the action force on the surface.

5.3 Mass and weight

1. Mass and weight are different concepts in science, although they might be used interchangeably in everyday language. ***Mass*** is the amount of matter in an object, measured in the SI unit kilograms (kg). In other words, your mass relates to the amount and type of particles (atoms and molecules) in you.
Weight is another name for gravitational force, which is the force of attraction between any two objects with mass, measured in newtons (N). It is the force that produces acceleration due to gravity, which accelerates masses toward each other. The symbol g is used to represent acceleration due to gravity, and $g = 9.8$ m s^{-2} on average on Earth's surface. Acceleration due to gravity g varies on different planets and moons, because gravitational force increases as the mass of the objects increases. Gravitational force also decreases as the distance between objects increases.
The symbol W is often used to represent weight and F_G is often used to represent gravitational force. W and F_G are the same thing, but the symbol W is more commonly used for gravitational force on Earth's surface.
When the force accelerating an object is weight, the equation $F = ma$ can be written as:

$$W = mg$$

where: W = weight (N)

m = mass (kg)

g = acceleration due to gravity (m s^{-2})

2.

Concept	Mass	Weight
Definition	the amount of matter in an object	gravitational force on an object $W = mg$
Unit	kg	N

3. a) $W = mg = 50 \times 9.8 = 490$ N
   b) $W = mg = 50 \times 1.625 = 81.25$ N

4. The statement is true in saying you have less weight on the Moon, as the Moon has less mass than Earth, and weight is the same thing as gravitational force. However, most people trying to lose weight are actually trying to lose mass – the amount of matter in their body. Your mass would not change on the Moon. Therefore customers probably won't be satisfied.
5. a) circumference $= 2\pi r = 57$ cm
   $r = 9.07 \ldots$ cm
   area $= \pi r^2 = 258$ cm^2
   mass $= 1.03 \times 258 = 265.74$ kg
   $W = mg = 265.74 \times 9.8 = 2604$ N
   b) $W = mg = 90 \times 9.8 = 882$ N
   The air exerts more weight than a washing machine.
6. In everyday language, we often say we use scales to "weigh" objects, but the unit on the scales is grams or kilograms, not newtons. How scales actually work is they measure force, not mass. If you stand on a bathroom scale and jump up and down, the reading on the scale changes, even though your mass hasn't. So the bathroom scale is actually measuring the force you exert on the scale, which is equal and opposite to the normal force it exerts on you. The labelling on the scale assumes that you are using it on Earth to measure your mass, where acceleration due to gravity is 9.8 m s^{-2} and you are standing still on it. It is labelled so that the force measured is converted to mass using $W = mg$.
7. The force they exert upwards must equal the weight of the 100 kg, so $W = mg = 100 \times 9.8 = 980$ N
8. $F = ma = 4.8 \times 9.8 = 47$ N downwards
9. On the moon, the astronaut's weight is:
   $W = mg = 75 \times 1.625 = 121.875$ N
   The label assumes $g = 9.8$, so it will incorrectly give the astronaut's mass as:
   $W = mg$
   $121.875 = m \times 9.8$
   $m = 12.4$ kg
10. The rock's correct mass is 53 g, as measured on Earth. On Mars, the 20 g reading tells us the weight of the rock on Mars is:
   $W = mg = 0.02 \times 9.8 = 0.196$ N.
   This means the acceleration due to gravity on Mars is:
   $W = mg = 0.053 \times g$
   $g = 3.7$ m s^{-2}
11. When a mass is hung at the end of a spring balance, the weight of the mass stretches the spring down. The tension in the spring pulls the mass up with an equal and opposite reaction force. The spring stops stretching further when the weight on the mass and tension are balanced. The length of the spring is proportional to the tension in it. To convert this tension to the mass of the object in kilograms, the tension, measured in newtons, is divided by acceleration due to gravity of 9.8 ms^{-2}.
   The lower acceleration due to gravity causes the scale to give a lower reading on Mars.
12. There is gravity in space, keeping the Moon orbiting around the Earth. If this gravity disappeared suddenly, the Moon would continue moving in a straight line, tangential to its original orbit, due to inertia. Gravity also keeps the planets in our Solar System, including Earth, orbiting around the Sun, and human-made satellites orbiting around Earth. The amount of gravity isn't small either. For example, the International Space Station, which orbits Earth at an altitude of around 400 km, experiences an acceleration due to gravity of 8.7 m s^{-2}, compared to 9.8 m s^{-2} on Earth's surface.
   So the sensation of "weightlessness" doesn't come from the absence of gravity. At amusement parks on a "drop tower" ride, riders seated on chairs are lifted up high on a tower, then the chairs are dropped, plummeting down the sides of the tower. The riders experience "weightlessness" because their chair is no longer pressing up on them, so there is no normal force. In fact, the only force acting is gravity (if we ignore air resistance)! The riders and the astronauts in the International Space Station are described as being in free fall, which leads to the sensation of "weightlessness".
   In everyday life, we are used to experiencing the normal force from the ground or our seat, and this normal force is usually the same magnitude as our weight. One way to describe this force is 1 g. If you are in a rocket where the normal force is pushing up on you twice as hard, you experience 2 g. In free fall, you would experience zero g. Using g allows us to compare the force we experience, even though we are different masses.
   Astronauts typically experience 4 g during a rocket launch, sometimes more. As our bodily fluids experience inertia while our bodies accelerate, blood can pool in parts of the body and the heart has difficulty pumping, so shortness of breath, blurred vision and loss of consciousness are common effects, making higher g dangerous.
13.

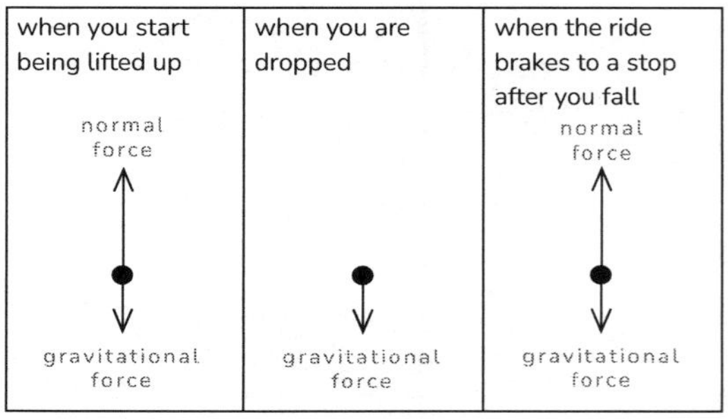

5.4 Forces and constant acceleration

1. $v = u + at$
   $27 = 0 + a \times 7.5$
   $a = 3.6$ m s^{-2}
   $F = ma = 1400 \times 3.6 = 5040$ N (in direction of motion)
2. $F = ma$
   $11565 = 1500 \times a$
   $a = 7.71$ m s^{-2}
   $v = u + at$
   $27 = 0 + 7.71 \times t$
   $t = 3.5$ s
3. Let direction of motion be positive.
   $v^2 = u^2 + 2as$
   $0 = 7^2 + 2a \times 4.55$
   $a = -5.38$ m s^{-2}
   $F = ma = 50 \times -5.38 = -269$ N (opposing direction of motion)
4. t = 0-3s: Moving north and accelerating with net force:
   $F = ma = m\frac{v-u}{t} = 92 \times \frac{(30\div3.6)-0}{3} = 256$ N (north)
   t = 3-6s: Coasting at constant velocity of 30 km h^{-1}. No net force is acting.
   t = 6-10s: Moving north and braking to slow down with net force:
   $F = ma = m\frac{v-u}{t} = 92 \times \frac{0-(30\div3.6)}{4} = -192$ N (192N south)

5. a) $F = ma = 75 \times 1.2 = 90$ N up
   b) $W = mg = 75 \times 9.8 = 735$ N down
   c) $F_{net} = T - W$
   d) $F_{net} = T - W$
   $90 = T - 735$
   $T = 825$ N up

6. Let down be positive.
   a) zero
   b) $v^2 = u^2 + 2as = 0 + 2 \times 9.8 \times 30$
   $v = 24.2$ m s^{-1}
   c) $v = u + at$
   $4.0 = 24.2 + a \times 2.0$
   $a = -10.1$ m s^{-2}
   d) $F_{net} = ma = W - N$
   $60 \times -10.1 = 60 \times 9.8 - N$
   $N = 1194$ N up
   The rider feels "weightless" when in free fall, because the normal force is zero, and heavier when the ride slows, because normal force is greater than their weight of $60 \times 9.8 = 588$ N.
7. a) $s = ut + \frac{1}{2}at^2$
   $0.1 = 0 + \frac{1}{2} \times a \times 0.4^2$
   $a = 1.25$ m s^{-2} to the right
   b) $F_{net} = ma = 0.04 \times 1.25 = 0.05$ N to the right
   c) $F_{resistive} = 1.5 - 0.05 = 1.45$ N to the left
8. In flight, the only force acting on both objects is weight. As object A has more mass, it has more weight according to $W = mg$, so there is more force slowing it down. However, this force also acts on a larger mass, resulting in the same acceleration due to gravity as object B. As velocity is zero at maximum height, the maximum height of both objects is given by:
   $v^2 = u^2 - 2gs$
   $s = \frac{u^2}{2g}$

5.5 Harder force problems – Set 1

1. a) In physics, both trapeze artists together is called the ***system***, since they will move together. As they stay rest, there must be no net force on them, according to the Law of Inertia. Therefore the tension in the top person's legs, T_1, pulling them up, must balance the combined weight of the people, $W_1 + W_2$, pulling them down.
   The net force on each person is also zero, meaning the forces acting on each of them is balanced.
   Between the people, the force the bottom person pulls the top person down with, T_2, with must equal the force the top person pulls the bottom person up with, since this is an action-reaction pair.
   The forces on the top person are the tension from the trapeze bar, T_1, pulling them up, and their own weight, W_1, pulling them down. In addition, they are in contact with the bottom person, who also pulls down on them with tension T_2. As the top person is stationary, the tension from the trapeze bar, T_1, must balance their weight, W_1, and tension from the bottom person, T_2. This means T_1 is greater than T_2, and the top person's legs must be stronger than their arms.
   The forces on the bottom person are their own weight, W_2, pulling them down, and tension from the top person's arms, T_2, pulling them up. As there is no net force on the bottom person, W_2 and T_2 are equal in magnitude. The trapeze bar exerts no force on the bottom person, since it is not in contact with them.
   Another way to think about why T_1 is greater T_2 is that the tension from the trapeze bar, T_1, must support the combined weight of the system. On the other hand, T_2 only needs to support the weight of the bottom person.
   b) $F_{net} = T_2 - W_2 = 0$
   $T_2 = W_2 = mg = 60 \times 9.8 = 588$ N
   c) $F_{net} = T_1 - T_2 - W_2 = 0$
   $T_1 = T_2 + W_2 = 588 + 80 \times 9.8 = 1372$ N
2. a) The forces acting on the system are the combined weight of the acrobats, pulling down, and normal force from the ground, pushing them up. These forces are balanced if they are stationary.
   The force in the bottom acrobat's legs pushing down into the ground is equal in magnitude to the normal force pushing up from the ground, as this is an action-reaction pair.
   The forces acting on the top acrobat are their own weight, pulling down, and normal force from the bottom acrobat's shoulders, pushing up.
   b) Let up be positive
   $F_{net} = N_1 - W_1 = 0$
   $N_1 = W_1 = mg = 60 \times 9.8 = 588$ N

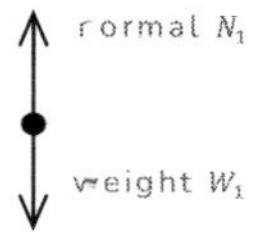

   c) $F_{net} = N - W = 0$
   $N = W = mg = (80 + 60) \times 9.8 =$ 1372 N up

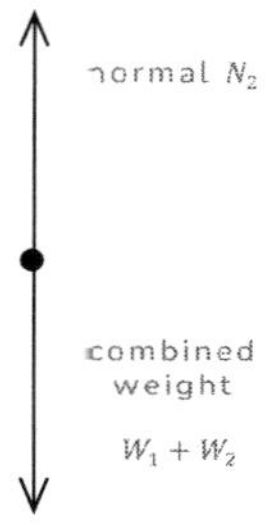

3. a) $v = u + at$
   $9.0 = 0 + a \times 18$
   $a = 0.5$ m s^{-2} forwards
   b) $F_T = ma = 70000 \times 0.5 = 35000$ N forwards
   c) $F_{net} = F_T - T_1 = ma$
   $35000 - T_1 = 10000 \times 0.5$
   $T_1 = 30000$ N
   d) $F_{net} = T_1 - T_2 = ma$
   $30000 - T_2 = 30000 \times 0.5$
   $T_2 = 15000$ N
   e) $T_2 = ma$
   $15000 = 30000 \times a$
   $a = 0.5$ m s^{-2} forwards
4. Let right be positive.
   a) $F_{net} = W_2 - W_1 = ma$
   $2200 \times 9.8 - 1200 \times 9.8 = 3400 \times a$
   $a = 2.88$ m s^{-2} in the direction of the counterweight
   b) $F_{net} = T - W_1 = m_1 a$
   $T - 1200 \times 9.8 = 1200 \times a$
   $T = 15200$ N
   c) $F_{net} = W_2 - T = m_2 a$
   $T = 2200 \times 9.8 - 2200 \times a = 15200$ N
5. 0 to 2 s: The elevator is stationary.
   The scales read 50 kg, as the normal force read by the scales will be the same

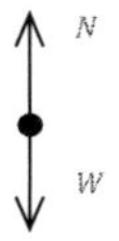

magnitude as weight. This is because the forces on the passenger are balanced.

2 to 9 s: The acceleration of the elevator is given by the gradient: $a = 1$ m s^{-2} upwards.

The net force on the passenger is:

$F_{net} = ma = 50 \times 1 = 50$ N up

The normal force on the passenger is given by:

$F_{net} = N - W = 50$

$N = 50 \times 9.8 + 50 = 540$ N up

The reading on the scales is $540 \div 9.8 = 55$ kg

9 to 12 s: The elevator moves up at a constant velocity of 7 m s^{-2}. The scales read 50 kg, again because forces are balanced and normal force is the same magnitude as weight.

12 to 19 s: The acceleration of the elevator is given by the gradient: $a = -1$ m s^{-2}. So it is going up but slowing down.

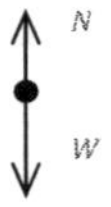

The net force on the passenger is:

$F_{net} = ma = 50 \times -1 = -50$ N (50 N down)

The normal force on the passenger is given by:

$F_{net} = N - W = -50$

$N = 50 \times 9.8 - 50 = 440$ N up

The reading on the scales is $440 \div 9.8 = 45$ kg

5.6 Harder force problems – Set 2

1. Let up be positive.

   $F_{net} = ma = T - W$

   $30 \times 0.25 = T - 30 \times 9.8$

   $T = 301.5$ N up

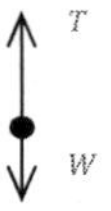

2. a) Let the weight of the child by W_1 and older sibling be W_2. Let the direction the older sibling descends be positive.

   $F_{net} = W_2 - W_1$

   $90 \times a = 50 \times 9.8 - 40 \times 9.8$

   $a = 1.09$ m s^{-2}

   b) $F_{net} = T - W_1$

   $40 \times 1.09 = T - 40 \times 9.8$

   $T = 436$ N up

3. a) $v = u + at$

   $97 = 0 + a \times 8.5$

   $a = 11.4$ m s^{-2} forwards

   b) Force from engine doesn't change:

   $F = ma = 2220 \times 11.4 = 25308$ N

   Acceleration with trailer: $F = ma$

   $25308 = (2220 + 1000) \times a$

   $a = 7.86$ m s^{-2} forwards

   c) Car:

   Trailer:

   d) The net force on the trailer is tension:

   $$T = ma = 1000 \times 7.86 = 7860 \text{ N}$$

4. a) $F = ma = 3 \times 32 \times 1 = 96$ N forwards

   b) Front trolley 1:

   Middle trolley 2:

   Back trolley 3:

   c) Trolley 1: $F_{net} = N_{2\ on\ 1} = ma = 32 \times 1 = 32$ N

   $N_{1\ on\ 2} = N_{2\ on\ 1} = 32$ N backwards, as this is an action-reaction pair

   Trolley 3: $F_{net} = N_{3\ on\ 2} - N_{1\ on\ 2} = ma$

   $N_{3\ on\ 2} - 32 = 32 \times 1$

   $N_{3\ on\ 2} = 64$ N forwards

5. Let down be positive.

   a) $F_{net} = ma = 50 \times 0.5 = 25$ N down

   b) $F_{net} = W - T$

   $120 \times 0.5 = 120 \times 9.8 - T$

   $T = 1116$ N up

   c) $F_{net} = W - N$

   $50 \times 0.5 = 50 \times 9.8 - N$

   $N = 465$ N up

6. Let moving to the right be positive.

   a) The net force on the system is provided solely by the 2 kg mass:

   $F_{net} = ma = W_{2kg}$

   $5 \times a = 2 \times 9.8$

   $a = 3.9$ m s^{-2} to the right

   b) The tension is the only unbalanced force on the 3 kg mass:

   $T = ma = 3 \times 3.9 = 12$ N

7. a)

Just after ignition, but before the rocket leaves the ground	After ignition, when the rocket starts accelerating upward	When the rocket is in the air, just after the first stage has run out of fuel and before the second stage is ignited
 (F_T can be smaller with normal force to balance W)		

   b) Net force on the rocket is given by:

   $$F_{net} = F_T - mg = ma$$

   Acceleration of the rocket is given by:

   $$a = \frac{F_T - mg}{m}$$

   Since thrust is constant, this equation shows that as the fuel is burned, total mass m decreases and acceleration a increases upwards. When fuel runs out, there is no thrust, so acceleration is just due to gravity downwards.

   c) While fuel is burned, as acceleration increases, velocity increases at an increasing rate. When fuel is used up and there is no thrust, the only acceleration is due to gravity, so velocity decreases until the rocket slows to a momentary stop, before changing direction to fall back towards Earth. The slope of the graph is equal to acceleration due to gravity.

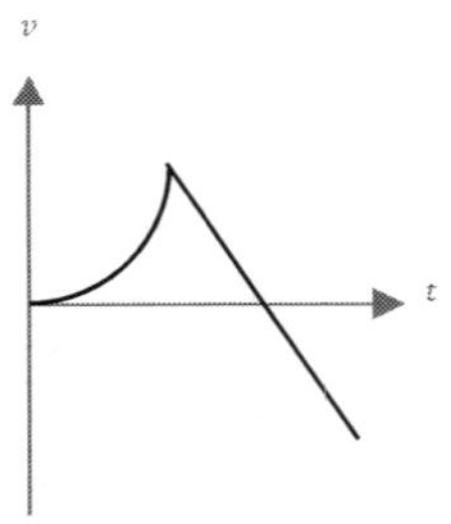

   d) $F = ma = N - W$

   $80 \times 49 = N - 80 \times 9.8$

   $N = 4700$ N

5.7 Linear regression

1. The universe appears to follow mathematical rules, which science tries to discover.
   Experiments can give us qualitative conclusions like "as force on an object increases, its acceleration increases".
   However, if we can also obtain an equation for the trend from the results of the experiment, we can use it to make quantitative predictions.
   Some real world trends are ***linear*** and their equation takes the form of:
   $$y = cx + b$$
   In words, this equation means that y is ***proportional*** to x.
   Graphing this equation gives a straight line, where c is the gradient and b is the y-intercept.
   An example of a linear trend is obtained when changing the net force F on an object with mass m to see how it affects the acceleration a produced:
   $$F = ma$$
   The graph for this is shown to the right, with the axes following the usual convention of putting the independent variable, net force F, on the x-axis and the dependent variable, acceleration a, on the y-axis.
   The graph shows that as the force on an object increases, its acceleration increases proportionally.
   Rearranging the equation $F = ma$ gives:
   $$a = \frac{1}{m} \times F$$
   This is in the form of $y = cx + b$:

y	$=$	c	x	$+$	b
a	$=$	$\frac{1}{m}$	F	$+$	0

   Now we can see that the gradient of the graph should be $\frac{1}{m}$, or the inverse of mass. The line of best fit also passes through the origin, since the y-intercept is zero.
   To make it easier to see the relationship with mass, we can switch the axes and let go of the convention of putting the independent variable on the x axis, as shown to the right. Graphing this would give the gradient as mass.
   Note that when comparing graphs where the axes are switched, the gradient of one will be the inverse of the other.
2. An example of a relationship that is non-linear is a quadratic relationship, which takes the form:
   $$y = ax^2 + bx + c$$
   where a, b and c are constants. For example, the relationship between the displacement of an object undergoing uniform acceleration and time is given by:
   $$s = ut + \frac{1}{2}at^2$$
   If the object starts at rest, this becomes $s = \frac{1}{2}at^2$, so graphing s on the y-axis and t on the x-axis gives the graph on the right. It shows that as time increases, displacement increases at an increasing rate.
   If a trend looks like this, it is hard to tell whether the relationship is in the form of $y = x^2$ or $y = x^3$ or some other power as they have a similar shape. It also hard to tell what constants there are in the equation.
   To determine an equation for the relationship, one technique used is a ***linear regression***. This involves choosing what to graph on each axis to obtain a linear relationship. Something calculated from the independent variable goes on one axis, while something calculated from the dependent variable goes on the other, rather than the variables on their own.
   For example, if the displacement s of an object dropped from rest as time t increases is measured, to verify that $s = \frac{1}{2}at^2$ actually describes the trend, we linearise the equation by writing it in the form $y = cx + b$:

y	$=$	c	x	$+$	b
s	$=$	$\frac{1}{2}a$	t^2	$+$	0

   If we graph s on the y-axis and t^2 on the x-axis, and the results are consistent with $s = \frac{1}{2}at^2$, then the graph will be a straight line with the gradient equal to $\frac{1}{2}a$. Since a dropped object accelerates due to gravity, theoretically $a = 9.8$ m s^{-2} so the gradient should be 4.9 m s^{-2}.
   Note that since t^2 is graphed on the x-axis, the unit is also squared, to give seconds squared (s^2).
3. a)

y	$=$	c	x	$+$	b
m	$=$	D	V	$+$	0

   b)

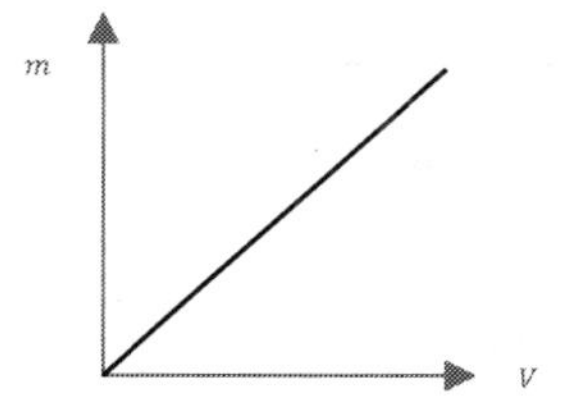

   c) D
4. a)

y	$=$	c	x	$+$	b
A	$=$	π	r^2	$+$	0

   b) r^2, the unit is m^2

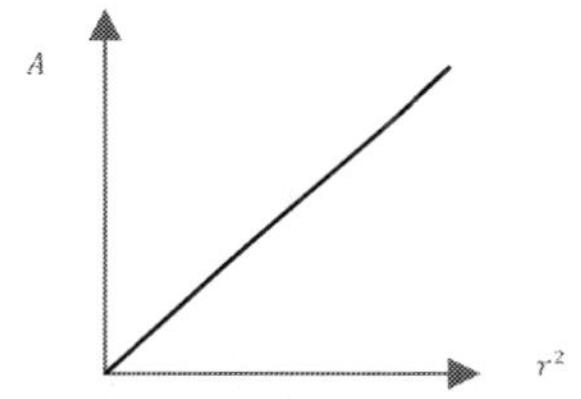

   c) π
5. a)

y	$=$	c	x	$+$	b
v	$=$	a	t	$+$	u

   b)

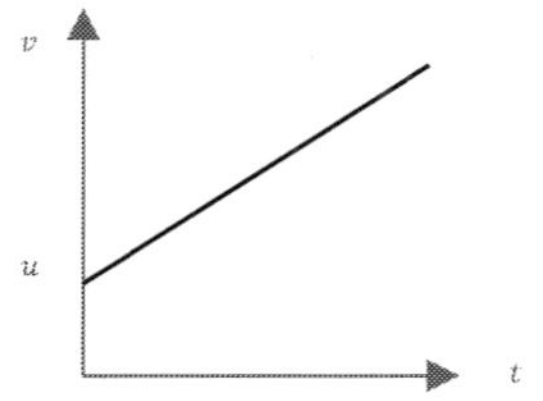

   c) The gradient is acceleration, a, and y-intercept is initial velocity, u.

6. a) $v^2 = u^2 + 2as$
   b) If the hypothesis is correct, then graphing v^2 on the y-axis and s on the x-axis should give a straight line graph, where the gradient is $2a$ and the y-intercept is u^2.

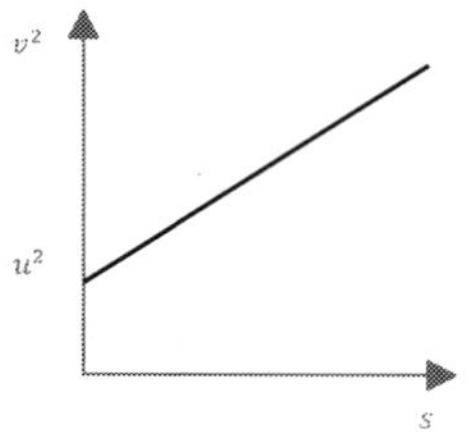

7. The equation can be rearranged as:

$$T^2 = \frac{4\pi^2}{g}L$$

If the data is consistent with the equation, then graphing T^2 on the y-axis and L on the x-axis should give a straight line graph, where the gradient is $\frac{4\pi^2}{g}$.

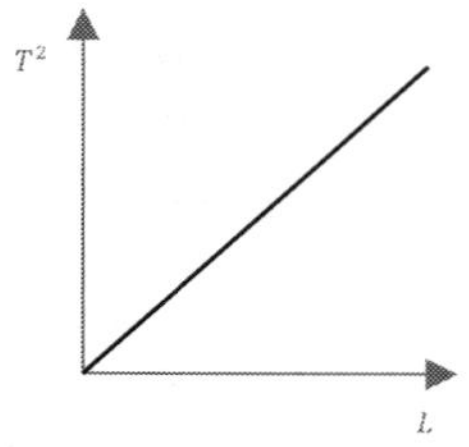

(Note: Graphing T vs $\sqrt{L}$ would also be correct, with a gradient of $\frac{2\pi}{\sqrt{g}}$.)

5.8 Experimental skills – Forces

1. Force acting on the system OR mass hanging from pulley
2. Acceleration of the cart OR system
3. Total mass of the system, same bench, same cart, same path travelled by cart, initial speed of cart
4. To investigate the effect of increasing the force acting on a system on the acceleration of the system.
5. As the total hanging mass pulling the system increases, the acceleration will increase. This is because $F = ma$ predicts that as the force F increases, acceleration a increases, if the total mass m being accelerated is constant.
6. a) An accelerometer is used to measure acceleration.
   b) Three trials are taken of each measurement.
   c) The controlled variables listed in question 3 are kept the same. Six values of hanging mass are tested, to see if there is a consistent trend.
7. Although controlling everything else apart from the force acting on the system is good practice, the length of the string and the distance the trolley is pulled back would not affect the acceleration, as distance is not a variable in $F = ma$. These distances only affect the final velocity the trolley reaches the end of the bench with, assuming that the string is not so long that the hanging mass ends up on the floor.
8. a) The net force is provided by the weight of the hanging mass carrier and the masses on it, $F_{net} = m_2 g$, if friction is ignored. All other forces acting on the system are balanced.
   b)

m_2 (kg)	a (ms^{-2})	F_{net} (N)
0.05	0.61	0.49
0.1	1.23	0.98
0.15	1.84	1.47
0.2	2.45	1.96
0.25	3.06	2.45
0.3	3.68	2.94

   c)

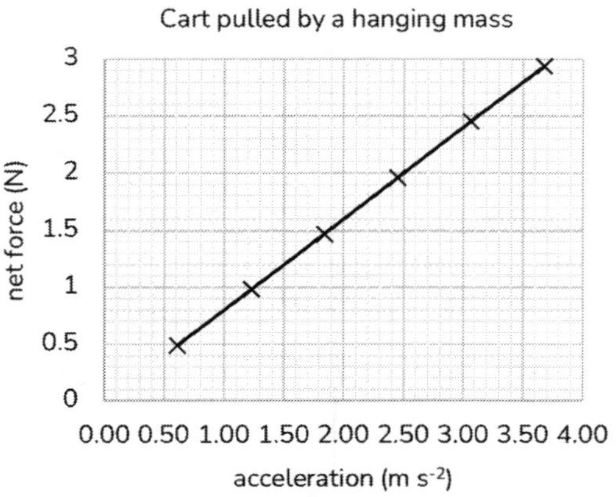

   d) Gradient = 0.8 kg. This is the total mass of the system, since $F_{net} = (m_1 + m_2)a$

6.2 Friction

1. Friction is a force which resists movement between two surfaces. Friction always opposes the direction of motion and is parallel to the surface.
   The magnitude of friction on an object moving on a surface depends on the coefficient of friction and the normal force between the object and surface.
   Normal force is the force which is perpendicular to the surface in contact with the object, and is the reaction force to the object pushing into the surface. If an object is sitting on a horizontal surface and is not pressed or lifted, the magnitude of the normal force is equal to its weight.
   The coefficient of friction μ depends on the roughness of the surfaces in contact. If a surface is slippery, the coefficient of friction will decrease. If a surface is rougher, the coefficient of friction will increase. The coefficient of friction also depends on the electrostatic forces between the surfaces.
2.

Symbol	Meaning	Unit
$\vec{F_f}$	frictional force	N
μ	coefficient of friction	none
$\vec{N}$	normal force	N

3. The coefficient of static friction μ_s applies when the rubber is not yet sliding on the concrete. This is higher than coefficient of kinetic friction μ_k, as less force is required to keep rubber sliding once it is already moving. Both coefficients are lower on wet concrete compared to dry, since the floor is more slippery. Due to this, floors should be kept dry to prevent people slipping.
4. Frictional force is given by $F_f = \mu N$. The normal force is equal to the component of weight pressing perpendicularly into the slippery dip. As the adult weighs more, there is more normal force, so the adult experiences more friction.
5. a) $F_f = \mu_s N = 0.45 \times 0.5 \times 9.8 = 2.2$ N forwards
   b) $F_f = \mu_k N = 0.38 \times 0.5 \times 9.8 = 1.9$ N forwards
6. a)

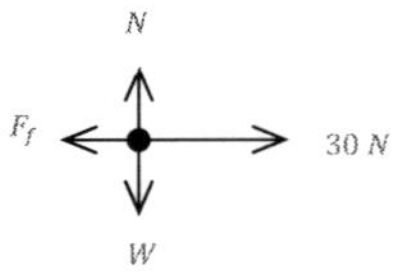

   b) $F_f = \mu N = \mu mg = 0.3 \times 5 \times 9.8 = 14.7$ N to the left

c) $F_{net} = ma = 30 - F_f$

$5 \times a = 30 - 14.7$

$a = 3.1$ m s^{-2} to the right

7. $F_{net} = ma = 75 - \mu_s N$

$0 = 75 - \mu_s \times 40 \times 9.8$

$\mu_s = 0.19$

$F_{net} = ma = 60 - \mu_k N$

$0 = 60 - \mu_k \times 40 \times 9.8$

$\mu_s = 0.15$

8. $F_f = ma = \mu_k N$

$80 \times a = 0.03 \times 80 \times 9.8$

$a = 0.294$ m s^{-2} backwards

Let the forward direction be positive.

$v^2 = u^2 + 2as$

$0 = 5^2 + 2 \times -0.294 \times s$

$s = 42.5$ m

9. $F_{net} = 0 = F_f - mg$

$F_f = \mu N = mg$

$0.7 \times N = 60 \times 9.8$

$N = 840$ N into the wall

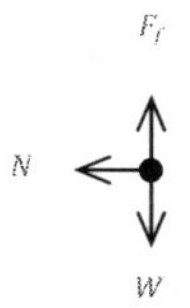

10. $s = ut + \frac{1}{2}at^2$

$4.5 = 0 + \frac{1}{2}a \times 1.5^2$

$a = 4$ m s^{-2} down

$F_{net} = ma = mg - \mu N$

$45 \times 4 = 45 \times 9.8 - 0.46 \times N$

$N = 567$ N perpendicular to the pole's surface

6.3 Experimental skills – Friction

1. Total mass of the box OR normal force on the box
2. Force needed to move the box at constant velocity
3. Pulling velocity, surface the box is moving on, same box
4. As the total mass of the box and contents increases, the force needed to keep the box moving at a constant velocity increases proportionally. This is because as weight increases, the normal force increases, so the friction that needs to be overcome by the pulling force increases, according to $\vec{F_f} = \mu\vec{N}$.
5. 1 Set up as shown:

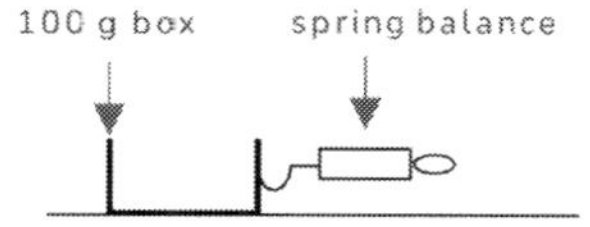

2 Drag the spring balance horizontally so that the box just starts to move across the bench top. Record the force required on the spring balance.

3 Repeat step 2 two more times.

4 Repeat steps 1 to 3, after adding 100 g, 200 g, 300 g, and 400 g in the box respectively.

6. a)

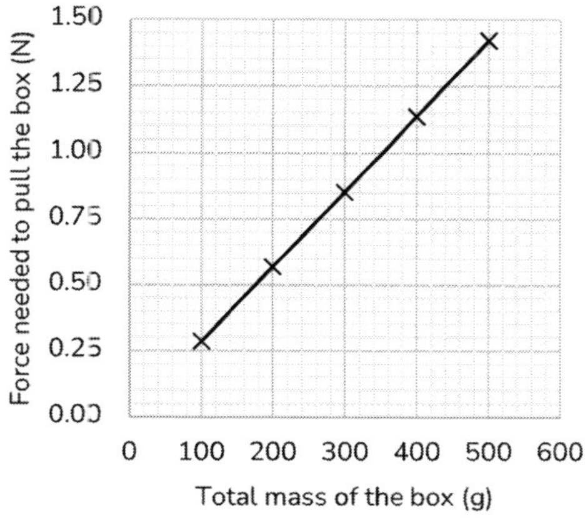

b) The force needed to pull the box is equal in magnitude to the frictional force F_f on the box:

$F_f = \mu N = \mu mg = \mu \times 9.8 \times$ mass in grams $\div 1000$

As F_f is plotted on the y-axis and mass in grams on the x-axis, the gradient $= \mu \times 0.0098 = 0.0028$ from the graph. So the coefficient of friction μ is 0.29.

6.5 Air resistance

1. When a skydiver jumps out of the plane, they accelerate downwards at 9.8 m s^{-2}. As their velocity increases, air resistance increases, so acceleration decreases. When acceleration is zero, and forces are balanced, the skydiver has reached terminal velocity.

   Opening a parachute at this time will increase air resistance so that it is greater than gravitational force. This means net force is upwards, so acceleration is upwards. The velocity is still downwards, but it decreases.

   The decrease in velocity means that air resistance decreases, until it is the same magnitude as weight. At this point, the net force on the skydiver is zero and they reach a new, slower terminal velocity.

2.

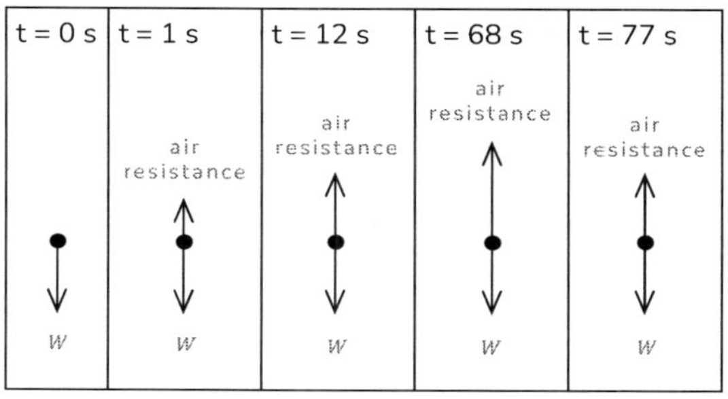

3.

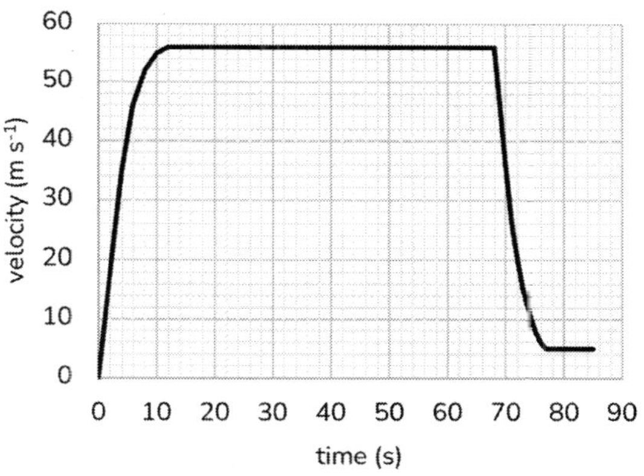

4. a) When just stepping out of the plane
   b) After the parachute opens
   c) When terminal velocity is reached
5. a) False – acceleration is not constant.
   b) True
   c) False – the area gives the displacement downwards from the plane.
   d) False – air resistance is bigger than weight when the parachute opens, to decelerate the skydiver.
   e) False – there is no net force at terminal velocity. Even if terminal velocity were not reached, gravitational force on the skydiver does not change.

Part 2: Dynamics - Answers

7.2 Forces in 2D

1. A vector is a quantity that has both magnitude and direction, such as force, acceleration, velocity or displacement. Vectors are represented by arrows. When several forces act on an object, the net force is the vector sum of all of these forces. Another term for net force is resultant force. To add vectors, you draw them head to tail. You can arrange the vector arrows any way to do this as long as you don't change their magnitude and direction. The net force is then drawn from the "tail of the tails" to the "head of the heads", creating a closed polygon. Rules of geometry, such as Pythagoras' theorem and trigonometry, can then be used to solve for the magnitude and direction of vectors.
If the forces are in equilibrium, then the net force is zero. This means forces are balanced and the object is either not moving or moving at constant velocity. If the net force is not zero, then the object will accelerate in the same direction as the net force. The net force is equal to mass times acceleration.

2. a) $F_{net} = \sqrt{5000^2 + 10000^2}$
$= 11180$ N
$\tan\theta = \dfrac{5000}{10000}$
$\theta = 27°$
The net force exerted is 11180 N S27°E.

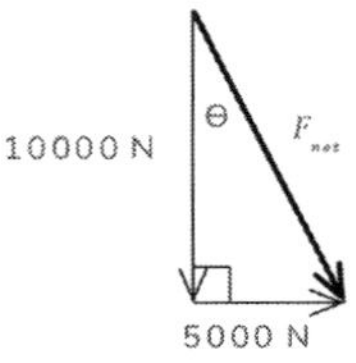

b) $F = ma$
$a = \frac{F}{m} = \frac{11180}{1000} = 11.18$ m s^{-2} S27°E

3. $\cos 45° = \frac{F_{tug}}{25000}$
$F_{tug} = 25000\cos 45° = 18000$ N

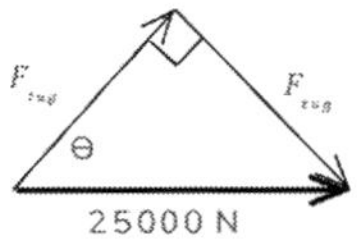

4.
a) $\tan 35° = \frac{F}{W} = \frac{F}{mg}$
$F = mg\tan 35°$
$= 28 \times 9.8 \times \tan 35° = 190$ N
The parent pulls the child horizontally with 190 N.
b) $\cos 35° = \frac{W}{T} = \frac{mg}{T}$
$T = \dfrac{mg}{\cos 35°} = \dfrac{28 \times 9.8}{\cos 35°} = 340$ N
The tension in each rope is 170 N at 35° to the vertical.

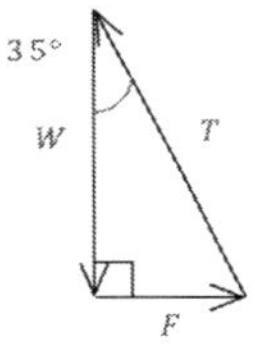

5. $\cos 45° = \frac{T}{W}$
$T = mg\cos 45° = 40 \times 9.8 \times \cos 45°$
$= 280$ N
The tension in each arm is 280 N at 45° to the horizontal bars.

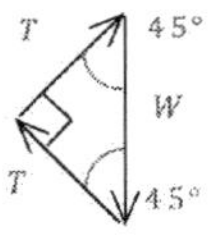

6. $\cos 45° = \frac{T}{W}$
$T = mg\cos 45° = 60 \times 9.8 \times \cos 45°$
$= 420$ N
The tension in the top arm is 420 N and the push in the bottom arm is 420 N at 45° to the vertical.

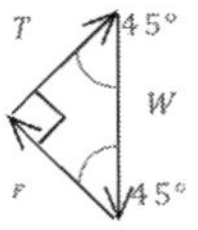

7. $\dfrac{T}{\sin 30°} = \dfrac{25000}{\sin 120°}$
$T = 14000$ N
The tension in each rope is 14000 N.

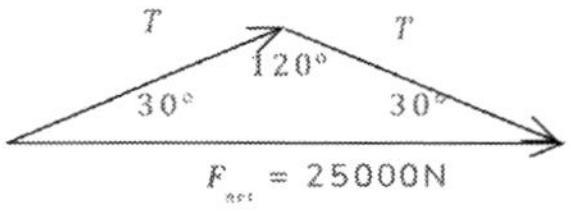

8. a) $\dfrac{F}{\sin 80°} = \dfrac{65\times9.8}{\sin 20°}$
$F = 1800$ N
The force in each arm is 1800 N at 80° to their torso.

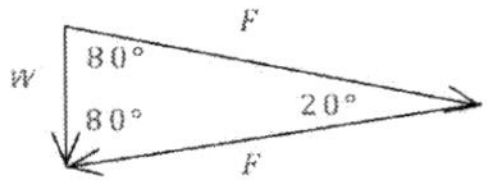

b) $W = F + F = 65 \times 9.8$
$F = 320$ N
The force in each arm is 320 N up in the support position, compared to the force of 1800 N in the iron cross, which nearly 6 times bigger.
c) As the angle of the arms approaches horizontal, the force in the arms would need to keep increasing to an infinite force, as the vertical component reduces to zero.

9. $\dfrac{T}{\sin 20°} = \dfrac{W+F_{net}}{\sin 140°} = \dfrac{35\times9.8+35\times2}{\sin 140°}$
$T = 220$ N
The tension in each bungee cord is 220 N at 20° to the vertical.

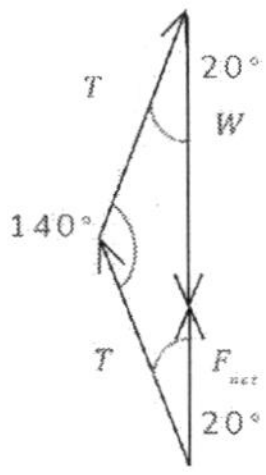

7.3 Graphical method of adding vectors

1. Weight is 170 N so the mass of the pot plant is 17 kg.

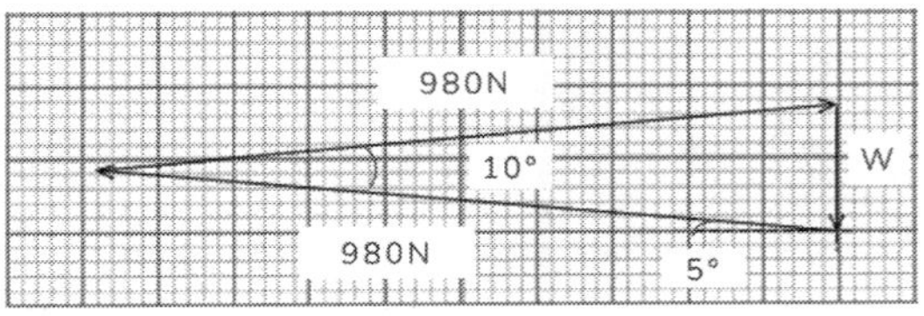

2.

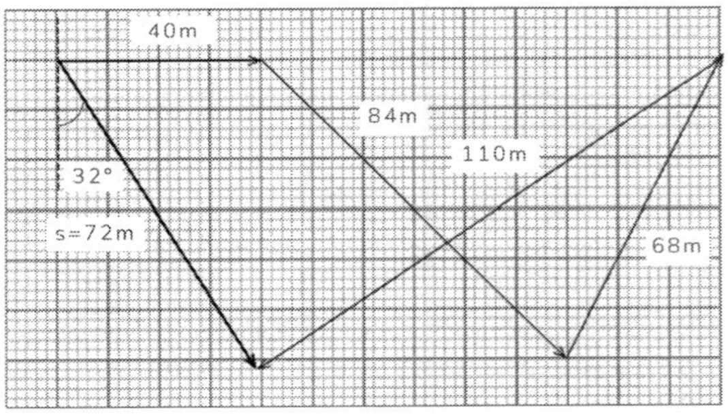

a) 72 m S32°E
b) $distance = 40 + 84 + 68 + 110 = 302\ m$
$average\ speed = \dfrac{302}{25} = 12\ m\ s^{-1}$
c) $v = \frac{72}{25} = 2.9\ m\ s^{-1}$

3. The tension in the bowstring is 75 N.

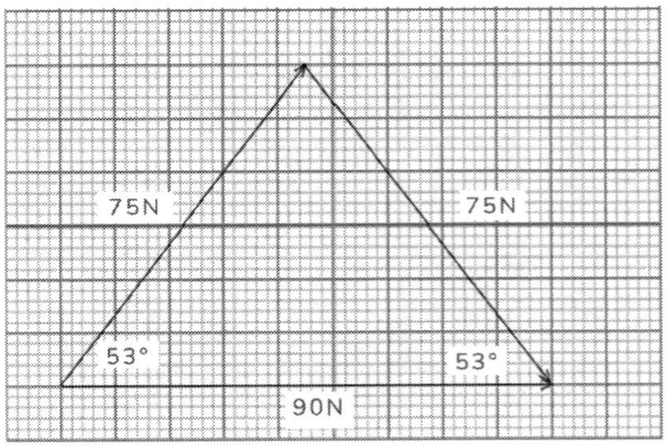

7.4 Subtracting vectors

1.

a)

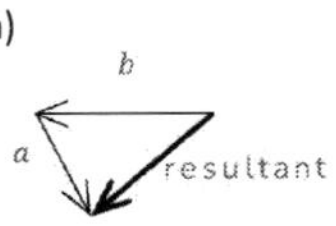

b)

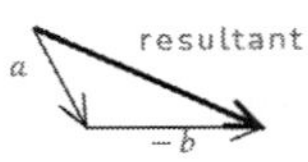

c)

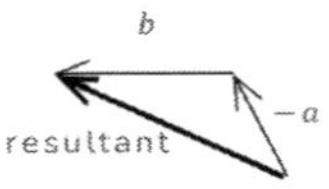

2.

a)

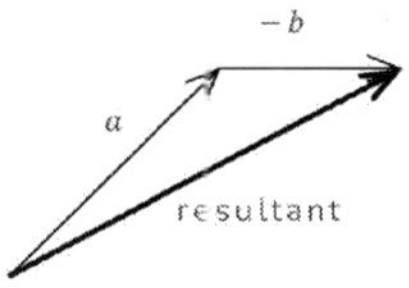

b)

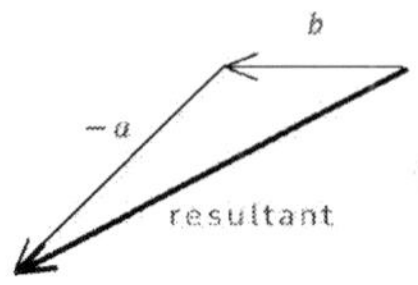

3. a) $\Delta\vec{v} = \vec{v} - \vec{u} = 2.0$ m s^{-1}

b) $\vec{a} = \frac{\Delta\vec{v}}{t} = \frac{2.0}{0.5} = 4.0$ m s^{-2} up

c) tension in the string

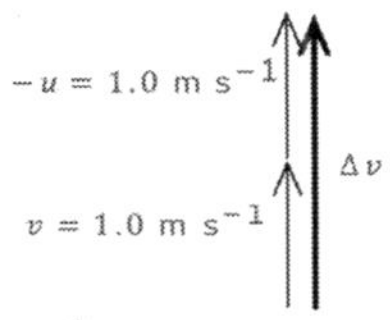

4. a) $\Delta v = \sqrt{7.63^2 + 7.63^2}$

$\vec{a} = \frac{\Delta\vec{v}}{t}$

$= \frac{\sqrt{7.63^2 + 7.63^2}}{0.25}$

$= 43.2$ m s^{-2}

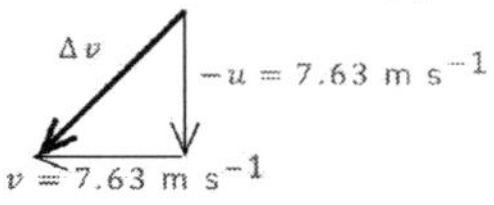

b) $F = ma = 7.26 \times 43.2 = 313$ N

c) tension in the wire

7.5 Resolving vectors

1. Resolving a vector into components means breaking it up into two parts that are perpendicular to each other. It is like working out the x and y coordinates of a point on a number plane. If you use Pythagoras' theorem to calculate $\sqrt{x^2 + y^2}$, you get the distance from the origin to the point. Similarly, using Pythagoras' theorem on the components of a vector gives the magnitude of the vector.
To find the direction of a point on a number plane, you can use trigonometry to find an angle. Similarly, if you know the components, you can use trigonometry to find the direction of a vector.
Resolving a vector is useful when there is something that only affects one component of the vector. For example, if you throw a ball forwards, gravity will pull the ball downwards. However, gravity will not stop the ball moving horizontally, since this component of the ball's velocity is at right angles to gravitational force. Gravity will only change the vertical velocity of the ball.

2. a) Let the height of the cliff be s_y.

$\tan 20° = \frac{s_y}{96}$

$s_y = 96 \tan 20° = 35$ m

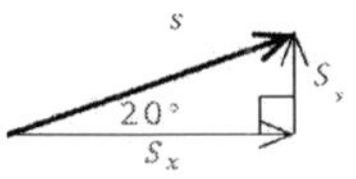

b) Let the length of the zipline be s.

$\cos 20° = \frac{96}{s}$

$s = \frac{96}{\cos 20°} = 102$ m

3. a) Let the horizontal component of force be F_x and vertical component be F_y.

$\cos 30° = \frac{F_x}{200}$

$F_x = 200 \cos 30° = 173$ N forward

$\sin 30° = \frac{F_y}{200}$

$F_y = 200 \sin 30° = 100$ N down

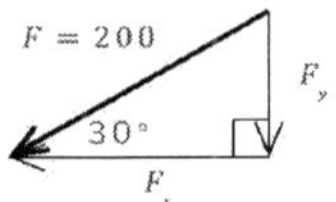

b) $F_x = ma$

$173 = 60a$

$a = 2.88$ m s^{-2} forward

c) $N = W + F_y = 60 \times 9.8 + 100 = 688$ N up

4. $\cos 68° = \frac{v_N}{v}$

$v_N = v \cos 68°$

$= 40 \cos 68° = 15$ km h^{-1}

$v_N = \frac{s_N}{t}$

$s_N = v_N t = 15 \times \frac{10}{60} = 2.5$ km

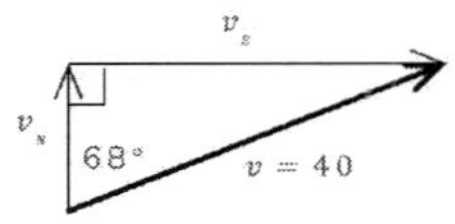

5. a) $s_y = ut + \frac{1}{2}at^2$

$= 0 + \frac{1}{2} \times 9.8 \times 3.346^2 = 54.86$ m

$\tan\theta = \frac{3.836}{54.86}$

$\theta = 4.000°$

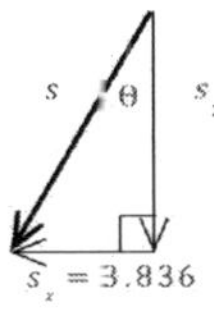

b) $\sin\theta = \frac{3.836}{s}$

$s = \frac{3.836}{\sin 4.000} = 54.99$ m

6. The hiker would be correct if there was no elevation, as they have measured the horizontal component of the displacement:

$v = \frac{s}{t}$

$t = \frac{s}{v} = \frac{1000}{4 \div 3.6} = 900$ s

$= 15$ min

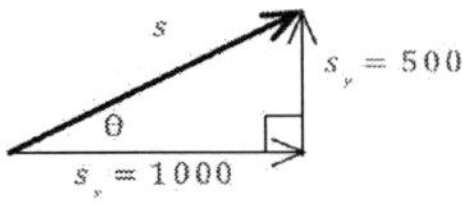

However, the hiker has not considered the vertical component of displacement of 500m, giving the total displacement of:

$s = \sqrt{1000^2 + 500^2} = 1118$ m

This would give a total time of:

$t = \frac{s}{v} = \frac{1118}{4 \div 3.6} = 1006.2$ s $= 17$ min

assuming they could maintain a speed of 4 km h^{-1} uphill. Therefore the hiker's calculation is invalid.

7. $F_x = 882 \cos 45° = 624$ N

$F_x = ma = m\frac{v_x - u}{t}$

$624 = 65 \times \frac{v_x - 0}{0.35}$

$v_x = 3.36$ m s^{-1}

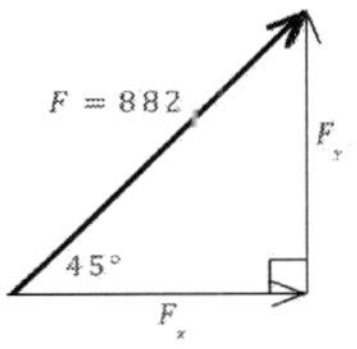

8. With starting blocks, the normal force on the runner from the blocks is perpendicular to the blocks, so the horizontal force on the runner is the horizontal component of normal force.

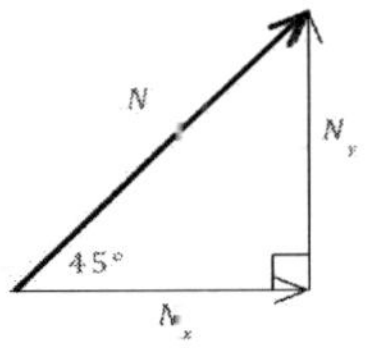

Without starting blocks, the normal force on the runner is vertically up, perpendicular to the ground, so horizontal force moving the runner forward is provided by friction only, $F_f = \mu N$. A successful start would therefore rely on the runner not slipping.

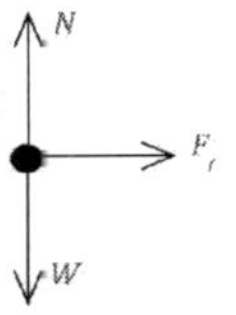

9. When pulling the nozzle, the vertical component of the force you exert upwards reduces the normal force on the nozzle. So friction is lower, since $F_f = \mu N$.

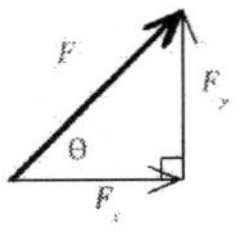

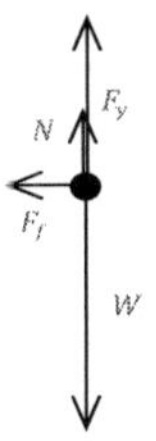

Pushing the nozzle means the vertical component of force you exert is downwards, increasing normal force. So you need to push against a larger frictional force than pulling.

10. In the vertical direction, net force is zero.

$F \sin 30° + N = W$

$N = mg - F \sin 30°$ ...(1)

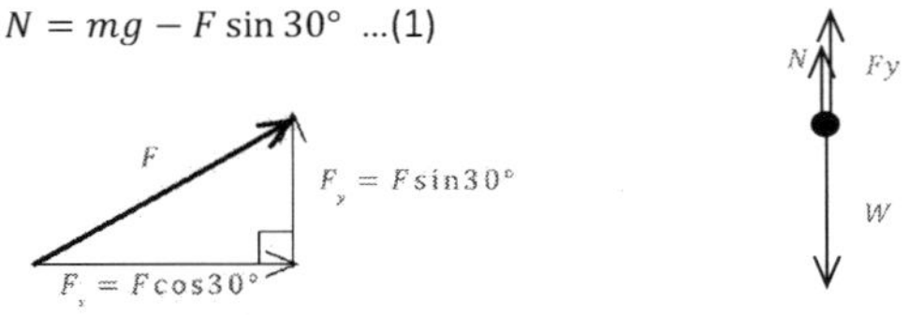

In the horizontal direction, net force is zero.

$F \cos 30° = \mu N = \mu(mg - F \sin 30°)$ (sub in (1))

$F \cos 30° + \mu F \sin 30° = \mu mg$

$F(\cos 30° + \mu \sin 30°) = \mu mg$

$F = \dfrac{\mu mg}{\cos 30° + \mu \sin 30°} = \dfrac{0.75 \times 12 \times 9.8}{\cos 30° + 0.75 \times \sin 30°} = 71$ N

7.6 Adding more than two vectors

1. a) Breaking up a vector into two vectors at right angles that add up to give the original vector is called resolving vectors into components.

   b) If any number of vectors are added together, then the vector sum of their horizontal components will equal the horizontal component of the resultant and the vector sum of their vertical components will equal the vertical component of the resultant.

2.

Force	x	y
T	$10.5 \cos 45°$	$-10.5 \sin 45°$
W	0	-0.25×9.8
Lift	0	F_L
Drag	$-F_D$	0
Total	0	0

$-10.5 \sin 45° - 0.25 \times 9.8 + F_L = 0$

$F_L = 9.9$ N

$10.5 \cos 45° - F_D = 0$

$F_D = 7.4$ N

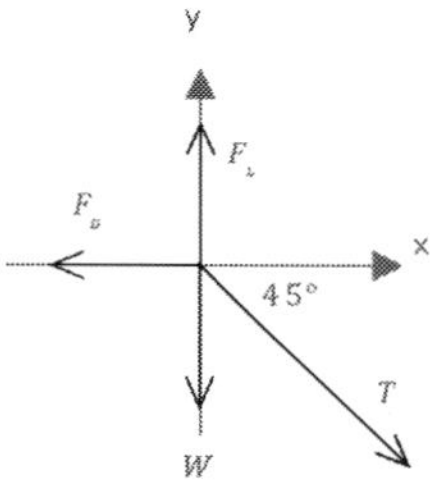

7.7 Inclined planes

1. An inclined plane is a ramp. An object on the ramp will tend to slide down, due to its weight. However, because the object cannot fall vertically through the ramp, its acceleration is restricted to being downwards along the ramp. So resolving weight into components, with one component along the ramp, is a useful way to find the acceleration of the object.
   We can rotate our frame of reference so that the x-axis is along the ramp, in the direction of the net force on the object, and the y-axis is perpendicular to the ramp, in the direction of normal force on the object.
   If the angle the inclined plane makes to the horizontal is θ, then the component of weight along the ramp is $mg \sin \theta$. If there is no other force with a component in the x-direction, such as friction, then the acceleration of the object down the ramp can be found using:

$$F = ma = mg \sin \theta$$
$$a = g \sin \theta$$

   This means that acceleration is independent of the mass of the object.
   The component of weight perpendicular to the ramp is $mg \cos \theta$. If there is no other force with a component in the y-direction, perpendicular to the ramp, then the normal force on the object is:

$$N = mg \cos \theta$$

   This is the reaction force from the object pushing against the ramp. Normal force has a magnitude equal to weight if the ramp is horizontal and zero if the ramp is vertical.

2.

Angle	Magnitude	Reasoning
a	90°	Weight is vertically down
b	60°	Sum of angles in a triangle is 180°
c	60°	Corresponding angles on parallel lines are the same
d	30°	Components of weight are at 90° to each other

3. $F = ma = mg \sin \theta$

   $a = g \sin \theta$

   $\theta = \sin^{-1}\frac{a}{g} = \sin^{-1}\frac{6.8}{9.8} = 44°$ for maximum angle

   $\theta = \sin^{-1}\frac{a}{g} = \sin^{-1}\frac{4.0}{9.8} = 24°$ for minimum angle

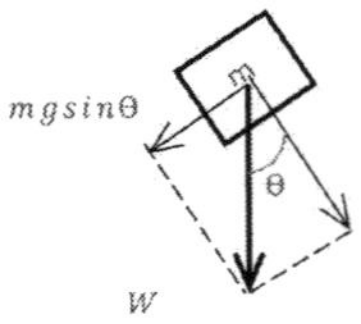

4. The student has correctly identified the direction of two forces on the object, weight and normal force, and shown that the net force is the vector sum of these forces. They have also correctly identified the angle between the normal force and weight. However, normal force is equal to $mg\cos\theta$ not mg, as it is the reaction force to the component of weight perpendicular to the ramp. Net force also needs to be down the ramp, not horizontal, since the object only accelerates along the ramp.

5. Just before the shoe slides, the component of weight down the ramp equals friction up the ramp, so net force is zero.

   $F_{net} = 0 = mg \sin \theta - \mu mg \cos \theta$

   $\mu \cos \theta = \sin \theta$

   $\mu = \frac{\sin \theta}{\cos \theta} = \tan \theta = \tan 14° = 0.25$

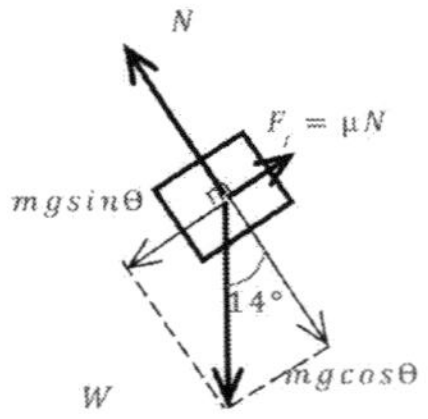

6. a) $s = ut + \frac{1}{2}at^2$

   $10 = 0 + \frac{1}{2}a \times 2.36^2$

   $a = 3.59$ m s^{-2} down ramp

   b) $F_{net} = ma = mg \sin \theta - \mu mg \cos \theta$

   $25 \times 3.59 = 25 \times 9.8 \sin 30° - \mu \times 25 \times 9.8 \cos 30°$

   $\mu = 0.15$

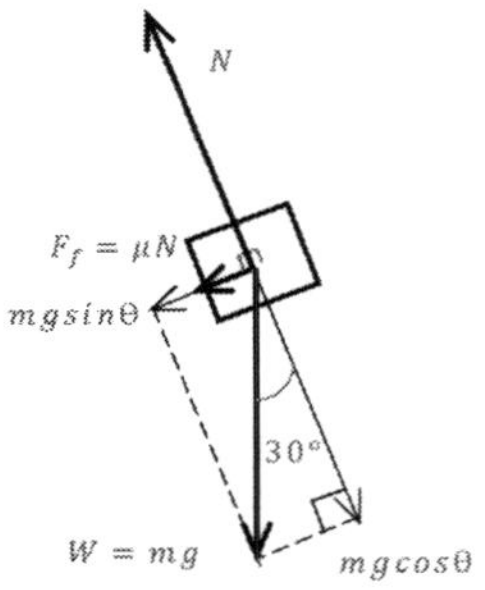

7. a) $F_{net} = ma = mg\sin\theta + \mu mg\cos\theta$

$a = g\sin\theta + \mu g\cos\theta$

$= 9.8\sin 30° + 0.065 \times 9.8\cos 30°$

$= 5.45$ m s^{-2} down the slope

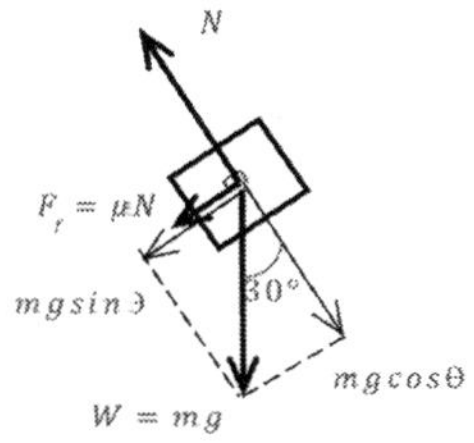

b) Let down the slope be positive.

$v^2 = u^2 + 2as$

$0 = (-15)^2 + 2 \times 5.45s$

$s = -20.6$ m

Skier travels 20.6 m up the slope

7.8 Experimental skills – Inclined plane

1. angle of the ramp
2. reading on the spring balance OR force down the ramp
3. same ramp, same trolley
4. As the angle of the ramp θ increases, the tension on the spring balance T increases according to $T = mg\sin\theta$

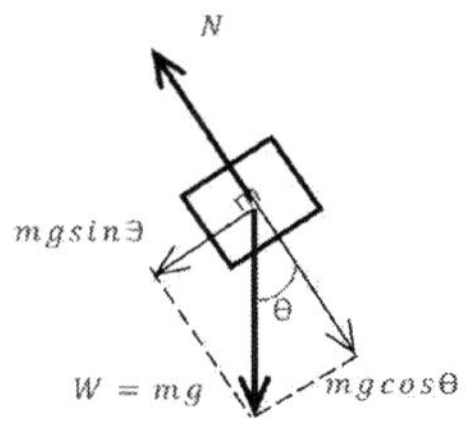

5. a) T on the y-axis, $\sin\theta$ on the x-axis. Gradient is mg.

b)

angle (°)	tension (N)	sinΘ
15	0.2	0.26
30	0.4	0.50
45	0.6	0.71
60	0.7	0.87
75	0.8	0.97

c)

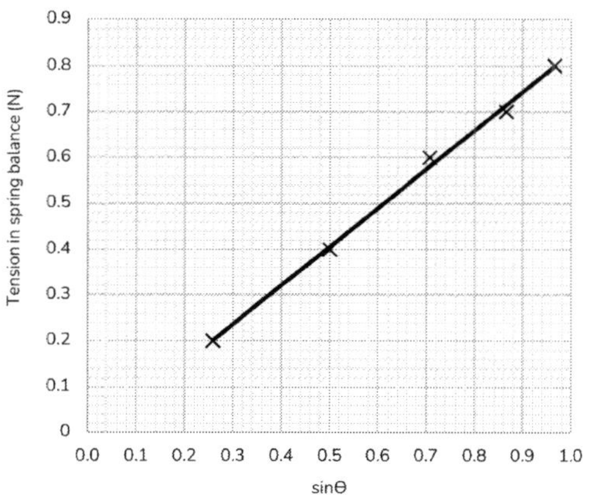

6. Tension is 0.5 N when sinΘ=0.6, giving an angle of 37°.
7. The line of best fit can be extrapolated to sinΘ=1, since this would give T=mg. This gives around T=0.82 N and mass of 84 g.

The gradient of the line of best fit is mg. From the line of best fit, gradient is 0.84, giving a mass of 86 g.

(Answers may vary depending on accuracy of drawing line of best fit.)

8. a) 0.2 ± 0.025 N

b) 0.2 N $\pm 12.5\%$

c) The maximum reading on this spring balance is 2.5 N but the maximum reading in the results is 0.8 N so we can use a smaller scale.

A better spring balance would have a smaller limit of reading, for example a 1 N scale with increments of 0.02 N, giving a limit of reading of 0.01 N.

9. The results were not reliable as only one trial was conducted, so there is no evidence that repeating the same experiment would obtain similar results.
10. The trend shown by the results is that as tension increases sinΘ increases proportionally.

The trend shows that the hypothesis of $T = mg\sin\theta$ is correct.

11. The number of data points measured was five. Compared to the line of best fit, most of these data points were close to it. Therefore, overall the trend was valid. However, validity could be improved by adding more measurements of angles smaller than 30° because there is only one data point for sinΘ≤0.5.

8.2 Relative motion in 1D

1. We are used to thinking about the ground as not moving and being the <u>absolute</u> frame of reference. So if someone is driving a truck at 60 km h^{-1}, we assume this is measured <u>relative</u> to the ground. "Relative to x" means x is the frame of reference of the <u>observer</u>. So "relative to the ground" means as observed by someone at <u>rest</u> on the ground. We can denote the velocity of the truck relative to the ground as $v_{t/g} = 60$ km h^{-1}, choosing "t" to represent "<u>truck</u>" and "g" to represent "<u>ground</u>".

Now an observer on the ground sees the police drive up behind the truck at $v_{p/g} = 70$ km h^{-1}, where "p" represents "<u>police</u>". The observer sees the police driving in the same direction as the <u>truck</u> and catching up to it. However, the police would see the <u>observer</u>, and everything else at rest on the <u>ground</u>, moving backwards at $v_{g/p} = -70$ km h^{-1}, where $v_{g/p}$ is the velocity of the ground relative to the <u>police</u>. In other words, in the frame of reference of the police, the ground is moving <u>backwards</u>. Notice that the velocity of the police relative to the ground is the <u>same</u> magnitude but in an <u>opposite</u> direction to the velocity of the ground relative to the police.

Therefore to determine how fast the truck moves relative to the police $v_{t/p}$, we find the <u>vector</u> sum of the velocity of the truck relative to the <u>ground</u> plus the velocity of the ground relative to the <u>police</u>:

$$\overrightarrow{v_{t/p}} = \overrightarrow{v_{t/g}} + \overrightarrow{v_{g/p}} = 60 - 70 = -10 \text{ km h}^{-1}$$

The resultant is negative, meaning that the police would see the truck moving <u>backwards</u> towards them at 10 km h^{-1}.

On the other hand, to determine how fast the police move relative to the truck $v_{p/t}$, we find the vector sum of the velocity of the <u>police</u> relative to the ground plus the velocity of the ground relative to the <u>truck</u>:

$$\overrightarrow{v_{p/t}} = \overrightarrow{v_{p/g}} + \overrightarrow{v_{g/t}} = 70 - 60 = 10 \text{ km h}^{-1}$$

So the truck driver would see the police moving <u>forwards</u> towards them at 10 km h^{-1}. This is the same <u>magnitude</u> as the velocity of the truck relative to the police, but in the opposite <u>direction</u>.

2. a) $\overrightarrow{v_{w/b}} = \overrightarrow{v_{w/g}} + \overrightarrow{v_{g/b}} = 110 - 90 = 20$ km h^{-1} forwards

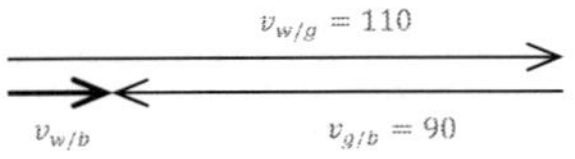

b) $\overrightarrow{v_{b/w}} = \overrightarrow{v_{b/g}} + \overrightarrow{v_{g/w}} = 90 - 110 = -20$ km h^{-1}

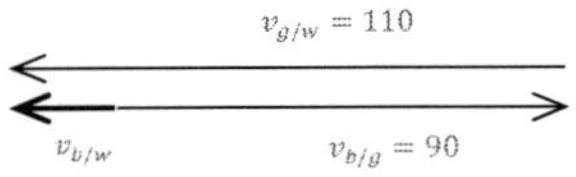

Relative velocity is 20 km h^{-1} backwards.

c) $v = \frac{s}{t}$

$20 \div 3.6 = \frac{100}{t}$

$t = 18$ s

3. a) By doing a run up, the cricketer gives the ball an initial velocity of <u>6.7 m s^{-1}</u>, even if they drop the ball, due to the ball having <u>inertia</u>. The ball's final velocity of 28 m s^{-1} is the <u>resultant</u> velocity of the ball relative to the cricketer plus the velocity of the cricketer relative to the <u>ground</u>.

b) $\overrightarrow{v_{b/g}} = \overrightarrow{v_{b/c}} + \overrightarrow{v_{c/g}}$

$28 = \overrightarrow{v_{b/c}} + 6.7$

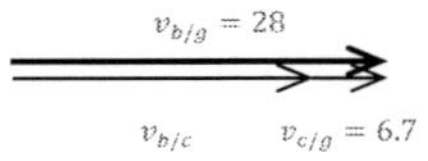

$\overrightarrow{v_{b/c}} = 21.3$ m s^{-1}

4. a) $\overrightarrow{v_{w/c}} = \overrightarrow{v_{w/g}} + \overrightarrow{v_{g/c}} = 5 - 15 = -10$ m s^{-1}

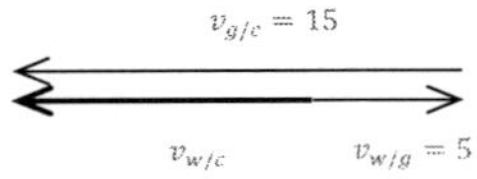

b) $\overrightarrow{v_{w/c}} = \overrightarrow{v_{w/g}} + \overrightarrow{v_{g/c}} = -15 - 5 = -20$ m s^{-1}

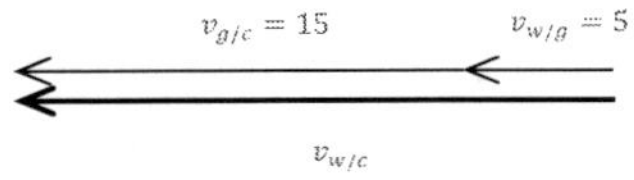

5. Let downstream be positive

a) 2.0 m s^{-1} upstream

b) 2.0 m s^{-1} downstream

c) 1.0 m s^{-1} upstream

d) $\overrightarrow{v_{a/c}} = \overrightarrow{v_{a/r}} + \overrightarrow{v_{r/c}} = -2 + 1 = -1$ m s^{-1} (upstream)

e) $\overrightarrow{v_{b/c}} = \overrightarrow{v_{b/r}} + \overrightarrow{v_{r/c}} = 2 + 1 = 3$ m s^{-1} (downstream)

f) $\overrightarrow{v_{a/b}} = \overrightarrow{v_{a/r}} + \overrightarrow{v_{r/b}} = -2 - 2 = -4$ m s^{-1} (upstream)

g) In still water, Ash swims 60 m in total, at 2.0 m s^{-1}:

$v = \frac{d}{t}$

$2 = \frac{60}{t}$

$t = 30$ s

In the river, Ash swims upstream 30 m at 1.0 m s^{-1}:

$v = \frac{d}{t}$

$1 = \frac{30}{t}$

$t = 30$ s

Going downstream, Ash swims 30 m at 3.0 m s^{-1}:

$v = \frac{d}{t}$

$3 = \frac{30}{t}$

$t = 10$ s

Ash takes 40 s to swim in the river, which is 10 s longer than in still water.

6. a) Using the second diagram with the signpost:
$v_{d/g} = \frac{s}{t} = 0.05 \div \frac{1}{24} = 1.2$ m s^{-1} eastward

b) Using the first diagram with the actor:
$v_{a/d} = \frac{s}{t} = 0.025 \div \frac{1}{24} = 0.6$ m s^{-1} eastward

$\overrightarrow{v_{a/g}} = \overrightarrow{v_{a/d}} + \overrightarrow{v_{d/g}} = 0.6 + 1.2 = 1.8$ m s^{-1} eastward

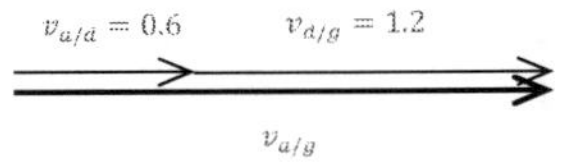

7. a) $v_{c/y} = -10$ m s^{-1} (north)

b) $t = 2$ s

c) $\overrightarrow{v_{c/y}} = \overrightarrow{v_{c/g}} + \overrightarrow{v_{g/y}} = 10 - 15 = -5$ m s^{-1} (south)

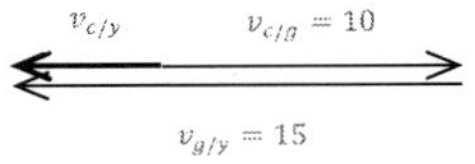

d) Acceleration of you relative to the ground is given by the gradient of the graph:

$a_{y/g} = \frac{\Delta v}{\Delta t} = \frac{15}{3} = 5$ m s^{-2} (north)

so $a_{g/y} = -5$ m s^{-2} (south)

The cyclist is not accelerating relative to the ground, so they have the same acceleration as the ground relative to you of 5 ms^{-2} south.

e)

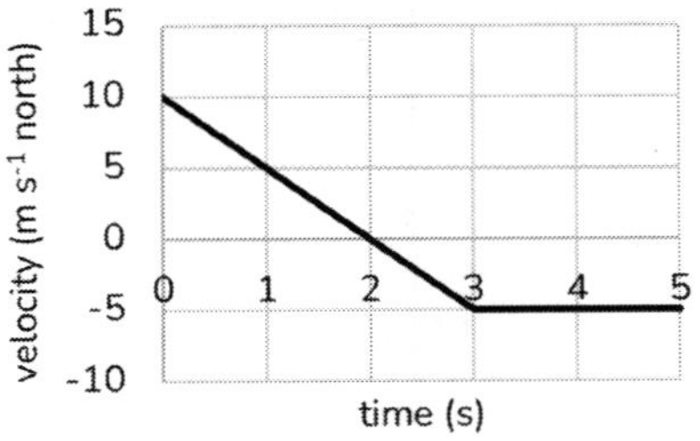

8.3 Relative motion in 2D

1. a)

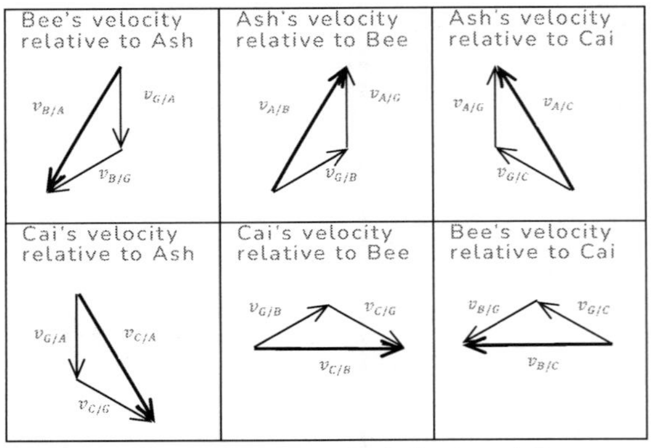

b) The 120° angle is measured relative to a stationary observer. As each skater is moving, the angle they see the others skate at will differ from this, as shown in the diagrams.

2. When the train was moving, the velocity of the rain as seen on the train was the vector sum of the velocity of the ground relative to the train and the velocity of the rain relative to the ground. The velocity of the rain relative to the ground is given by:

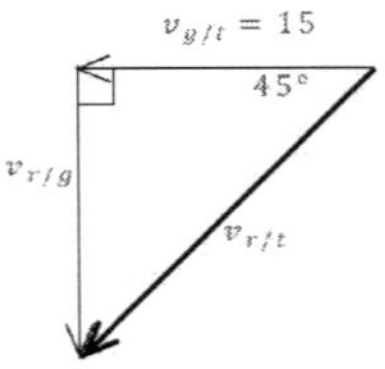

$\overrightarrow{v_{r/t}} = \overrightarrow{v_{r/g}} + \overrightarrow{v_{g/t}}$

$\tan 45° = \frac{v_{r/g}}{15}$

$v_{r/g} = 15$ m s^{-1} down

The velocity of the rain relative to the train is faster:

$\overrightarrow{v_{r/t}} = \overrightarrow{v_{r/g}} + \overrightarrow{v_{g/t}} = \sqrt{15^2 + 15^2} = 21$ m s^{-1} at 45° to the vertical downwards

3. $\overrightarrow{v_{b/w}} = \overrightarrow{v_{b/g}} + \overrightarrow{v_{g/w}} = \sqrt{60^2 + 60^2 - 2 \times 60 \times 60 \cos 45°}$
$= 46$ km h^{-1}
$\frac{\sin\theta}{60} = \frac{\sin 45°}{46}$
$\theta = 67°$
The black car moves at 46 km h^{-1} at 247° T relative to the white car

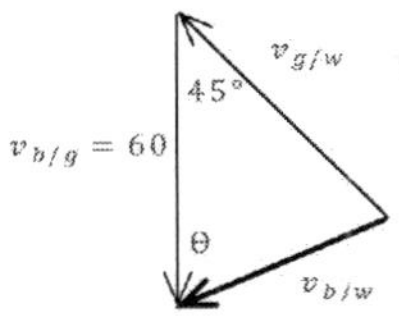

4. a) As Bee just floats on the water, their velocity is zero relative to the water. However, Bee's velocity is 1.0 m s^{-1} south relative to Ash. Ash's velocity relative to Bee is the same magnitude but the opposite direction, 1.0 m s^{-1} north.
Since Cai's swimming velocity is directed eastward, their swimming will not affect their velocity north or south, as components that are perpendicular have no effect on each other. Similarly, the river cannot slow down Cai's velocity to the east, since the current is perpendicular to it. However, the current will make Cai drift south at the same velocity as Bee. The displacement of Bee and Cai downstream will therefore be the same as each other. However, Cai will have a velocity eastwards at 2.0 m s^{-1} relative to Bee.
Ash observes Cai moving diagonally downstream, with a total velocity that is the resultant of their swimming velocity plus the velocity of the current.

b)

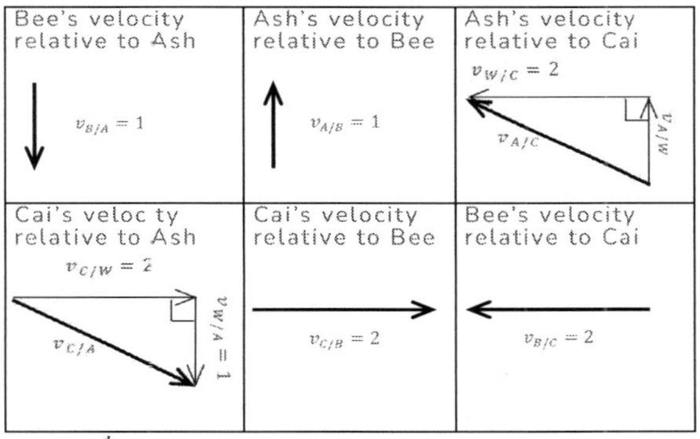

c) $v = \frac{d}{t}$
$2 = \frac{5}{t}$
$t = 2.5$ s

d) $v = \frac{d}{t}$
$1.0 = \frac{d}{2.5}$
$d = 2.5$ m

e) $\overrightarrow{s_{C/A}} = \sqrt{5^2 + 2.5^2} = 5.6$ m
$\tan\theta = \frac{5}{2.5}$
$\theta = 63°$
Cai's total displacement is 5.6 m at 63° to the bank.

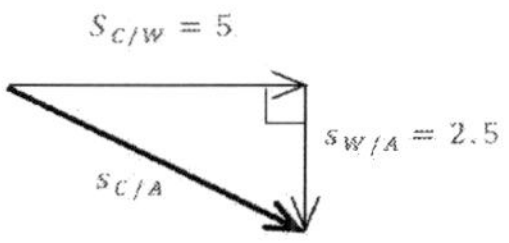

f) Method 1: $v = \frac{d}{t} = \frac{5.6}{2.5} = 2.2$ m s^{-1} at 63° to the bank
Method 2: Using the vector diagram in part (b)
$\overrightarrow{s_{C/A}} = \sqrt{2^2 + 1^2} = 2.2$ m s^{-1}
$\tan\theta = \frac{2}{1}$
$\theta = 63°$
Cai's velocity is 2.2 m s^{-1} at 63° to the bank, either way.

5. a) Because Cai's displacement needs to be directly east of Ash, their resultant velocity also needs to be towards the east. This resultant velocity, relative to Ash, is the vector sum of their swimming velocity and the velocity of the water south. However, this time Cai is going to need a component of their swimming velocity to be north at 1.0 m s^{-1}, to counteract the river flowing south. So Cai will need to direct their 2.0 m s^{-1} swimming velocity diagonally upstream, relative to the water. This means the component of their swimming velocity to the east is less than 2.0 m s^{-1}, so it will take more time to cross the river, compared to allowing the river to carry them downstream while swimming across.
In addition, Cai appears to move diagonally upstream relative to Bee at 2.0 m s^{-1}, who is in the same frame of reference as the water.

b)

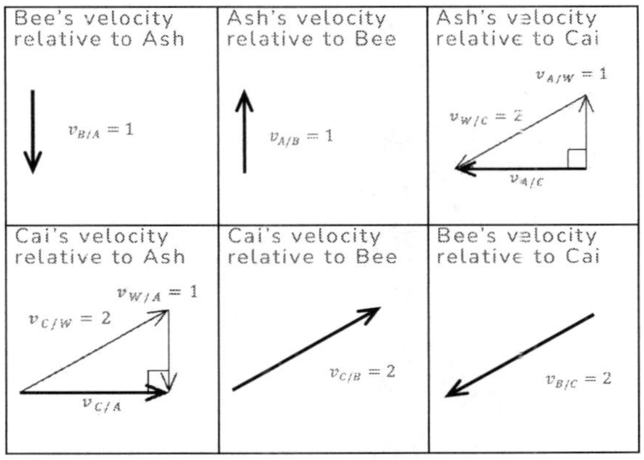

c) Using the vector diagram in part (b) for Cai's velocity relative to Ash:
$\overrightarrow{v_{C/A}} = \sqrt{2^2 - 1^2} = 1.7$ m s^{-1}
Cai's velocity is 1.7 m s^{-1} east.

d) $v = \frac{d}{t}$
$1.7 = \frac{5}{t}$
$t = 2.9$ s

6. a) $\overrightarrow{v_{P/G}} = \sqrt{30.9^2 + 12.5^2} = 33.3$ m s^{-1}
$\tan\theta = \frac{12.5}{30.9}$
$\theta = 22°$
The plane will fly at 33.3 m s^{-1} at 22° T.

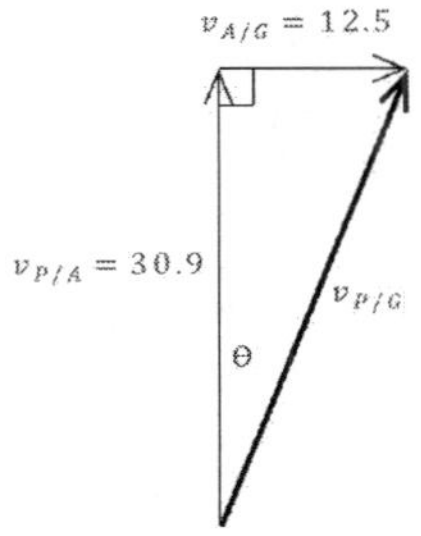

b) $\overrightarrow{v_{P/G}} = \sqrt{30.9^2 - 12.5^2} = 28.3$ m s^{-1}
$\sin\theta = \frac{12.5}{30.9}$
$\theta = 24°$
The plane needs to steer at N24°W relative to the air, to give a resultant velocity of 28.3 m s^{-1} north.

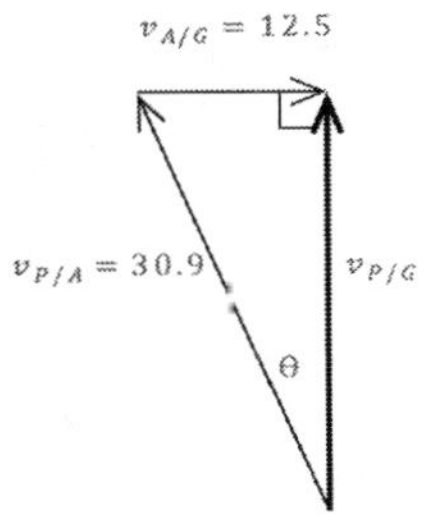

c) $v = \frac{d}{t}$
$28.3 = \frac{215 \times 10^3}{t}$
$t = 7597$ s $= 2.11$ h

7. a) Your swimming speed may be less than the speed of the rip current. If so, your resultant velocity will be away from the beach. In the example given, the swimmer would move further away from the beach at 0.4 m s^{-1}.

b) $\overrightarrow{v_{S/G}} = \sqrt{2.0^2 + 2.4^2} = 3.1$ m s^{-1}

$\tan\theta = \frac{2.0}{2.4}$

$\theta = 40°$

The swimmer's velocity is 3.1 m s^{-1} at 40° to the rip.

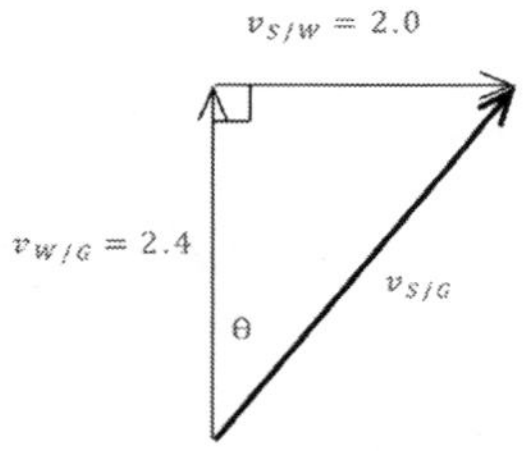

8. a) In still air, a boat moving with velocity $v_{b/g}$ will perceive a headwind $v_{g/b} = -v_{b/g}$ as the ground and air moves backwards relative to it. When the velocity of this headwind is added as a vector to the velocity of the actual wind relative to the ground $v_{a/g}$, the resultant velocity is the apparent velocity of the air observed by the boat. If the boat is sailing at the same velocity as the wind, the resultant is zero, so the apparent velocity of the air observed by the boat is

zero.

b) The apparent wind, or wind relative to the boat, is the resultant of the actual wind relative to the ground $v_{a/g}$ and the headwind created by the motion of the boat through the air $v_{g/b} = -v_{b/g}$. As seen in the vector diagram, when the boat's velocity relative to the ground $v_{b/g}$ is directed at an angle into the wind, the resultant is larger in magnitude than the actual wind, so there is more force on the sail.

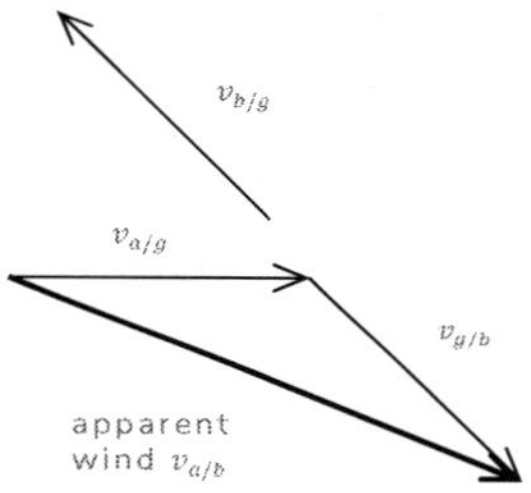

9.2 Work

1. Energy is measured in joules and is the ability to do work. Work can be calculated by multiplying the displacement an object travels through by the force which acts on it in the same direction as the displacement.
The energy possessed by a moving object is called kinetic energy. Heat or thermal energy involves moving particles with kinetic energy. Sound energy also involves kinetic energy, as it consists of vibrating particles. Electrical energy is also a special case of kinetic energy, as it involves moving charges.
The kinetic energy of moving objects often comes from some type of potential energy, or stored energy. A ball that is lifted off the floor has gravitational potential energy. This gets transformed into kinetic energy when the ball is released. The higher the ball is lifted, the more gravitational potential energy it has, and the more kinetic energy it will reach the ground with once released. So gravitational potential energy is determined by the position of an object in a gravitational field.
A battery has electrical potential energy. The positive and negative terminals of the battery exert an electrostatic force on the electrons in a circuit connected to the battery, giving those electrons kinetic energy. So electrical potential energy is transformed into electrical energy when there is a closed circuit allowing electrons to flow.
Chemical potential energy is stored in the chemical bonds between atoms. This is the type of energy stored in food, as well as fuel combusted to power vehicles and generators. Actually, chemical bonds are made from electrostatic attraction between negative electrons and positive nuclei, so they are a special case of electrical potential energy.
Elastic potential energy is stored in objects that are stretched, such as a rubber band, or compressed, such as a spring.
If we calculate the total amount of energy we start with and end with, it appears that this total amount stays the same. This is the Law of Conservation of Energy, which means that energy is not created or destroyed. It just gets transferred from one object to another or transformed from one type to another.
Some types of energy may not be useful in a situation. For example, using your laptop transforms some electrical energy into heat, but it is not the function of a laptop to be a heating device. The more wasted energy, the less energy efficient a device is.
2. $W = \vec{F}.\vec{s} = 6.5\cos 45° \times 25 = 115$ J

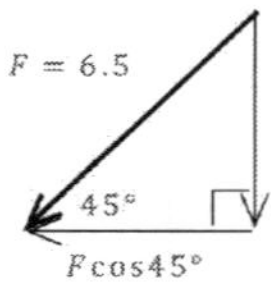

3. $v^2 = u^2 + 2as$
$4.5^2 = 0^2 + 2 \times a \times 20$
$a = 0.51$ m s^{-2}
$W = \vec{F}.\vec{s}$
$= mas$
$= 75 \times 0.51 \times 20$
$= 770$ J
4. $W = \vec{F}.\vec{s}$
$= mg\sin\theta \times s$
$= 75 \times 9.8 \times \frac{80}{2000} \times 2000$
$= 58800$ J
$= 59$ kJ
This is around one fifth of the energy contained in the apple.

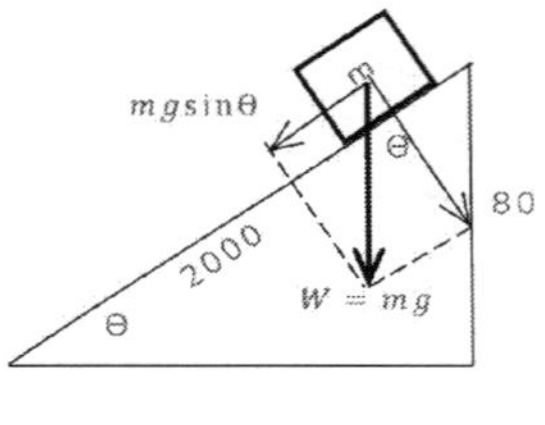

5. $W = \vec{F}.\vec{s}$
$= \mu mgs$
$= 0.45 \times 52 \times 9.8 \times 1.2$
$= 280$ J

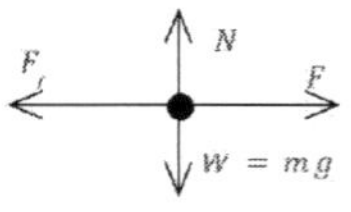

6. a) As the length of the band is increased, the force required to lengthen the band increases proportionally.
b) You cannot just multiply the distance 0.2 m by 5 N, as the band requires less force to stretch at shorter distances, and therefore less energy. So the area under the graph must be used, since work equals force times displacement.
W = area under the graph $= \frac{1}{2} \times 0.2 \times 5 = 0.5$ J

9.3 Gravitational potential energy

1. To lift an object of mass m upwards, we must apply a force equal to at least its weight, mg. If we change the object's height by Δh, we also change its gravitational potential energy by ΔU. Since work means energy

transformed, change in gravitational potential energy is given by:

$\Delta U = W = \vec{F}.\vec{s} = mg\Delta h = mg(h_f - h_i)$

where: ΔU = change in gravitational potential energy (J)

W = work done (J)

m = mass of object (kg)

g = acceleration due to gravity (m s^{-2})

h_i, h_f = initial and final heights above ground (m)

Δh = change in height (m)

Lifting 1 kg from 1 m to 2 m above the ground, the change in gravitational potential energy is:

$\Delta U = W = mg\Delta h = mg(h_f - h_i) = 1 \times 9.8 \times (2 - 1)$
$= +9.8$ J

Here, we do the work to lift the object, since our muscles exert the force to move it. The positive sign means we increased the gravitational potential energy of the object. We have stored more energy in the object, so if we release it, it will have more energy to transform into kinetic energy and hit the ground with more speed.

If we drop the same object from 2 m to 1 m instead, the change in gravitational potential energy is:

$\Delta U = W = mg\Delta h = mg(h_f - h_i) = 1 \times 9.8 \times (1 - 2)$
$= -9.8$ J

The negative sign means the gravitational potential energy of the object decreased. This time we aren't doing the work – the gravitational field is.

So energy and work can have a sign, even though they are scalar quantities. It's like a bank account can have positive and negative changes, even though money is not a vector.

2. $W = \Delta U = mg\Delta h = 150 \times 9.8 \times 1.2 = 1764$ J
3. $W = mg\Delta h = mg(h_f - h_i)$
$= 5000 \times 9.8 \times -1000$
$= -49$ MJ (negative sign means gravity does the work)
4. a) $F = mg\sin\theta$
$\sin 15° = \frac{1.2}{s}$
$W = \vec{F}.\vec{s}$
$= mg\sin\theta \times s$
$= (65 + 18) \times 9.8 \times \sin 15° \times \frac{1.2}{\sin 15°}$
$= 980$ J
b) $W = mg\Delta h = (65 + 18) \times 9.8 \times 1.2 = 980$ J
5. We often assume that something on the ground has zero gravitational potential energy because it has zero height. If so, lifting an object off the ground by height h gives it gravitational potential energy U equal to:

$\Delta U = mg\Delta h$

$U_f - U_i = mg(h_f - h_i)$

$U_f - 0 = mg(h_f - 0)$

$U = mgh$

So lifting a 1 kg object 1 m above the ground gives it gravitational potential energy of +9.8 J. 2 m above the ground puts it at +19.6 J. The change in gravitational potential energy between these points is +9.8 J.

Actually, we can choose to set the point of zero potential energy anywhere. If we set $h = 0$ to be the final position of the object, 2 m above the ground, and still let up be positive, the change in gravitational potential energy required to lift the object from 1 m to 2 m above the ground now looks like this:

$\Delta U = W = mg\Delta h = mg(h_f - h_i) = 1 \times 9.8 \times (0 - (-1))$
$= +9.8$ J

This is the same change in potential energy, so it doesn't matter what frame of reference we use to calculate change. It's like measuring the length of a line - it's easier if you put your ruler with zero at one end of the line, but you don't have to and you will still measure the same length.

Sometimes it is inconvenient to assume the ground has zero potential energy. If we drop something from the ground floor down the stairwell to the basement, it is easier to think of the basement as having zero gravitational potential energy. Spacecraft might not even return to Earth, so there is often a better frame of reference to set as zero potential energy than Earth's surface.
6. a) $W = mg\Delta h = mg(h_f - h_i)$
$= 92 \times 9.8 \times (22.5 - 10) = 11270$ J
b) $\Delta U = mg\Delta h = mg(h_f - h_i)$
$= 92 \times 9.8 \times (10 - 22.5) = -11270$ J
The gravitational field does the work.
7. a) Change in gravitational potential energy is given by the area under the graph times mass.
$\Delta U = 9500 \times \frac{1}{2} \times (9.8 + 8.8) \times 400 \times 10^3$
$= 3.5 \times 10^{10}$ J
b) First half:
$\Delta U = 9500 \times \frac{1}{2} \times (9.8 + 9.2) \times 200 \times 10^3$
$= 1.8 \times 10^{10}$ J
Second half:
$\Delta U = 9500 \times \frac{1}{2} \times (9.2 + 8.8) \times 200 \times 10^3$
$= 1.7 \times 10^{10}$ J
There is less change in gravitational potential energy in the second half of the journey, so it takes less energy to lift the rocket through the second half.
8. a) Number of pulleys
b) Effort, distance the string is pulled
c) Same mass being lifted, no acceleration, same type of string and pulley type (radius and material)
d) The student is wrong. There is no Law of Conservation of Force. Force and energy are different concepts – work done, or energy transformed, is given by $W = \vec{F}.\vec{s}$. For the one pulley system, the work done by pulling the spring balance was:
$W = \vec{F}.\vec{s} = 2.45 \times 0.5 = 1.23$ J
For the two pulley system, the work done was:
$W = \vec{F}.\vec{s} = 1.23 \times 1.0 = 1.23$ J
These are the same value. This work increased the gravitational potential energy of the mass:
$W = mgh = 0.25 \times 9.8 \times 0.5 = 1.23$ J
This is the same amount of energy, so energy is conserved.
e) Since the mass goes up at constant velocity, net force is zero. For the one pulley system:
$T = mg = 0.25 \times 9.8 = 2.45$ N
In the two pulley system, there are two strings lifting the mass, so each string has half this tension:
$T = 1.23$ N

9.4 Kinetic energy

1. Kinetic energy is energy possessed by a moving object. If a net force acts on an object, there will be a change in its kinetic energy. The net force also means work is done on the object, equal to the change in kinetic energy:
$\Delta K = W = \vec{F}.\vec{s} = m\vec{a}.\vec{s}$ ... (1)
If the acceleration is constant, we can find an expression for $\vec{a}.\vec{s}$ using:
$\vec{v}^2 = \vec{u}^2 + 2\vec{a}.\vec{s}$
$\vec{a}.\vec{s} = \frac{1}{2}(\vec{v}^2 - \vec{u}^2)$ ...(2)
Substituting $\vec{a}.\vec{s}$ from equation (2) into (1) gives:
$\Delta K = m \times \frac{1}{2}(\vec{v}^2 - \vec{u}^2)$
$K_f - K_i = \frac{1}{2}m\vec{v}^2 - \frac{1}{2}m\vec{u}^2$
As both sides of this equation involve a final quantity minus an initial quantity, we deduce that for any given mass and velocity, kinetic energy is:
$K = \frac{1}{2}m\vec{v}^2$
where: K = kinetic energy (J)
m = mass of object (kg)
$\vec{v}$ = velocity of object (m s^{-1})
2. a) $\Delta K = W = \vec{F}.\vec{s} = 295 \times 5 = 1475$ J
b) $\Delta K = \frac{1}{2}mv^2 - \frac{1}{2}mu^2$
$1475 = \frac{1}{2} \times 50 \times v^2 - 0$
$v = 7.7$ m s^{-1}
3. a) $K = \frac{1}{2}mv^2 = \frac{1}{2} \times 0.025 \times 15^2 = 2.8125$ J
b) $W = \Delta K$
$\vec{F}.\vec{s} = \frac{1}{2}mv^2 - \frac{1}{2}mu^2$
$F \times 0.1 = \frac{1}{2} \times 0.025 \times 15^2 - 0$
$F = 28$ N
4. Let the object's original kinetic energy be: $K = \frac{1}{2}mv^2$
Tripling the speed to $v' = 3v$ gives the new kinetic energy:
$K' = \frac{1}{2}mv'^2 = \frac{1}{2}m \times (3v)^2 = 9 \times \frac{1}{2}mv^2 = 9K$
The object has nine times the kinetic energy.
5. Rearranging the equation in the hypothesis gives:
$v^2 = 2gh$
If the hypothesis is correct, then a linear graph should be obtained by plotting v^2 on the y-axis and h on the x-axis, with a gradient of $2g = 19.6$. The graph plotted matches these criteria, so the results support the hypothesis.

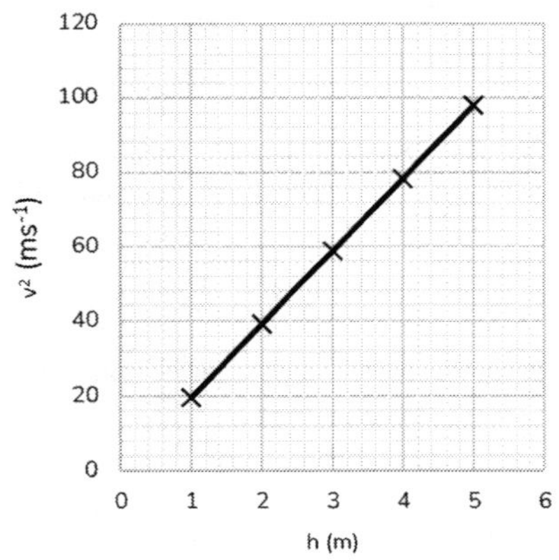

9.5 Energy transformations

1. When an object falls, it loses gravitational potential energy and gains kinetic energy. At any point in the object's journey, if we add up all the energy involved, the total energy E must stay the same, according to the Law of Conservation of Energy. In other words, the total initial energy E_i, comprising the initial gravitational potential energy U_i plus initial kinetic energy K_i, must equal the total final energy E_f, comprising the final gravitational potential energy U_f plus final kinetic energy K_f:

$E_f = E_i$

$K_f + U_f = K_i + U_i$

We can rearrange this formula to show that any gain in kinetic energy must equal the amount of gravitational potential energy transformed:

$K_f - K_i = -(U_f - U_i)$

$\Delta K = -\Delta U$

We can also use the formula to derive an expression for the final velocity of an object of mass m dropped from rest, after travelling through a distance h to the ground:

$K_f - K_i = -(U_f - U_i)$

$\frac{1}{2}mv^2 - 0 = -(0 - mgh)$

$\frac{1}{2}v^2 = gh$

$v = \sqrt{2gh}$

2. a) The ball starts with gravitational potential energy equal to mgh. According to the Law of Conservation of Energy, all the gravitational potential energy becomes kinetic energy at the bottom of the ramp, before transforming back to the same amount of gravitational potential energy mgh on the other side. Therefore, the ball reaches a height h on the other side.
b) The gravitational potential energy of the ball is transformed into kinetic energy, but now that the ramp is flattened, it is not transformed back into gravitational potential energy and stays as kinetic energy. Hence the ball will keep going forever, if there is no net force on the ball. This matches the Law of Inertia, which states that an object in motion stays in motion unless acted upon by a net force.
3. $E_i = E_f$
$\frac{1}{2}mu^2 = \frac{1}{2}mv^2 + mg\Delta h$
$\frac{1}{2} \times 10^2 = \frac{1}{2}v^2 + 9.8 \times (4 - 2)$
$v = 7.8$ m s^{-1}
4. a) The gymnast leaves the trampoline when the trampoline is no longer compressed. At this point, they will have their maximum kinetic energy. At maximum height above the trampoline, all of this initial kinetic energy is converted to gravitational potential energy:
$\frac{1}{2}mv^2 = mgh$
$v = \sqrt{2gh} = \sqrt{2 \times 9.8 \times 7.5} = 12$ m s^{-1}
b) $Fs = mgh$
$F \times 0.5 = 59 \times 9.8 \times 7.5$
$F = 8673$ N up
5. a) $F_{net} = ma$
$2 \times 9.8 = 5 \times a$
$a = 3.92$ m s^{-2}
$v^2 = u^2 + 2as = 0 + 2 \times 3.92 \times 1$
$v = 2.8$ m s^{-1}
b) $m_1gh = \frac{1}{2}(m_1 + m_2)v^2$
$2 \times 9.8 \times 1 = \frac{1}{2} \times 5 \times v^2$
$v = 2.8$ m s^{-1}

6. The hypothesis is wrong. The student has the same gravitational potential energy of mgh at the top of both slides, which will be transformed into the same amount of kinetic energy $\frac{1}{2}mv^2$ at the bottom, giving the student the same speed. It is true that the force due to gravity down Slide 1 is $mg\sin 30°$, which is less than this force down Slide 2 of $mg\sin 60°$. However, Slide 1 has a length of:
$\frac{4}{\sin 30°} = 8$ m
while Slide 2 has a shorter length of:
$\frac{4}{\sin 60°} = 4.6$ m
As work done is $W = F.s$ the same amount of work is done on both slides.
7. Let the height above the lowest point in the swing be h and the vertical component of the displacement from the pivot be x, so $h + x = 4$ m.
$\cos 60° = \frac{x}{4}$
$x = 2$ m so $h = 2$ m
$\frac{1}{2}mv^2 = mgh$
$v = \sqrt{2gh} = \sqrt{2 \times 9.8 \times 2} = 6.3$ m s^{-1}
8. a) $W = Fs = mgh = 1 \times 9.8 \times 10 = 98$ J
   b) zero
   c) While the work done is different in each frame of reference, this does not mean that energy was created or destroyed within each frame of reference, so it is still consistent with the Law of Conservation of Energy.
9. If there is friction or air resistance acting on a moving object, some of its kinetic energy will be transformed to heat, possibly even sound or light.
For example, a meteorite falling through the Earth's atmosphere will transform gravitational potential energy into kinetic energy. However, air resistance from the atmosphere will transform some of the kinetic energy into heat and light. If we are lucky, we might get to see the meteorite as a shooting star.
Friction often means the speeds measured in our motion experiments might be less than what we predict if we ignore friction. This is because some of the kinetic energy is lost as heat, possibly sound if you can hear the object moving. We can try to minimise the effect of friction by using smooth surfaces with a low coefficient of friction.
Friction and air resistance can sometimes waste energy by converting it into forms we do not find useful. For example, a child on a swing is most interested in moving, therefore they want kinetic energy. The Law of Conservation of Energy predicts that the child would swing back and forth forever, with energy being transformed from gravitational potential energy to kinetic energy, then back to gravitational potential energy. Instead, friction where the chains are anchored to the pivot, and air resistance as they move through the air means they eventually stop, because some of the kinetic energy is wasted as heat and sound. We can use an energy flow diagram to represent the energy transformations.
10. The parachute starts with gravitational potential energy:
$U = mgh = 0.015 \times 9.8 \times 5 = 0.735$ J
It reaches the ground with kinetic energy:
$K = \frac{1}{2}mv^2 = \frac{1}{2} \times 0.015 \times 1.28^2 = 0.0123$ J
The difference is heat produced: $0.735 - 0.0123 = 0.723$ J
11. a) There are 4 boxes shown after the one that just reached the bottom of the ramp, so a box takes 1 s to travel 1.6 m. As they are equally spaced apart, the velocity must be constant.
$v = 1.6$ m s^{-1}
$K = \frac{1}{2}mv^2 = \frac{1}{2} \times 0.25 \times 1.6^2 = 0.32$ J
   b) As the boxes are equally spaced apart, there is no acceleration and no change in kinetic energy. However, there is still work done, since there is a change in gravitational potential energy. The component of gravitational force down the ramp causes the displacement of the box, so work done is:
$W = \vec{F}.\vec{s} = 0.25 \times 9.8 \times \sin 15° \times 1.6 = 1.0$ J
As there must be friction on the ramp stopping the boxes from accelerating, the gravitational potential energy must have transformed into heat or sound. The height of the ramp h is given by:
$\sin 15° = \frac{h}{1.6}$
$h = 1.6 \sin 15°$
So gravitational potential energy transformed into heat is the same as work done:
$$U = mgh = 0.25 \times 9.8 \times 1.6 \sin 15° = 1.0 \text{ J}$$

9.6 Power

1. It is easier to walk up a gently sloping ramp than climb up steeper stairs to get to the same point. However, because you gain the same amount of gravitational potential energy, you do the same work. You do this work in less time though, if you take the stairs rather than the ramp. ***Power*** is the rate at which energy is transformed or used. In other words, it is the rate at which work is done. It is measured in ***watts*** (W):
$$P = \frac{W}{t}$$
where: P = power (W)
W = work done (J)
t = time in which the work was done (s)
Since energy is conserved, and power is energy transformed divided by time, that means for a given time period, power is also conserved. For example, if you ride a bike that generates electricity for a light bulb, the more power you exert with your legs, the more electrical power you generate, and the brighter the light bulb glows, putting out light energy at a faster rate.
2.

Feature	Work	Power
Unit	joule (J)	watt (W)
Definition	amount of energy transformed	rate at which energy is transformed
Formula	$W = \Delta E = \vec{F}.\vec{s}$	$P = \frac{W}{t}$

3. $P = \frac{W}{t} = \frac{\Delta K}{t} = \frac{mv^2}{2t} = \frac{145 \times 2.2^2}{2 \times 2} = 175$ W
4. $P = \frac{W}{t} = \frac{mgh}{t}$
$63.5 = \frac{20 \times 9.8 \times 12}{t}$
$t = 37$ s
5. $v^2 = u^2 + 2as$
$0 = 3^2 + 2a \times 2.50$
$a = -1.8$ m s^{-2}

$v = u + at$
$0 = 3 - 1.8t$
$t = \frac{3}{1.8}$ s
$P = \frac{W}{t} = \frac{\Delta K}{t} = \frac{K_f - K_i}{t} = \frac{0 - mu^2}{2t} = \frac{-75 \times 3^2}{2 \times \frac{3}{1.8}} = -203$ W

6. $P = \frac{W}{t} = \frac{Fs}{t}$
   Since $v = \frac{s}{t}$ then $P = Fv$
7. a) $F_f = \mu N$
   $mg \sin\theta = \mu mg \cos\theta$
   $\mu = \tan 30° = 0.58$
   b) $F_f = mg \sin\theta = 2 \times 9.8 \times \sin 30° = 9.8$ N
   $P = Fv = 9.8 \times \frac{3}{1} = 29.4$ W
8. Velocity is constant, so work is done by friction, converting gravitational potential energy to heat at the rate: $P = Fv$
   The diagram shows 5 time intervals, which is:
   $t = 5 \times \frac{1}{20} = 0.25\ s$
   The distance between the first and last image is:
   $s = 14 \times 0.05 = 0.7$ m
   $v = \frac{s}{t} = \frac{0.7}{0.25} = 2.8$ m s^{-1}
   As friction must balance the force down the ramp:
   $F_f = mg \sin\theta = 30 \times 9.8 \times \sin 11° = 56$ N
   Work is done at the rate of:
   $P = Fv = 56 \times 2.8 = 157$ W

9.7 SI units, dimensional analysis and prefixes

1. kg, m, s
2. A joule is the unit for work. The formula for work is:
   $W = Fs$
   Putting units in gives:
   J=N×m=kg×m×s^{-2}×m=kg m^2 s^{-2}
3. From question 2, the joule is equivalent to:
   J=kg m^2 s^{-2}
   $\frac{1}{2}mv^2$ has units equivalent to:
   kg×(m s^{-1})^2=kg m^2 s^{-2}
   which is the same as joules.
4. A watt is the unit for power. The formula for power is:
   $P = \frac{W}{t}$
   This has the SI unit of:
   $\frac{J}{s} = \frac{kg\ m^2\ s^{-2}}{s}$=kg m^2 s^{-3}
5. a) megajoule, 10^6 J, energy
   b) gigawatt, 10^9 W, power
   c) nanometre, 10^{-9} m, distance
   d) microgram, 10^{-6} g, mass
   e) millisecond, 10^{-3} s, time
6. a) 8700 kJ or 8.7 MJ
   b) 160 MW
   c) 650 nm
   d) 17 µm to 180 µm
7. $1200 \times 4.18 \times 10^3 = 5016000$ J
8. 1600×746 W $= 1193600$ W $= 1194$ kW

9.8 Rolling

1. If an object slides across a smooth surface, all the points on the object will have the same velocity. This motion in a straight line is called ***translational motion***.
   However, if a wheel moves at constant velocity v, but rolls without sliding, then only the centre of the wheel moves at constant velocity.
   In the frame of reference of the wheel's centre, the rim undergoes ***rotational motion***. Each point on the rim has a velocity that is tangential to the wheel. The bottom of the wheel moves backwards relative to the wheel's centre at velocity $-v$, and experiences friction from the surface. Meanwhile, the top of the wheel moves forwards relative to the wheel's centre at velocity v.
   In the frame of reference of an observer on the ground, the bottom of the wheel is stationary for the moment it comes into contact with the ground. The top of the wheel appears to move at velocity $2v$, which is the resultant of its velocity relative to the wheel's centre and the velocity of the wheel's centre relative to the ground. Basically we add v to the velocity of all the points on the wheel to change the frame of reference from the wheel's centre to the ground.
   If we track the path of a point on the rim, it looks semicircular from the ground's frame of reference.
   Where does the extra energy for points on the rim come from? The Law of Conservation of Energy means some of the kinetic energy moving the wheel forward in a straight line, or translational kinetic energy, must be traded for rotational kinetic energy instead.
2. When a ball moves down a ramp, it may slide or roll. So gravitational potential energy is transformed into a combination of translational kinetic energy and rotational kinetic energy. If the ball rolls, its acceleration down the ramp will be less than predicted by the component of gravity acting down the ramp, so it is as if there is a force opposing its motion. This is called ***rolling resistance***.
   Friction causes objects to roll. Cars on slippery ground experience less friction, so tyres can skid rather than roll, making it hard to accelerate by speeding up, braking or turning corners.
   Rolling resistance also depends on variables such as the shape and size of the object, how easily deformed the object and surface are, and the contents of the rolling object. For example, compared to an empty jar rolling, if the same jar is half full of rice, some energy must be transferred to move the rice grains, with less energy available for the kinetic energy of the jar itself. If friction between the rice grains is not overcome, the rice may remain stationary, if the component of weight down the ramp is less than rolling resistance.
3. When descending down the string, the yo-yo also spins, so some of its gravitational potential energy is converted to rotational kinetic energy, resulting in less translational kinetic energy.
4. a) $K_f - K_i = -(U_f - U_i)$
   $\frac{1}{2}mv^2 = mgh$
   $v = \sqrt{2gh} = \sqrt{2 \times 9.8 \times 0.45} = 2.97$ m s^{-1}
   b) Total initial energy: $mgh = 0.02 \times 9.8 \times 0.45 = 0.0882$ J
   Actual translational kinetic energy:
   $\frac{1}{2}mv^2 = \frac{1}{2} \times 0.02 \times 2.5^2 = 0.0625$ J
   Percentage transformed into rotational kinetic energy:
   $\frac{0.0882 - 0.0625}{0.0882} \times 100\% = 29.1\%$
5. The car must exert a force equal to rolling resistance in order to maintain a constant velocity. At 20 psi rolling resistance is 230 N:
   $P = Fv = 230 \times 100 \div 3.6 = 6389$ W
   At 50 psi rolling resistance is 130 N:
   $P = Fv = 130 \times 100 \div 3.6 = 3611$ W
   It would be better to pump the tyres to 50 psi as it would require 57% less power to maintain a 100 km h^{-1} speed, and therefore use less fuel.

10.2 Momentum and impulse

1. The momentum of an object is given by its mass multiplied by its velocity:
$\vec{p} = m\vec{v}$
where: $\vec{p}$ = momentum (kg m s^{-1})
m = mass (kg)
$\vec{v}$ = velocity (m s^{-1})
It can be thought of as "mass in motion" in classical physics. Since velocity is a vector, momentum is too, with the same direction as the velocity.
Impulse is the change in momentum. In other words, impulse is the final momentum minus the initial momentum:
$\vec{I} = \Delta\vec{p} = \vec{p_f} - \vec{p_i} = m\vec{v} - m\vec{u}$
where: $\vec{I}$ = impulse (kg m s^{-1})
$\vec{v}$ = final velocity (m s^{-1})
$\vec{u}$ = initial velocity (m s^{-1})
Impulse is a vector, with the same direction obtained by vector subtraction of the final and initial momentum.
2. $\vec{p} = m\vec{v} = 0.17 \times 6.5 = 1.1$ kg m s^{-1}
3. $\Delta\vec{p} = \vec{p_f} - \vec{p_i} = m\vec{v} - m\vec{u}$
$= 1200 \times \frac{60}{3.6} - 1200 \times \frac{40}{3.6} = 6700$ kg m s^{-1}
4. Change in momentum:
Child: $\Delta\vec{p} = \vec{v_f} - \vec{p_i} = m\vec{v} - 0 = m\vec{v}$
Adult: $\Delta\vec{p} = \vec{p_f} - \vec{p_i} = 2m \times (\frac{1}{2}\vec{v}) - 0 = m\vec{v}$
The child and the adult have the same change in momentum.
Gain in kinetic energy:
Child: $\Delta K = K_f - K_i = \frac{1}{2}mv^2$
Adult: $\Delta K = K_f - K_i = \frac{1}{2} \times (2m) \times (\frac{1}{2}v)^2 = \frac{1}{4}mv^2$
The adult gained half the kinetic energy of the child.
5. Let up be positive.

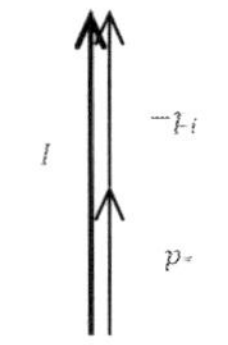

$\vec{I} = \Delta\vec{p} = \vec{p_f} - \vec{p_i} = 52 \times 4.2 - (52 \times -4.8) = 468$ kg m s^{-1}
6. $\vec{I} = \Delta\vec{p} = \vec{p_f} - \vec{p_i}$
$= \sqrt{(0.3135)^2 + (0.3135)^2 - 2 \times (0.3135)^2 \cos 30°}$
$= 0.162$ kg m s^{-1}
$\frac{\sin\theta}{0.165 \times 1.9} = \frac{\sin 30°}{0.162}$
$\theta = 75°$
Impulse is 0.162 kg m s^{-1} at 90° to the side of the snooker table.

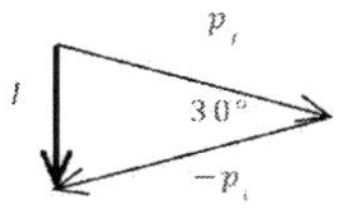

7. a)

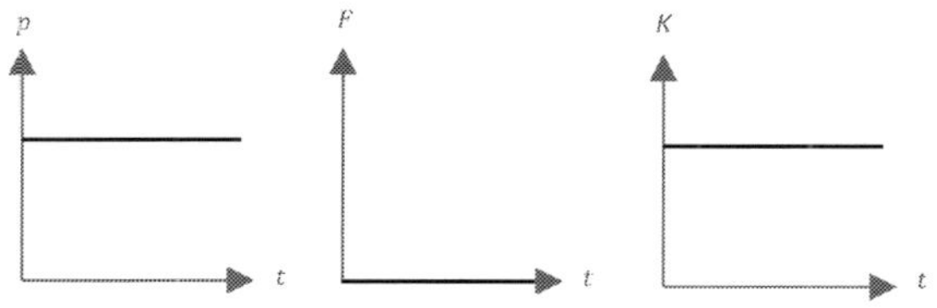

b)

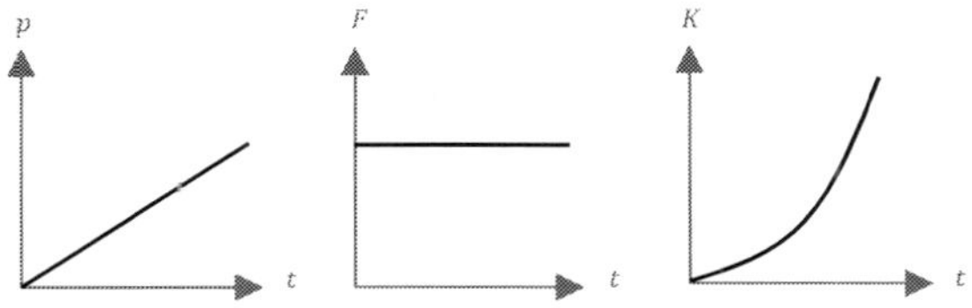

8. Recall that a net force acting on an object causes an acceleration that is proportional to the force, and that acceleration is a change in velocity over time. Combining these gives:
$\vec{F} = m\vec{a} = \frac{m(\vec{v}-\vec{u})}{t} = \frac{m\vec{v}-m\vec{u}}{t} = \frac{\Delta\vec{p}}{t} = \frac{\vec{I}}{t}$
This formula tells us that impulse is also force multiplied by time:
$\vec{I} = \vec{F}t$
where: $\vec{I}$ = impulse (N s)
$\vec{F}$ = force applied (N)
t = time the force is applied to the object (s)
So impulse is increased if a larger force acts on an object, or if the force acts for a longer time. For example, pressing the accelerator of a car harder or for a longer time will increase the change in momentum of the car.
9. Impulse can be calculated using:
$\vec{I} = \Delta\vec{p} = \vec{p_f} - \vec{p_i} = m\vec{v} - m\vec{u} =$ kg m s^{-1}
Impulse can also be calculated using:
$\vec{I} = \vec{F}t = m\vec{a}t =$ kg m s^{-2}s $=$ kg m s^{-1} same as above
10.

Feature	Impulse	Force	Kinetic energy
Formula	$\vec{I} = \vec{F}t$	$\vec{F} = m\vec{a}$	$K = \frac{1}{2}mv^2$
Unit	kg m s^{-1} or N s	N	J
Scalar or vector?	Vector	Vector	Scalar

11. When an acrobat falls, they come to a stop. If the initial velocity u they reach the ground at is kept the same, and they come to a stop with final velocity $v = 0$ after hitting the ground, the acrobat's change in momentum stays the same, regardless of the surface they fall on. This change in momentum, or impulse, is given by the equation:
$\vec{I} = \Delta\vec{p} = \vec{p_f} - \vec{p_i} = 0 - m\vec{u}$
However, using a crash mat increases the time taken t in which the impulse takes place. This decreases the force acting on the acrobat from the ground, hence making it safer, according to the equation: $\vec{I} = \vec{F}t$
12. When a soccer player wants a ball to travel as fast as possible, the maximum strength of their muscles determines the maximum force they can exert on the ball. However, by following through, they increase the time that the force acts on the ball.
Therefore, according to the equation:
$\vec{I} = \Delta\vec{p} = \vec{p_f} - \vec{p_i} = m\vec{v} - m\vec{u} = \vec{F}t$ this means that the final momentum and final velocity of the ball will be increased.
13. $\vec{I} = \vec{F}t$
$m\vec{v} - m\vec{u} = \vec{F}t$
$86 \times \vec{v} - 86 \times 1.25 = 18.2 \times 1.18$
$\vec{v} = 1.50$ m s^{-1}
14. The impulse on each car is the same, because both have the same mass and change in velocity:
$\vec{I} = \Delta\vec{p} = m\vec{v} - m\vec{u} = 0 - 2000 \times \frac{60}{3.6} = -33000$ kg m s^{-1}
The work done on each car is also the same, because both cars have the same change in kinetic energy:
$W = \Delta K = 0 - \frac{1}{2}mu^2 = -\frac{1}{2} \times 2000 \times \left(\frac{60}{3.6}\right)^2 = -280000$ J
The force acting on Car A is:
$\vec{F} = \frac{\vec{I}}{t} = \frac{0 - 2000 \times \frac{60}{3.6}}{0.03} = -1100000$ N $= -1100$ kN (backwards)
The force acting on Car B is:
$\vec{F} = \frac{\vec{I}}{t} = \frac{0 - 2000 \times \frac{60}{3.6}}{0.14} = -240000$ N $= -240$ kN (backwards)
The force acting on Car B is around 22% that of Car A, so Car B is much safer. Bee is correct.

10.3 Impulse graphs

1. The equation $\vec{I} = \vec{F}t$ can be used to calculate impulse if the force acting remains constant over time, for example if the engine of a car applies constant thrust, or if the only force on a falling object is gravitational force.
   If the force acting on an object changes, the impulse can be calculated by finding the area between a force vs time graph and the x-axis. The area will have units equivalent to the unit of the x-axis multiplied by the y-axis, in this case newton-seconds (N s). Whenever the force decreases, the area under the graph will be smaller too, representing a smaller change in momentum during that time period.
2. a)

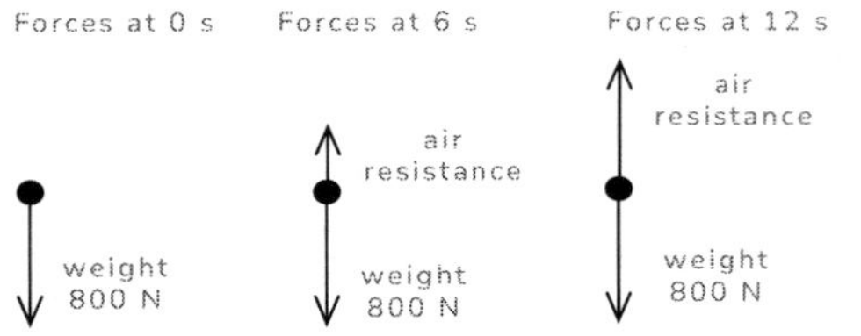

   b) Terminal velocity is reached when net force is zero, at 12 s.
   $\Delta\vec{p} = m\vec{v} - m\vec{u}$
   $\frac{1}{2} \times 12 \times 800 = \frac{800}{9.8} \times \vec{v} - 0$
   $\vec{v} = 58.8 \text{ m s}^{-1}$
3. a) $\Delta\vec{p} = \frac{1}{2} \times 3 \times 9000 + 1 \times 9000 = 22500$ N s
   b) $\Delta\vec{p} = m\vec{v} - m\vec{u}$
   $22500 = 1250 \times \vec{v} - 0$
   $\vec{v} = 18 \text{ ms}^{-1} = 64.8 \text{ km h}^{-1}$
   c) The first sentence is invalid. The equation cannot be used to calculate impulse for 0 to 1 s because the force is not constant. However, the second statement is correct, since force is constant for 3 to 4 s.
4. a) $\Delta\vec{p} = m\vec{v} - m\vec{u}$
   $\frac{1}{2} \times 35 \times 10^{-3} \times 500 = 0.165 \times 32 - 0.165\vec{u}$
   $\vec{u} = -21.0 \text{ m s}^{-1}$
   The initial velocity was 21.0 m s^{-1} towards the player.
   b) $\vec{F} = \frac{\Delta\vec{p}}{t} = \frac{\frac{1}{2} \times 35 \times 10^{-3} \times 500}{35 \times 10^{-3}} = 250$ N away from the player
5. Punch A has a higher peak force, so it would do more damage than Punch B.
   Calculating the change in momentum from Punch A from the area under the graph:
   $\Delta\vec{p} = \frac{1}{2} \times 80 \times 10^{-3} \times 1750 = 70$ N s
   And Punch B:
   $\Delta\vec{p} = \frac{1}{2} \times 100 \times 10^{-3} \times 1400 = 70$ N s
   Both punches exert the same change in momentum on the object, so the object's velocity will change by the same amount. This is because Punch B is in contact with the object for a longer time than Punch B.
6. a)

0 – 0.2	0.2 – 0.4	0.4 – 1.0	1.0 – 1.4
Squatting at rest	Pushing against the plate to go up	In the air	Landing and slowing down
normal force 750 N weight 750 N	normal force 2250 N weight 750 N	weight 750 N	normal force 1375 N weight 750 N

   b) $W = mg$
   $m = \frac{750}{9.8} = 76.5$ kg
   c) $s = ut + \frac{1}{2}at^2$
   $0 = u \times 0.6 - \frac{1}{2} \times 9.8 \times 0.6^2$
   $u = 2.94 \text{ m s}^{-1}$
   d) $\vec{I} = \Delta\vec{p} = \vec{p_f} - \vec{p_i} = 76.5 \times 2.94 - 0 = 225$ N s up
   e) The average net force gives the same impulse in the same time as the varying force.
   $\Delta\vec{p} = \vec{F}t$
   $\vec{F} = \frac{\Delta\vec{p}}{t} = \frac{225}{0.2} = 1125$ N up
   f) Work done is the change in kinetic energy.
   $W = \frac{1}{2}mv^2 - 0 = \frac{1}{2} \times 76.5 \times 2.94^2 = 331$ J
   g) Work is done by the net force on the athlete over the distance from the squat to jumping.
   $W = \vec{F}\vec{s}$
   $\vec{s} = \frac{331}{1125} = 0.29$ m
   h) Area under the graph from 0.2 s to 0.4 s minus the impulse from gravity gives the change in momentum:
   $\Delta\vec{p} = \frac{0.2+0.1}{2} \times 1500 + 750 \times 0.2 - 750 \times 0.2$
   $= 225$ N s up
   Area under the graph from 1.0 s to 1.4 s minus the impulse from gravity gives the change in momentum:
   $\Delta\vec{p} = \frac{2000 + 625}{2} \times 0.1 + \frac{1}{2} \times 625 \times 0.3$
   $= 225$ Ns up
   The impulse given by the first part of the graph is the same as part (d), as the force and time the athlete is pushing against the force plate for must produce the velocity used to calculate (d). As the athlete lands with the same speed, but going down, the same impulse is required to slow the athlete to rest, so the change in momentum from the second area under the graph is the same as the first.

10.4 Conservation of momentum

1. If two objects exert force on each other, either because they are in contact or because there is a field exerting force between them, then the force the first object exerts on the second $\overrightarrow{F_{12}}$ will be equal in magnitude but in an opposite direction to the force the second object exerts on the first $\overrightarrow{F_{21}}$. These forces are an action-reaction pair:
   $\overrightarrow{F_{12}} = -\overrightarrow{F_{21}}$
   If these forces act for time t, then the impulse is given by:
   $\overrightarrow{F_{12}}t = -\overrightarrow{F_{21}}t$
   Impulse means the same thing as change in momentum, so:
   $\Delta\overrightarrow{p_2} = -\Delta\overrightarrow{p_1}$
   Change in momentum is given by the final momentum minus the initial momentum, and momentum is given by mass times velocity, so:
   $m_2\overrightarrow{v_2} - m_2\overrightarrow{u_2} = -(m_1\overrightarrow{v_1} - m_1\overrightarrow{u_1})$
   This equation can be rearranged to show that the total final momentum is equal to the total initial momentum:
   $m_1\overrightarrow{v_1} + m_2\overrightarrow{v_2} = m_1\overrightarrow{u_1} + m_2\overrightarrow{u_2}$
   $\Sigma\overrightarrow{p_f} = \Sigma\overrightarrow{p_i}$
   In other words, in a closed system, where there is no other unbalanced force acting on the objects, total momentum is conserved. This concept is known as the Law of Conservation of Momentum.
2. $m_1\overrightarrow{v_1} + m_2\overrightarrow{v_2} = m_1\overrightarrow{u_1} + m_2\overrightarrow{u_2}$
   $0.25 \times 1.5 - m_2 \times 0.21 = 0$
   $m_2 = 1.8$ kg
3. $m_1\overrightarrow{v_1} + m_2\overrightarrow{v_2} = m_1\overrightarrow{u_1} + m_2\overrightarrow{u_2}$
   $4.5 \times 3.1 + 1.5 \times \overrightarrow{v_2} = 4.5 \times 6.2 + 0$
   $\overrightarrow{v_2} = 9.3 \text{ m s}^{-1}$ in the same direction as the bowling ball

4. a) $(m_1 + m_2)\vec{v} = m_1\vec{u_1} + m_2\vec{u_2}$
$(170 + 50)\vec{v} = 170 \times 1.5 + 0$
$\vec{v} = 1.2 \text{ m s}^{-1}$ north
b) When Cai is released, Cai will still have the same horizontal velocity as the boat relative to the water due to the Law of Inertia, as no net force acts on them in the horizontal direction. Therefore the boat will continue at the same velocity, since the horizontal momentum of every body stays unchanged.
5. a) Let the direction the bullet moves be positive.
$m_B\vec{v_B} + m_G\vec{v_G} = m_B\vec{u_B} + m_G\vec{u_G}$
$1.67 \times 21 + 100 \times \vec{v_G} = 0$
$\vec{v_G} = -0.351$ m s^{-1} (opposite direction to the bullet)
b) $\Sigma K_f = \frac{1}{2}m_B v_B{}^2 + \frac{1}{2}m_G v_G{}^2$
$= \frac{1}{2} \times 0.00167 \times 21^2 + \frac{1}{2} \times 0.1 \times 0.351^2 = 0.374$ J
0.374 J of elastic potential energy was transformed into kinetic energy.
6. As energy is conserved, the gain in gravitational potential energy equals the loss in kinetic energy:
$\frac{1}{2}(m_D + m_P)v^2 = (m_D + m_P)gh$
$v = \sqrt{2gh} = \sqrt{2 \times 9.8 \times 0.01} = \sqrt{0.196}$
According to the Law of Conservation of Momentum:
$m_D\vec{u_D} + m_P\vec{u_P} = (m_D + m_P)\vec{v}$
$24 \times \vec{u_D} + 0 = (24 + 450) \times \sqrt{0.196}$
$\vec{u_D} = 8.7$ m s^{-1}
7. $0.165 \times \vec{v_1} + 0.165 \times \vec{v_2} = 0.165 \times \vec{u_1} + 0$
Since the vector triangle is isosceles, and the balls are the same mass, $v_1 = v_2$.

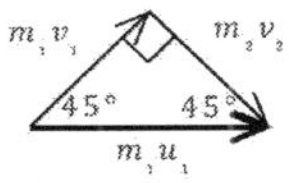

Since the angles of a triangle add to 180°, the other angle is a right angle.
$(m_1v_1)^2 + (m_2v_2)^2 = (m_1u_1)^2$
$2 \times (0.165 \times v_1)^2 = (0.165 \times 3.2)^2$
$v_1 = v_2 = 2.3$ m s^{-1}
8. To jump 10 cm off the ground, the initial velocity u would be given by:
$v^2 = u^2 + 2as$
$0 = u^2 - 2 \times 9.8 \times 0.1$
$u = 1.4$ m s^{-1}
Before the people jump, from Earth's frame of reference, both the Earth and people start at rest. Total momentum is conserved so as the people jump up, the velocity of the Earth is:
$m_P\vec{u_P} + m_E\vec{u_E} = m_P\vec{v_P} + m_E\vec{v_E}$
$0 = 144000 \times 80 \times 1.4 + 6 \times 10^{24} \times \vec{v_E}$
$\vec{v_E} = -2.7 \times 10^{-18}$ m s^{-1}
The Earth moves away from its starting position by a miniscule amount. However, as the people fall back due to Earth's gravity pulling them, the reaction force is they pull the Earth towards them with the same force. As they land back on the ground, in order for total momentum to be conserved, it must equal zero, so the Earth stops moving relative to the people.
(The local seismometer may pick up bigger vibrations as the Earth is not rigid.)

10.5 Elastic and inelastic collisions

1. A Newton's cradle is a toy made of identical balls hanging on strings in a line. If one ball at the end is pulled aside and released, only one ball at the other end moves out, and not the others. Why?
Let the mass of each ball be m. In a collision, momentum must be conserved, so:
$\Sigma\vec{p_f} = \Sigma\vec{p_i}$
Since $\vec{p} = m\vec{v}$, the Law of Conservation of Momentum is satisfied if one ball is pulled aside then collides with velocity $\vec{v}$ and causes any of the following scenarios:

- one ball at the other end exits with the same velocity $\vec{v}$;
- two balls at the other end exit with the velocity $\frac{1}{2}\vec{v}$;
- three balls at the other end exit with the velocity $\frac{1}{3}\vec{v}$; or
- all four balls move with the velocity $\frac{1}{4}\vec{v}$.

However, if the Law of Conservation of Energy is also true, then the total initial energy must equal the total final energy. The ball initially pulled aside starts with gravitational potential energy, which becomes kinetic energy when released. When it collides with the next ball, that energy must either be transferred or transformed. If no energy is transformed as heat, sound or deformation in the collision, then the total initial kinetic energy must equal the total final kinetic energy. This type of collision, involving only a transfer of kinetic energy, is called an ***elastic collision***.

The initial kinetic energy of the ball pulled aside, just before collision, is:

$$K_i = \frac{1}{2}m\vec{v}^2$$

If only one ball exits at the other end with velocity $\vec{v}$, then its final kinetic energy is:

$$K_f = \frac{1}{2}m\vec{v}^2 = K_i$$

Since kinetic energy was conserved, this is an elastic collision.

However, if two balls exit at the other end with velocity $\frac{1}{2}\vec{v}$ then final kinetic energy is:

$$K_f = \frac{1}{2} \times (2m) \times (\frac{1}{2}\vec{v})^2 = \frac{1}{4}m\vec{v}^2 = \frac{1}{2}K_i$$

This implies that half of the initial energy had to be lost as sound or heat, or the Law of Conservation of Energy would be violated. Since this much energy loss does not happen, two balls cannot exit. Repeating the same calculation while assuming that three or four balls exit would result in a bigger loss in kinetic energy. Therefore, to satisfy both the Law of Conservation of Momentum and the Law of Conservation of Energy, one ball being pulled aside must lead to one ball exiting the other side.

Collisions where kinetic energy is not conserved are called ***inelastic collisions***. In inelastic collisions, the initial kinetic energy is transformed into heat or sound, as the colliding objects are deformed instead of bouncing apart in their original form. A completely inelastic collision happens when two objects collide and are stuck together. Momentum is still conserved in these collisions, just not kinetic energy.

2. Let the original direction of Ash and Bee be positive.
a) $\Sigma\vec{p_f} = \Sigma\vec{p_i}$
$m_A\vec{v_A} + m_B\vec{v_B} = m_A\vec{u_A} + m_B\vec{u_B}$
$100 \times 1 + 100 \times \vec{v_B} = 100 \times 4 + 100 \times 1$
$\vec{v_B} = 4$ m s^{-1}
b) $\Sigma K_i = \frac{1}{2}m_A u_A^2 + \frac{1}{2}m_B u_B^2$
$= \frac{1}{2} \times 100 \times 4^2 + \frac{1}{2} \times 100 \times 1^2 = 850$ J
$\Sigma K_f = \frac{1}{2}m_A v_A^2 + \frac{1}{2}m_B v_B^2 = \frac{1}{2} \times 100 \times 1^2 + \frac{1}{2} \times 100 \times 4^2$
$= 850$ J

As kinetic energy is conserved, this is an elastic collision.

3. a) Initial velocity just before touching the ground u is given by:

$\frac{1}{2}mu^2 = mgh_i$

$u = \sqrt{2 \times 9.8 \times 1} = \sqrt{19.6}$ m s^{-1} down

Final velocity just after touching the ground v is given by:

$\frac{1}{2}mv^2 = mgh_f$

$v = \sqrt{2 \times 9.8 \times 0.85} = \sqrt{16.66}$ m s^{-1} up

$\Delta\vec{p} = m\vec{v} - m\vec{u}$

$= 0.075 \times \sqrt{16.66} - (-0.075 \times \sqrt{19.6})$

$= 0.64$ kg m s^{-1} up

b) $K_i = \frac{1}{2}mu^2 = \frac{1}{2} \times 0.075 \times 19.6 = 0.735$ J

$K_f = \frac{1}{2}mv^2 = \frac{1}{2} \times 0.075 \times 16.66 = 0.62475$ J

Percentage lost $= \frac{0.735-0.62475}{0.735} \times 100\% = 15\%$

so this is an inelastic collision.

4. a) Let the original direction of the car be positive.

$\Sigma\vec{p_f} = \Sigma\vec{p_i}$

$m_C\vec{u_C} + m_T\vec{u_T} = (m_C + m_T)\vec{v}$

$1.6 \times 80 + 3.6 \times -60 = (1.6 + 3.6) \times \vec{v}$

$\vec{v} = -17$ km h^{-1}

b) $K_i = \frac{1}{2}m_C{u_C}^2 + \frac{1}{2}m_T{u_T}^2$

$= \frac{1}{2} \times 1600 \times \left(\frac{80}{3.6}\right)^2 + \frac{1}{2} \times 3600 \times \left(\frac{60}{3.6}\right)^2 = 895062$ J

$K_f = \frac{1}{2}(m_C + m_T)v^2$

$= \frac{1}{2} \times (1600 + 3600) \times \left(\frac{17}{3.6}\right)^2 = 57978$ J

Percentage lost $= \frac{895062-57978}{895062} \times 100\% = 94\%$

The energy is probably lost as heat, sound and work done to deform the car and truck.

5. a) $\vec{p_f} = \Sigma\vec{p_i}$

$(m_A + m_B)v_{AB} = \sqrt{(m_Av_A)^2 + (m_Bv_B)^2}$

$(35 + 42) \times v_{AB} = \sqrt{(35 \times 3)^2 + (42 \times 2.5)^2}$

$v_{AB} = 1.9$ m s^{-1}

$\tan\theta = \frac{p_B}{p_B} = \frac{m_Av_A}{m_Bv_B} = \frac{35\times3}{42\times2.5}$

$\theta = 45°$

Ash and Bee move at 1.9 m s^{-1} north east.

b) $K_i = \frac{1}{2}m_A{u_A}^2 + \frac{1}{2}m_B{u_B}^2$

$= \frac{1}{2} \times 35 \times 3^2 + \frac{1}{2} \times 42 \times 2.5^2 = 290$ J

$K_f = \frac{1}{2}(m_A + m_B){v_{AB}}^2$

$= \frac{1}{2} \times (35 + 42) \times 1.9^2 = 140$ J

This is an inelastic collision as kinetic energy is not conserved.

6. a) $m_A\vec{u_A} + m_B\vec{u_B} = m_A\vec{v_A} + m_B\vec{v_B}$

$85 \times 5 + m_B \times -5 = 85 \times -3.6 + m_B \times 6.3$

$m_B = 65$ kg

b) $\Delta\vec{p} = m_B\vec{v_B} - m_B\vec{u_B} = \vec{F}t$

$65 \times (6.3 - (-5)) = \vec{F} \times 0.3$

$\vec{F} = 2400$ N north

c) 884 N south, since this is an action-reaction pair with the force Ash exerts on Bee.

d) $K_i = \frac{1}{2}m_A{u_A}^2 + \frac{1}{2}m_B{u_B}^2$

$= \frac{1}{2} \times 85 \times 5^2 + \frac{1}{2} \times 65 \times 5^2 = 1875$ J

$\Sigma K_f = \frac{1}{2}m_Av_A^2 + \frac{1}{2}m_Bv_B^2$

$= \frac{1}{2} \times 85 \times 3.6^2 + \frac{1}{2} \times 65 \times 6.3^2 = 1841$ J

This collision was almost elastic, as very little kinetic energy was lost.

7. As momentum is conserved:

$\Sigma\vec{p_i} = \Sigma\vec{p_f}$

$mu_W + 0 = mv_W + mv_R$

$u_W = v_W + v_R$ ...(1)

As kinetic energy is conserved:

$\Sigma K_i = \Sigma K_f$

$\frac{1}{2}mu_W^2 = \frac{1}{2}mv_W^2 + \frac{1}{2}mv_R^2$

$u_W^2 = v_W^2 + v_R^2$ ...(2)

Substitute (1) in (2)

$(v_W + v_R)^2 = v_W^2 + v_R^2$

$v_W^2 + 2v_Wv_R + v_R^2 = v_W^2 + v_R^2$

$2v_Wv_R = 0$ so either v_W or v_R is zero. Since the collision transfers energy to the red ball, $v_W = 0$.

Substituting this into (1), $v_R = u_W$.

This results in an elastic collision, where both momentum and kinetic energy are conserved.

8. a) $T = \frac{1}{f} = \frac{1}{40} = 0.025$ s

b) $2 \times 0.025 = 0.05$ s

c) 6 cm

d) $\vec{u_L} = \frac{s}{t} = \frac{0.06}{0.05} = 1.2$ m s^{-1} right

e) $\vec{p_L} = m_L\vec{u_L} = 0.01 \times 1.2 = 0.012$ kg m s^{-1} right

f) $\vec{v_S} = \frac{s}{t} = \frac{0.09}{2\times0.025} = 1.8$ m s^{-1} right

g) $\vec{v_L} = \frac{s}{t} = \frac{0.065}{4\times0.025} = 0.65$ m s^{-1} right

h) $\Sigma\vec{p_f} = \Sigma\vec{p_i}$

$m_L\vec{v_L} + m_S\vec{v_L} = m_L\vec{u_L} + m_S\vec{u_S}$

$0.01 \times 0.65 + m_S \times 1.8 = 0.01 \times 1.2 + 0$

$m_S = 0.0031$ kg $= 3.1$ g

i) $\Sigma K_i = \frac{1}{2}m_L{u_L}^2 = \frac{1}{2} \times 0.01 \times 1.2^2 = 0.0072$ J

$\Sigma K_f = \frac{1}{2}m_L{v_L}^2 + \frac{1}{2}m_S{v_S}^2$

$= \frac{1}{2} \times 0.01 \times 0.65^2 + \frac{1}{2} \times 0.0031 \times 1.8^2 = 0.0071$ J

Taking into account limit of reading of the graph paper and rounding, total initial and final kinetic energies are the same, so this is an elastic collision.

9. a) $\vec{u_W} = \frac{s}{t} = \frac{0.068}{2\times0.05} = 0.68$ m s^{-1} to the right

$\vec{p_W} = m_W\vec{u_W} = 0.16 \times 0.68 = 0.1088$ kg m s^{-1} to the right

b) $\vec{v_W} = \frac{s}{t} = \frac{0.058}{2\times0.05} = 0.58$ m s^{-1} at 31°

$\vec{p_W} = m_W\vec{v_W} = 0.16 \times 0.58 = 0.0928$ kg m s^{-1} at 31°

c) $\vec{v_B} = \frac{s}{t} = \frac{0.07}{4\times0.05} = 0.35$ m s^{-1} at 59°

$\vec{p_B} = m_B\vec{v_B} = 0.16 \times 0.35 = 0.056$ kg m s^{-1} at 59°

d) $\Sigma\vec{p_f} = \vec{p_{Wf}} + \vec{p_{Bf}}$

$= \sqrt{0.0928^2 + 0.056^2}$

$= 0.1084$ kg m s^{-1}

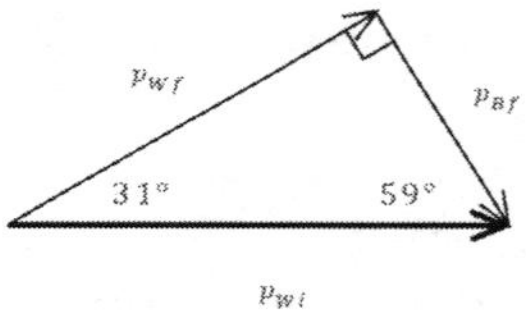

This is the same magnitude as the initial momentum, calculated in part (a), allowing for errors due to limit of reading and rounding. Therefore momentum was conserved in this collision.

e) $\Sigma K_i = \frac{1}{2}m_W{u_W}^2 = \frac{1}{2} \times 0.16 \times 0.68^2 = 0.036992$ J

$\Sigma K_f = \frac{1}{2}m_W{v_W}^2 + \frac{1}{2}m_B{v_B}^2$

$= \frac{1}{2} \times 0.16 \times 0.58^2 + \frac{1}{2} \times 0.16 \times 0.35^2 = 0.036712$ J

Taking into account limit of reading of the graph paper and rounding, total initial and final kinetic energies are the same, so this is an elastic collision.

Part 3: Electricity and magnetism - Answers

11.2 Charging objects

1. For a long time, people have known that rubbing some materials together could leave materials with a property called charge, which results in either attraction or repulsion between them. The force between charges is called ***electrostatic force***. A charge can exert a force on other charges without touching them, due to the electric field around a charge.
   To explain both attraction and repulsion, two types of charge are needed, called positive and negative. Like charges repel each other, so two positive objects will repel and so will two negative objects. Unlike charges attract, so a positive and negative object will attract each other.
   Where does the charge come from? Atoms in objects already have charges in them – positive protons in the nucleus and negative electrons outside the nucleus. Protons and electrons have the same magnitude of charge, but opposite signs. A neutral atom has equal numbers of protons and electrons, so it has no charge overall.
   Charging objects only involves moving charges to different locations, not creating or destroying them, as charge is conserved. Generally, only electrons are removed or added to atoms, as their mass is much lighter than protons and they are found on the outside of the nucleus. So if rubbing an object leaves the object with a positive charge, then the object has had electrons removed, leaving the object with more positive charge than negative charge. Similarly, a negatively charged object has had electrons added, resulting in more negative charge than positive charge in the object.
   If there are too many electrons on one object compared to another object nearby, the excess electrons can jump from one object to the other, resulting in sparks of lightning. You may see a spark and feel an electric shock if you shuffle across carpet, then touch a metal doorknob – that is due to you building up charge as you shuffle, then electrons jumping between you and the doorknob!
   The electricity running through the circuits of your house is actually a flow of electrons.
2. Charging by ***friction*** involves an object being rubbed by a different material. This can either add or remove electrons from the object, depending on the nature of the material. Some objects have a greater tendency to gain electrons than others. For example, wool is more likely to lose electrons than rubber, so if you rub a rubber balloon with wool, the wool will become positively charged, while the balloon becomes negatively charged. The more you rub, the more electrons are transferred. However, since both wool and the rubber balloon are insulators, charge is only transferred between the areas being rubbed, and does not flow throughout the whole object.
   Charging by ***induction*** involves bringing a charged object close to a neutral one, for example a negatively charged balloon near some tiny pieces of paper with no charge overall. The electrons in the paper will be repelled by the balloon to the far side of the paper, leaving the near side of the paper positively charged. The near side of the paper is then attracted to the balloon.
   Charging by ***contact*** involves touching a neutral object with a charged one. If one of the objects is a good electrical conductor, such as a metal, then charge can travel easily through the material and become evenly distributed throughout it. However, if the metal object is to retain its charge, then it needs to be insulated from its surroundings and supported by a material that does not allow charge to move through easily. Otherwise, if the metal is also in contact with other objects, especially conductors, then it will not retain its charge, since it can easily flow to the other objects too.
   A metal strip in contact with the ground can transfer excess charge through it to the Earth. The process of removing excess charge from an object by transferring it to another large object is called ***earthing*** or ***grounding***. This process is used as a safety feature in case of lightning strikes or electrical surges.
3. a) B is positive, C is negative, D is negative.
   b) B and D would attract, since they have opposite charges.
4. a) Neutral, because there are equal numbers of protons and electrons.
   b) Electrostatic force, because negative electrons are attracted to positive protons as they are unlike charges.
   c) As positive protons would repel each other, the nucleus would be unstable unless there was also another force that attracted them.
5. Electrons in the stream of lightning strike the earth wire and are conducted through the metal support tower to the ground, where they spread out across the whole Earth. The electrons cannot travel as easily through the insulating discs to the power lines, since charge does not flow easily through insulators.
6. a) Charge on cloth: neutral
   Charge on pipe: neutral
   Force between them: none
   b) Type of charging: friction
   Which way the electrons moved: from cloth to pipe
   c) Charge on cloth: positive
   Charge on pipe: negative
   Force between them: attractive
   d) Charge on can: neutral
   e) Type of charging: induction
   Electrons in the can are repelled away from the negative pipe. The near side of the can is attracted to the pipe, so the can will roll towards the pipe.
7. a) According to the triboelectric series, hair is more likely to lose electrons than silk. Therefore when hair is rubbed by silk, friction transfers electrons from your hair to the silk, leaving your hair positive and the silk negative. As your hair and silk shirt are the opposite charges, they attract. The positive hairs repel each other, so your hair stands up.
   b) i)

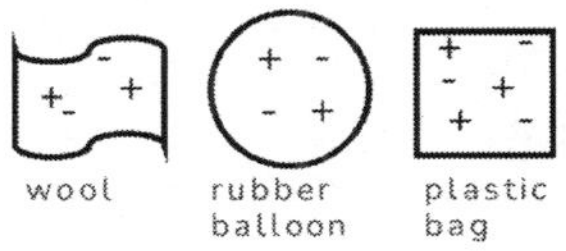

   ii)

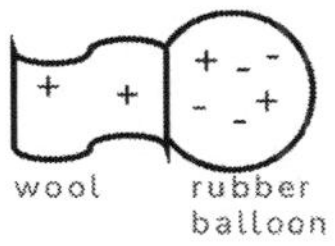

   iii)

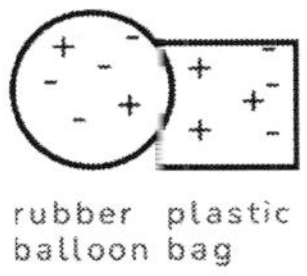

8.

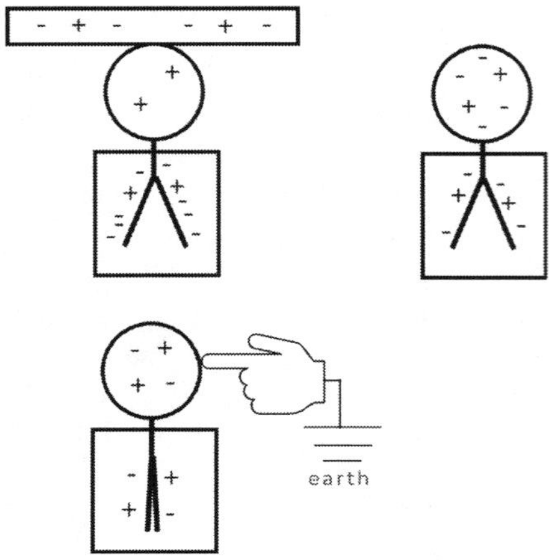

11.3 Measuring charge

1. Charge is a property of matter which causes objects to experience a force in an electric field. Charge is a scalar quantity, but can have a positive or negative sign. The unit used to measure charge is the ***coulomb*** (C). This unit was named a long time before electrons were discovered, so it does not match a whole number of electrons. However, once electrons were discovered in 1897, the charge of one electron was found to be -1.602 x 10^{-19} C. This means the charge of one proton is +1.602 x 10^{-19} C.
   Electricity experiments show that charge is conserved. That means that in a closed system, the total amount of charge stays the same. In other words, the total number of protons stays the same, and the total number of electrons too – they just get transferred between objects.
   Current is the flow of charge. In other words, it is the rate at which charge passes a point:
   $$I = \frac{q}{t}$$
   where: I = current (A)
   q = amount of charge passing a point (C)
   t = time in which charge passes a point (s)

   The unit for current is the ***ampere*** (A), which is one coulomb passing a point per second.
2. $\frac{1.5\times10^{-10}}{1.602\times10^{-19}} = 936329588 = 9.4 \times 10^8$ electrons lost
3. a) Total charge $= +5.0\ \mu C - 8.0\ \mu C = -3.0\ \mu C$
   Charge on each sphere $= -\frac{3.0}{2} = -1.5\ \mu C$
   b) $-1.5 - 5.0 = -6.5\ \mu C$ from B to A
   c) $\frac{6.5\times10^{-6}}{1.602\times10^{-19}} = 4.1 \times 10^{13}$ electrons
4. Ampere second (As)
5. Number of electrons $= \frac{4.0}{1.602\times10^{-19}} = 2.5 \times 10^{19}$ C

11.4 Electrostatic force

1. In 1785, Charles de Coulomb worked out an experimental formula for the electrostatic force between two charges. He brought a charged ball close to another charged ball stuck on the end of a horizontal beam that could rotate. The electrostatic force could be measured using the deflection of the beam. This force was directly proportional to the product of the magnitudes of the charges and inversely proportional to the square of the distance between them:
   $$F_E = \frac{1}{4\pi\varepsilon_0}\frac{q_1 q_2}{r^2}$$
   where: F_E = electrostatic force between charge 1 and 2 (N)
   q_1 and q_2 = charges 1 and 2 respectively (C)
   r = distance between charge 1 and 2 (m)
   $\varepsilon_0 = 8.854 \times 10^{-12}$ $A^2\ s^4\ kg^{-1}\ m^{-3}$
   = electric permittivity constant for a vacuum

   Electric permittivity ε means how much a medium interferes with an electric field. A ***medium*** is the material the electric field is going through. If the medium is a vacuum, there are no charges in it, so it has the lowest possible electric permittivity, sometimes called the ***permittivity of free space ε_0***. Any other medium will have atoms with charges in it, which will experience a force and therefore store some electrical potential energy.
   To simplify calculations, we usually assume air has the same permittivity as a vacuum. However, water has a high electric permittivity compared to a vacuum, so on humid days, static electricity experiments don't work as well, since the force between charged objects is reduced.
2. a) Let the original force be $F_E = \frac{1}{4\pi\varepsilon_0}\frac{q_1q_2}{r^2}$
   New force is $F_E' = \frac{1}{4\pi\varepsilon_0}\frac{(2q_1)q_2}{r^2} = 2 \times \frac{1}{4\pi\varepsilon_0}\frac{q_1q_2}{r^2} = 2F_E$
   Force doubles.
   b) New force is $F_E' = \frac{1}{4\pi\varepsilon_0}\frac{q_1q_2}{(2r)^2} = \frac{1}{4} \times \frac{1}{4\pi\varepsilon_0}\frac{q_1q_2}{r^2} = \frac{1}{4}F_E$
   c)

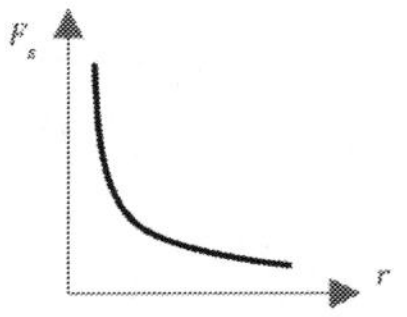

   d)

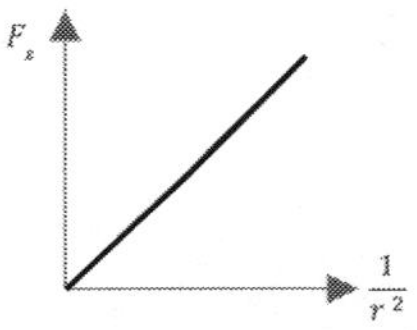

3. a) $F_E = \frac{1}{4\pi\varepsilon_0}\frac{q_1q_2}{r^2}$
   $= \frac{1}{4\pi\times8.854\times10^{-12}} \times \frac{3.0\times10^{-6}\times6.0\times10^{-6}}{0.01^2}$
   = 1600 N repulsive
   b) No, the charges exert the same magnitude of force on each other as the calculation is the same. The forces are also an action-reaction pair.
4. C
5. Let the charge of each balloon be q.
   $\tan 10° = \frac{F_E}{mg}$
   $F_E = \frac{1}{4\pi\varepsilon_0}\frac{q_1q_2}{r^2} = mg\tan 10°$
   $\frac{1}{4\pi\times8.854\times10^{-12}}\frac{q^2}{0.30^2} = 0.0035 \times 9.8\tan 10°$
   $q = 2.46 \times 10^{-7}$ C
   This charge could either be positive or negative.

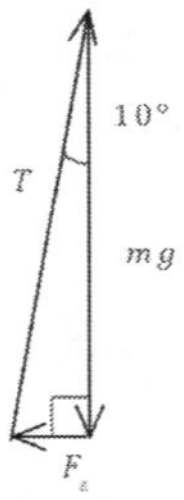

6. Substituting units for all symbols in the equation $F_E = \frac{1}{4\pi\varepsilon_0}\frac{q_1q_2}{r^2}$:
   $N = \frac{1}{\varepsilon_0}\frac{C^2}{m^2}$
   From $F = ma$:
   $kg\ m\ s^{-2} = \frac{1}{\varepsilon_0}\frac{C^2}{m^2}$
   $\varepsilon_0 = \frac{C^2\ s^2}{kg\ m^3} = \frac{A^2\ s^2\ s^2}{kg\ m^3} = A^2\ s^4\ kg^{-1}\ m^{-3}$

7. a) $F_E = \frac{1}{4\pi\varepsilon_0}\frac{q_1 q_2}{r^2}$

$= \frac{1}{4\pi \times 8.854\times10^{-12}} \times \frac{8\times1.602\times10^{-19}\times1.602\times10^{-19}}{(9.57\times10^{-11})^2}$

$= 2.01 \times 10^{-7}$ N repulsive

b) $F_E = \frac{1}{4\pi\varepsilon_0}\frac{q_1 q_2}{r^2}$

$= \frac{1}{4\pi \times 8.854\times10^{-12}} \times \frac{1.602\times10^{-19}\times1.602\times10^{-19}}{(0.5\times9.57\times10^{-11})^2}$

$= 1.01 \times 10^{-7}$N attractive

c) $F_E = \frac{1}{4\pi\varepsilon_0}\frac{q_1 q_2}{r^2}$

$= \frac{1}{4\pi \times 8.854\times10^{-12}} \times \frac{8\times1.602\times10^{-19}\times1.602\times10^{-19}}{(0.5\times9.57\times10^{-11})^2}$

$= 8.06 \times 10^{-7}$N attractive

d) The hydrogen and oxygen nuclei repel each other with a force of 2.01×10^{-7}N. However, the shared pair of electrons would together exert a combined force of around 2.02×10^{-7}N attracting the hydrogen nucleus, as well as a combined force of 16.1×10^{-7} N attracting the oxygen nucleus to themselves, which exceeds the repulsive force between the hydrogen and oxygen nuclei.

e) According to the calculations in parts (b) and (c), the shared electrons are much more strongly attracted to the oxygen nucleus than the hydrogen nucleus. Therefore they will actually be positioned closer to the oxygen nucleus, making the oxygen end more negative and the hydrogen end more positive.

11.5 Visualising electric fields

1. A field is a region of space where a body experiences a non-contact force. An electric field is a field in which a charge experiences a force. Electric field is a vector quantity, since it has both magnitude and direction.
The direction of an electric field is defined by the direction in which a positive charge would experience a force in this field. In other words, each electric field line tells us the direction a positive charge would accelerate, if it were placed on that field line and were free to move.
Electric field lines never cross – if there is more than one source of electrostatic force on a charge, then the direction of the electric field line shows the direction of the resultant force on the charge.
Electric field lines are perpendicular to the surface of a charged conductor, since the charge is free to move throughout the object. This means that any components of the electric field that are not perpendicular to the surface will be balanced, making the resultant force at any point on the surface perpendicular to it.
2. a)

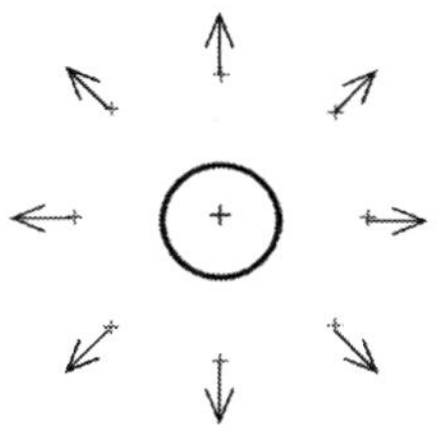

b)

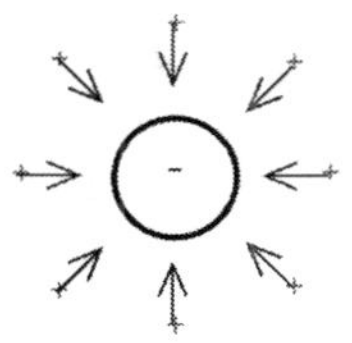

3. A is negative, as the electric field lines point towards the charge. B is positive, as the electric field lines point away from the charge. The electric fields lines around both charges have the same spacing, so their charges have equal magnitude.
4. A is positive, B is negative.
5. The ***electric field strength***, or magnitude of an electric field, tells us how much force per coulomb of charge an object would experience in the electric field. The distance between electric field lines represents electric field strength. Field lines closer together represents a stronger electric field. So the field lines near a point charge (a single charge at a point) are closer together, and get further apart away from the point charge, representing that the electric field gets weaker. This is because the force F_E experienced by a charge q_1 due to another charge q_2 decreases inversely proportionally with the square of the distance between the charges r^2, according to:

$$F_E = \frac{1}{4\pi\varepsilon_0}\frac{q_1 q_2}{r^2}$$

In a ***uniform electric field***, the field lines are the same distance apart, indicating that the electric field strength is the same throughout.
6. a) Both A and B are negative.
b) The electric field in the middle is zero. Both A and B would attract a positive charge in opposite directions with the same magnitude of force, so they balance.
7. a)

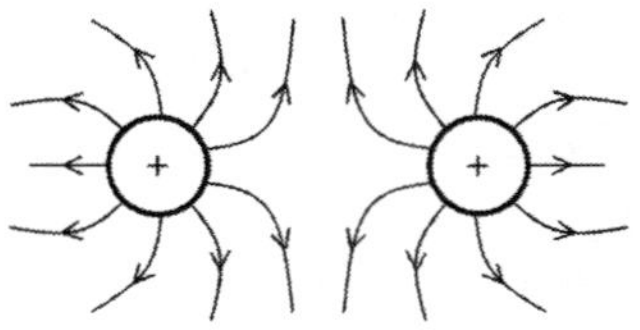

b)

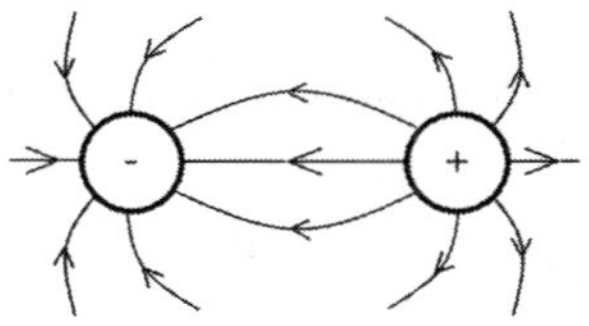

8. a) left
b) The force is equally strong at a and b, since parallel plates create a uniform electric field.
9. When placed in an external electric field, the electrons in a conductor are free to move. Therefore electrons in the metal box will be forced to the left, leaving the right side positive. There is now an induced electric field inside the metal box, in the opposite direction to the external electric field. These two fields cancel each other out, so the resultant electric field is zero.
10. The positively charged balloon creates an electric field radiating out from the balloon. When water molecules are in this electric field, they rotate so that the negative oxygen end is attracted to the positive balloon. This induced negative side of the water stream is attracted to the positive balloon as they have positive charge, therefore the stream of water bends towards the balloon.

11.6 Electric field strength

1. Electrostatic force between two charges increases as the magnitude of both charges increases. Likewise, gravitational force, or weight, between two masses increases with mass. It is therefore useful to have some way to compare field strength due to an object in a way that does not depend on the properties, such as charge or mass, of another object you put in that field.
For example, ***gravitational field strength*** is the gravitational force $\vec{F_G}$ per unit of mass m. Since gravitational force $\vec{F_G}$, or weight, is given by:

$$\vec{F_G} = m\vec{g}$$

we can rearrange this to see that numerically, acceleration due to gravity $\vec{g}$ is the same thing as gravitational field strength:

$$\vec{g} = \frac{\vec{F_G}}{m}$$

Similarly, ***electric field strength*** is the electrostatic force $\vec{F_E}$ acting on an object per coulomb of charge q:

$$\vec{E} = \frac{\vec{F_E}}{q}$$

where: $\vec{E}$ = electric field strength at a point (N C^{-1})
$\vec{F_E}$ = electrostatic force exerted on a charge at that point (N)
q = charge (C)

This equation is often rearranged to make the electrostatic force the subject:

$$\vec{F_E} = q\vec{E}$$

2. $m\vec{a} = q\vec{E}$
$6.645 \times 10^{-27} \times \vec{a} = q \times 120$
$q = +3.20 \times 10^{-19}$ C
3. Let up be positive
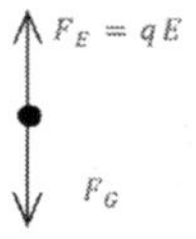

$F_E = F_G$
$qE = mg$
$-1.5 \times 10^{-6} \times E = 1.2 \times 10^{-4} \times 9.8$
$E = -784$ N C^{-1} down
4. $\tan\theta = \frac{qE}{mg}$
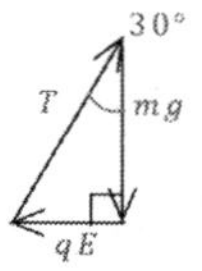

$qE = mg\tan\theta$
$-3.0 \times 10^{-6} \times E$
$= 4.8 \times 10^{-5} \times 9.8 \tan 30°$
$E = -91$ N C^{-1} (to the right)
5. Since the electrostatic force between two charges is:
$$F_E = \frac{1}{4\pi\varepsilon_0}\frac{q_1 q_2}{r^2}$$
the electric field due to q_1 on q_2 is:
$$\vec{E} = \frac{\vec{F_E}}{q_2} = \frac{1}{4\pi\varepsilon_0}\frac{q_1}{r^2}$$
As can be seen from the equation, this value does not depend on the second charge, since it cancels out.
6. $\vec{E} = \frac{1}{4\pi\varepsilon_0}\frac{q}{r^2}$
$\vec{E_1} = \frac{1}{4\pi\varepsilon_0}\frac{3q}{r^2} = 3\vec{E}$
$\vec{E_2} = \frac{1}{4\pi\varepsilon_0}\frac{q}{(3r)^2} = \frac{1}{9}\vec{E}$
Therefore tripling the charge triples the electric field strength, but tripling the distance reduces the field strength by 1/9.
7. a)
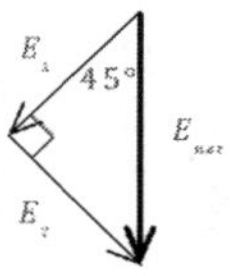

b) Using geometry, the displacement of the electron from each proton is in a direction 45° to the midline and distance between them r is:
$$r = \sqrt{\left(\frac{7.4\times10^{-11}}{2}\right)^2 + (3.7 \times 10^{-11})^2} = 5.2 \times 10^{-11} \text{ m}$$
Net electric field due to both protons is along midline, away from midpoint:
$$E_{net} = \sqrt{\left(\frac{1}{4\pi\varepsilon_0}\frac{q_1}{r^2}\right)^2 + \left(\frac{1}{4\pi\varepsilon_0}\frac{q_2}{r^2}\right)^2}$$
$$= \sqrt{2 \times \left(\frac{1}{4\pi \times 8.854\times10^{-12}} \times \frac{+1.602\times10^{-19}}{(5.2\times10^{-11})^2}\right)^2}$$
$= 7.5 \times 10^{11}$ N C^{-1} up
8. The electron will accelerate in the opposite direction to the electric field as it is negative, so it moves towards the midpoint. At the midpoint, there is no electric field, but it will keep moving due to inertia, to the other side past the midpoint. However, the electric field is reversed on the other side, so the electron's acceleration also reverses, slowing down the electron until it stops momentarily. Then the electron is accelerated back towards the midpoint. The pattern continues so the electron oscillates along the midline.

11.7 Electrical potential energy

1. Recall that work is done whenever a force $\vec{F}$ causes a displacement $\vec{s}$:
$$W = \vec{F}.\vec{s}$$
Energy is required to do work, so work is also the amount of energy transformed from one type to another. For example, you do work to lift an object off the ground, as you need to apply a force at least as big as gravitational force to displace it upwards. In doing so, you transform kinetic energy into gravitational potential energy. The work you do can also be calculated from the change in one form of energy, such as change in potential energy:
$$W = \Delta U$$
If you let the object go, it loses gravitational potential energy and gains kinetic energy. This time, the gravitational field does the work, so the work you do is negative.
Similarly, an electric field exerts a force on a positive charge in the same direction as the field. The charge gains kinetic energy by losing electrical potential energy, and the electric field does the work. Since the electrostatic force $\vec{F_E}$ on a charge q in an electric field $\vec{E}$ is given by:
$$\vec{F_E} = q\vec{E}$$
the work done on a charge in an electric field is given by:
$$W = \vec{F}.\vec{s} = qEd$$
where d is the displacement of the charge from its starting position.
On the other hand, to make a positive charge move in the opposite direction to the electric field, something other than the electric field needs to do the work and the charge gains electrical potential energy. ***Electrical potential energy*** is the energy stored due to the position of charges relative to each other.
For a negative charge, work is done by the electric field to move it in the opposite direction to the field, whilst work is done by something else to move it in the same direction to the field.
The terminals of a battery provide a source of electrical potential energy, due to a chemical reaction inside the battery. The electrical potential energy is converted into kinetic energy if the battery is attached to an electric circuit where charges are able to move, producing an electric current.
2. a) downwards
b) $\vec{F_E} = q\vec{E} = 1.602 \times 10^{-19} \times 1 \times 10^8$
$= 1.602 \times 10^{-11}$ N downwards
c) $W = q\vec{E}d = 1.602 \times 10^{-19} \times 1 \times 10^8 \times 2$
$= 3.204 \times 10^{-11}$ J
d) $W = \Delta K$
$\frac{1}{2}mv^2 = \frac{1}{2} \times 1.673 \times 10^{-27} \times v^2 = 3.204 \times 10^{-11}$
$v = 2.0 \times 10^8$ m s^{-1} downwards
3. $W = \Delta K$
$q\vec{E}d = \frac{1}{2}mv^2$
$-1.602 \times 10^{-19} \times \vec{E} \times 0.15$
$= \frac{1}{2} \times 9.109 \times 10^{-31} \times (1.2 \times 10^8)^2$
$\vec{E} = -2.7 \times 10^5$ NC^{-1} in the opposite direction to the electron's motion

11.8 Electrical potential and potential difference

1. Recall electric field strength is a measure of electrostatic force per unit charge, and it is a useful quantity to define because it is independent of the charge put into the field. Similarly, it is useful to define ***electrical potential***, or ***voltage*** at a point, as the electrical potential energy per unit charge, measured in volts (V):

   $$V = \frac{U}{q}$$

   where: V = electrical potential or voltage at a point (V)
   U = electrical potential energy (J)
   q = charge (C)

   Electrical potential is a scalar quantity, but it can have a positive or negative sign, depending on whether voltage drops or increases.
   Just as the ground is often the reference point for zero gravitational potential energy, the negative or earthed terminal that positive charges accelerate towards is taken as zero electrical potential.
   So between a positive and negative plate, the positive plate would be at the highest potential and the negative plate would be at the lowest potential. In other words, a positive charge would have the most electrical potential energy at the positive plate and the least at the negative plate, because it would be lost as kinetic energy.
   Conversely, a negative charge would have the least electrical potential energy at the positive plate and the most at the negative plate. A negative charge would accelerate from low electrical potential to high electrical potential. The value and sign of electrical potential at each point does not change because the charge is negative. This is because for a negative charge, in the formula:

   $$V = \frac{U}{q}$$

   both U and q would be negative, giving a positive V.
   Equipotential lines are imaginary lines connecting points at the same electrical potential. As an analogy, on a flight of steps, all points on each step are at the same gravitational potential. No work needs to be done to move at constant velocity horizontally along a step, due to inertia. However, work must be done to go up one step. Similarly, in an electric field, no work is done when a charge moves along an equipotential line. Therefore equipotential lines are always perpendicular to electric field lines.
2. a) No. Since field lines show the direction of force on a positive charge, an electron would accelerate in the opposite direction, losing potential energy to gain kinetic energy.
   b) No. Since $F = qE$ and the electric field is uniform, the electron experiences the same force.
   c) Yes. A positive charge would have more potential energy per unit charge at A, since it would accelerate towards B, so electrical potential is also higher, according to $V = \frac{U}{q}$. It does not change for a negative charge since both U and q are negative.
   d) No. They have the same electrical potential because no force is required to move a positive charge from B to C, since the electric field only exerts force parallel to the field lines. Hence no work needs to be done on the charge.
3. a)

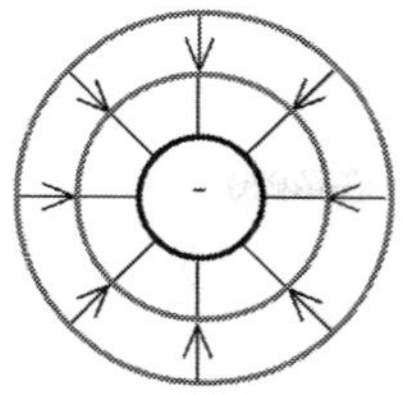

   b)

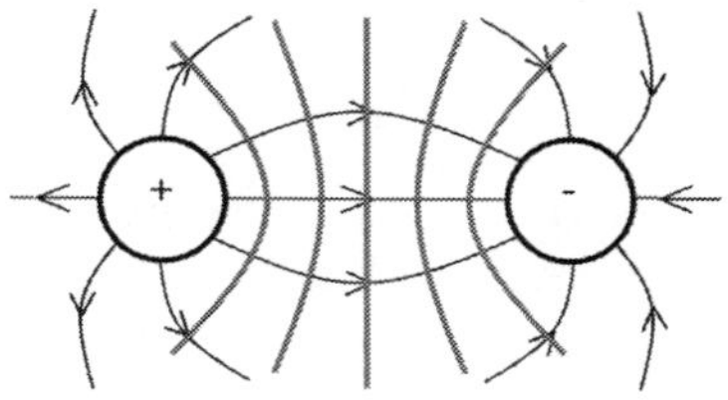

4. Potential difference is related to work done on a charge by:

   $$\Delta V = \frac{\Delta U}{q} = -\frac{W}{q}$$

   where: ΔV = potential difference between two points (V)
   ΔU = change in electrical potential energy (J)
   W = work done on the charge (J)
   q = charge (C)

   This formula is often rearranged to give:

   $$W = -q\Delta V$$

   If the electrical potential decreases between two points, then the change in electrical potential ΔV is negative. The magnitude is also known as ***voltage drop*** V, so the formula is written as:

   $$W = qV$$

5.

A to B	B to C	A to C	A to D	B to D
-6 V	0	-6 V	-2 V	4 V
1.5 J	0	1.5 J	0.5 J	-1 J
Field	Neither	Field	Field	External

6. $W = qV = \Delta K = \frac{1}{2}mv^2$

   $1.6 \times 10^{-19} \times 20 \times 10^3 = \frac{1}{2} \times 2.2 \times 10^{-25} \times v^2$

   $v = 170000$ m s^{-1}

11.9 Potential difference and electric field strength

1. Recall that ***electric field strength*** E is defined as the electrostatic force F_E acting on an object per coulomb of charge q:

   $$E = \frac{F_E}{q}$$

   Recall that work done W moving a charge through a distance d in an electric field E is:

   $$W = \vec{F}.\vec{s} = qEd \text{ ...(1)}$$

   Also, work done on a charge can be calculated using the potential difference V between the two points that the charge moves between:

   $$W = qV \text{...(2)}$$

   Combining equations 1 and 2 gives:

   $$qV = qEd$$
   $$V = Ed$$

This gives a second way to define ***electric field strength***, as the potential difference per <u>metre</u>:

$$E = \frac{V}{d}$$

where: E = <u>electric field strength</u> (V m^{-1})
V = potential difference between two points (V)
d = distance between the same two points, <u>parallel</u> to the electric field (m)

This equation can only be used for a <u>uniform</u> electric field, since it is based on an assumption that the force was the same throughout the displacement over which work was done. This equation provides a practical way to determine electric field strength experimentally, since the voltage between two plates can be measured with a <u>voltmeter</u> and the distance between them can be measured with a <u>ruler</u>.

2. a) $E = \frac{V}{d} = \frac{12}{0.06} = 200$ V m^{-1} so the electric field strength would double.
   b) As $\vec{F} = q\vec{E}$ and the electric field strength has doubled, the force on a charge also doubles.
   c) As the voltage between the plates has not changed, and the gain in kinetic energy is given by $W = qV$, which is independent of distance between the plates, reducing the distance will not change the gain in kinetic energy. Even though the force on the charge has increased, it has half the distance over which to accelerate the charge, so the work done on the charge, $W = \vec{F}.\vec{s}$, does not change.
3. Using $E = \frac{V}{d}$ gives V m^{-1} as the unit. Using $\vec{E} = \frac{\vec{F_E}}{q}$ gives N C^{-1} as the unit.
4.

Quantity	Definition	Formula	Unit	Vector or scalar?
Electrostatic force	Force between charges	$F = qE$	N	Vector
Electrical potential energy	Energy stored due to the position of charges relative to each other	$W = Fs$ $= -\Delta U$	J	Scalar
Electric field	Electrostatic force acting on an object per coulomb of charge OR Potential difference per metre	$E = \frac{F}{q} = \frac{V}{d}$	NC^{-1} Vm^{-1}	Vector
Potential difference	Voltage, or change in electrical potential energy per unit charge	$V = \frac{W}{q}$	V	Scalar

Electric field and potential difference are independent of charge.

5. C
6. a) $E = \frac{V}{d} = \frac{500}{0.25} = 2000$ V m^{-1}
   b) The dipoles will align themselves in the electric field so that the negative end of the dipole is attracted to the positive plate, and the positive end of the dipole is attracted to the negative plate. Therefore, the dipoles will create an electric field in the opposite direction to the electric field from the plates, reducing the total electric field between the plates.

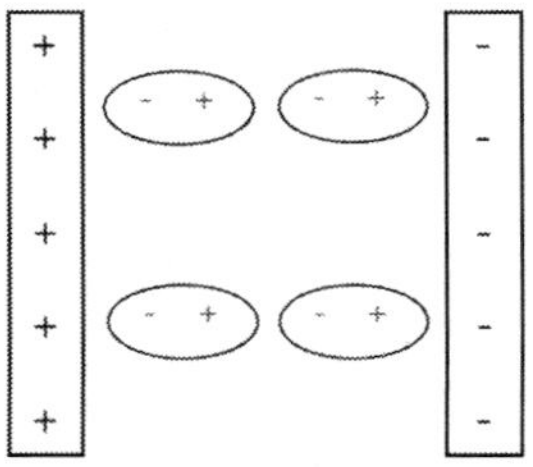

   c) No for two reasons. The given formula for electrostatic force is between two point charges, not two parallel plates. The constant ε_0 is the permittivity of free space, where there is a vacuum between the charges, and in a capacitor the space is filled with some material.
7. a) $\Delta K = W$
   $\frac{1}{2}mv^2 = qV$
   $v = \sqrt{\frac{2qV}{m}}$
   b) Let the original initial speed be v and new final speed be v'.
   Substitute voltage of $2V$ to find new final speed:
   $v' = \sqrt{\frac{2q \times 2V}{m}} = \sqrt{2}v$
   The final speed will increase by a factor of $\sqrt{2}$.
8. a)

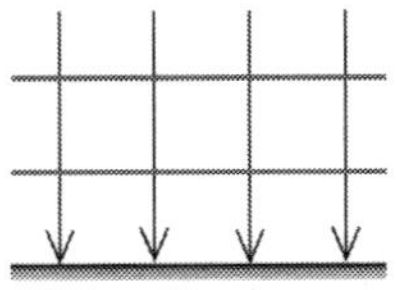

   b) $V = Ed = 100 \times 1.5 = 150$ V
   c) Let up be positive: $F_{net} = F_E - F_G$
   $ma = qE - mg$
   $0.2 \times 10^{-6} \times a = 20 \times 10^{-9} \times 100 - 0.2 \times 10^{-6} \times 9.8$
   $a = 0.2$ m s^{-2} up

12.2 Circuit basics

2. When objects rub together, charges may be transferred from one object to another, causing a <u>charge imbalance</u>. However, if there is a continuous flow of charge, then an <u>electric current</u> is produced. ***Current*** is defined as the amount of charge flowing past a point in a circuit per second. The unit used to measure current is the <u>ampere</u>. An instrument called an <u>ammeter</u> is used to measure current.
Charges flow easily through ***conductors***, such as <u>metals</u>. Charge does not flow easily through ***insulators***, such as air, rubber or <u>plastic</u>. In other words, <u>conductors</u> have low resistance compared to <u>insulators</u>. ***Resistance*** is how well something <u>resists</u> the flow of charge, therefore doing work by transforming electrical energy into some other form. The unit used to measure resistance is the <u>ohm</u>.
If there is an <u>unbroken</u> loop of conducting material, then current can flow, and this is called a ***closed circuit***. However, if the path is <u>broken</u>, for example by an air gap, then current cannot flow, and this is an ***open circuit***. So if we use copper wires to connect components in a circuit, then we provide a <u>low resistance</u> path for current to flow through. However, to switch off a circuit, we can create an air gap that has <u>high resistance</u> and does not allow charges to flow across.
Charges in a circuit also require <u>electrical potential</u> energy to transform into kinetic energy, in order to produce a current. In other words, electrical energy in a circuit is really the <u>kinetic</u> energy of moving charges. The electrical potential energy can be provided in the form of a battery or electrical powerpoint, which produces an <u>electric field</u> between the positive and negative terminals. The electrical potential energy used per coulomb of charge

between two points in a circuit is also known as potential difference or *voltage*. The unit used to measure voltage is the volt. An instrument called a voltmeter is used to measure voltage.

3. Circuits are often drawn in rectangular paths, with components placed on the sides of the rectangles, but never the corners. The current flows from the positive terminal, represented by the long line on the battery, through the circuit, then to the negative terminal, represented by the short line. In Circuit A, shown to the right, starting from the positive terminal, current passes through a wire, ammeter, wire, light bulb, wire, closed switch, then a final wire.
   All the components in Circuit A are connected in ***series***, meaning that current passes through the same path to go through them.
   An ammeter must always be connected in series with the component it is measuring the current through, in order to work.
   In Circuit B, shown to the right, starting from the positive terminal, current passes through a wire, but then splits into two different paths. The first path consists of two resistors in series, joined by wires together then back to the battery. The second path has a voltmeter, connected in parallel with the first resistor. ***Parallel*** means current splits up to pass through more than one path to reach the components.
   A voltmeter must always be connected in parallel between the points it is measuring the potential difference across. Therefore this voltmeter is only measuring the voltage across the first resistor and not the second.
4. a) A battery is connected to two light bulbs in parallel with each other. The first is in series with an open switch. The second is in series with a closed switch.
   b) Only the bottom one
5. a) A battery is connected to a variable resistor, light bulb and ammeter in series. A voltmeter is in parallel with the light bulb.
   b) Voltage across the light bulb
6.

a)

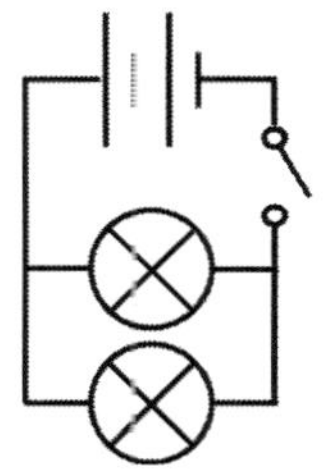

b)

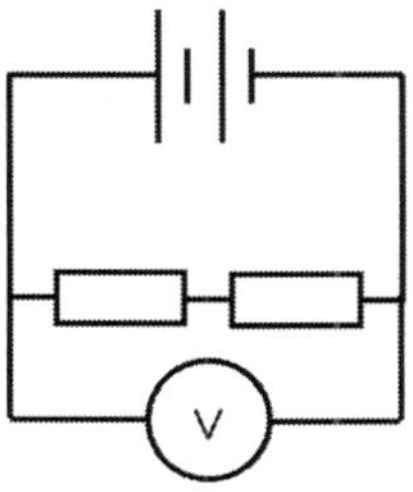

c)

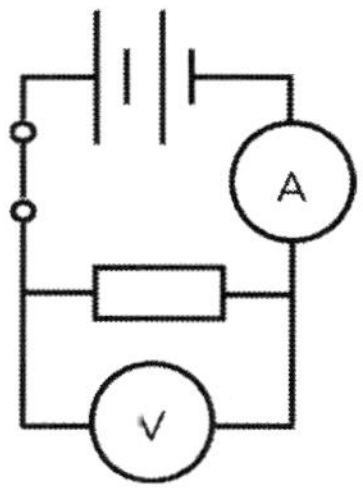

d)

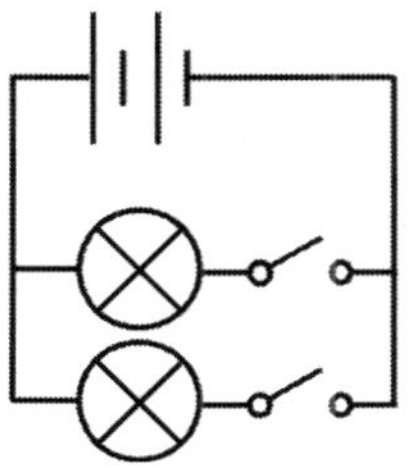

12.3 Voltage and current

1. ***Voltage***, or potential difference, is the electrical potential energy per unit charge and is measured in volts:

$$V = \frac{W}{q}$$

where: V = voltage across two points (V)
W = work done or change in electrical potential energy (J)
q = charge (C)

This formula can also be written with work done as the subject:

$$W = qV$$

Inside a battery, chemical reactions transform chemical potential energy into electrical potential energy, resulting in a positive charge at one terminal, called the ***cathode***, and a negative charge at the other terminal, called the ***anode***. An electric field exists from the cathode to the anode, in the same direction as the flow of positive charge.
Current is the rate at which positive charge flows:

$$I = \frac{q}{t}$$

where: I = current (A)
q = amount of charge passing a point (C)
t = time in which charge passes a point (s)

By definition, the ***cathode*** is the terminal the current leaves a power source from, and the ***anode*** is the terminal the current enters the power source.
Historically, current was defined before anyone knew what kind of charges moved, but after J. J. Thomson discovered the electron in 1897, we now know that it is in fact the negative electrons that move in metals. We are still stuck with the same definition of current though, so current is still taken to flow from the positive to negative terminal when electrons flow the opposite way. In other words, if one coulomb of electrons pass a point in a circuit per second, then we say that one ampere of current flowed the other way.
When current only travels in one direction, it is called ***direct current***, or DC. For example, batteries are a DC source.
The current from an electric socket, on the other hand, switches direction periodically, and is called ***alternating current***, or AC. In Australia, the top two holes switch between being the positive and negative terminals 50 times per second, or at a frequency of 50 Hz, with a peak voltage of around 240 V. The third hole is earthed as a safety feature.

2.

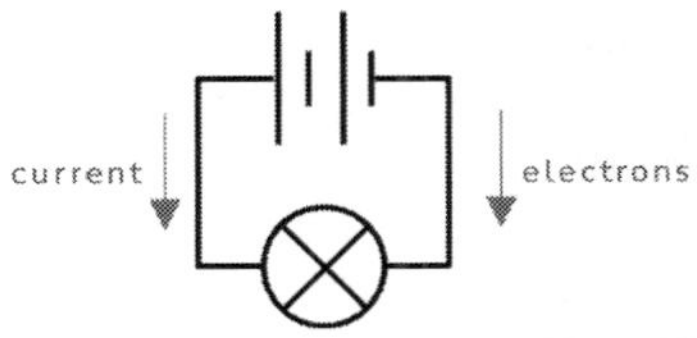

3. The input electricity supply is 50 to 60 Hz alternating current, meaning the current direction switches back and forth 50 to 60 times per second, whereas the output is direct current, meaning current goes in one direction. The input voltage of 100-240 V is higher than output voltage of 5 V, meaning the output has less electrical potential energy per unit charge output. However, the input current of 0.18 A is lower than output current of 2 A, meaning that more charge passes a point per second in the output circuit.

4. $I = \frac{q}{t}$

   $q = It = 1 \times 1 = 1\ \text{C} = \frac{1}{1.602\times10^{-19}}$ electrons

   $= 6.24 \times 10^{18}$ electrons

5. $I = \frac{q}{t}$

   $q = It = 2 \times 45 \times 60 = 5400\ \text{C}$

6. $I = \frac{q}{t}$

   $q = It = 60 \times 10^{-3} \times 0.25 = 0.015\ \text{C}$

   $= \frac{0.015}{1.602\times10^{-19}}$ electrons $= 9.36 \times 10^{16}$ electrons

7. a) $I = \frac{q}{t}$

   $q = It = 1.8 \times 35 \times 60 = 3780\ \text{C}$

   b) $W = qV = 3780 \times 1.5 = 5670\ \text{J}$

8. a) $W = qV$

   $q = \frac{W}{V} = \frac{0.16}{2} = 0.08\ \text{C}$

   b) $I = \frac{q}{t} = \frac{0.08}{5} = 0.016\ \text{A}$

12.4 Resistance, ohmic and non-ohmic conductors

1. The ***resistance*** of any component in an electric circuit is measured in ohms and is defined as the ratio of voltage applied across it to the current through it:

$$R = \frac{V}{I}$$

where: R = resistance (Ω)
V = voltage between two points (V)
I = current (A)

This equation tells us that the higher the voltage, the higher the current, and that this ratio is the same if the resistance of the material stays the same. This equation also predicts that if the same voltage is applied across different materials, conductors with low resistance will have a higher current through them than insulators with high resistance.

Resistance in a circuit will transform electrical energy into some other form. For example, a light bulb transforms electrical energy into light. Metal wires transform electrical energy into heat.

2.

Voltage	Current	Resistance
Electrical potential energy per unit charge	Rate at which charge passes a point	Ratio of applied voltage to current
$V = \frac{W}{q}$	$I = \frac{q}{t}$	$R = \frac{V}{I}$
volt, V	ampere, A	ohm, Ω
Voltmeter 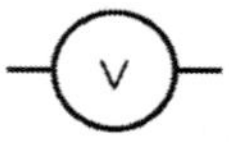	Ammeter 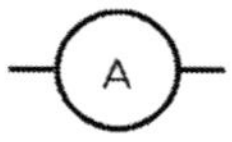	Resistor 

3. $R = \frac{V}{I}$

   $V = IR = 5 \times 2 = 10\ \text{V}$

4. a) silicon 2600 Ω, wet wood 4000 Ω

   b) Same applied voltage, same shape and size of material

   c) Silicon, as it has a lower resistance, and therefore will conduct more current given the same voltage.

5. $R = 50\ \text{m} \times \frac{8.75\ \Omega}{1000\ \text{m}} = 0.4375\ \Omega$

   $R = \frac{V}{I}$

   $I = \frac{V}{R} = \frac{12.9}{0.4375} = 29.5\ \text{A}$

6. If a component stays the same resistance when different voltages are applied to it, then it is described as ***ohmic***. This means that if you experimentally measure the current through a material when different voltages are applied, the graph of voltage on the y-axis and current on the x-axis will be a straight line, with voltage proportional to current. The gradient of the graph will be constant and equal to the resistance. In other words, ohmic conductors obey an experimental trend called Ohm's Law:

$$V = IR$$

where: R = constant resistance (Ω)
V = voltage between two points (V)
I = current (A)

Ohm's Law looks like the same equation as the definition of resistance, but here it is used to describe how the resistance of a component behaves if different voltages are applied to it.

Resistors are designed to be ohmic – that is, they stay the same resistance over a large range of applied voltages. If you switch the direction of the voltage but keep it the same magnitude, the resistance stays the same, meaning that current reverses but stays the same magnitude.

Metal wires are ohmic to some extent – they stay a constant resistance if they can stay at the same temperature, but if they get hot, their resistance increases. For example, a filament light bulb, which contains a thin tungsten wire, is close to ohmic if only switched on for short periods of time, but does not obey Ohm's Law if left on to get hot.

A light emitting diode (LED) is an example of a ***non-ohmic*** conductor – that is, its resistance depends on the applied voltage. LEDs will only light up and let current through in one direction. If the voltage is applied the wrong way, the resistance is enormous, so almost no current flows. If a large enough voltage is applied the wrong way, the LED will break.

Because non-ohmic conductors do not have constant resistance, the gradient of a voltage-current graph cannot be used to calculate the resistance. Instead, you need to use the equation that defines resistance, substituting the actual measurement of voltage and current to calculate resistance at that voltage.

7. a) $R = \frac{V}{I} = \frac{1.5}{2\times10^{-3}} = 750\ \Omega$

   b) $R = \frac{V}{I} = \frac{3}{40\times10^{-3}} = 75\ \Omega$

   c) $R = \frac{V}{I} = \frac{4}{100\times10^{-3}} = 40\ \Omega$

   d) The LED is non-ohmic. As the graph is not linear, resistance does not stay constant and depends on the voltage applied. When the voltage is below 2 V, current is very small, so resistance is huge. As the voltage increases, current increases, so resistance decreases.

8. a) Method:

   1. Set up as shown in the diagram.
   2. Using 2 V on the power pack, switch on the circuit, measure and record the reading on the voltmeter and ammeter. Switch the circuit off.
   3. Repeat step 2 twice more and find the average.
   4. Repeat steps 2 to 3, using 4 V, 6 V, 8 V, 10 V and 12 V.
   5. Graph the voltage vs current to determine whether the graph is linear. If so, the resistor is ohmic.
   6. Calculate the gradient of the graph. This is equal to its resistance.

b)

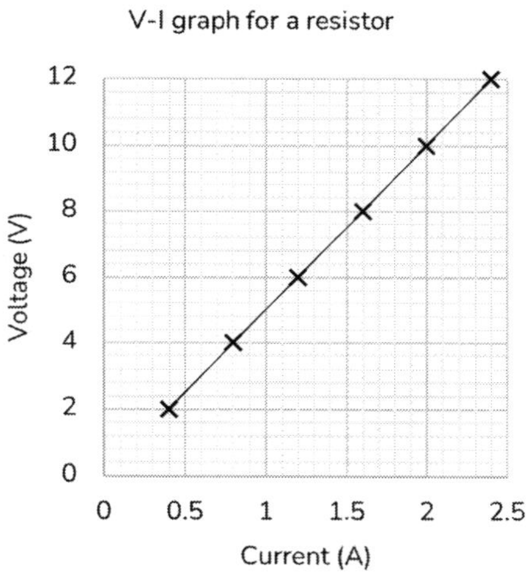

c) $R = gradient = \frac{12-2}{2.4-0.2} = 5\ \Omega$

12.5 Experimental skills – Resistance and resistivity

1. The resistance of a straight conductor depends on the material, its length and cross-sectional area:

$$R = \frac{\rho L}{A}$$

where: R = resistance (Ω)
ρ = resistivity (Ωm)
L = length of the conductor (m)
A = cross-sectional area of the conductor (m^2)

Resistivity is a property of the material and describes how well the material allows current to flow, independent of its size and shape. Since current is the rate at which charge moves past a point, resistivity therefore depends on how much charge is free to move in a material and what obstacles it encounters when it moves. In metals, the electrons in the outermost energy shell of the atom, known as ***valence electrons***, are the most free to move, since they are furthest from the positive nucleus, and therefore experience the least electrostatic force from it. Metals therefore have low resistivity. In non-metals, valence electrons are more strongly attracted to the nucleus, therefore less free to move, so they have high resistivity compared to metals.

As electrons move through a conductor, their movement is hindered by the other parts of the atoms, causing resistance. Collisions between the electrons and other parts of the atoms transform the kinetic energy of the electrons into heat, and this is called ***resistive heating***. Therefore, work is done by resistance in circuits. Resistive heating is useful in a toaster, but wasteful in power lines that distribute electricity between power stations and homes.

If a conductor gets hotter, the atoms in it will have more kinetic energy and are able to collide more with the flow of electrons. This makes it harder for electrons to flow, therefore increasing resistivity.

2. The width of the corridor models how increasing cross-sectional area of a conductor decreases resistance, so current can more easily flow through. The length of the corridor models how increasing the length of a conductor increases resistance.
3. In the formula $R = \frac{\rho L}{A}$, in the numerator unit for resistivity, Ωm, multiplied by the unit for length, m, gives Ωm^2. The m^2 cancels with the unit for cross-sectional area in the denominator.
4. a) Carbon, copper and tungsten are conductors. Glass and sulfur are insulators.
   a) copper
   b) sulfur
5. a) Resistive heating in the nichrome wire leads to risk of burns. Using a switch so that current runs through the nichrome wire for only short times will reduce the amount of heat produced.
   b) Method:
   1. Set up as shown in the diagram.
   2. Clip the two wires with alligator clips so that 20 cm of wire is between them. Record the length of the wire.
   3. Turn on the switch. Measure and record the reading on the voltmeter and ammeter. Turn off the switch.
   4. Calculate the resistance using $R = \frac{V}{I}$.
   5. Repeat steps 3 and 4 two more times and calculate the average resistance.
   6. Repeat steps 2 to 5 using lengths of 40 cm, 60 cm, 80 cm and 100 cm.

   c) As length increases, resistance increases proportionally, because this is predicted by $R = \frac{\rho L}{A}$.
   d) Graphing $R = \frac{\rho L}{A}$ with R on the y-axis and L on the x-axis should give a straight line with the equation $y = mx$, where the gradient is $\frac{\rho}{A}$.
6. a) Length, temperature of the wire
   b) Resistance on the y-axis and $\frac{1}{A}$ on the x-axis. Gradient will be ρL.

12.6 Electrical energy and power

1. Recall that work done on charges in a circuit, measured in joules, is given by:

$$W = qV \quad ...(1)$$

where: W = work done or change in electrical potential energy (J)
q = charge (C)
V = voltage between two points (V)

It is often impractical to measure the amount of charge in a circuit though, and much easier to measure current, using an ammeter. Recall the definition of current is:

$$I = \frac{q}{t} \quad ...(2)$$

where: I = current (A)
q = amount of charge passing a point (C)
t = time in which charge passes a point (s)

Equation (2) can be written as:

$$q = It \quad ...(3)$$

Substituting equation (3) into equation (1) gives:

$$W = VIt \quad ...(4)$$

Therefore, to calculate how much energy a component uses in an electric circuit, you just need to measure the voltage across the component with a voltmeter, the current through it with an ammeter and the time for which the circuit was on.

Recall that power is the rate at which energy is transformed, measured in watts:

$$P = \frac{W}{t} \quad ...(5)$$

where: P = power (W)

W = work done (J)

t = time in which the work was done (s)

Substituting equation (4) into equation (5) gives:

$$P = VI$$

Therefore, the power from an electrical power supply increases with both voltage and current.

2. a) $P = VI = 5 \times 2 = 10$ W
   b) $W = Pt = 10 \times 30 \times 60 = 18000$ J $= 18$ kJ
3. a) $P = VI$
   $120 = 12I$
   $I = 10$ A
   b) $W = Pt$
   $83800 = 120t$
   $t = 698$ s $= 11.6$ min
4. a) $R = \frac{V}{I}$ so $I = \frac{V}{R}$ and $V = IR$
   $P = VI = V \times \frac{V}{R} = \frac{V^2}{R}$
   $P = VI = IR \times I = I^2R$
   b) $P = I^2R$
   $1100 = I^2 \times 75$
   $I = 3.8$ A
5. a) $V = IR$
   $4 = I \times 0.07$
   $I = 57$ A The circuit breaker is triggered.
   b) $P = VI = 4 \times 57 = 230$ W
6. For the 20 Ω resistor: $P = VI = \frac{V^2}{R} = \frac{12^2}{20} = 7.2$ W
   For the 100 Ω resistor: $P = VI = \frac{V^2}{R} = \frac{12^2}{100} = 1.44$ W
   Therefore the same voltage does not mean the same power supplied.
7. a) 1 kWh = 1 kW × 1 h $= 1000 \text{ W} \times 60 \times 60$ s
   $= 3.6 \times 10^6$ J
   b) W = 2 kW × 5 h = 10 kWh
   $= 10 \times 3.6 \times 10^6 \text{ J} = 3.6 \times 10^7$ J
8. a) $V = IR$
   $I = \frac{V}{R} = \frac{1.5}{5} = 0.3$ A Same current for each battery.
   b) $W = Pt = VIt$
   AA battery: $4.05 = 1.5 \times 0.3 \times t$
   $t = 9$ h
   C battery: $12 = 1.5 \times 0.3 \times t$
   $t = 27$ h
   D battery: $18 = 1.5 \times 0.3 \times t$
   $t = 40$ h

12.7 Series circuits

1. In series, current passes through only one path to go through all components. The number of electrons entering a point in a circuit equals the number of electrons leaving it, as charge is conserved. So in the circuit shown, the charge q_1 passing through resistance R_1 equals the charge q_2 passing through resistance R_2:

$$q_1 = q_2 \quad ...(1)$$

Since current is given by $I = \frac{q}{t}$, in any given time t the charge passing a point is $q = It$, so equation (1) also means:

$$I_1 t = I_2 t$$

Therefore the current from the battery I_T and the current through each light bulb respectively, I_1 and I_2, must be the same:

$$I_T = I_1 = I_2 \quad ...(2)$$

This means it would not matter where you put an ammeter in a series circuit, as you would measure the same current.

The change in electrical potential energy in the battery, W_T, equals the sum of the work done by the light bulbs W_1 and W_2, as energy is conserved:

$$W_T = W_1 + W_2$$

Since $W = qV$:

$$qV_T = qV_1 + qV_2$$

As charge is conserved, the voltage supplied to the circuit is equal to the sum of the voltages across each component in series:

$$V_T = V_1 + V_2 \quad ...(3)$$

The resistance of a component is the voltage across it divided by the current through it: $R = \frac{V}{I}$ or $V = IR$. Expressing equation (3) in terms of current and resistance:

$$I_T R_T = I_1 R_1 + I_2 R_2 \quad ...(4)$$

From equation (2), the current throughout the circuit is the same, so:

$$R_T = R_1 + R_2 \quad ...(5)$$

This means that the total resistance in a series circuit is equal to the resistances of the components in series added together.

2. 35 Ω
3. 180 Ω
4. a) $R_T = R_1 + R_2 = 5 + 25 = 25$ Ω
   b) $V = IR$
   $8 = I \times 25$
   $I = 0.32$ A
   $V_1 = IR_1 = 0.32 \times 5 = 1.6$ V
   $V_2 = IR_2 = 0.32 \times 20 = 6.4$ V
   c) The sum of the voltages equals the voltage supplied of 8 V. This is due to the Law of Conservation of Energy, which states that the total electrical potential energy per unit charge lost over each resistor must equal the total amount supplied.
   d)

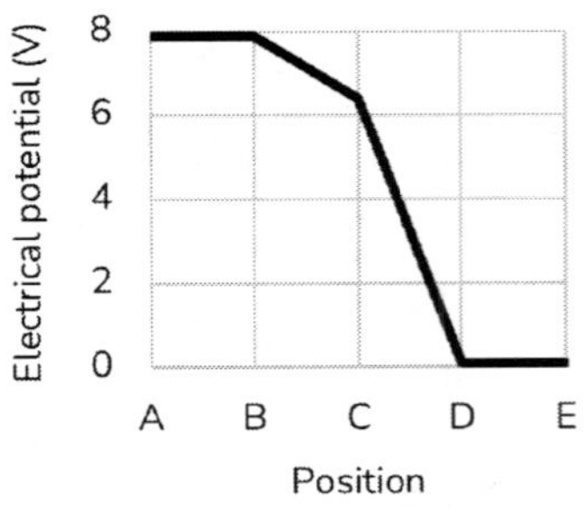

5. a) Voltmeter reading is $V = 12$ V
   $V = IR$
   $12 = I \times 5$
   $I = 2.4$ A
   Power is $P = VI = 12 \times 2.4 = 28.8$ W

b) $R_T = R_1 + R_2 = 5 + 5 = 10\ \Omega$
$V = IR$
$12 = I \times 10$
$I = 1.2$ A
Voltage across light bulb: $V = IR = 1.2 \times 5 = 6\ V$
Power from light bulb: $P = VI = 6 \times 1.2 = 7.2$ W

c) With two light bulbs, the current and voltage both halved and the power was a quarter of the original single light bulb. Since brightness is related to light energy produced, the brightness with two light bulbs also becomes a quarter of the original, with both light bulbs in series having the same brightness.

d) In part (a), the power is 28.8 W. In part (b), both light bulbs use 7.2 W of power, so the total power drawn from the power supply is 7.2 + 7.2 = 14.4 W. This is half the original power used.

e) Total resistance is: $R_T = 5 + 5 + 5 = 15\ \Omega$
Current is: $V = IR$
$12 = I \times 15$
$I = 0.8$ A
Voltage across one light bulb: $V = IR = 0.8 \times 5 = 4\ V$
Power from one light bulb: $P = VI = 4 \times 0.8 = 3.2$ W
Voltage and current are both one-third the original value and power is one-ninth the original value.

f) The more light bulbs in series, the dimmer they are. If one light bulb breaks, there is an open circuit and none of the others will work either. You cannot switch each light bulb on individually, because the same current runs through all of them.

6. a) $R_T = R_1 + R_2 = 5 + 7.5 = 12.5\ \Omega$
$V = IR$
$10 = I \times 12.5$
$I = 0.8$ A
Power of R_1 is: $P_1 = V_1 I_1 = I^2 R_1 = 0.8^2 \times 5 = 3.2$ W
Power of R_2 is: $P_2 = V_2 I_2 = I^2 R_2 = 0.8^2 \times 7.5 = 4.8$ W
R_2 is brighter than R_1.

b) R_2 is brighter than R_1, regardless of which position they are in, since the calculations are independent of their position.

7. a) $V = IR$
$1.8 = 0.02 \times R$
$R = 90\ \Omega$

b) $V_T = IR_T$
$3 = 0.02 \times R_T$
$R_T = 150\ \Omega = 90 + R$
$R = 60\ \Omega$

8. As the resistance of the variable resistor increases, the total resistance of the circuit increases, therefore current decreases according to $V = IR$, since the voltage of the power source has not changed. The voltage across the light bulb will also decrease according to $V = IR$, due to the current decreasing and the resistance of the light bulb itself staying constant. Therefore the power used by the light bulb also decreases, according to $P = VI$, leading to the light bulb being dimmer.

9. a) $R_T = R_1 + R_2 = 5 + 1.2 = 6.2\ \Omega$
$V = IR$
$10 = I \times 6.2$
$I = 1.94$ A

b) Without cable: $P = VI = 12 \times (\frac{12}{5}) = 28.8$ W
With cable: $P = VI = I^2 R = \left(\frac{12}{6.2}\right)^2 \times 5 = 18.7$ W
The light bulb emits less power when the extension cable is used.

c) $P = VI = I^2 R = \left(\frac{12}{6.2}\right)^2 \times 1.2 = 4.50$ W

12.8 Parallel circuits

1. In ***parallel***, current splits into multiple paths to go through components.
Charge is conserved, so in the circuit diagram shown, the total charge q_T entering point X from the positive terminal must equal the sum of q_1 passing through resistance R_1 and q_2 passing through resistance R_2, giving:
$$q_T = q_1 + q_2 \quad ...(1)$$
Since current is given by $I = \frac{q}{t}$, if we divide both sides by the amount of time passed t, equation (1) gives:
$$I_T = I_1 + I_2 \quad ...(2)$$
This means the reading on ammeter A_T equals the sum of the readings on ammeters A_1 and A_2.
If we assume that wires have minimal resistance, all the electrical potential energy from the battery is transformed in the resistors. This means that the voltage, or electrical potential energy per unit charge, measured by each of the voltmeters is the same:
$$V_T = V_1 = V_2 \quad ...(3)$$
The total electrical power supplied by the battery is:
$$P_T = I_T V_T$$
So multiplying equation (2) by V_T tells us what the total power is:
$$I_T V_T = I_1 V_T + I_2 V_T$$
From equation (3), we can see all the voltages are equal, so:
$$I_T V_T = I_1 V_1 + I_2 V_2$$
This means that the total power from the battery is equal to the sum of the power used by each resistor:
$$P_T = P_1 + P_2 \quad ...(4)$$
Resistance of a component is the voltage across a component divided by the current through it: $R = \frac{V}{I}$.
Rearranging this gives $I = \frac{V}{R}$.
Expressing equation (2) in terms of voltage and resistance then gives:
$$\frac{V_T}{R_T} = \frac{V_1}{R_1} + \frac{V_2}{R_2}$$
Recall from equation (3) that the voltage across each resistor is the same, so:
$$\frac{1}{R_T} = \frac{1}{R_1} + \frac{1}{R_2} \quad ...(5)$$
This means that the more resistors are added in parallel, the smaller the total resistance will be.

2. $\frac{1}{R_T} = \frac{1}{R_1} + \frac{1}{R_2} = \frac{1}{5} + \frac{1}{10}$
$R_T = 3.3\ \Omega$

3. $\frac{1}{R_T} = \frac{1}{R_1} + \frac{1}{R_2} + \frac{1}{R_3} = \frac{1}{5} + \frac{1}{5} + \frac{1}{5}$
$R_T = 1.7\ \Omega$

4. $\frac{1}{R_T} = \frac{1}{R_1} + \frac{1}{R_2}$
$\frac{1}{1} = \frac{1}{2} + \frac{1}{R_2}$
$R_2 = 2\ \Omega$

5. $\frac{1}{R_T} = \frac{1}{R_1} + \frac{1}{R_2} + \frac{1}{R_3}$
$\frac{1}{5} = \frac{1}{20} + \frac{1}{10} + \frac{1}{R_3}$
$R_3 = 20\ \Omega$

6. a) $\frac{1}{R_T} = \frac{1}{R_1} + \frac{1}{R_2} = \frac{1}{10} + \frac{1}{7.5}$
$R_T = 4.3\ \Omega$

b) $V_T = I_T R_T$
$12 = I_T \times 4.3$
$I_T = 2.8$ A
$V_1 = I_1 R_1$
$12 = I_1 \times 10$
$I_1 = 1.2$ A
$V_2 = I_2 R_2$

$12 = I_2 \times 7.5$

$I_2 = 1.6$ A

7. a) $V = IR$

$12 = I \times 5$

$I = 2.4$ A

Power is $P = VI = 12 \times 2.4 = 28.8$ W

b) $V_1 = I_1R_1$

$12 = I_1 \times 5$

$I_1 = 2.4$ A

$V_2 = I_2R_2$

$12 = I_2 \times 5$

$I_2 = 2.4$ A

$I_T = I_1 + I_2 = 2.4 + 2.4 = 4.8$ A

The same current goes through each light bulb. The total current from the battery doubles.

c) Since the current and voltage across each light bulb in part (a) and (b) is the same, they use the same power and have the same brightness.

d) Power in part (b) is:

$P_T = P_1 + P_2 = 28.8 + 28.8 = 57.6$ W.

This is double the power in part (a).

e) Adding light bulbs in parallel does not affect the brightness of the other light bulbs in the circuit, but adding light bulbs in series makes them dimmer. If one light bulb breaks in parallel, there is still a closed circuit to the other light bulbs, so they will still work, but in a series circuit they would all stop working. If a switch is connected in series to each light bulb in parallel, you can switch each one on individually, but you cannot do this with light bulbs in series.

8. a) $I_T = I_1 + I_2 = 2.5 + 2 = 4.5$ A

b) $P_1 = I_1V_1 = 2.5 \times 9 = 22.5$ W

$P_2 = I_2V_2 = 2 \times 9 = 18$ W

Light bulb 1 is brighter as it uses more power.

c) $R_1 = \frac{V_1}{I_1} = \frac{9}{2.5} = 3.6\ \Omega$

$R_2 = \frac{V_2}{I_2} = \frac{9}{2} = 4.5\ \Omega$

$R_T = \frac{V_T}{I_T} = \frac{9}{4.5} = 2\ \Omega$

9. a)

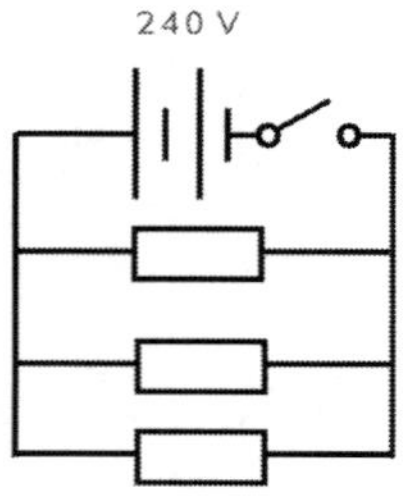

b) Total power available before circuit breaker trips:

$P = VI = 240 \times 12 = 2880$ W

Total power used by appliances:

$800 + 2000 + 150 = 2950$ W

The circuit breaker will be tripped.

10. a) $\frac{1}{R_T} = \frac{1}{R_1} + \frac{1}{R_2} = \frac{1}{5} + \frac{1}{0.025}$

$R_T = 0.025\ \Omega$

$V_T = I_TR_T$

$12 = I_T \times 0.025$

$I_T = 480$ A

This will trip the circuit breaker.

b) A short circuit produces an extremely high current and power. This can lead to a large amount of resistive heating, which can burn out appliances and even set fire to them.

12.9 Mixed circuit problems

1. $\frac{1}{R_{23}} = \frac{1}{R_2} + \frac{1}{R_3} = \frac{1}{5} + \frac{1}{10}$

$R_{23} = 3.3\ \Omega$

$R_T = R_1 + R_{23} = 5 + 3.3 = 8.3\ \Omega$

2. $R_{12} = R_1 + R_2 = 5 + 5 = 10\ \Omega$

$\frac{1}{R_T} = \frac{1}{R_{12}} + \frac{1}{R_3} = \frac{1}{10} + \frac{1}{10}$

$R_T = 5\ \Omega$

3. a)

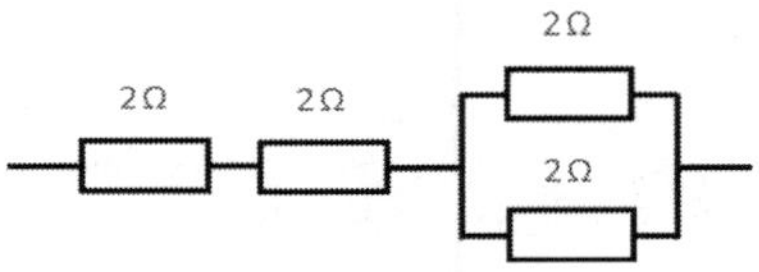

b)

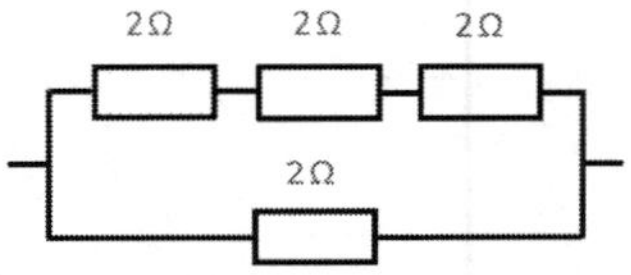

4. For R_3: $V_3 = 48$ V (part (a))

$V_3 = I_3R_3$

$48 = I_3 \times 30$

$I_3 = 1.6$ A (part (b))

For R_1 and R_2 combined: $V_{12} = I_{12}R_{12}$

$48 = I_{12} \times 30$

$I_{12} = I_1 = I_2 = 1.6$ A (part (b))

Total current: $I_T = I_{12} + I_3 = 1.6 + 1.6 = 3.2$ A (part (c))

For R_1: $V_1 = I_1R_1 = 1.6 \times 20 = 32$ V (part (a))

For R_2: $V_2 = I_2R_2 = 1.6 \times 10 = 16$ V (part (a))

For R_1: $P_1 = V_1I_1 = 32 \times 1.6 = 51.2$ W (part (d))

For R_2: $P_2 = V_2I_2 = 16 \times 1.6 = 25.6$ W (part (d))

For R_3: $P_3 = V_3I_3 = 48 \times 1.6 = 76.8$ W (part (d))

$W = VIt = 48 \times (1.6 + 1.6) \times 5 \times 60 = 46080$ J (part (e))

5. Voltage over 75 Ω resistor is:

$V = IR = 0.178 \times 75 = 13.35$ V

Current through the 300 Ω and 100 Ω resistors respectively is:

$I = \frac{V}{R} = \frac{13.35}{300} = 0.0445$ A

$I = \frac{V}{R} = \frac{13.35}{100} = 0.1335$ A

Ammeter A_2 measures the total current:

$I_T = 0.178 + 0.0445 + 0.1335 = 0.356$ A

Total resistance of the circuit:

$R_T = 30 + 1 \div \left(\frac{1}{300} + \frac{1}{100} + \frac{1}{75}\right) = 67.5\ \Omega$

Voltmeter measures total voltage across circuit:

$V_T = I_TR_T = 0.356 \times 67.5 = 24.0$ V

6. $R_T = 1 \div \left(\frac{1}{5} + \frac{1}{5}\right) + 5 = 2.5 + 5 = 7.5\ \Omega$

$I_T = \frac{V_T}{R_T} = \frac{12}{7.5} = 1.6$ A

$V_3 = I_3R_3 = 1.6 \times 5 = 8$ V

$P_3 = V_3I_3 = 8 \times 1.6 = 12.8$ W

$V_1 = V_2 = V_T - V_3 = 12 - 8 = 4$ V

$I_1 = I_2 = \frac{V_1}{R_1} = \frac{4}{5} = 0.8$ A

$P_1 = P_2 = V_1I_1 = 4 \times 0.8 = 3.2$ W

Light bulb R_3 is 4 times brighter than light bulbs R_1 and R_2.

7. Low heat: Close S_1 and S_3

$R_T = R_1 + R_2 = 20 + 20 = 40\ \Omega$

$I = \frac{V}{R_T} = \frac{12}{40} = 0.3$ A

$P = VI = 12 \times 0.3 = 3.6$ W

Medium heat: Close S_1 and S_4 (S_2 and S_4 also work)

$R_T = 20\ \Omega$

$I = \frac{V}{R_T} = \frac{12}{20} = 0.6$ A

$P = VI = 12 \times 0.6 = 7.2$ W

High heat: Close S_1, S_2 and S_4

$\frac{1}{R_T} = \frac{1}{R_1} + \frac{1}{R_2} = \frac{1}{20} + \frac{1}{20}$

$R_T = 10\ \Omega$

$I = \frac{V}{R_T} = \frac{12}{10} = 1.2$ A

$P = VI = 12 \times 1.2 = 14.4$ W

8. a) A and C
   b) B and D
   c) The circuit shown effectively has the three light bulbs in parallel, since current splits into three paths at points A and C, then rejoins at points B and D. Removing the middle wires turns the circuit into a series circuit.
   In parallel, the total resistance is: $\frac{1}{R} = \frac{1}{2} + \frac{1}{2} + \frac{1}{2}$
   $R = 0.67\ \Omega$
   Total power drawn: $P = VI = \frac{V^2}{R} = \frac{6^2}{0.67} = 54$ W
   So each light bulb draws 18 W.
   In series, the total resistance is: $R = 2 + 2 + 2 = 6\ \Omega$
   Total power drawn: $P = VI = \frac{V^2}{R} = \frac{6^2}{6} = 6$ W
   So each light bulb draws 2 W. The light bulbs are 9 times brighter with the wires in place.

13.2 Magnetic fields around permanent magnets

1. A field is a region where an object experiences a non-contact force. A magnetic field is a region that will exert a force on magnets and magnetic materials. For example, a fridge magnet will exert a magnetic force on objects containing iron, such as a paper clip. The fridge magnet only needs to be close enough to the paper clip, but does not need to touch it.
   Magnets have a north pole and a south pole. If the north pole of one magnet is brought close to the south pole of another, they will attract. If like poles of two magnets are brought close together, they will repel.
   Magnetic fields are invisible, but sprinkling iron filings in the region will reveal their shape. We can model the magnetic field by drawing field lines that follow this shape.
   The strength of the magnetic field is represented by the density of the field lines. In other words, if field lines are closer together, the magnetic field is stronger. Therefore, field lines around a bar magnet will be closer together near the poles than away from the poles.
   The direction of a magnetic field is defined as the direction that the north pole of a compass will point. Since a compass is actually a magnet, this means that the north pole of a compass will be repelled from the north pole of a bar magnet, and attracted towards the south pole of a bar magnet.
   This means that the direction of magnetic field lines around a bar magnet go from its north pole to its south pole.
   Since the direction of the field lines shows the direction a compass will point, this means the compass direction is given by the tangent to the magnetic field line at each position.
   It also means that magnetic field lines cannot cross, because the compass will not point in two different directions.
   Because a magnetic field has both strength and direction, it is a vector quantity. So if there is a magnetic field exerted by two different magnets, the strength and direction of the magnetic field at one point is given by the resultant of the magnetic fields from each magnet.

2. a)

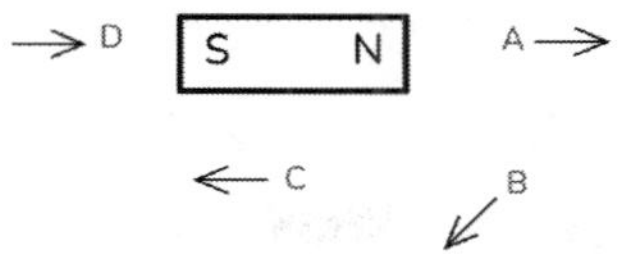

   b) The arrows show the direction of the magnetic field lines and are tangential to them.
   c) D, because the magnetic field lines are closer together.
   d)

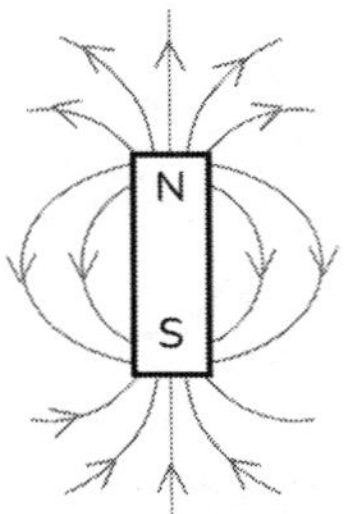

3. a) X – south, Y – north
   b) X – north, Y – north
4. a)

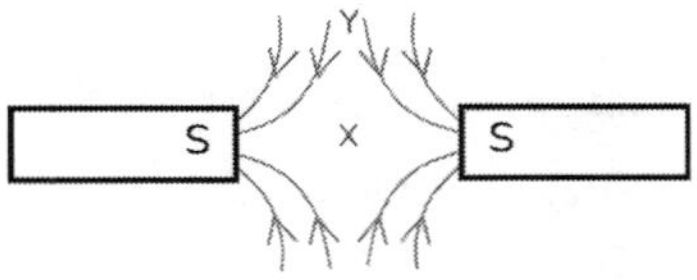

   X has zero magnetic field, as the magnetic field from both magnets cancel each other out. Y therefore has a stronger magnetic field, since the net magnetic field will be downwards.

   b)

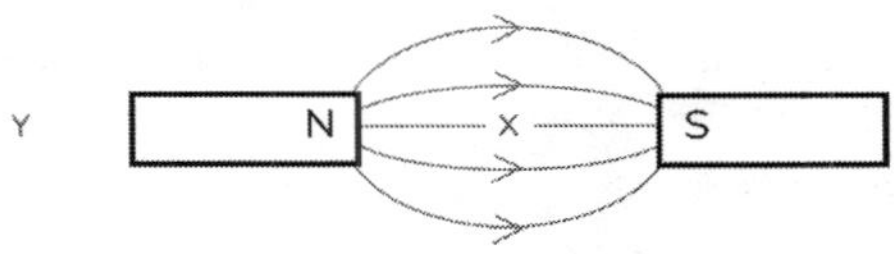

   X has a stronger magnetic field than Y, since both magnets give a magnetic field to the right at X and the net magnetic field is the vector sum of these. Y is much further than the magnet on the right so the field from it is much weaker.

5.
   a)

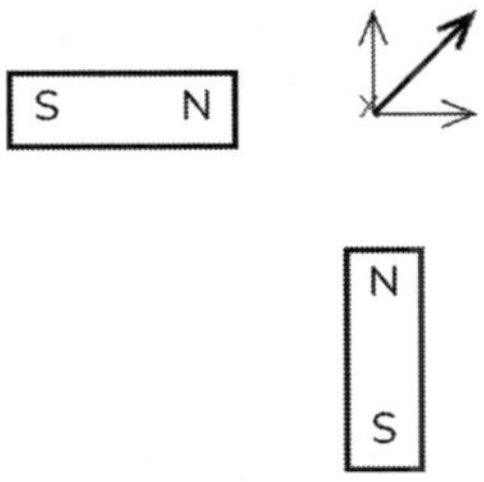

   b)

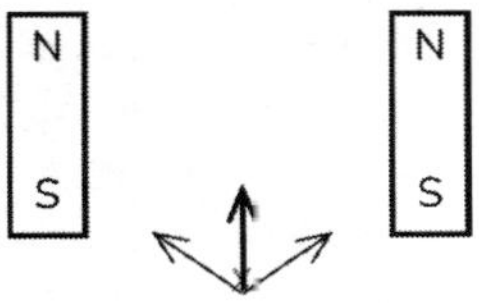

   c)

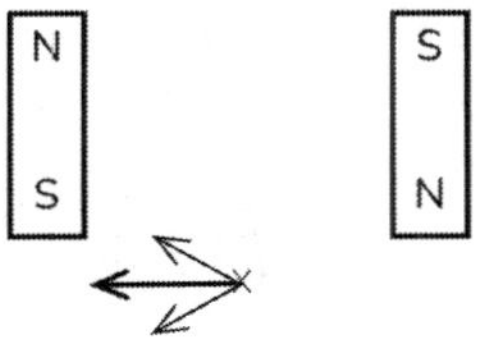

6.

a)

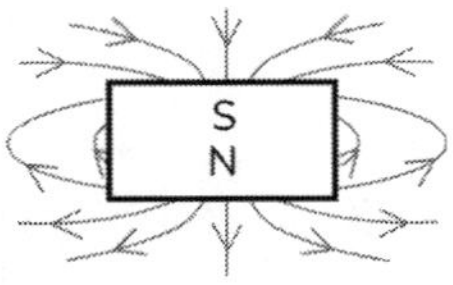

b)

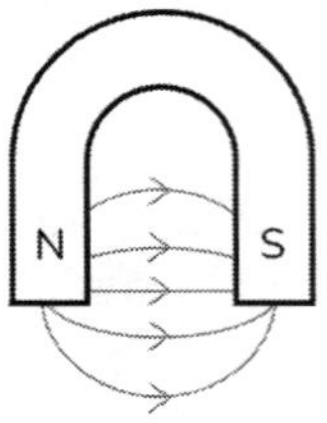

c)

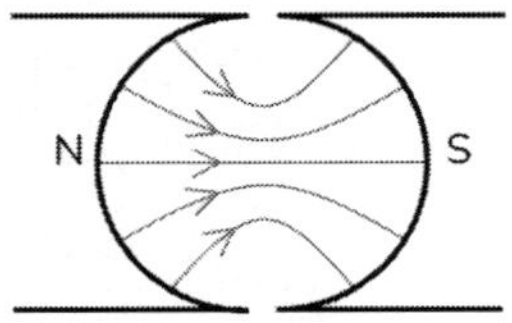

d)

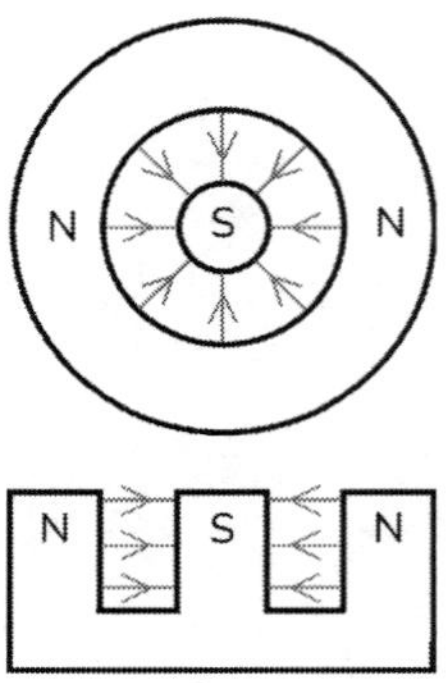

13.3 Magnetic fields around currents

1. In 1820, Hans Christian Oersted showed that when a current flowed through a straight wire, it could make a compass needle deflect, so that it pointed at right angles to the wire. When the current was switched off, the compass went back to pointing in the direction of the Earth's magnetic field.
   This implies that an electric current can produce a magnetic field. It shows that electricity and magnetism are linked, opening up a new area of study in physics called electromagnetism.
   One way to remember the observed direction of the magnetic field around a straight conductor is the "***right hand grip rule***". Point your right thumb in the direction of the current. Then, the direction your fingers would close if you were gripping something gives the direction of the magnetic field. This direction goes in a circle all the way around the current-carrying conductor. Recall that current direction is defined as the direction positive charge flows, which is the opposite direction to the flow of electrons.
   As fields extend in all three dimensions, we need new symbols to represent them on a two dimensional page. We can represent the direction out of the page with a dot – imagine an arrow is coming towards you and what you see closest to you is the arrow tip. We can represent the direction that is into the page with a cross – imagine the arrow is going away from you and what you see closest to you is the tail feathers.
   Therefore, to represent the magnetic field around a current going up the page through a straight conductor, we can use dots on the left of it, to show the magnetic field is out of the page, and crosses on the right of it, to show the magnetic field is into the page.
   We can also use dots and crosses to represent current direction, so if the current in the conductor is travelling away from you, into the page, you use a cross. The right hand grip rule would tell you that the magnetic field around this current is clockwise.
   The strength of the magnetic field decreases as you move away from the conductor. Therefore, you need to draw the magnetic field lines further apart.

2.

a)

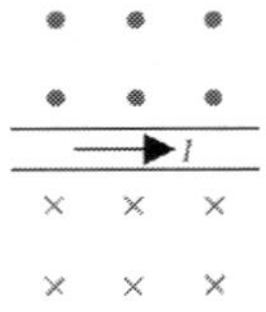

b)

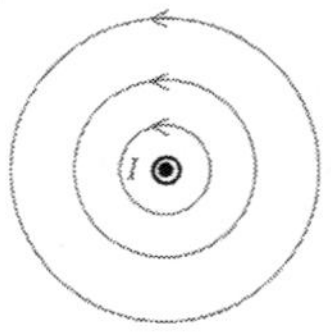

c)

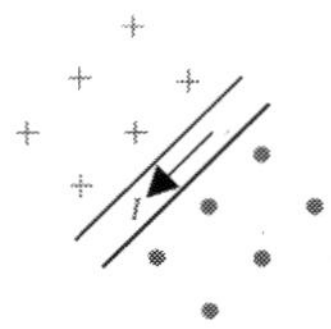

3. a) into page
   b) into page
   c) into page
   d) into page
   e) out of page
   f) out of page
4. a) X: zero; Y: up
   b) X: into page; Y: out of page
5. a) left
   b) into page
   c) out of page
6. When the current travels to the right, the magnetic field at X is into the page, as given by the "right hand grip rule". As the current changes to travel to the left, the magnetic field at X is out of the page. This continues periodically.
7. When current flows along a conductor that is not straight, it still produces a magnetic field. The magnetic field at a point will be the vector sum of the magnetic field from each bit of the conductor.
   The magnetic field inside a circular coil with a current travelling clockwise in it will therefore be into the page. This is because if we use the "right hand grip rule" and stick our thumb on each point on the circular coil in the current direction, our fingers always point into the page in the middle of the coil.
   To work out the direction of the magnetic field in a coil faster, we can use the "***right hand coil rule***". This time, your fingers wrap in the direction of the current in the coil and your thumb represents the direction of the magnetic field inside the coil.
   A ***solenoid*** consists of a long wire looped around many turns around a cylinder. Each turn contributes a magnetic field inside the solenoid the same way a single coil does. Therefore, the more turns, the stronger the magnetic field inside the solenoid will be, with the direction given by the right hand coil rule. Solenoids can often have hundreds of turns.
   Outside the solenoid, there is still a magnetic field, which gets weaker away from the ends of the solenoid. The shape of the magnetic field outside the solenoid is similar to the magnetic field around a bar magnet. Unlike a bar magnet, we can turn the magnetic field from the solenoid on and off, by switching the current on and off. This is how an ***electromagnet*** works, and these are found in automated locking systems. The magnetic field lines outside the solenoid go from the north pole to the south pole of the electromagnet. However, inside the solenoid, they go from the south pole to the north pole, since this is the direction that a compass inside the solenoid would point.
8. a) Anti-clockwise
   b) From Y to X
   c) From Y to X

9.

a)

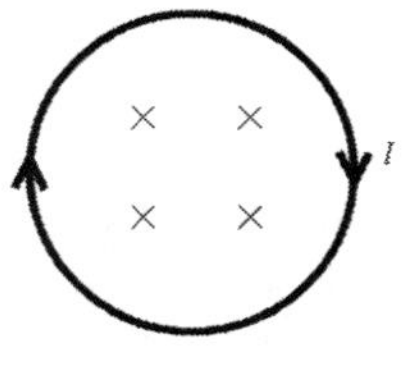

b)

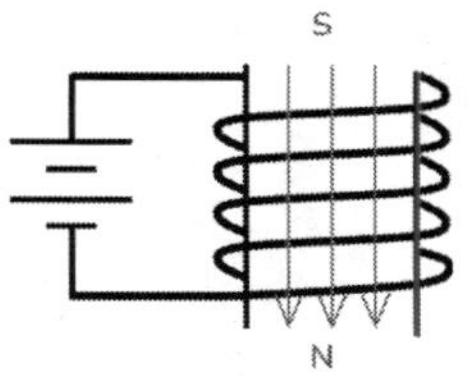

c)

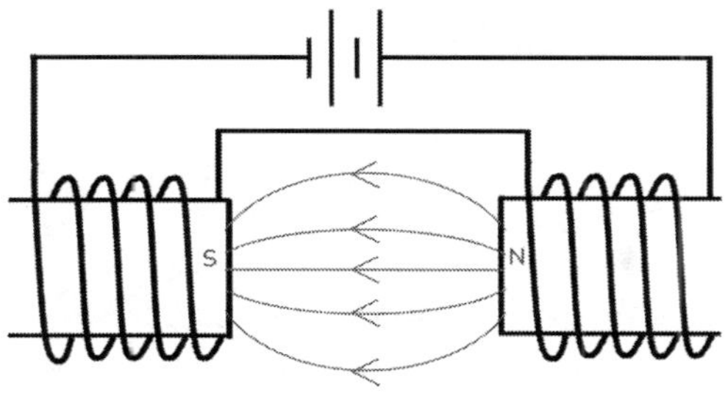

10. A big coil of wire underground, carrying a current running clockwise around the laboratory, would produce a south pole at the surface of the ground, which would attract the north pole of the compass.

11.

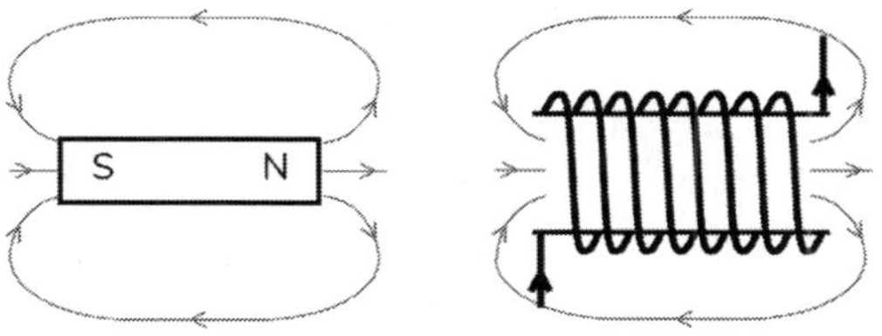

12. A solenoid has many turns, with the total magnetic field inside it being the sum of the magnetic fields from each single turn, whereas a straight conductor would only produce a magnetic field around the size of each single turn at most.

13. When the switch is closed, current flows through the solenoid, creating a magnetic field upwards. This turns the top of the solenoid into a north pole, which will attract the south pole of the bar magnet, so the bar magnet moves down towards the solenoid.

13.4 Magnetic field strength

1. Magnetic field strength is related to how much force is exerted on magnetic objects by an external magnetic field, and is measured in tesla (T).
   The magnetic field strength at a point due to a current-carrying conductor is given by the vector sum of the magnetic field strength contributed by each bit of the conductor. If you change the shape and size of the conductor, you change the magnetic field strength. For a long, straight current-carrying conductor, the magnetic field strength is given by:
   $$\vec{B} = \frac{\mu_0 I}{2\pi r}$$
   where: $\vec{B}$ = magnetic field strength (T)
   $\mu_0 = 4\pi \times 10^{-7}$ (N A^{-2}) (permeability of free space)
   I = current through the conductor (A)
   r = perpendicular distance from the conductor (m)

   μ_0 is a constant called the ***permeability of free space***. ***Magnetic permeability*** means how much a material affects the strength of a magnetic field passing through it. The permeability of free space is defined for a magnetic field passing through a vacuum. For a magnetic material such as iron, the permeability would be much higher.
   This formula assumes that r is small compared to conductor length.
   Magnetic field strength is a vector, so it has a magnitude and a direction, given by the right hand grip rule for a straight conductor.
2. Toaster in use 5 x 10^{-7} T
   Earth's magnetic field at equator 3.1 x 10^{-5} T
   Fridge magnet 5 x 10^{-3} T
3. $\vec{B} = \frac{\mu_0 I}{2\pi r} = \frac{\text{N A}^{-2} \times \text{A}}{\text{m}} = \frac{\text{kg} \times \text{m} \times \text{s}^{-2}}{\text{A} \times \text{m}} = \text{kg s}^{-2}\ \text{A}^{-1}$
4. a) At X: $\vec{B} = \frac{\mu_0 I}{2\pi r} = \frac{4\pi \times 10^{-7} \times 2.5}{2\pi \times 1 \times 10^{-2}} = 5 \times 10^{-5}$ T into page
   At Y: $\vec{B} = \frac{\mu_0 I}{2\pi r} = \frac{4\pi \times 10^{-7} \times 2.5}{2\pi \times 2 \times 10^{-2}} = 2.5 \times 10^{-5}$ T out of page

   b)

   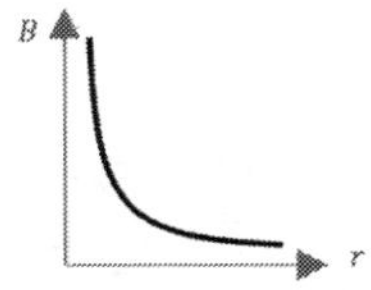

5. At X from left wire: $\vec{B} = \frac{\mu_0 I}{2\pi r} = \frac{4\pi \times 10^{-7} \times 2}{2\pi \times 1 \times 10^{-2}} = 4 \times 10^{-5}$ T up
   At X from right wire: $\vec{B} = \frac{\mu_0 I}{2\pi r} = \frac{4\pi \times 10^{-7} \times 3}{2\pi \times 1 \times 10^{-2}} = 6 \times 10^{-5}$ T up
   Total at X: $\vec{B} = 4 \times 10^{-5} + 6 \times 10^{-5} = 1 \times 10^{-4}$ T up
   At Y from left wire: $\vec{B} = \frac{\mu_0 I}{2\pi r} = \frac{4\pi \times 10^{-7} \times 2}{2\pi \times 3 \times 10^{-2}} = 1.3 \times 10^{-5}$ T up
   At Y from right wire: $\vec{B} = \frac{\mu_0 I}{2\pi r} = \frac{4\pi \times 10^{-7} \times 3}{2\pi \times 1 \times 10^{-2}} = 6 \times 10^{-5}$ T down
   Total at Y: $\vec{B} = 6 \times 10^{-5} - 1.3 \times 10^{-5} = 4.7 \times 10^{-5}$ T down
6. Magnetic field from the current must be east, so current must run from north to south through the wire.
   $$\tan 45° = \frac{B}{50 \times 10^{-6}}$$
   $$B = 50 \times 10^{-6}\ \text{T}$$
   $$\vec{B} = \frac{\mu_0 I}{2\pi r}$$
   $$50 \times 10^{-6} = \frac{4\pi \times 10^{-7} \times I}{2\pi \times 1.5 \times 10^{-2}}$$
   $$I = 3.75\ \text{A}$$
7. A coil with many turns is called a solenoid. If a solenoid is much longer than its diameter, and the coils are very close together, the magnetic field inside the solenoid is uniform. Its direction is given by the right hand coil rule.
   The magnetic field strength inside a solenoid is given by:
   $$\vec{B} = \frac{\mu_0 N I}{L}$$
   where: $\vec{B}$ = magnetic field strength (T)
   $\mu_0 = 4\pi \times 10^{-7}$ (N A^{-2}) (permeability of free space)
   I = current through the conductor (A)
   N = number of turns (m)
   L = length of solenoid (m)

   The formula tells us that the magnetic field strength is proportional to the current, which means that the strength of an electromagnet made using a solenoid can be varied by changing the current.
   The magnetic field strength can also be increased by increasing the number of turns or decreasing the length of the solenoid. This means that packing the coils tightly together, so there are more turns per length, will increase the magnetic field.
8. $I = \frac{V}{R} = \frac{12}{131} = 0.0916$ A
   $\vec{B} = \frac{\mu_0 N I}{L} = \frac{4\pi \times 10^{-7} \times 2160 \times 0.0916}{0.06} = 4.14 \times 10^{-3}$ T
9. Since $\vec{B} = \frac{\mu_0 N I}{L}$ the gradient of the graph is $\frac{\mu_0 N}{L}$.
   $\frac{\mu_0 N}{L} = \frac{4\pi \times 10^{-7} \times N}{L} = \frac{15 \times 10^{-3}}{3}$
   $\frac{N}{L} = 4000$ turns per m $= 40$ turns per cm

10. The magnetic field increases if the number of turns per length, $\frac{N}{L}$, increases. For a given number of turns, this increases if the turns are closer together, reducing L.
11. a)

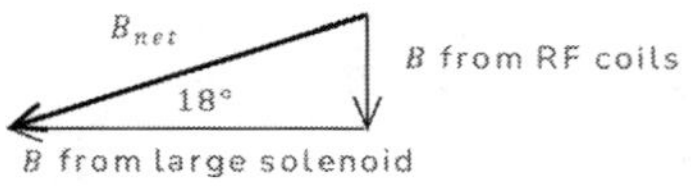

b) $B = \frac{0.5}{\tan 18°} = 1.54$ T

$B = \frac{\mu_0 NI}{L}$

$1.54 = \frac{4\pi \times 10^{-7} \times 9550 \times I}{1}$

$I = 128$ A

13.5 Ferromagnetism

1. Atoms have negatively charged electrons that orbit around a nucleus and spin on their axes. Moving charges create currents, and currents create magnetic fields. So a moving electron creates a tiny magnetic field, called a ***magnetic moment***, meaning it will rotate to align itself with an external magnetic field.
   In many atoms, the magnetic moments of all the electrons cancel out, so overall the atom has no magnetic moment. Elements with atoms like this, such as copper, are not attracted to magnets.
   However, ***paramagnetic*** materials, such as aluminium and oxygen, have atoms with a permanent magnetic moment. The magnetic moments of all the atoms in the material are in random directions, so they cancel each other out, leaving no magnetic field overall.
   Magnetic moments in paramagnetic materials will align themselves in the same direction as an external magnetic field though. So paramagnetic materials are weakly attracted to magnets.
   Ferromagnetic materials, such as iron, nickel and cobalt, also have atoms with a permanent magnetic moment. Unlike paramagnetic materials, in ferromagnetic materials, the atoms group into regions called ***domains***, in which neighbouring atoms have their magnetic moments aligned in the same direction. Normally, the magnetic fields of the domains are in different random directions.
   However, the magnetic fields of the domains will align themselves in the same direction as an external magnetic field. This causes ferromagnetic materials to be strongly attracted to magnets. The total magnetic field is increased, because the magnetic field from the ferromagnetic material is added to the external magnetic field.
   When the external magnetic field is removed, the domains retain the same alignment for a short time. This is why stroking an iron nail many times with the same pole of a magnet will temporarily magnetise the nail too, because the magnetic fields of its domains are now in the same direction. The nail is demagnetised when the domains return to random directions by being agitated, for example by hitting or heating the nail. This type of iron is called ***soft iron***, where the overall magnetic field is not permanent.
   Permanent magnets are made of ***hard iron***, with added impurities that stop the domains returning to random alignments.
   Alloys containing iron are often also ferromagnetic. For example, strong neodymium magnets contain a mixture of the elements iron, neodymium and boron.
2. The magnetic field direction can be determined using the right hand coil rule, where the fingers represent the movement of positive charge and the thumb represents the magnetic field produced. If the electron were a positive charge, this would give a magnetic field pointing down, but this is flipped upwards because the electron is negative.
3. Both have atoms with a permanent magnetic moment, which align themselves with an external magnetic field. However, in ferromagnetic materials, the atoms are grouped in domains, with the same magnetic alignments, even when there is no external field. This makes ferromagnetic materials much more strongly attracted to magnets than paramagnetic materials.
4. Soft iron is temporarily magnetised by an external magnetic field, but loses its magnetism after some time or when agitated. Hard iron has had impurities added so that it retains magnetism permanently.
5. Copper will not be attracted to the magnet.
   This is because the magnetic moments of its electrons cancel each other out, leaving copper atoms with no overall magnetic moment.
   Soft iron will be attracted to the magnet.
   This is because the magnetic moments of its atoms are grouped into domains, which align with the external magnetic field of the bar magnet and are therefore attracted to it.
   A compass will rotate so that its north pole points away from the north pole of the bar magnet.
   This is because it is made of magnetised iron, with its domains permanently aligned in the same direction, so the whole compass will rotate so its magnetic field is aligned with the bar magnet's field.
6. a) The unmagnetised pin

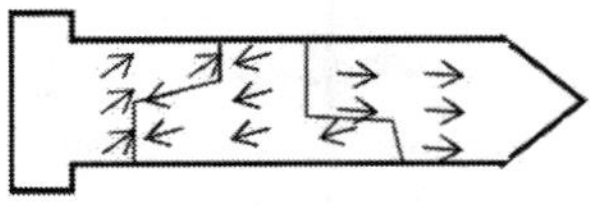

The bar magnet and pin after stroking

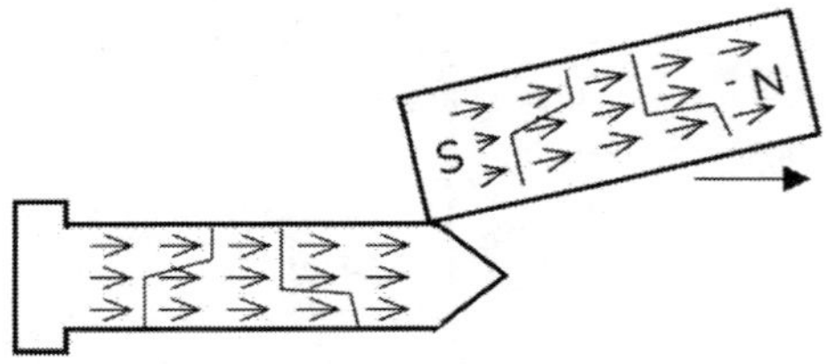

   b) The magnetic moments of the domains in the magnetised pin are all aligned in the same direction, so they don't cancel out. So the magnetised pin can act as a magnet to align the domains of the unmagnetised pin, thus attracting that pin.
   c) The magnetised pin acts like a compass. It will rotate to point north so that its domains align with the Earth's magnetic field.
7. a) X is a north pole.
   b) A bar magnet is made of hard iron, so there is a permanent magnetic field. However, an electromagnet consists of a conducting coil wound around a soft iron core, with no permanent magnetisation. This means that the electromagnet can be switched on and off. The strength of the electromagnet can also be varied, by varying the current, according to $\vec{B} = \frac{\mu_0 NI}{L}$, whereas the strength of the bar magnet cannot be changed easily.
   c) $I = \frac{V}{R} = \frac{120}{50} = 2.4$ A

   $\vec{B} = \frac{\mu_0 NI}{L} = \frac{4\pi \times 10^{-7} \times 1000 \times 2.4}{0.25} = 0.012$ T

d) A soft iron core would increase the total magnetic field inside the coil, as the magnetic field produced by the current would make the domains in the soft iron core align in the same direction, adding their magnetic field to the one from the coil. This would create a stronger total magnetic field than that calculated in part (c). In a hard iron core, the domains are not free to rotate, so it has a fixed magnetic field, which could either increase or decrease the total magnetic field, depending which way the core was aligned.

8. a) False – most metals are not, for example copper.
   b) True – the magnetic moments of the domains in each are still aligned in the same direction.
   c) False – soft or hard here refers to whether the iron is permanently magnetised, not its strength.
   d) False – copper is not ferromagnetic.

9. a) Since the direction of the magnetic field is defined by the direction a compass points, the field lines go from the geographic south pole towards the geographic north pole, making X a magnetic south pole.
   b) At high temperatures, the atoms in the ferromagnetic material move around more, so they are no longer aligned in domains. However, the domains are aligned in the same direction in a bar magnet, therefore this model cannot explain why compasses point north.
   c) Currents in coils also produce a magnetic field with a similar shape. If this were a better model for Earth's magnetic field, Earth's core would need to contain conducting material and some energy source capable of moving charges in it.

Part 4: Waves and thermodynamics - Answers

14.2 Describing waves

1. A ***wave*** is something that transfers energy but not matter. For example, when someone talks, sound is transferred from their mouth to your ear, but air is not. Heat and light are forms of energy that reach us from the Sun, but this process does not involve particles being moved from the Sun to Earth.
   In waves, particles may still be moving, but they are vibrating on the spot.
   Mechanical waves need a medium to travel through, which contains particles of matter. For example, sound is a mechanical wave and can travel through solids, liquids and gases. However, sound cannot travel through a vacuum, which is a region of space with no particles.
   Electromagnetic waves can travel through a vacuum. For example, light can travel through space, even though there are very few particles in it.
2. A wave is defined as something that transfers energy but not matter. Similar to a wave, the rock does transfer kinetic energy to the water. However, unlike a wave, the rock has moved downwards, transferring its matter downwards. Therefore, a rock is not a wave.
3. a) The ripples transferred kinetic energy from the rock to the leaf, and waves involve transfer of energy. However, water molecules did not get transferred from the rock to the leaf, they just vibrate up and down, as shown by the leaf's bobbing motion. Therefore the ripples transferred energy but not matter, fitting the definition of a wave.
   b) The ripples involve vibrating water molecules, which are matter, therefore they are mechanical waves. Electromagnetic waves do not need a medium to travel through.
4. All waves involve vibrations, resulting in repeating patterns in the medium the wave travels through.
   In ***transverse waves***, these vibrations are perpendicular to the direction in which energy travels. Transverse waves often look like a sine wave. Water waves behave approximately like transverse waves.
   The ***amplitude*** is the furthest distance a particle vibrates away from its normal position, or equilibrium position, when there is no wave. In transverse waves, ***crests*** are the highest points displaced by the amplitude and ***troughs*** are the lowest points.
   Points on the wave which are doing the same thing at the same time are described as ***in phase***. For example, all the points on the wave that are at a crest at the same time are in phase. This means they have the same displacement, velocity and acceleration. All the points that are at a trough are also in phase with each other. The distance between two consecutive points in phase is called the ***wavelength***. Crests and troughs are half a wavelength out of phase with each other.
   In ***longitudinal waves***, vibrations are along the direction in which energy travels. Sound is a longitudinal wave, so when you hear a sound, it means the particles between the source of the sound and your ear are moving towards then away from your ear repeatedly. This causes ***compressions***, where the particles bunch up, and ***rarefactions***, where particles spread out. In other words, compressions are high pressure regions, while rarefactions are low pressure.
   The particles in the middle of each compression will all be in phase, meaning that neighbouring compressions are all a wavelength apart. However, particles in the middle of a compression are half a wavelength out of phase with particles in the middle of a neighbouring rarefaction.
   The furthest distance a particle vibrates away from its equilibrium position is still called the amplitude in a longitudinal wave.
5. a) Longitudinal
   b) Transverse
6. a) 3 cm
   b) 6 cm
   c) A and E, B and F, C and G
   d) C and G
   e) Transverse, because the points are vibrating perpendicular to the direction of travel.
7.

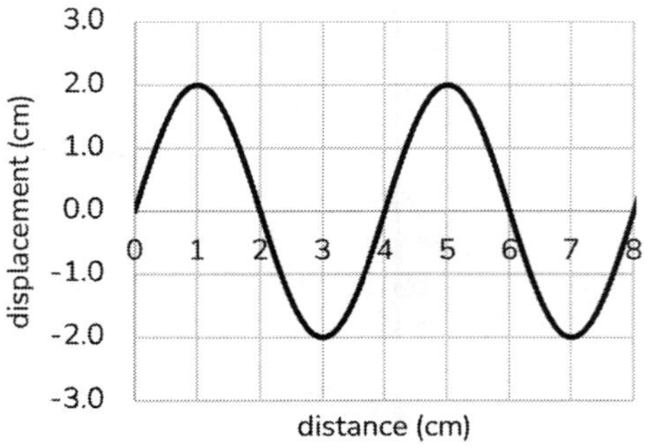

8. a) G
   b) A
   c) 12 cm
   d) D, moved 1 cm to the right, this is the amplitude
   e) The particle that is at the 15 cm equilibrium position, which has moved to the 16 cm position when the wave travels through, is in phase with D. This is because it is one wavelength, or 12 cm, from D. Both particles have moved the same distance, 1cm to the right.
   f) The particle that is at the 9 cm equilibrium position, which has moved to the 8 cm position when the wave travels through it is half a wavelength out of phase with D. While D moved 1 cm to the right, this particle moved 1 cm to the left, so it is doing the exact opposite to D.

14.3 Timing waves

1. When there is a recurring event, the ***period*** is the time between one event and another. For example, the period of Earth's rotation on its axis is one day.
   The ***frequency*** of a recurring event is how often it occurs in a given time period. For example, the frequency of a teenager's heart beat might be 60 beats per minute. This would be the same thing as 1 beat per second. Therefore the period of this teenager's heart beat would be 1 second.
   As waves involve recurring vibrations, the ***period*** is the time taken for a particle to undergo one complete vibration and return to the same spot. For example, when a surfer bobs up and down due to ocean waves, the period is the time between one crest of a wave until the

next crest. Therefore in one period, one wavelength of ocean wave has passed the surfer.

The frequency of a wave is the number of vibrations per second. For example, the frequency of the ocean waves would be the number of times the surfer goes down, then back up, per second. The frequency is also the number of wavelengths that has passed the surfer per second. The unit for frequency is the hertz (Hz).

The equation for the relationship between period and frequency is:

$$T = \frac{1}{f} \text{ or } f = \frac{1}{T}$$

where: T = period (s)
f = frequency (Hz)

2. $f = \frac{1}{T} = \frac{1}{5} = 0.2$ Hz
3. $f = \frac{120}{60} = 2$ Hz
$T = \frac{1}{f} = \frac{1}{2} = 0.5$ s
4. $T = \frac{1}{f} = \frac{1}{50} = 0.02$ s
5. $f = \frac{1}{4} = 0.25$ Hz

14.4 Wave speed

1. The speed of a wave, like the speed of anything, is the distance it moves per second in the direction of travel. Specifically, wave speed refers to the speed of the energy transfer, and not the speed of any individual particle, which would have an average velocity of zero over one period.
One way in which wave speed can be calculated is to pick a part of the wave, for example the crest, and measure how far it travels per second:
$$v = \frac{s}{t} \dots (1)$$
where: v = wave speed (m s^{-1})
s = distance travelled by the wave (m)
t = time in which the distance is travelled (s)
The time taken for one wavelength pass a point is also the period. Substituting this into equation (1):
$$v = \frac{s}{t} = \frac{\lambda}{T} \dots (2)$$
where: v = wave speed (m s^{-1})
λ = wavelength (m) (Note: λ is a Greek symbol, pronounced "lambda")
T = period (s)
Recall the frequency of a wave is related to period by:
$$f = \frac{1}{T} \dots (3)$$
Substituting this into equation (2) gives the ***wave equation***:
$$v = f\lambda$$
where: v = wave speed (m s^{-1})
f = frequency (Hz)
λ = wavelength (m)
2. a) $v = f\lambda$
$4 \times 1000 = 6 \times \lambda$
$\lambda = 670$ m
b) P waves travel faster than S waves, so they would pass through the plantation first. As P waves are longitudinal, they will cause the trees to vibrate back and forth along the direction the earthquake wave travels, resulting in areas where trees are close together and far apart. As S waves are transverse, they will cause the trees to vibrate perpendicular to the direction the earthquake travels. This can either be up and down, or side to side, in a pattern that looks like it has crests and troughs.
3. $v = f\lambda = \frac{5}{3} \times 3.5 = 5.8$ m s^{-1}
4. a) $v = f\lambda$
$331 = f \times 0.78$
$f = 424$ Hz
b) $v = f\lambda$
$349 = f \times 0.78$
$f = 447$ Hz
5. In mechanical waves, there must be some ***restoring force*** to return particles to their equilibrium position in order for a wave to travel. For example, when a ukelele string is plucked, tension in the string pulls it back towards its equilibrium position. However, inertia keeps it moving past this point, so it vibrates. Increasing tension means the string can vibrate faster, increasing wave speed.
Mechanical waves involve continuous energy transformations. For example, stretching a ukelele string gives it elastic potential energy, which transforms into kinetic energy when released, which transforms back into elastic potential energy when it stretches out on the other side. The medium affects how fast this energy transformation occurs.
6. a) Move the end a bigger distance up and down, to increase the maximum displacement from the equilibrium position.
b) Move the end faster up and down, so it undergoes one vibration in a shorter period of time.
c) Move the end slower up and down, so frequency reduces. This reduces wavelength since $v = f\lambda$ and the velocity does not change if the spring stays stretched with the same tension.
d) Move back to stretch the spring further and increase tension. This increases the restoring force bringing the spring back to equilibrium position, therefore increasing the speed at which energy is transferred.
7. a) Glass has the fastest speed of sound because it is a solid, so particles are packed closely together, allowing more frequent collisions between them to transfer energy. Sea water is a liquid, with particles a bit further apart, so the speed of sound is slower in it. Air is a gas, so particles are furthest apart, with the least frequent collisions, so the speed of sound is the slowest.
b) $v = f\lambda$
Glass: $\lambda = \frac{v}{f} = \frac{5100}{100} = 51$ m
Sea water: $\lambda = \frac{v}{f} = \frac{1530}{100} = 15.3$ m
Air: $\lambda = \frac{v}{f} = \frac{343}{100} = 3.43$ m
As speed of sound decreases, wavelength decreases.
8. Ocean waves can be generated by wind or earthquakes. The waves travel along the surface of the ocean due to the restoring forces of surface tension and gravity acting on the water molecules.
Changing the water depth changes the medium in which ocean waves travel. Generally, waves travel slower in shallower water as they come closer to shore, because water molecules drag along the bottom of the ocean as they vibrate. Due to conservation of energy, as waves slow down, their amplitude increases. When the water shallow enough to interrupt the water molecules' vibrations, the wave breaks and it looks like the top of the wave forms a peak. A surfer on a breaking wave can move

towards the shore instead of bobbing up and down, as it is not behaving like a wave anymore.

9. Deep area: $v = f\lambda$

   $4.4 = f \times 1$

   $f = 4.4$ Hz

   Shallow area: $v = f\lambda = 4.4 \times 0.7 = 3.1$ m s^{-1}

10. a) As the period increases, the wavelength increases directly proportionally.

    b) As $v = f\lambda = \frac{\lambda}{T}$

    $\lambda = vT$

    So the gradient represents the wave velocity.

    c) $v = gradient = \frac{0.5}{0.1} = 5$ m s^{-1}

    d) By altering the medium, that is, the depth of the water.

14.5 Graphing transverse waves

1. a) 2 m

   b) $v = f\lambda$

   $1 = f \times 2$

   $f = 0.5$ Hz

   c)

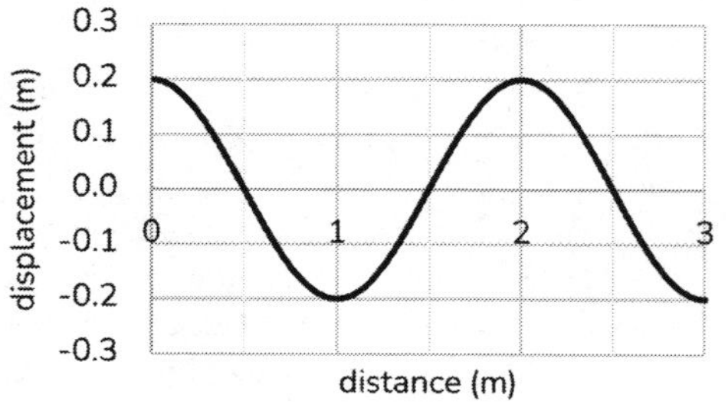

2. a) 4 m

   b) The smallest speed corresponds with the smallest distance travelled by the wave, which is 2 m in 1.5 s.

   $v = \frac{s}{t} = \frac{2}{1.5} = 1.3$ m s^{-1}

   c) In 1.5 s, the crest has moved half a wavelength, so it moves one wavelength in 3.0 s and this is its period.

   d) Every 3 s after 1.5 s

3. A displacement-time graph for a wave tracks the position of one particle in the wave from its equilibrium position over time.

   If you were to graph the displacement of a surfer bobbing up and down on ocean waves over time, you would get a displacement-time graph. This would be the same graph for any other surfer bobbing up and down on the same waves, just shifted in time to account for when crests reach them.

   As for any displacement-time graph, the gradient of this graph for a wave still represents the velocity of the object. This is the velocity of the individual particle, not the wave speed.

4. a) 0.75 m

   b) In a trough at a displacement of -0.75 m

   c) $T = 4$ s, $f = \frac{1}{T} = \frac{1}{4} = 0.25$ Hz

   d) There is no information about distance in order to find wavelength. Therefore it is also not possible to find velocity, as $v = f\lambda$.

   e) 0 s, then every 2 s after this

   f) 1 s, then every 2 s after this

5. a) $T = 2.5$ s, $f = \frac{1}{T} = \frac{1}{2.5} = 0.4$ Hz

   b) 8 m

   c) $v = f\lambda = 0.4 \times 8 = 3.2$ m s^{-1}

   d) Every 8 m, or multiples of the wavelength, from this boat

6. a)

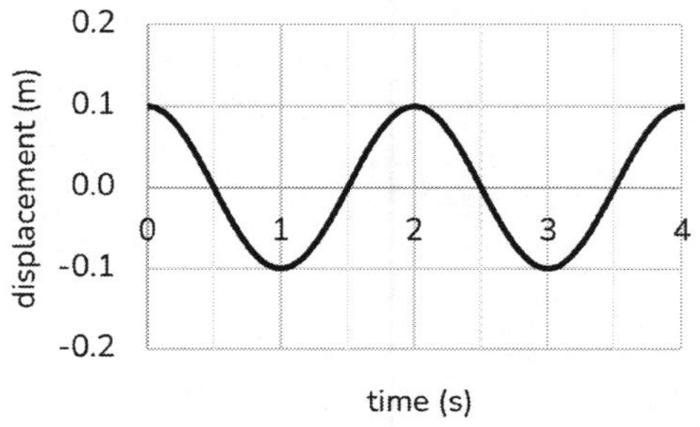

   b)

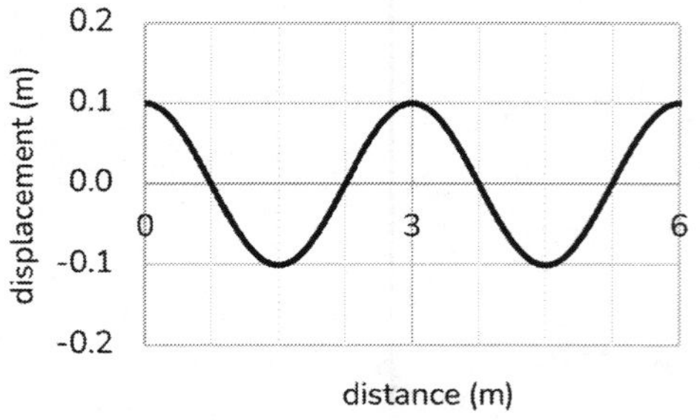

14.6 Sound waves

1. Sound is a type of wave, so it transfers energy without transferring matter. When you hear a sound, energy is transferred to your ear, but air particles are not, so you do not feel wind hitting your ear.

   Sound is a longitudinal wave. This means that it makes particles in the medium it travels through vibrate along the direction of travel. This causes a higher density of particles in some areas, called compressions. In between are areas with a lower density of particles, called rarefactions. The compressions can also be described as areas of high pressure, while the rarefactions are areas of low pressure.

   Higher frequency sound waves have a higher pitch. This means that the vocal cords of a soprano vibrate with a higher frequency than the vocal cords of a bass when they sing. Lower frequency sounds have a longer wavelength in the same medium, because they travel at the same speed. This is how musical instruments can play at different pitches – a double bass will play deeper notes of a low frequency because the longer strings allow longer wavelengths to vibrate on it than a violin with shorter strings.

   As the amplitude of sound waves increases, the volume increases, so the sound is louder. For example, to make a louder sound on a guitar, you need to pluck the string harder, increasing the amplitude of the string's vibration. Increasing the amplitude therefore also increases the energy transferred by the wave.

   When you tune a stringed instrument, for example a guitar, you tighten or loosen the string to change its tension. This changes the medium that sound waves travel through, therefore changing the speed of waves. Higher tension means higher speed, therefore higher frequency and pitch, according to the wave equation $v = f\lambda$.

2. a) $v = f\lambda$

   $340 = 440 \times \lambda$

   $\lambda = 0.77$ m

b) $v = f\lambda$

$340 = f \times 1.54$

$f = 221$ Hz.

This is around half the frequency and at a lower pitch that the "A" used for tuning.

3. a) $v = f\lambda$

$340 = 20 \times \lambda$

$\lambda = 17$ m

b) $v = f\lambda$

$340 = 20 \times 1000 \times \lambda$

$\lambda = 0.017$ m

4. a) $T = 4.5$ ms

$f = \frac{1}{T} = \frac{1}{4.5\times10^{-3}} = 220$ Hz

b) $v = f\lambda$

$340 = 220 \times \lambda$

$\lambda = 1.5$ m

c) The second note is louder, as a larger voltage is produced. The second note has twice the frequency, so it sounds higher pitched:

$T = \frac{4.5}{2} = 2.25$ ms

$f = \frac{1}{T} = \frac{1}{2.25\times10^{-3}} = 440$ Hz

5. a)

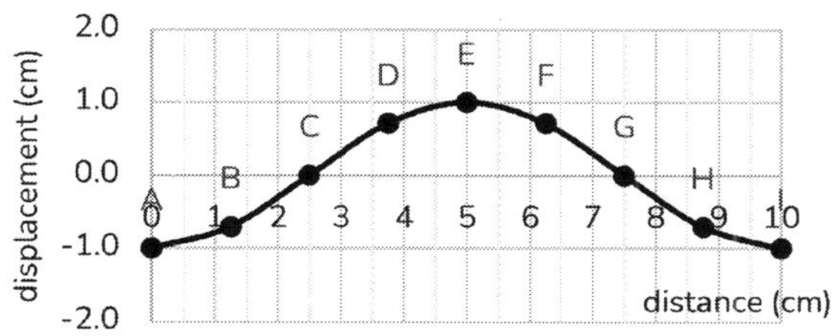

b) 10 cm

c) 1 cm

d) G, zero displacement

e) C, zero displacement

f) Particles in the compressions and rarefactions both have zero displacement.

g) 2.5 cm

h) $v = \frac{s}{t} = \frac{0.025}{0.5} = 0.05$ m s^{-1}

6. a)

Displacement-distance graph	Pressure-distance graph
Displacement of each particle from their equilibrium position	Pressure, or how dense the medium is
Maximum displacement from equilibrium position in the positive direction	High pressure, or compressions
Maximum displacement from equilibrium position in the negative direction	Low pressure, or rarefactions
Distance between two points at maximum displacement from equilibrium position in the same direction	Distance between two points of high pressure, or two points of low pressure

b)

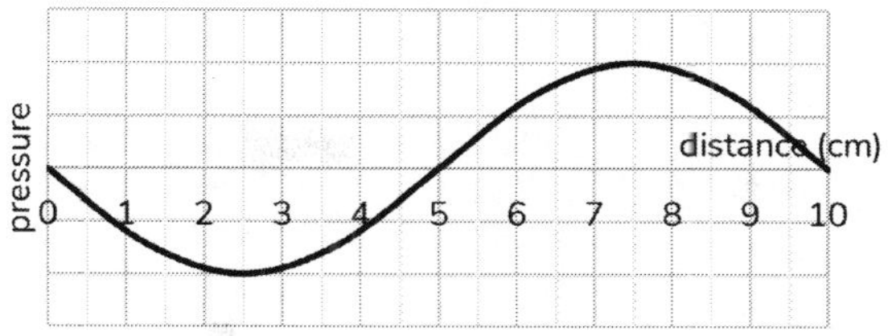

14.7 Light and electromagnetic waves

1. Sunlight reaches Earth, but matter is not transferred in the process, consistent with the definition of waves. Light is an example of an ***electromagnetic wave***, and the whole range of electromagnetic waves is called the electromagnetic spectrum. Each type of electromagnetic wave has a different wavelength and frequency.
The Sun produces all types of electromagnetic waves. Those with the highest frequencies - gamma rays, x-rays and ultraviolet rays - cause mutations in living cells, leading to cancer. Fortunately, the Earth's atmosphere blocks the gamma rays and x-rays and the ozone layer partially filters the ultraviolet rays. Light is the part of the electromagnetic spectrum that is visible to the human eye. Infrared is perceived by us as heat. Microwaves and radio waves have many uses, including communications.
Unlike mechanical waves, electromagnetic waves do not need a medium consisting of particles to travel through. However, like all waves, electromagnetic waves still involve vibrations. These vibrations consist of changing electric fields and magnetic fields. The fields change in magnitude and direction in a pattern like a sine wave, that repeats every wavelength. These changing fields are perpendicular to the direction the wave travels, so electromagnetic waves are transverse waves.
Electromagnetic waves can be created by changing electric fields, for example by turning an electric circuit on and off, which accelerates electrons in the circuit.
Electromagnetic waves can be blocked using metal shielding, which has electrons that can move. The electric field makes electrons accelerate, turning electromagnetic energy into the kinetic energy of the electrons, then into heat due to the resistance of the metal.
The electric fields are accompanied by magnetic fields, since electric currents produce magnetic fields.
All electromagnetic waves travel at the same speed in a vacuum, 3 x 10^8 m s^{-1}, which is also taken as its speed in air. Nothing we know of in the universe travels faster than this!
2. They all travel at the same speed.
3. a) $v = f\lambda$

$3 \times 10^8 = f \times 650 \times 10^{-9}$

$f = 4.6 \times 10^{14}$ Hz

b) $v = f\lambda$

$3 \times 10^8 = f \times 475 \times 10^{-9}$

$f = 6.3 \times 10^{14}$ Hz

4. $v = f\lambda$

$3 \times 10^8 = 105.7 \times 10^6 \times \lambda$

$\lambda = 2.8$ m

5. a) $v = f\lambda$

$3 \times 10^8 = 850 \times 10^6 \times \lambda$

$\lambda = 0.35$ m

b) Radio wave

6. a) $v = f\lambda$
   $3 \times 10^8 = f \times 259 \times 10^{-9}$
   $f = 1.2 \times 10^{15}$ Hz
   b) Ultraviolet
7. A wavelength is the distance between two consecutive points in phase. For light, this would be the distance between two consecutive points with the same magnitude and direction of electric field and magnetic field. The frequency is the number of times the electric field or magnetic field returns to the same magnitude and direction at a point in one second.
8.

Feature	Sound	Light
Is energy transferred?	Yes	Yes
Is matter transferred?	No	No
Can it travel through a vacuum?	No	Yes
Type of wave	Mechanical	Electromagnetic
Describe the vibrations involved	Longitudinal – particles vibrate along the direction of travel	Transverse – electric and magnetic fields vary periodically perpendicular to the direction of travel
Speed in air	340 m s^{-1}	3 x 10^8 m s^{-1}

15.1 Inverse square law

1. The further from a light bulb, the dimmer the light. In other words, less power, or energy per second, is received. However, due to the Law of Conservation of Energy, the power emitted at the source must be equal to the total power that has radiated outwards in all directions. If we could track each bit of energy that is radiating outwards from the source, in the same amount of time, those bits of energy will have spread out over the area of a sphere centred on the light bulb.
   The surface area of a sphere is $4\pi r^2$ so the power is spread out across this area.
   The amount of power per unit area, reaching a distance r away from the source of energy, is called the intensity. So the intensity of light observed is given by:
   $$I = \frac{P}{4\pi r^2}$$
   where: I = intensity at a distance r away from the source (W m^{-2})
   P = power emitted by the source (W)
   r = distance from the source (m)
   If the same light bulb is used, then $\frac{P}{4\pi}$ is a constant, so intensity is inversely proportional to the square of the distance from the source. This type of relationship is called an inverse square relationship.
   The inverse square relationship applies to all types of waves, including sound waves. The inverse square relationship also describes many other phenomenon, including electrostatic and gravitational fields.
   Comparing intensities at two different distances, r_1 and r_2, from the same source with power P, gives:
   $$I_1 = \frac{P}{4\pi r_1^2} \text{ and } I_2 = \frac{P}{4\pi r_2^2}$$
   Rearranging to make P the subject then equating the formulae gives:
   $$4\pi r_1^2 I_1 = 4\pi r_2^2 I_2$$
   $$I_1 r_1^2 = I_2 r_2^2$$
   Any unit for intensity or distance can be used in this equation, so long as the same unit is used on both sides. The SI unit for light intensity is the candela. When shopping for light bulbs, the brightness of the light bulb is often stated in lumens and the intensity of light received at a distance is measured in lux.
2. Electric field strength is defined as $E = \frac{F}{q}$ so the electric field around a point charge q is given by $E = \frac{1}{4\pi\varepsilon_0}\frac{q}{r^2}$. Since $\frac{1}{4\pi\varepsilon_0}q$ is a constant, as the distance r from charge q increases, the electric field strength decreases according to the inverse square of r. Therefore this is an example of the inverse square law.
3. Let $r_2 = 3r_1$
   $I_1 r_1^2 = I_2 r_2^2$
   $I_1 r_1^2 = I_2 \times (3r_1)^2 = 9I_2 r_1^2$
   $I_2 = \frac{I_1}{9}$
4. Let $I_2 = 2I_1$
   $I_1 r_1^2 = I_2 r_2^2$
   $I_1 r_1^2 = 2I_1 \times r_2^2$
   $r_2^2 = \frac{r_1^2}{2}$
   $r_2 = \frac{r_1}{\sqrt{2}} = 0.7r_1$
   They should move the lamp closer to the subject so it is $\frac{1}{\sqrt{2}}$ or 70% of the original distance.
5. $I = \frac{P_1}{4\pi r_1^2} = \frac{4\times 650}{4\pi \times 8^2} = \frac{650}{64\pi}$
   $I = \frac{P_2}{4\pi r_2^2}$
   $\frac{650}{64\pi} = \frac{P_2}{4\pi \times 12^2}$
   $P_2 = 5850 \text{ W} = \frac{5850}{650}$ lights $= 9$ lights
6. a) Since $I = \frac{P}{4\pi r^2}$, graphing radius on the x-axis would give a curved line of best fit, resembling a hyperbola. While this graph would illustrate a qualitative trend, it cannot confirm whether the inverse square law is satisfied. However, graphing radius squared should give a straight line with gradient $\frac{P}{4\pi}$.
   b) While the line of best fit is a straight line, consistent with the inverse square law, the y intercept is 40 lux, where it should be zero. The inconsistency could be due to background light of 40 lux.
   c) gradient $= 1194 = \frac{P}{4\pi}$
   $I = \frac{P}{4\pi r^2} = 1194 \times \frac{1}{1^2} 1194$ lux
7. $I_1 r_1^2 = I_2 r_2^2$
   $100000 \times 1^2 = I_2 \times 30^2$
   $I_2 = 111$ lux
   Percentage of Earth's intensity $= \frac{111}{100000} \times 100 = 0.111\%$
8. Let the intensity of light from Alpha Centauri and Sirius be I_1 and I_2 respectively.
   $I_2 = 3.8I_1$
   $\frac{P_2}{4\pi r_2^2} = \frac{3.8P_1}{4\pi r_1^2}$
   $\frac{25}{r_2^2} = \frac{3.8 \times 1.5}{4.3^2}$
   $r_2 = 9.0$ ly

9. a) $I = \frac{P}{4\pi r^2} = \frac{3.86\times10^{26}}{4\pi\times(1.50\times10^{11})^2} = 1370$ W m^{-2}

b) $P = \frac{\Delta E}{\Delta t}$

$\Delta E = P\Delta t = 1370$ W $\times$ 9 h $\times$ 60 min $\times$ 60 s

$= 4.44 \times 10^7$ J

or $\Delta E = P\Delta t = 1.370$ kW $\times$ 9 h $= 12.3$ kWh

c) Area $= \frac{15}{12.3} = 1.2$ m^2

10. a) $I_1 r_1^2 = I_2 r_2^2$

$1 \times 10^{-10} \times 1^2 = 1 \times 10^{-12} \times r_2^2$

$r_2 = 10$ m from the whisperers

They need to move 9 m further away.

b) $I_1 r_1^2 = I_2 r_2^2$

$1 \times 10^{-6} \times 1^2 = I_2 \times 100^2$

$I_2 = 1 \times 10^{-10}$ W m^{-2}

The talking is within the threshold of hearing and would sound like a whisper at 20 dB.

15.2 Reflection

1. Reflection is the change in direction of a wave at the boundary between two <u>media</u>, so that the wave bounces back into the medium it started in. You see <u>light</u> reflected from your face when you look in a mirror. An echo is a reflection of a <u>sound</u> wave. Reflection is a <u>property</u> of all waves, but particles can also reflect. For example, a <u>ball</u> will bounce off hard ground.

   Reflection changes the <u>direction</u> of a wave, but not its speed, intensity, wavelength or frequency. Reflection can however change the phase of a wave. If you send a wave through a spring or rope so a crest reaches a fixed end, a <u>trough</u> will return, so the wave reflects half a wavelength <u>out of phase</u> back towards you.

   However, if you send the same wave through the same spring towards an end that is free to move perpendicular to the direction of travel, then if a crest reaches the end, a <u>crest</u> will return. In other words, the wave will reflect <u>in phase</u> back towards you.

2. $v = \frac{d}{t}$

   $d = vt = 1500 \times \frac{3.3}{2} = 2500$ m

3. Speed when stretched 5.0 m: $v = \frac{s}{t} = \frac{10}{1.6} = 6.3$ m s^{-1}

   Speed when stretched 7.0 m: $v = \frac{s}{t} = \frac{14}{1.9} = 7.4$ m s^{-1}

   The speed of the pulse in the slinky spring increased when tension increased, because this changes the medium. Higher tension means bigger restoring force, leading to a higher wave speed.

4. When an incident ray travels perpendicular to a reflective boundary, the <u>reflected ray</u> is also perpendicular to the boundary. The line that is perpendicular to a boundary is called the <u>normal</u>.

   When an <u>incident ray</u> travels towards a boundary at an angle to the normal, it reflects back on the other side of the normal, so that the angle of incidence i <u>equals</u> the angle of reflection r:

   $$i = r$$

   This is called the Law of Reflection, which was first described by the Arab scientist, Alhazen. The angles i and r are measured between the normal and the ray, not between the ray and the <u>boundary</u>.

5. a) 20°

   b)

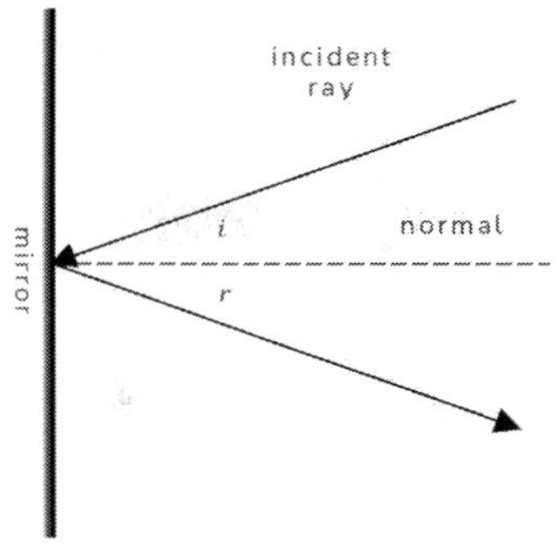

6. a)

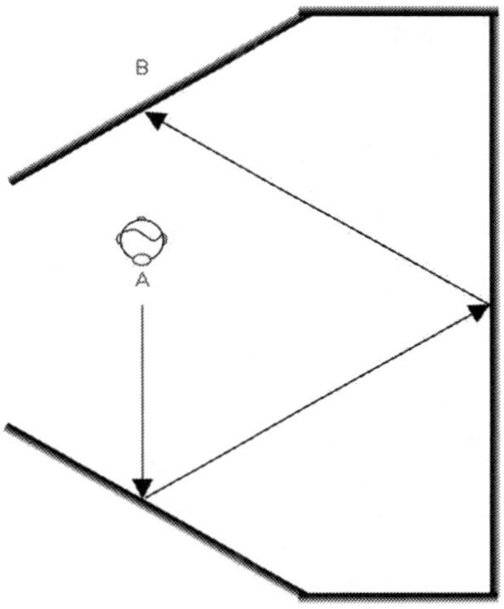

   b) Your back

7. We only see if <u>light</u> reaches our eye. To see an object, light must either be transmitted or <u>reflected</u> by the object. In a mirror, an object would appear to be in the <u>same</u> direction as the light ray reaching our eye travelled. This means that we may be able to see our feet in a mirror, even if the mirror did not reach the <u>ground</u>. The mirror just needs to reach low enough so that a light ray from our feet could be reflected by the mirror in such a way that the angle of <u>incidence</u> equalled the angle of reflection. The point on the mirror at which a light ray from our feet would reflect towards our eye would be <u>halfway</u> up between our feet and eyes, so the bottom of the mirror only needs to be as low as this to let us see our feet.

8. The Law of Reflection also applies to curved reflective surfaces, if angles are measured using a <u>tangent</u> to the surface at the point an incident ray strikes it.

   A curved surface that caves in towards the source of waves is described as <u>concave</u>.

   An example of a concave surface is a satellite dish, used to collect <u>radio waves</u> for communication or TV. The satellite dish is in the shape of a <u>parabola</u>, which in mathematics has the formula $y = ax^2$. The parabolic shape means that all rays that are <u>incident</u> on the dish parallel to its axis of symmetry will be reflected through a point, called the focal point or focus. The receiver for the radio waves is placed at the <u>focus</u>, which is at the tip of the antenna, to collect a strong signal from all of the reflected rays.

   A curved surface that bends away from the source of waves is described as <u>convex</u>.

   Convex mirrors do not <u>reflect</u> waves towards a focus. Instead, incident rays that are parallel to the axis of symmetry will reflect so the rays <u>diverge</u>, meaning they spread and go in different directions.

Convex mirrors provide a wider field of view, because light from larger angles to the axis of symmetry can be reflected towards your eye. This is why convex mirrors are often at the end of narrow driveways, so drivers can see hazards around corners.

9. If the light bulb is placed at the focus, then all of the light incident on the mirror will reflect back in a beam that is parallel to the axis of symmetry of the mirror, according to the Law of Reflection. This produces a much brighter beam of light going forwards than the light bulb would give without the mirror.

15.3 Superposition

1. If two particles meet, such as two balls coming from opposite directions, they collide and change velocity, in a way that conserves momentum.
   However, if two waves meet, they do not change velocity. Instead, they produce a resultant wave, so that the displacement of the medium at each point is given by the vector sum of the two waves meeting there. The waves themselves pass through each point unaffected. This overlapping of waves is called ***superposition***. The production of the resultant wave is called ***interference***. Interference is a property of waves, but not particles.
   Constructive interference occurs when the resultant displacement of a particle due to two waves gives it a bigger displacement than it would have had due to either wave.
   Destructive interference occurs when the resultant displacement of the particle due to two waves becomes zero, because each separate wave would have displaced it by an equal magnitude in opposite directions.
2. The spring will appear as a flat line, as each pulse displaces the spring in opposite directions by the same magnitude, giving a resultant displacement of zero and destructive interference.
3. a)

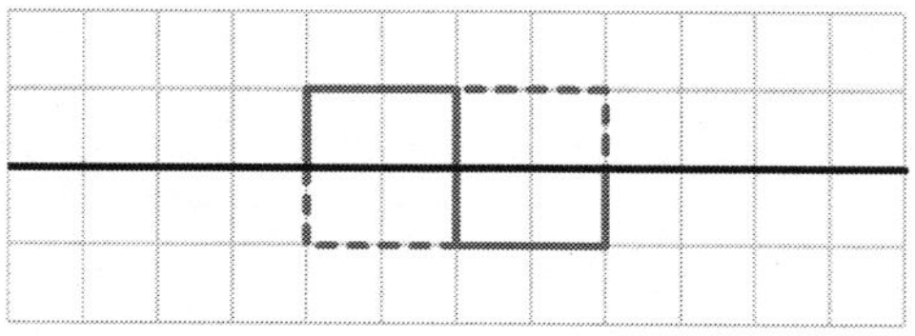

b)

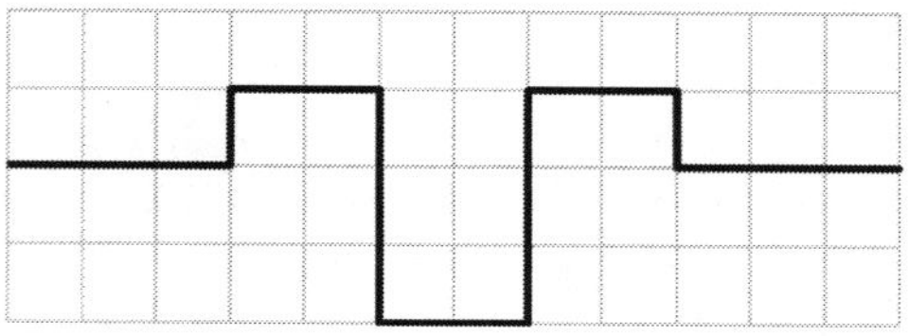

or

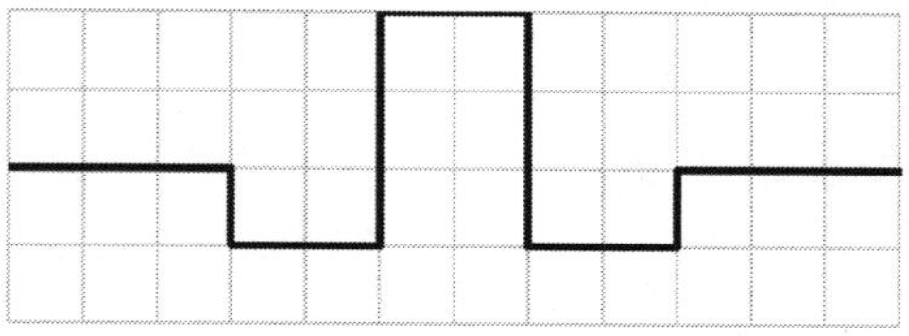

15.4 Standing waves on strings

1. Musical notes are due to resonant frequencies and the standing waves that they make.
   If you hold the end of a string and vibrate it repeatedly to send a continuous wave through it, the wave would reflect half a wavelength out of phase at the other end if it is a fixed end.
   The resultant wave would be the superposition of the incident and reflected waves. For most frequencies of vibration, the resultant would look chaotic. It would feel difficult to keep vibrating your end up and down at the same frequency, since the reflected wave would return energy to you, making your end vibrate with a different displacement to what you are giving the string.
   However, for certain frequencies, called ***resonant frequencies***, a repeating pattern is produced, with the most basic pattern making the string move like a bit like skipping rope.
   These frequencies need to produce wavelengths on the string which satisfy the boundary conditions of having a fixed end, called a node, at each end, since both ends are being held. A ***node*** is a point on the string where the total displacement due to the wave is zero. For example, the ends of guitar strings are nodes, as they do not move when the string is plucked.
   The lowest resonant frequency, called the ***fundamental frequency***, f_1, will produce nodes at each end with half a wavelength between them.
   The wavelength λ that a particular frequency produces on the string is given by the speed of waves through the string, v, divided by the frequency f, according to the wave equation $v = f\lambda$. The speed is determined by the medium, so using the same string at the same tension will mean that the waves travel through the string at the same speed. The longer the wavelength on the string, the lower the frequency that you must shake the end with.
   In other words, if the distance between the ends is L, the wavelength for this fundamental frequency will be $\lambda_1 = 2L$. Using the wave equation, the fundamental frequency will be equal to:
   $$f_1 = \frac{v}{\lambda_1} = \frac{v}{2L}$$
2. When an incident wave on a string reaches the other end as shown, so there is a crest in the middle, the reflected wave will also have a crest in the middle of the string, because it is reflected half a wavelength out of phase.
   The resultant wave will be produced by the superposition of the incident and reflected waves. The amplitude is therefore increased, with constructive interference occurring.
   At time $t = \frac{1}{4}T$, one quarter of a period later, the crest of the incident wave will have reached the other end. The reflected wave will therefore have a trough at the same end. The resultant wave at the other end still has zero displacement, producing a node. In fact, the resultant wave everywhere on the string experiences destructive interference.
   Another quarter of a period later, at time $t = \frac{1}{2}T$, there will be a trough in the middle of the string due to both the incident and reflected waves, so the resultant will therefore have a larger trough and exhibit constructive interference.
   Then another quarter of a period later, at time $t = \frac{3}{4}T$, the string will again experience destructive interference.

One period later, the pattern repeats, with constructive interference producing a large crest in the middle. The middle of the string undergoes the largest displacements compared to other points on the string, due to the fundamental frequency, and is called an ***antinode***. Shaking the string at a resonant frequency feels easy compared to other frequencies, as the resultant wave produces a node near your hand. The energy you put in will keep increasing the amplitude of vibration. However, there is no energy transferred along the string overall, since the incident and reflected waves move energy in opposite directions. The waves produced by resonant frequencies are therefore called ***standing waves***.

3. As well as the fundamental frequency f_1, there are higher resonant frequencies that will also produce standing waves. Collectively, these resonant frequencies are also known as ***harmonics***. The second longest wavelength λ_2 of a standing wave on this string of length L will still have a node at each end, but also a node in the middle, making the wavelength of the second harmonic $\lambda_2 = L$. The wavelength of the second harmonic is therefore half the wavelength of the fundamental frequency.
   The velocity of the wave v is still the same, because the medium has not changed. Using the wave equation, the frequency of the second harmonic is:
   $$f_2 = \frac{v}{\lambda_2} = \frac{v}{L}$$
   The frequency of the second harmonic is therefore double the fundamental frequency, $f_1 = \frac{v}{2L}$. When playing the second harmonic on a musical instrument, its pitch will sound one octave higher in compared to the fundamental frequency.
4. a)

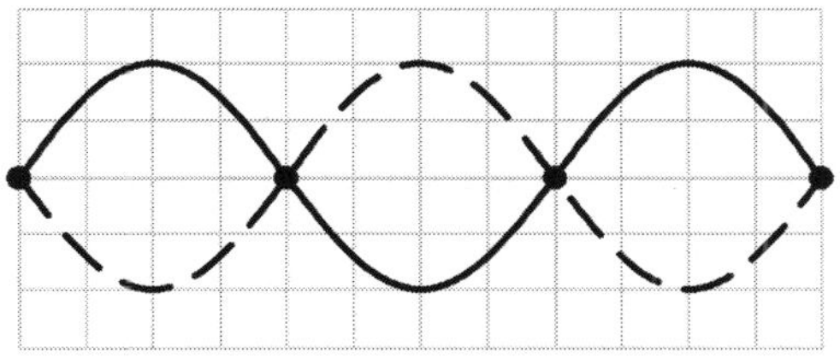

   b) $\lambda_3 = \frac{2}{3}L$
   c) $f_3 = \frac{v}{\lambda_3} = \frac{3v}{2L}$
5. a)

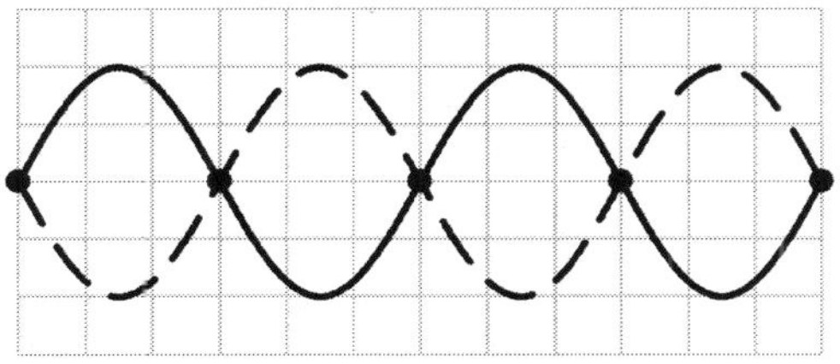

   b) $\lambda_4 = \frac{1}{2}L$
   c) $f_4 = \frac{v}{\lambda_4} = \frac{2v}{L}$
6. B
7. a) $v = f_1\lambda_1 = 310 \times 0.41 = 127 \text{ m s}^{-1}$
   b) $\lambda_2 = L = 0.205 \text{ m}$
   $f_2 = \frac{v}{\lambda_2} = \frac{127}{0.205} = 620$ Hz (Notice this is double the fundamental frequency.)
8. Reducing the string length reduces the wavelength of the standing wave produced by the fundamental frequency, given by $\lambda_1 = 2L$. The fundamental frequency therefore increases, according to $f_1 = \frac{v}{\lambda_1} = \frac{v}{2L}$, so the pitch increases.
9.

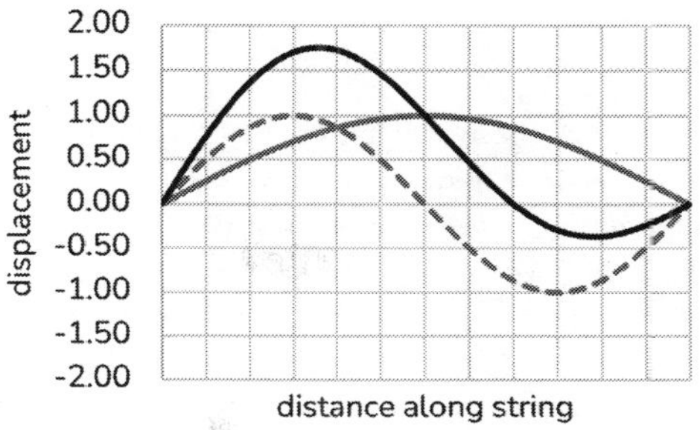

15.5 Standing waves in pipes

1. In wind instruments such as flutes, the medium through which waves travel is the air column. Unlike waves moving through strings, which are transverse, sound waves moving through air columns are longitudinal. Standing waves are still able to form though, so the notes we hear from the musical instruments are the resonant frequencies. In other words, sounds at the resonant frequencies are amplified by the musical instrument. If we visualise the waves as displacement-distance graphs, with the distance representing the position along the air column, we can analyse standing waves in pipes in a similar way as standing waves in strings.
   At an open end of a pipe, air particles are free to move to the maximum displacement, so the boundary condition for open ends is an antinode. A flute is open at both ends, so there would be an antinode at each end.
   At a closed end, air particles are not free to move, so the boundary condition is a node. If you blow across the top of an empty bottle to produce a note, there would be a node at the bottom of the bottle and an antinode at the mouth.
   The fundamental frequency produces the standing wave with the longest wavelength that satisfies the boundary conditions of the pipe. All other resonant frequencies must also satisfy the boundary conditions, but will have shorter wavelengths and higher frequencies than the fundamental frequency.
2. a) Standing waves are produced by resonant frequencies, for which the incident and reflected waves interfere in a way so that there is no energy transferred in the medium overall, since the incident and reflected waves move energy in opposite directions.
   b) It increases the amplitude of the standing wave.
3. a) One quarter of a wavelength is shown in the pipe, so the full wavelength is $4L$. Using the wave equation:
   $f_1 = \frac{v}{\lambda_1} = \frac{v}{4L} = \frac{340}{4 \times 0.58} = 147$ Hz
   b) $\lambda_2 = \frac{4L}{3}$, $f_2 = \frac{v}{\lambda_2} = \frac{3v}{4L}$

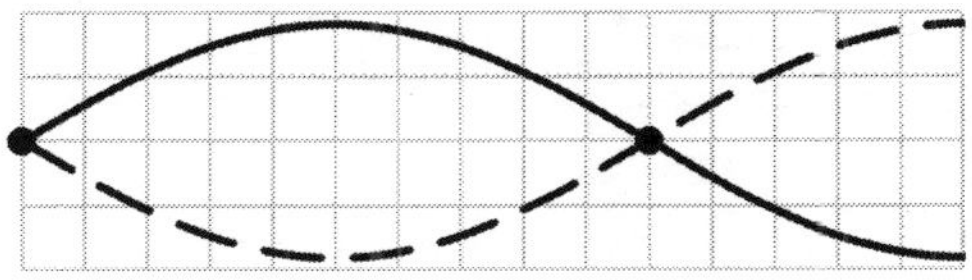

   c) $\lambda_3 = \frac{4L}{5}$, $f_3 = \frac{v}{\lambda_3} = \frac{5v}{4L}$

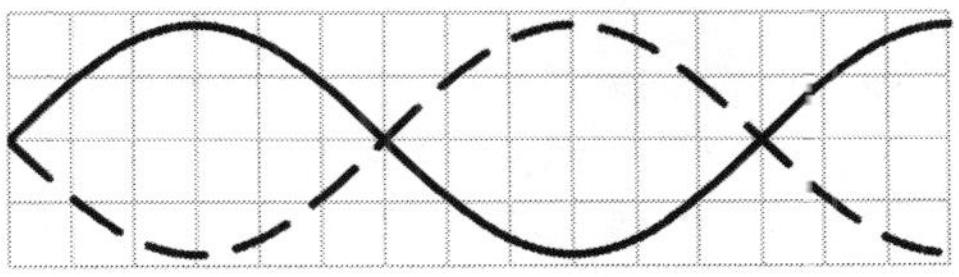

   d) D

4. a) $f_1 = \frac{v}{4L} = \frac{340}{4\times0.025} = 3400$ Hz
   b) The baby's most sensitive frequency is:
   $$f_1 = \frac{v}{4L} = \frac{340}{4 \times 0.015} = 5700 \text{ Hz}$$
   As the baby grows up, the ear canal lengthens, so the most sensitive frequency decreases. Their hearing becomes more sensitive at lower pitches as they age.
5. a) $\lambda_1 = 2L$
   $f_1 = \frac{v}{\lambda_1} = \frac{v}{2L} = \frac{340}{2\times1.2} = 140$ Hz
   b) $\lambda_2 = L$
   $f_2 = \frac{v}{\lambda_2} = \frac{v}{L}$
   c) $\lambda_3 = \frac{2}{3}L$
   $f_3 = \frac{v}{\lambda_3} = \frac{3v}{2L}$
   d) $\lambda_n = \frac{2L}{n}$
   $f_n = \frac{nv}{2L}$

15.6 Interference from two sources

1. Interference is a property of waves, but not particles. Interference occurs between a wave and its reflection returning towards the wave source. Interference also occurs when waves from different sources meet.
   If two loudspeakers play the same single frequency in phase with each other, what can you hear as you walk around in front of them? Constructive interference occurs when the part of the wave reaching you from one loudspeaker is in phase with the wave reaching you from the other loudspeaker. For example, if a compression reaches you from both loudspeakers at the same time, the sound you hear will seem louder.
   Destructive interference occurs when the parts of the waves from each loudspeaker reaching you are half a wavelength out of phase with each other. For example, this happens if a compression reaches you from one loudspeaker and a rarefaction reaches you from the other at the same time. At these points, the sound you hear will seem softer. If you try this in a room where there are no echoes from the walls, at the points of destructive interference there would be silence.
   The type of interference that occurs at a point depends on the difference in distance between that point and each source, also known as the ***path difference***. The path difference is zero at all points an equal distance from both sources. The waves reaching these points from both sources will always be in phase, so constructive interference occurs. This will also happens at points which are one wavelength further from one source than the other. In fact, constructive interference will happen at all points where the path difference is an integer number of wavelengths, or $n\lambda$, where $n = 0, 1, 2$ and so on.
   If the path difference is half a wavelength at a point, the waves reaching that point are out of phase, so destructive interference occurs. In fact, destructive interference will happen at all points where the path difference is $(n + \frac{1}{2})\lambda$, where $n = 0, 1, 2$ and so on.
   All other points will experience an amplitude somewhere in between the amplitude from constructive interference and zero.
   Interference occurs with water waves too. Where constructive interference occurs, the water has a greater amplitude. Where destructive interference occurs, the water will look flat.
   Evidence of light interference was only gathered in 1801, since the wavelength of visible light is on the nanometre scale. This means that the path differences for light interference are tiny compared to sound.
2. a) $v = f\lambda$
   $$\lambda = \frac{v}{f} = \frac{340}{261} = 1.3 \text{ m}$$
   b) The path difference is 0.65 m, which is half a wavelength. Therefore, the waves will be out of phase and destructive interference occurs, leading to a lower volume.
   c) The path difference is 1.3 m, which is a wavelength. Therefore, the waves will be in phase and constructive interference occurs, leading to a higher volume.
3. a) There would be a pattern of loud spots, where constructive interference occurs, and soft spots, where destructive interference occurs. Constructive interference would occur wherever the path difference between the point on the line and each speaker was an integer number of wavelengths. Destructive interference would occur in between these points, wherever the path difference was some integer number of wavelengths plus half a wavelength.
   b) A higher frequency note would have a shorter wavelength. Therefore, the points where constructive interference would occur would be closer together, as the path difference leading to the same number of integer wavelengths' difference would be smaller.

15.7 Beats

1. Beats are an interesting phenomenon that occurs when two sound waves of slightly different frequencies, f_1 and f_2, interfere to produce a resultant wave. This resultant wave has an amplitude that varies with time, with the peaks occurring with a frequency f_{beat} that is the difference between f_1 and f_2:
   $$f_{beat} = f_2 - f_1$$
   The frequency of these peaks is called the ***beat frequency***.
   The resultant wave also sounds like it has a frequency that is the average of f_1 and f_2:
   $$f_{resultant} = \frac{f_1 + f_2}{2}$$
   The resultant wave sounds like it has a pitch that is in the middle of f_1 and f_2 but with a volume that rises and falls rhythmically at the beat frequency.
2. a) From the first graph, there are 15 periods in 1 second, so $f_1 = 15$ Hz.
   From the second graph, there are 12 periods in 1 second, so $f_2 = 12$ Hz.
   b) $f_{resultant} = \frac{f_1+f_2}{2} = \frac{15+12}{2} = 13.5$ Hz
   c) $f_{beat} = f_2 - f_1 = 15 - 12 = 3$ Hz
3. a) $f_{beat} = f_2 - f_1 = 445 - 440 = 5$ Hz, therefore 5 beats per second.
   b) The beat frequency decreases, until there are no beats when the frequencies match.

15.8 Diffraction

1. As well as superposition, another behaviour that waves exhibit and particles do not is diffraction. ***Diffraction*** is the bending of waves around obstacles or through a gap, also known as an aperture. If you stand around a corner from someone and shout, they can still hear you, because sound is a wave and will diffract around the corner. However, if you try to throw a ball at them, the ball will be blocked by the corner, because particles do not diffract.

Diffraction explains why ocean waves can still reach a beach, even if the beach is sheltered by a headland. The wavefronts bend around the headland, producing circular wavefronts as they spread out. The part of the wavefront hitting the edge of the obstacle acts a bit like a source of energy, with the energy spreading out in all directions. However, some energy is lost from the wave as it is absorbed by the obstacle, so the amplitude of the wave is smaller as it diffracts around the obstacle. So the ocean wave bending around the headland will carry less energy than the wave before the headland.
How much diffraction occurs depends on the size of the wavelength and the obstacle or aperture. If an obstacle is small compared to the wavelength, then the wave will diffract right around it. For example, if you stand in the ocean, the oncoming waves will diffract around you, so the waves behind you do not look like they have been affected by your presence, because the distance between the crests of the ocean waves is likely to be bigger than your width.
This is also why you cannot see viruses using a light microscope. Viruses are typically 20 nm to 400 nm in diameter, which is smaller than the wavelengths of visible light, with a range of around 400 nm to 700 nm, so the light diffracts around the virus unaffected.
When wavefronts go through an aperture, diffraction is most noticeable when the wavelength is close to the size of the aperture. Diffraction still happens if the aperture is wider, but most of the wave will just pass through unaffected, so diffraction will only be observed close to the edges of the aperture and not the middle. If the aperture is too small compared to the wavelength, most of the wave's energy will be absorbed by the edges of the aperture.

2.

Feature	Reflection	Diffraction
Where it occurs	Boundary of media	Obstacles in the same medium
Direction wave changes	Determined by Law of Reflection $i = r$	Spreads out around obstacle
Does the wavelength change	No	No
Can particles have this property	Yes	No
Are observable effects affected by wavelength?	No	Yes –diffraction is most noticeable when the aperture or obstacle is similar in size to the wavelength

3. Visible light has a tiny wavelength compared to the width of a person, so diffraction of light around the person is not noticeable and most of the light energy is absorbed by the person. On the other hand, the wavelengths of sound humans can hear are between:
20 Hz: $\lambda = \frac{v}{f} = \frac{340}{20} = 17$ m
20 kHz: $\lambda = \frac{v}{f} = \frac{340}{20\times10^3} = 0.017$ m
The low frequencies have wavelengths much wider than the person, so will easily diffract around them to reach you.

4. The radio signals have the following wavelengths:
702 kHz: $\lambda = \frac{v}{f} = \frac{3\times10^8}{702\times10^3} = 427$ m
105.7 MHz: $\lambda = \frac{v}{f} = \frac{3\times10^8}{105.7\times10^6} = 2.838$ m
Since longer wavelengths diffract better than shorter ones, the 702 kHz signal is more likely to bend around the hill to reach the town.

5. Bee could be further away from the thunderstorm, with more obstacles between them and the thunderstorm. Longer wavelengths of sound with lower frequencies and pitches will reach Bee, as they diffract more easily around obstacles than shorter wavelengths. By contrast, Ash hears more of the frequencies present in the sound of the thunderstorm, as they are closer.

15.9 Doppler effect

1. If you walk down a street and pass a house every 5 seconds, a car driving down the same street at five times your speed will pass a house every second – in other words, at a higher frequency.
Similarly, if you and a source of sound move towards each other, then each wavelength reaches you faster, so frequency appears to increase and wavelength appears to decrease. Consequently, you hear a more high pitched sound.
If the same source moves away from you, then each wavelength takes longer to pass you, so the wavelength of sound that you observe will increase. Consequently, the frequency you observe will decrease and you hear a more low pitched sound.
This is called the ***Doppler effect*** - the change in frequency of a wave observed when there is relative motion between the source of the wave and the observer. The Doppler effect explains why the engine noise of aeroplanes flying by appears to increase in pitch as they approach and decrease as they recede, if we observe the aeroplanes from the ground, but the passengers do not observe a change in pitch.
When a source and observer move towards each other, the apparent frequency detected by the observer is given by:

$$f' = \frac{f(v_w + v_o)}{v_w - v_s}$$

where: f' = frequency detected by the observer (Hz)
f = frequency of the sound emitted at the source (Hz)
v_w = speed of the wave in the medium (m s^{-1})
v_o = speed of the observer relative to the ground (m s^{-1})
v_s = speed of the source relative to the ground (m s^{-1})
So the faster the speed of the source v_s towards from the observer, the higher the apparent frequency.
When a source and observer are moving away from each other, the apparent frequency detected by the observer is given by:

$$f' = \frac{f(v_w - v_o)}{v_w + v_s}$$

This formula looks like the formula used when they move towards each other, except that the plus and minus signs are swapped, so that the apparent frequency f' is lower than the original frequency f.
So the faster the speed of the source v_s away from the observer, the lower the apparent frequency.

2. The phone would have a lower pitch when it moves away from the observer and a higher pitch when it moves towards the observer. The highest pitch would occur at point X, when the phone has the highest velocity away from the observer, and the lowest pitch would occur at point Y, when the phone has the highest velocity towards the observer.

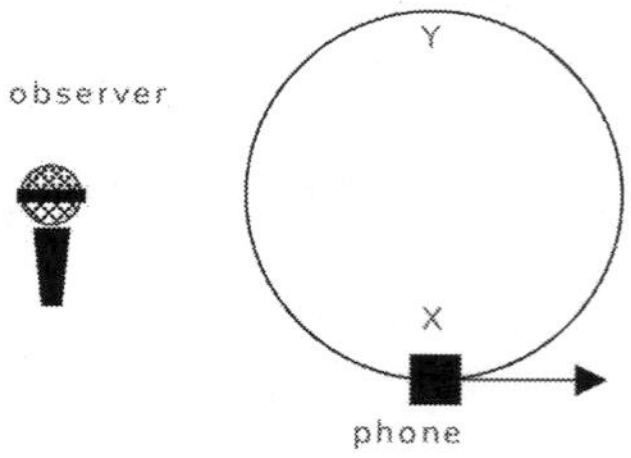

3. a) +
   b) –
   c) As the apparent frequency is larger, the source and observer must be approaching each other.
   d) Velocity has direction while speed does not, so treating these quantities as velocities would involve one of them having a minus sign in front, to represent that it is travelling in the opposite direction to the other.
4. a) $f' = \frac{f(v_w - v_o)}{v_w + v_s} = \frac{674(340-0)}{340+1.2} = 672$ Hz
   b) $f' = \frac{f(v_w + v_o)}{v_w - v_s} = \frac{674(340+0)}{340-1.2} = 676$ Hz
5. a) If the apparent wavelengths are longer, then the apparent frequency is lower. This means that galaxies must be moving away from Earth.
   b) Both A and B are receding from the Earth, as the apparent wavelengths in the hydrogen spectrum are longer than the actual wavelengths. However, B is receding faster than A, since the increase in wavelength is bigger. C is moving towards from Earth, as the apparent wavelengths in the hydrogen spectrum are shorter than the actual wavelengths.
6. a) $f' = \frac{f(v_w + v_o)}{v_w - v_s} = \frac{1.5\times10^6\times(1570+0.12)}{1570-0} = 1500115$ Hz
   b) $f' = \frac{f(v_w + v_o)}{v_w - v_s} = \frac{1500115\times(1570+0)}{1570-0.12} = 1500230$ Hz
   c) $f_{beat} = 1500230 - 1.5\times10^6 = 230$ Hz
7. $f' = \frac{f(v_w - v_o)}{v_w + v_s}$

   $1522 = \frac{1610(340-0)}{340+v_s}$

   $v_s = 19.66$ m s^{-1}

   $v = u + at$

   $19.66 = 0 + 9.8 \times t$

   $t = 2.01$ s

15.10 Refraction

1. A straw in a glass of water looks bent from most angles. The straw is not actually bent though – this phenomenon occurs because the light ray reaching us from the straw bends as it travels from water to air.

   The change in direction of travel of a wave as it moves from one medium to another is called ***refraction***. This bending occurs because waves travel at different speeds in different media.

   For example, light travels faster than anything we know in a vacuum, at 3 x 10^8 m s^{-1}. Light travels very close to this speed in air. However, as light enters other media, such as glass or water, it slows down. The ***refractive index*** of a medium gives the ratio of the speed of light in a vacuum to the speed of light in that medium:

   $$n = \frac{c}{v}$$

   where: n = refractive index (this is a number with no units)
   c = 3 x 10^8 m s^{-1} = speed of light in a vacuum
   v = speed of light in the medium (m s^{-1})

   The refractive index of a vacuum is equal to 1, and air is usually taken to be the same. Any other medium will have a refractive index greater than 1, since light will travel slower in it.

   If light travels perpendicular to a boundary between two media, it would only change speed, but not direction. In other words, a light ray travelling along the normal to a boundary would keep travelling through the second medium along the normal, but with a different speed. Another way of saying this is if the angle of incidence is zero (that is, the incident ray's angle to the normal is zero) then the angle of refraction is also zero.

   However, if the angle of incidence is greater than zero, then the change in speed will make the ray bend. For example, a light ray travelling from air into water would slow down, as the refractive index of water is greater than air. This makes light bend towards the normal, so the angle of incidence i is greater than the angle of refraction r. It is as if the light travels the path that would take it the least time to go from one point in the air to the water, rather than following a direct path which would cause it to travel further in the slower medium.

   As light emerges from water into air, it speeds up, as the refractive index of air is less than water. This makes the light bend away from the normal, so the angle of incidence is less than the angle of refraction. Again, light follows the path that takes the least time between two points.
2. a) Oil: $n = \frac{c}{v}$

   $v = \frac{c}{n} = \frac{3\times10^8}{1.43} = 2.10 \times 10^8$ m s^{-1}

   Water: $v = \frac{c}{n} = \frac{3\times10^8}{1.33} = 2.26 \times 10^8$ m s^{-1}

   b)

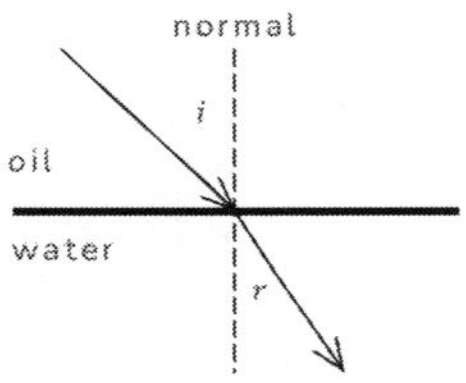

3. The unknown glass is not the same type of glass as the known glass. Since the laser bends away from the normal when travelling from the unknown to the known glass, light travels slower in the unknown glass than the known glass, and the unknown sample therefore has a refractive index that is higher than 1.50.
4. a) Air: $n = \frac{c}{v}$

   $v = \frac{c}{n} = \frac{3\times10^8}{1.00} = 3 \times 10^8$ m s^{-1}

   Water: $v = \frac{c}{n} = \frac{3\times10^8}{1.33} = 2.26 \times 10^8$ m s^{-1}

   Glass: $v = \frac{c}{n} = \frac{3\times10^8}{1.52} = 1.97 \times 10^8$ m s^{-1}

   b) The light from the fish would have refracted towards the normal when travelling from water to glass, since it would slow down due to the higher refractive index of glass.

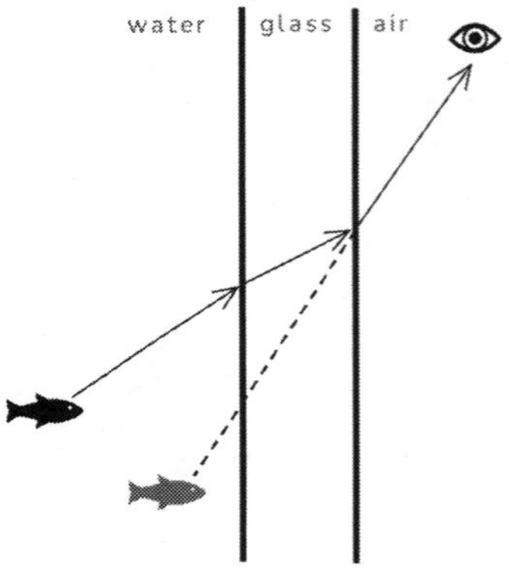

   It would then bend away from the normal into air, since air has the lowest refractive index and therefore light travels fastest there. So the fish would actually be further from the glass and vertically higher than what it is observed to be.

5. The refractive index of any medium light travels through is the speed of light in a vacuum c divided by its speed in the medium v:

$$n = \frac{c}{v}$$

This equation can be rearranged to make the speed of light in a vacuum the subject:

$$c = nv \quad ...(1)$$

If light travels between two media, with refractive indices n_1 and n_2 and wave speeds of v_1 and v_2 respectively, then substituting these into equation (1) gives a relationship between these speeds and refractive indices:

$$n_1 v_1 = n_2 v_2$$

$$\frac{v_1}{v_2} = \frac{n_2}{n_1} \quad ...(2)$$

Substituting the wave equation $v = f\lambda$ for each medium gives:

$$\frac{f_1\lambda_1}{f_2\lambda_2} = \frac{n_2}{n_1} \quad ...(3)$$

However, the frequency of a wave does not change as it passes from one medium to another, so:

$$f_1 = f_2$$

Substituting this into equation (3) and cancelling the frequencies gives a relationship between the wavelengths in the media and refractive indices:

$$\frac{\lambda_1}{\lambda_2} = \frac{n_2}{n_1} \quad ...(4)$$

Combining equations (2) and (4) together, the ratios of velocities, wavelengths and refractive indices when a wave travels between two media is given by:

$$\frac{v_1}{v_2} = \frac{\lambda_1}{\lambda_2} = \frac{n_2}{n_1}$$

This formula can also be used to describe waves other than light which travel between media, but the value of the refractive indices will depend on both the type of wave and the media.

6. a) The wavelength decreases as the water waves move from deep to shallow water. Since frequency stays the same, the speed of the wave must decrease in shallow water, according to $v = f\lambda$. The decrease in speed causes the wave to refract and the direction of the wave bends towards the normal.

   b) From part (a), the speed of the wave will increase when moving from shallow to deep water. Therefore the wavelength will increase, according to $v = f\lambda$. The increase in speed also causes the wave to refract and the direction of the wave bends away from the normal.

7. a) $n = \frac{c}{v}$

   $v = \frac{c}{n} = \frac{3 \times 10^8}{1.53} = 1.96 \times 10^8$ m s^{-1}

   a) $\frac{\lambda_1}{\lambda_2} = \frac{n_2}{n_1}$

   $\frac{365 \times 10^{-9}}{\lambda_2} = \frac{1.53}{1.33}$

   $\lambda_2 = 317$ nm

8. The speed of light in the cornea is:

   $v = \frac{c}{n} = \frac{3 \times 10^8}{1.376} = 2.180 \times 10^8$ m s^{-1}

   The speed of light in the aqueous humour is:

   $v = \frac{c}{n} = \frac{3 \times 10^8}{1.336} = 2.246 \times 10^8$ m s^{-1}

   So light travels faster in the aqueous humour compared to the cornea.

   The frequency of light remains the same in both media.

   The ratio of the wavelengths in the cornea to the aqueous humour is given by:

   $\frac{\lambda_1}{\lambda_2} = \frac{n_2}{n_1} = \frac{1.336}{1.376} = 0.971$

   So the wavelength in the cornea is slightly smaller than the wavelength in the aqueous humour.

15.11 Angle of refraction

1. $\frac{\sin i}{\sin r} = \frac{n_2}{n_1}$

   $\frac{\sin 75°}{\sin 44°} = \frac{n_2}{1}$

   $n_2 = 1.39$

   The concentration is around 35%.

2. Angle of refraction in the glass is given by:

   $\frac{\sin i}{\sin r} = \frac{n_2}{n_1}$

   $\frac{\sin 60°}{\sin r} = \frac{1.5}{1}$

   $r = 35°$

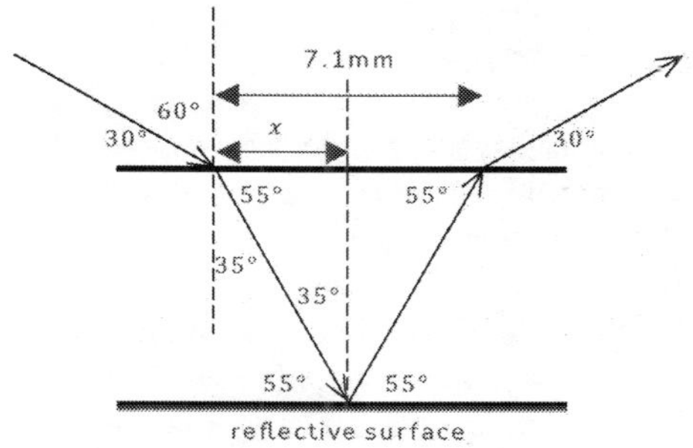

   Once the ray reaches the reflective surface, it reflects at an angle of reflection of 35°.

   For the triangle made by the ray in the glass and distance along the glass x, between where it enters the glass and the normal to the reflective surface:

   $\tan 35° = \frac{x}{5\text{mm}}$

   $x = 3.53$ ... mm

   Total distance between where the light enters and emerges from the glass is 7.1 mm.

3. a)

i (°)	r (°)	$\sin i$	$\sin r$
15	10	0.26	0.17
30	20	0.50	0.34
45	28	0.71	0.47
60	36	0.87	0.59
75	40	0.97	0.64

   b) $\sin r = \frac{n_1}{n_2} \sin i$

   This is a linear regression, in the form y=mx, with the gradient $\frac{n_1}{n_2}$.

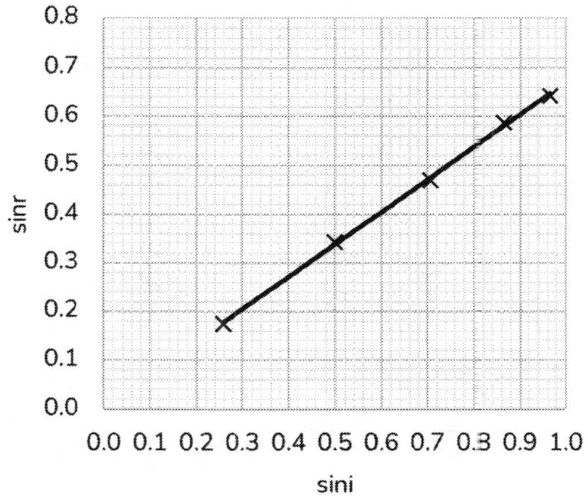

c) The line of best fit gives the average trend, which minimises the effect of random errors.

gradient $= 0.67 = \frac{n_1}{n_2} = \frac{1}{n_2}$

$n_2 = 1.5$

d) $n = \frac{c}{v}$

$1.49 = \frac{3 \times 10^8}{v}$

$v = 2.0 \times 10^8$ m s^{-1}

15.12 Lenses

1. Lenses have curved surfaces which can refract waves in order to make them converge (come together) or diverge (spread out).
   A surface on a lens that caves in towards the source of waves is described as concave, while a surface that curves away is convex.
   At any point on the curved surface, the normal is perpendicular to the tangent to the curved surface.
2. A magnifying glass is an example of a convex lens. In fact, it is a bi-convex lens, with both sides being parts of a sphere. An imaginary line through the middle of the lens, perpendicular to its surface is called the principal axis.
   Light rays that are parallel to the principal axis will bend towards the normal, because light slows down. This bending makes the rays converge in the glass. When they reach the boundary to travel back into air, the rays speed up and bend away from the normal. Because the glass is curved, this bending also makes the rays converge, so the rays meet at a focus.
   Sunlight passing through a magnifying glass can be directed to the focus, concentrating enough heat there to burn paper!
3. a) The angle of incidence is zero, so although the ray slows down in the glass, refraction does not make it change direction.
   b) Focus
   c) The image is bigger than the object, but otherwise appears the same.
4. They both direct light through a focus.
5. The light rays will bend towards the normal, because the lens has a higher refractive index than air, so light slows down. The shape of the concave lens makes the rays diverge. When they reach the boundary to travel back into air, the rays speed up and bend away from the normal, making the rays diverge again. Therefore the concave lens gives a wider beam of torch light.
6. The concave lens makes light rays from an object on one side of the peephole diverge on the other side, due to refraction. A person viewing these rays would perceive that they come from a smaller image. Therefore the concave lens makes it possible to see more of the object, giving a wider field of view.

15.13 Dispersion

1. What our eyes and brains interpret as white light is actually made of the visible wavelengths of the electromagnetic spectrum. However, refraction can separate white light into a rainbow, so that we can see the different wavelengths in it. Blue and violet light have wavelengths near the 400 nm end of the visible light spectrum, while red light has wavelengths near the 700 nm end.
   The rainbow occurs because refractive index depends on both the wavelength of the wave and the medium it is travelling through. This means that different wavelengths of light experience slightly different speeds in a medium other than a vacuum. So if a ray of white light travels from a medium with refractive index n_1 to another medium with refractive index n_2, the different wavelengths of the ray start with the same angle of incidence i but refract at different angles of refraction r. The relationship is given by Snell's Law:
   $$\frac{\sin i}{\sin r} = \frac{v_1}{v_2} = \frac{\lambda_1}{\lambda_2} = \frac{n_2}{n_1}$$
   When a ray of white light undergoes dispersion through glass, it is observed that blue light has a larger angle of refraction than red light, therefore blue light must have a slower speed in glass than red light, and a higher refractive index.
   This separation of different wavelengths into different angles is called ***dispersion***. Dispersion is a phenomenon that can occur for all waves, but is most easily observed with light. There is not much dispersion of sound waves in air, but it becomes more significant for sound waves in solids.
2.

Feature	Dispersion	Diffraction
Where it occurs	Boundary of two media	Obstacles in the same medium
Does wave direction change?	Yes, due to refraction	Yes, spreads out around obstacle
Does wave speed change?	Yes	No
Does the wavelength change?	Yes	No
Contrast the effect on shorter and longer wavelengths	Shorter wavelengths bend towards the normal more than longer wavelengths, when travelling into a slower medium	Diffraction is most noticeable when the aperture or obstacle is the same size as wavelength. Longer wavelengths diffract more.

3. a) As wavelength increases, refractive index decreases, so blue light has a higher refractive index than red light.
   b) Blue light of 450 nm has a refractive index of 1.3395:
   $n = \frac{c}{v}$
   $v = \frac{c}{n} = \frac{3 \times 10^8}{1.3395} = 2.24 \times 10^8$ m s^{-1}
   Red light of 670 nm has a refractive index of 1.331:
   $v = \frac{c}{n} = \frac{3\times10^8}{1.331} = 2.25 \times 10^8$ m s^{-1}
   so red light travels faster in water than blue.
   c) According to $\frac{\lambda_1}{\lambda_2} = \frac{n_2}{n_1}$ as the refractive index increases from air to water, the wavelength of light decreases.
   The wavelength of the blue light of 450 nm becomes:
   $\lambda_2 = \frac{\lambda_1}{n_2} = \frac{450\times10^{-9}}{1.339} = 336$ nm
   The wavelength of the red light of 670 nm becomes:
   $\lambda_2 = \frac{\lambda_1}{n_2} = \frac{670\times10^{-9}}{1.331} = 503$ nm

d) For $i = 25°$ and blue light: $\frac{\sin 25°}{\sin r} = \frac{1.339}{1}$ so $r = 18.4°$

For $i = 25°$ and red light: $\frac{\sin 25°}{\sin r} = \frac{1.331}{1}$ so $r = 18.5°$

For $i = 85°$ and blue light: $\frac{\sin 85°}{\sin r} = \frac{1.339}{1}$ so $r = 48.1°$

For $i = 85°$ and red light: $\frac{\sin 85°}{\sin r} = \frac{1.331}{1}$ so $r = 48.5°$

An angle of incidence of 25° only causes dispersion between blue and red light with a 0.1° difference in angle, but an angle of incidence of 85° causes a difference in angle of 0.4°. Therefore the larger angle of incidence causes the most noticeable dispersion.

e) Not to scale

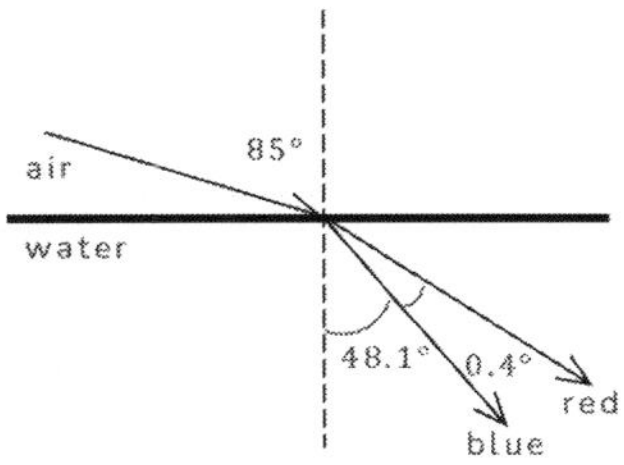

15.14 Total internal reflection

1. ***Total internal reflection*** occurs when a wave travelling from a slower medium towards a faster medium is reflected from the boundary, instead of refracting as it leaves the slower medium. In other words, the ray must travel from a medium with a higher refractive index towards the boundary of a medium with a lower refractive index. If this ray exits the slower medium, the angle of refraction will be greater than the angle of incidence, and it will bend away from the normal.
However, the refracted ray can only have an angle of refraction less than 90° and still be in the faster medium. The angle of incidence that gives an angle of refraction of 90° is called the ***critical angle***, i_C:
$$\frac{\sin i_C}{\sin 90°} = \frac{n_2}{n_1}$$
$$\sin i_C = \frac{n_2}{n_1}$$
If the angle of incidence is bigger than i_C, the ray will no longer exit to the faster medium, but will instead reflect from the boundary. The Law of Reflection still applies to total internal reflection, so the angle of reflection will be equal to the angle of incidence.
2. During the day, the sun warms the ground, so at night, heat rises from the ground through convection. This makes the air warmer as the height above the ground increases. Sound travels faster in warm air, since the air particles have more kinetic energy compared to cold air and collide faster. By contrast, the air just above a lake stays cooler since water does not heat up as fast as the ground during the day. Therefore sound travels slower in the cool air just above the lake.
As height above the ground increases at night, the temperature increases and the refractive index of sound decreases. Therefore sound travelling from the ground towards the warm air above will refract, bending the path of the sound downwards. If sound travels upwards so that the angle of incidence to a layer of warm air is greater than the critical angle, the sound will undergo total internal reflection and be directed back towards the ground. This means that if you hear a source of sound across a lake, the sound can travel to you both in a direct line as well as along a second route via refraction and reflection, amplifying the sound.
3. The largest angle of incidence possible on any boundary is 90°. Travelling to a medium with higher refractive index means that the wave will bend towards the normal in the second medium, with an angle of refraction below 90°, so the wave will never be bent back into the first medium.
4. a) $\sin i_C = \frac{n_2}{n_1} = \frac{1}{1.33}$

   $i_C = 48.8°$

   b) $\tan 48.8° = \frac{r}{1}$

   $r = 1.14$ m
5. a) $n = \frac{c}{v}$

   $1.5 = \frac{3 \times 10^8}{v}$

   $v = 2 \times 10^8$ m s^{-1}

   b) $v = \frac{s}{t}$

   $2 \times 10^8 = \frac{1000 \times 10^3}{t}$

   $t = 0.005$ s

   c) Without cladding: $\sin i_C = \frac{n_2}{n_1} = \frac{1}{1.50}$

   $i_C = 41.8°$

   With cladding: $\sin i_C = \frac{n_2}{n_1} = \frac{1.49}{1.50}$

   $i_C = 83.4°$

   d) Since the cladding increases the critical angle of the glass core significantly, only rays travelling almost parallel to the optical fibre will be totally internally reflected. This prevents rays from spending a longer time in the optical fibre reflecting from the edges and zigzagging in between, therefore reducing the time delay between different rays reaching the end.
6. Critical angle of diamond: $\sin i_C = \frac{n_2}{n_1} = \frac{1}{2.42}$

   $i_C = 24.4°$

   The rays in the diamond with the 60° base reach the bottom at an angle of incidence of 60°, which is bigger than the critical angle. Therefore the rays will totally internally reflect from the first side of the diamond. However, the rays reach the other side at an angle of incidence of 0°, so leave the diamond rather than totally internally reflect back through the top.

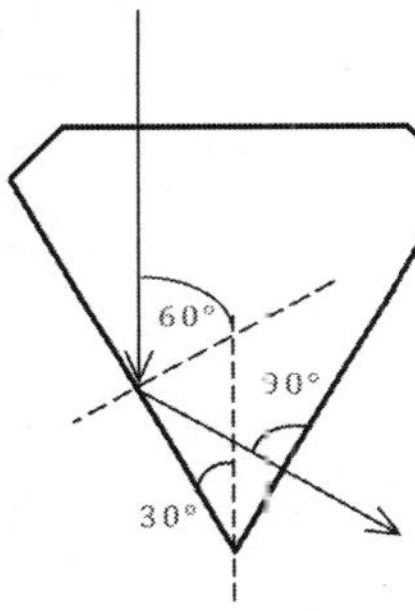

The rays in the diamond with the 90° base will reach the bottom at an angle of incidence of 45°, which is bigger than the critical angle. Therefore the rays will totally internally reflect from the first side of the diamond. They reach the other side at an angle of incidence of 45° too, so they also totally internally reflect from the second side, back through the top of the diamond.

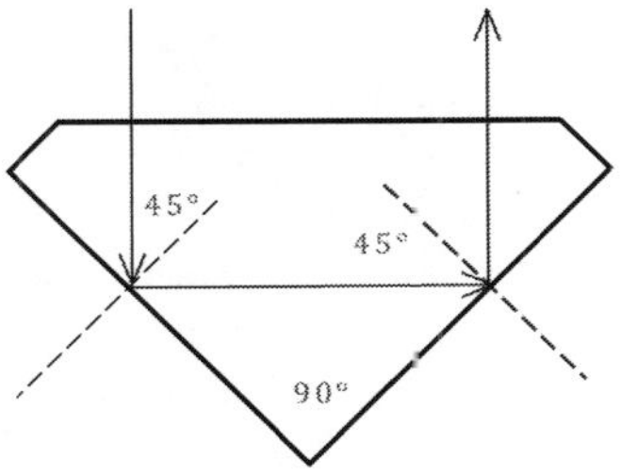

The rays in the diamond with the 140° base will reach the bottom at an angle of incidence of 20°, which is smaller than the critical angle. Therefore the rays will exit the diamond and not reflect.

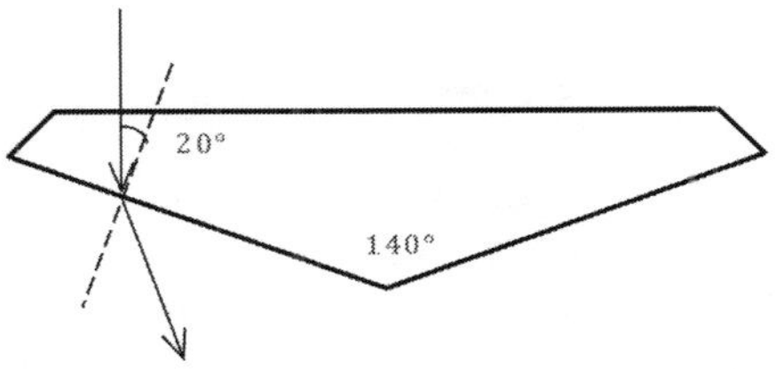

This means the 90° base is the most sparkly, due to the rays entering from the top being totally internally reflected back.

16.2 Heat transfer

1. When air is heated, the air particles have more kinetic energy to overcome electrostatic attraction to each other, so they move further apart. As the hot air expands, it becomes less dense and floats upwards. This means that the air closer to the ceiling is hotter than the air closer to the ground. The current of rising hot air pushes particles of smoke upwards, which can be toxic to inhale.
   This current of hot air also pushes air already near the ceiling sideways. As hot air moves away from the heat source, it cools, so the particles lose kinetic energy and are less able to overcome the electrostatic force of attraction between each other. The cooler air is more dense than hotter air and sinks due to gravitational force. This means that the ground is often the coolest place in a burning room.
   This process of hotter substances rising and cooler substances sinking is called ***convection***, and occurs in substances where particles are free to move, such as liquids and gases, but not in substances where particles are strongly bonded to others, such as solids.
2. As the cooler air closer to the ground warms up, it rises up inside the robe, because it becomes less dense. Hot air escapes from the top of the robe via convection.
3. On a hot day, heat from the Sun makes particles in the outermost pane of glass vibrate with more kinetic energy. Particles in glass can collide with neighbouring particles and transfer some of their kinetic energy to them. These particles in turn then collide with more particles to transfer kinetic energy. This process is called ***conduction*** - the transfer of heat through a substance from the hotter parts to the colder parts, as a result of collisions between neighbouring particles.
   Conduction happens more easily in solids than gases, because particles in a solid are closer together and are more likely to collide and transfer kinetic energy. By contrast, the gas layer in a double glazed window will not easily conduct heat, because gas particles are more spread out. In other words, the glass is a better conductor of heat than gas, which is a better insulator.
   This means double glazed windows will keep a house cool on hot days. They will also work on cold days to keep heat from inside escaping to the outside.
4. To conduct heat, a material must have particles that are able to transfer kinetic energy to other particles through collisions. However, to conduct electricity, a material must have charges that are free to move. If diamond is good conductor of heat but not electricity, there must be particles that can easily collide with others, but it must not have charges that are free to move.
5. The processes of convection and conduction both involve the kinetic energy of particles in transferring heat. However, heat travels to Earth from the Sun without particles present, via electromagnetic radiation. Heat in this form is also known as infrared. Infrared is invisible to our eyes, but can be detected by specialised human nerve cells called thermoreceptors in our skin.
   The wavelength of infrared ranges from around 700 nm to 1 mm, which is longer than the wavelength of visible light at 400 nm to 700 nm, but shorter than the wavelength of microwaves. Infrared travels at the same speed as light in a vacuum, 3 x 10⁸ m s⁻¹.
6. For 700 nm: $f = \frac{c}{\lambda} = \frac{3\times10^{8}}{700\times10^{-9}} = 4.3\times10^{14}$ Hz
   For 1 mm: $f = \frac{c}{\lambda} = \frac{3\times10^{8}}{1\times10^{-3}} = 3\times10^{11}$ Hz
7. a) A has more intensity at every wavelength. Since intensity is a measure of power per area around the light bulb, more power must have been supplied to it. Electrical power is given by $P = VI = \frac{V^2}{R}$, so the higher voltage supplied more power, producing graph A.
   b) Visible light is the useful form of energy from a light bulb, which is 400 nm to 700 nm, or 0.4 μm to 0.7 μm. The graph shows the intensity peaks at longer wavelengths than this, so much of the energy from this light bulb is produced as heat and it is not energy efficient.
   c) Both settings produce both visible light and heat, since there is intensity in both these wavelength ranges. Light bulb A on the 12 V setting is brighter and hotter, as intensity is higher at all wavelengths. Light bulb A would appear more bluish-white while Light bulb B appears more orange, as light bulb A's peak wavelength is shorter and closer to the blue end of the visible spectrum.

16.3 Skills – Conduction

1. $k = \frac{Q}{t}\frac{d}{A\Delta T} = \frac{W\,m}{m^2\,K} = W\,m^{-1}\,K^{-1}$
2. a) $\frac{Q}{t} = \frac{kA\Delta T}{d} = \frac{0.96\times3\times4\times(25-10)}{0.003} = 57600\text{ W} = 58\text{ kW}$
   b) $E = Pt = 58000\times60\times60 = 210\text{ MJ}$
3. a) proportional to
   b)

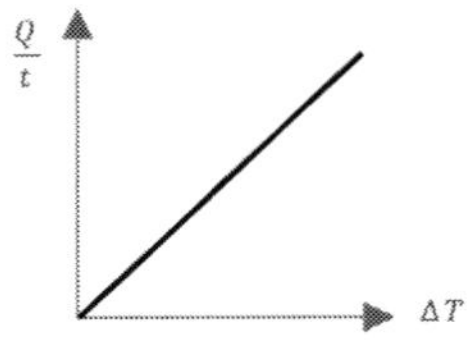

   c) $\frac{Q}{t} = \frac{kA\Delta T}{d}$
   $\frac{1}{2}\times58000 = \frac{0.96\times3\times4\times(T-10)}{0.003}$
   $T = 18°\text{C}$
4. a) The rate of heat loss is proportional to the area of the window.
   b)

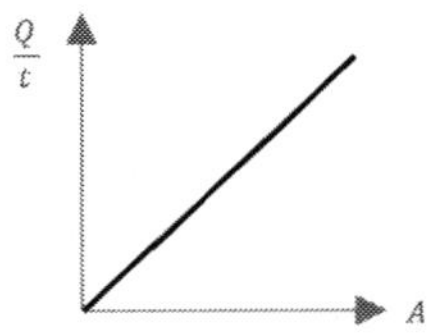

   c) As $\frac{Q}{t} = \frac{kA\Delta T}{d}$ and $\frac{Q}{t}$ is on the y-axis with A on the x-axis, this graph is in the form $y = mx$ with the gradient being $\frac{k\Delta T}{d}$. With the specifications given, the gradient will be $\frac{k\Delta T}{d} = \frac{0.96\times(25-10)}{0.003} = 4800\text{ W m}^{-2}$.
5. a) The rate of heat loss is inversely proportional to the thickness of the glass.

b)

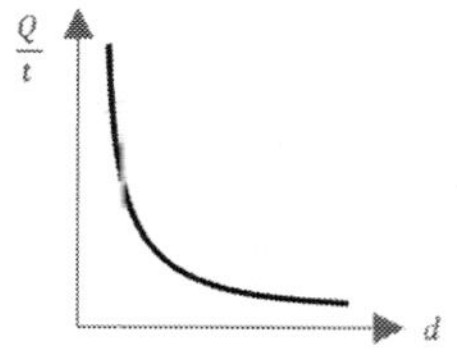

c) If $\frac{1}{d}$ is plotted on the x-axis, and the data shows a linear relationship, then the gradient should be equal to $kA\Delta T$. If the value of this gradient matches experimentally measured values, then the relationship $\frac{Q}{t} = \frac{kA\Delta T}{d}$ will be demonstrated.

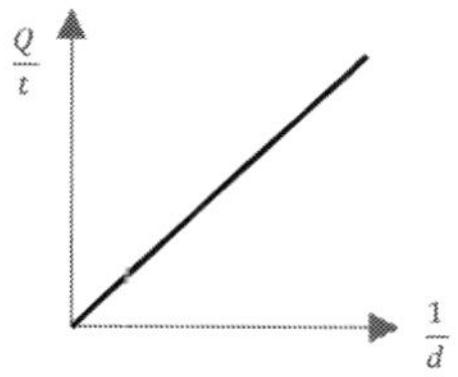

6. As $\frac{Q}{t} = \frac{kA\Delta T}{d}$, the thermal conductivity of the material k is proportional to the heat loss $\frac{Q}{t}$ as shown by the graph. Therefore choosing glass with a lower k will reduce heat loss, for example double glazing.

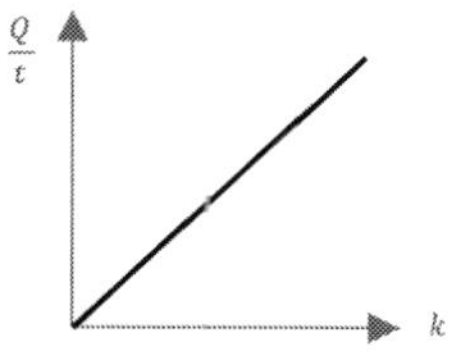

16.4 Temperature and heat

1. Heat can be transferred in matter via collisions of neighbouring particles to transfer kinetic energy, in a process known as conduction, as well as a process involving hotter, less dense matter rising and cooler, more dense matter sinking, known as convection. The ***heat*** contained in a substance is therefore the total kinetic energy of its particles. As heat is a form of energy, it is measured in joules.
The kinetic energy of each particle in a substance varies, so ***temperature*** is a measure of the average kinetic energy of particles in a substance. Two common units for temperature are degrees Celsius (°C) and Kelvin (K). The Kelvin scale is often used in science, and the lowest temperature on it, 0 K, is called absolute zero. Theoretically, 0 K is the temperature at which all particles in a substance stop moving and therefore have no kinetic energy. The Celsius scale is converted to Kelvin by:
$$K = °C + 273.15$$
Therefore, an increase in one degree Celsius is the same as an increase in one Kelvin.
2. a) 273.15 K
   b) 298.15 K
3. a) -273.15 °C
   b) 100 °C
4. On a hot day, the ocean may feel pleasantly cool, but the sand on the beach can feel painfully hot. Although the water and sand are subjected to the same sunlight and surrounding air temperature, the water can absorb heat better than sand, without its temperature being raised as much. This ability to absorb heat energy is called ***heat capacity***. The ***specific heat capacity*** is the heat energy required to raise the temperature of one gram of a substance by 1 °C or 1 K. So the specific heat capacity of water is much higher than sand, since the same amount of heat causes a bigger temperature increase in sand. The amount of energy (Q) needed to raise the temperature of a substance is given by:
$$Q = mC\Delta T$$
where: Q = amount of energy (J)
m = mass of substance (g)
C = specific heat capacity of substance (J g^{-1} K^{-1})
ΔT = increase in temperature (K or °C)
5. The working out is correct. If the mass of the water is in kilograms, which is 1000 grams, the unit for energy calculated will be multiplied by 1000 J, or kilograms. So using an alternate unit, the specific heat capacity of water is 4.18 kJ kg^{-1} K^{-1}. To calculate change in temperature, final minus initial temperature was used, giving a negative value for heat, which indicates heat is lost by the water.
6. For the cup of water:
$Q = mC\Delta T = 250 \times 4.18 \times (80 - 25) = 57475$
For the 2.0 L:
$Q = mC\Delta T$
$57475 = 2000 \times 4.18 \times (T_f - 25)$
$T_f = 32$ °C
7. a) Same container, same amount of heat added, same initial temperature, same mass of liquid.
   b) According to $Q = mC\Delta T$, if Q and m are controlled, then a smaller specific heat capacity C will lead to a larger change in temperature ΔT. Therefore sunflower oil has the smallest specific heat capacity, leading to the largest temperature change, while water has the largest specific heat capacity.
8. $VIt = mC\Delta T$
$12 \times 10t = 800 \times 4.18 \times (100 - 25)$
$t = 2090$ s = 35 min
9. $Mgh = mC\Delta T$
$20 \times 9.8h = 250 \times 4.18 \times 0.5$
$h = 2.7$ m
10. The high specific heat capacity of water helps maintain the human body at a constant temperature, even if the surrounding air temperature is hot or cold. This is because the high specific heat capacity means it takes a large heat gain or loss in order to change the temperature of water in the body.

16.5 Thermal equilibrium

1. When substances at different temperatures come into contact, heat energy is transferred from regions of high temperature to low temperature. This occurs via collisions between the particles in the substances. The amount of heat lost by the hotter substance must equal the amount of heat gained by the cooler substance, according to the law of conservation of energy. When both substances reach the same temperature, thermal equilibrium is reached.
At thermal equilibrium, collisions still occur, and individual particles may still transfer kinetic energy, but there is no net heat transfer between the two substances.
2. a) heat gain by milk = heat loss by water
$Q_m = -Q_w$
$m_m C_m \Delta T_m = -m_w C_w \Delta T_w$
$30 \times 3.93 \times (T_f - 4) = -220 \times 4.18 \times (T_f - 95)$
$117.9T_f - 471.6 = -919.6T_f + 87362$
$T_f = 85$ °C

b) The kinetic energy of the water particles is transferred to the kinetic energy of the milk particles through collisions.

3. a) heat gain by water = heat loss by gold

$Q_w = -Q_g$

$m_w C_w \Delta T_w = -m_g C_g \Delta T_g$

$10 \times 4.18 \times (40.72 - 25) = -10 \times C_g \times (40.72 - 550)$

$657.096 = 5092.8C_g$

$C_g = 0.13$ J g^{-1} K^{-1}

b) The specific heat capacity of gold is much lower than the specific heat capacity of water, so much more heat is required to raise the temperature of water, leading to the final temperature being closer to the initial temperature of water.

4. a) $density = \frac{mass}{volume}$

$2400 = \frac{m_c}{5 \times 5 \times 0.10}$

$m_c = 6000$ kg

b) $density = \frac{mass}{volume}$

$1.246 = \frac{m_c}{5 \times 5 \times 4}$

$m_a = 124.6$ kg

c) heat gain by air = heat loss by concrete

$Q_a = -Q_c$

$m_a C_a \Delta T_a = -m_c C_c \Delta T_c$

$124.6 \times 1.006 \times (T_f - 10) = -6000 \times 1.17 \times (T_f - 20)$

$125.3476T_f - 1253.476 = -7020T_f + 140400$

$T_f = 19.8$ °C

d) The concrete floor has both more mass and a higher specific heat capacity than the air in the room, so more heat is required to raise the temperature of the floor. Therefore, if the floor was shaded from sun during the day, for example using a curtain, then it will stay cooler than the air and absorb heat from the air.

16.6 Latent heat

1. When water at room temperature is heated, the kinetic energy of its particles increases. This means the temperature increases, because temperature is a measure of the average kinetic energy of the particles.
When water reaches its boiling point at 100 °C, all the liquid water turns into water vapour, its gaseous state. However, the temperature of the water no longer increases and stays at 100 °C, until the phase change is complete, meaning the average kinetic energy of the particles stays the same. The heat transferred to the water does not just disappear though, as this would violate the law of conservation of energy. Instead, it causes the phase change from liquid to gas, by overcoming the electrostatic attraction between the water particles and increasing the electrical potential energy between them. Because the heat does not cause a temperature change, it is called ***latent heat*** – latent means hidden.
The ***latent heat of fusion*** L_f is the amount of energy that must be absorbed by a substance for it to melt, turning it from solid to liquid. It is also the same amount of energy a liquid substance must lose to its surroundings in order to freeze.
The ***latent heat of vaporisation*** L_v is the amount of energy that must be absorbed by a substance for it to boil or evaporate, turning it from liquid to gas. It is also the same amount of energy a gaseous substance must lose to its surroundings in order to condense, turning it from gas to liquid.
The value of latent heat is experimentally determined using the formula:

$$Q = mL$$

where: Q = amount of energy required to change the state of a substance (J)
m = mass of the substance (g)
L = specific latent heat of fusion or vaporisation (J g^{-1})

2. a) The heat is transformed into electrical potential energy required to overcome the electrostatic attraction between water particles, so that the ice changes state. At 0 °C, the heat turns ice into water and at 100 °C, the heat turns water into ice.

b) Around 350 kJ kg^{-1}. (The latent heat of fusion is the heat added while the ice stays at 0 °C.)

c) Around 2300 kJ kg^{-1}. (The latent heat of vaporisation is heat added while the water stays at 100 °C.)

d) Around 400 kJ

e) $Q = mC\Delta T$

400 kJ $= 1$ kg $\times C \times (100 - 0)$

$C = 4$ kJ kg^{-1} K^{-1}

f) From the graph, for ice: $Q = mC\Delta T$

$100 = 1 \times C \times 50$

$C = 2$ kJ kg^{-1} K

From the graph, for water vapour: $Q = mC\Delta T$

$200 = 1 \times C \times 100$

$C = 2$ kJ kg^{-1} K

These are both around half of the specific heat capacity of water.

3. a) 55 °C

b) $Q = mL$

$4300 - 1300 = 15 \times L$

$L = 200$ J g^{-1}

c) The larger the latent heat of fusion, the more heat energy is needed to melt the candle, and the longer the candle will stay solid before it melts, so the candle will last longer.

d) From the graph, for solid paraffin: $Q = mC\Delta T$

$1300 - 0 = 15 \times C \times (55 - 20)$

$C = 2.5$ J g^{-1} K^{-1}

e) From the graph, for liquid paraffin: $Q = mC\Delta T$

$4900 - 4300 = 15 \times C \times (70 - 55)$

$C = 2.7$ J g^{-1} K^{-1}

Made in the USA
Monee, IL
10 November 2026

68383432R00201